T0190405

IOANNIS KE-
PLERI S. C. MAIEST.
MATHEMATICI
STRENA

Seu

De Niue Sexangula.

Cum Priuilegio S. Cæf. Maieft. ad annos xv.

FRANCOFVRTI AD MOENVM,
apud Godefridum Tampach.

Anno M. DC. XI.

Jeffrey C. Lagarias
Editor

The Kepler Conjecture

The Hales-Ferguson Proof

by

Thomas Hales
Samuel Ferguson

Including
A Special Issue of
Discrete & Computational Geometry

 Springer

Editor
Jeffrey C. Lagarias
Department of Mathematics
University of Michigan
Ann Arbor, MI 48109-1043
USA
lagarias@umich.edu

ISBN 978-1-4614-1128-4   e-ISBN 978-1-4614-1129-1
DOI 10.1007/978-1-4614-1129-1
Springer New York Dordrecht Heidelberg London

Library of Congress Control Number: 2011941118

Mathematics Subject Classification (2010): 52C17, 11H31, 05B40, 03B35, 68T15

Printed on acid-free paper

Springer is part of Springer Science+Business Media (www.springer.com)

Preface

The Kepler conjecture asserts that the densest packing of three-dimensional Euclidean space by equal spheres is attained by the "cannonball" packing, or face-centered-cubic (FCC) packing, which fills space with density $\frac{\pi}{\sqrt{18}} \approx 0.74048$. This conjecture, formulated by Kepler in his booklet *"Strena, seu, de Niue Sexangula,"* was published in 1611, exactly four hundred years ago. Notably, in 1900 Hilbert included the sphere packing problem in his famous problem list, as part of his 18th problem. More than a century later, in a landmark result, the Kepler conjecture was solved in work of Thomas C. Hales and Samuel P. Ferguson. An abridged version of their proof appeared in the *Annals of Mathematics* in 2005, followed a year later by the publication of a detailed proof.

This book presents the Hales-Ferguson proof of the Kepler conjecture, together with supporting material and commentary. It begins with an introductory overview chapter, followed by a chapter on the local density approach to sphere packing bounds. This is followed by the six papers of Hales and Ferguson giving their detailed proof, as published in 2006 in a special issue of *Discrete & Computational Geometry*. Next comes a 2010 paper by Hales (with five other authors) making a slight revision to the 2006 proof, and listing corrections. It concludes with two of Hales's initial papers on the problem, published in 1997. All chapters except for the first are papers reprinted from *Discrete & Computational Geometry*.

This book is divided into four parts, as follows.

Part I: Introduction and Survey

The editor has written the two chapters in this part. The first chapter is introductory and features a brief summary of the history of work on the problem, together with a description of Hales and Ferguson's 1998 preprints, details of the peer review process of the Hales and Ferguson papers (which took eight years), and subsequent developments. It also includes remarks on the reliability of the proof and the subsequent approach of Hales to obtain a formal proof of the Kepler Conjecture (in a formal logical system), entirely checkable by computer.

The second chapter describes the general program of obtaining sphere packing upper bounds using local density inequalities, an approach that can be applied to sphere packing in any dimension. It was published as a paper in 2003 in *Discrete & Computational Geometry*. After considering the general case, it specializes to the three-dimensional case, and desecribes the main features of the Hales-Ferguson local density inequality as presented in their 1998 preprints; the 2006 published proof established a slightly modified inequality.

Part II: Proof of the Kepler Conjecture
These six chapters comprise the heart of the volume. They reprint the six papers of Hales with Ferguson that together provide the detailed proof of the Kepler Conjecture. As noted earlier, these papers appeared as a special issue (2006) of *Discrete & Computational Geometry*. However, in this book the short index to definitions that appeared in this special issue is omitted; it is replaced by the two indexes at the end of the book.

In 2010 Hales published an addendum and a list of errata to the published proof; we have added asterisks in the margin of the reprinted papers locating these corrections, which are listed in Part III of the volume on pp. 361–374.

Part III: A Revision to the Proof of the Kepler Conjecture
This part presents an important follow-up paper of Hales (with five coauthors) that was published in *Discrete & Computational Geometry* in 2010. It explains Hales's program to obtain a formal proof of the Kepler Conjecture. The initial process of formalization uncovered one logical gap in the original proof, and this follow up paper provides a correction filling that gap. It also supplies a list of errata to the original papers.

Part IV: Initial Papers of the Hales Program
This part presents two early papers of Hales on Kepler's conjecture, which were published in *Discrete & Computational Geometry* in 1997. These papers give his original formulation of an approach to proving the Kepler Conjecture via a local density inequality, and carry out some initial steps of this approach. They explain and establish a basic framework followed in the subsequent proof. As it turned out, to obtain a proof, this approach required some modification. This modification included changes to the local density inequalities as described in Part III. The proofs given in Parts II and III are independent of these two papers.

In reading this volume, it should be helpful to start with the introductory Chapter 1. Next one might study Chapter 2, which describes the general framework for obtaining upper bounds on sphere packing density, in any dimension. At present it is not known in which dimensions optimal such inequalities (i.e., tight inequalities) may exist: they are known to exist in dimensions 1, 2, and now, by the Hales-Ferguson proof, in dimension 3. It seems likely that optimal inequalities might exist in dimensions 8 and 24 as well. One might then read the historical survey of Hales in Chapter 3, which also describes some features of the proof given in the next five chapters. (This chapter could alternatively be read before reading Chapter 2.) Next one could look at some of the details of the formulation of the proof (Chapter 4). It would also be useful to look at the Revision paper (Chapter 9) to see features of the ongoing work towards a formal proof of the Kepler conjecture. Finally the reader may consider the remaining chapters in the volume.

I thank many people who helped me with this editing project. Ricky Pollack gave useful general advice. Tom Hales made many useful comments regarding the introductory Chapter 1. JoAnn Sears (University of Michigan Library) helped obtain a phographic copy of Kepler's 1611 volume from Cornell University Library. Finally I thank Ann Kostant and John Spiegelman for work on the preparation of this volume. During the preparation of this work I received support from NSF grants DMS-0500555 and DMS-0801029.

Jeffrey C. Lagarias
Ann Arbor, MI
January 11, 2011

Contents

Introduction and Survey

Introduction to Part I

Part I contains two introductory chapters.

The first chapter in an introductory paper written for this volume. It gives a summary of the history of the problem, starting with the work of Kepler. It describes the difficulty of the problem, gives a description of Hales and Ferguson's 1998 preprints, including details of the peer review process of the Hales and Ferguson papers (which took 8 years), and subsequent developments. It also includes remarks on the reliability of the proof and the subsequent approach of Hales to obtain a formal proof of the Kepler Conjecture (in a formal logical system), entirely checkable by computer.

The second chapter provides a technical survey of local density approaches to obtain upper bounds on sphere packing density in any dimension. It then views the Hales-Ferguson approach and other approaches to the Kepler Conjecture within this context. It also provides an overview of the Hales-Ferguson proof, as given in the six 1998 preprints. This paper appeared in 2002, during the 1998–2006 peer reviewing process of the Kepler Conjecture proof, and affected the final revised form of the proof.

1

The Kepler Conjecture and Its Proof, by J. C. Lagarias

This introductory chapter gives a brief history of the Kepler problem and indicates sources of its difficulty. It describes various approaches to solving the problem, based on local density inequalities. It then discusses the proof of Hales and Ferguson and the peer-reviewing of their papers. It also discusses further developments, towards obtaining a formal proof of the Kepler conjecture, and other applications of the proof methodology.

Contents

1. The Kepler Problem for Sphere Packing
2. Why the Kepler Problem is Difficult
3. Local Density Approach: History
4. Hales Program and Hales-Ferguson Papers
5. Peer-Reviewing of the Hales-Ferguson Papers
6. Reliability of the Hales-Ferguson Proof
7. Formal Proof of the Kepler Conjecture
8. Applications of the Hales-Ferguson Proof Methodology
9. Contents of This Volume

Acknowledgments. I thank Gabor Fejes-Toth for editorial work on the DCG special issue in July 2006. I thank him, Tom Hales, Steven G. Krantz, and Richard Pollack for helpful comments on this paper. This work was supported in part by NSF grant DMS-0801029.

JOHANN KEPLER,

MATHEMATICIAN TO
HIS IMPERIAL MAJESTY

A NEW YEAR'S GIFT

or

On the Six-Cornered Snowflake.

Published by GODFREY TAMPACH at
FRANKFORT ON MAIN,
in the year 1611.

The Kepler Conjecture and Its Proof

Jeffrey C. Lagarias

Abstract This paper describes work on the Kepler conjecture starting from its statement in 1611 and culminating in the proof of Hales-Ferguson in 1998–2006. It discusses both the difficulty of the problem and of its solution.

1 The Kepler Problem for Sphere Packing

The Kepler conjecture asserts that a densest packing of three-dimensional Euclidean space by equal spheres is given by the "cannonball" packing, or face-centered-cubic (FCC) packing, which fills space with density $\frac{\pi}{\sqrt{18}} \approx 0.74048$. This conjecture was formulated by Kepler in 1611 [43, pp. 14–17], as follows. After discussing cubical packing in two-dimensional layers versus hexagonal packing in two-dimensional layers, Kepler says (see Figure 1):

> Now if you proceed to pack the solid bodies as tightly as possible, and set the files that are first arranged on the level on top of others, layer on layer, the pellets will be either squared (A in diagram), or in triangles (B in diagram). If squared, either each single pellet of the upper range will rest on a single pellet of the lower, or, on the other hand, each single pellet of the upper range will settle between every four of the lower. In the former mode any pellet is touched by four neighbors in the same plane, and by one above and one below, and so on throughout, each touched by six others. The arrangement will be cubic, and the pellets, when subjected to pressure, will become cubes. But this will not be the tightest pack. In the second mode, not only is every pellet touched by its four neighbors in the same plane, but also by four in the plane above and four below, so throughout one will be touched by twelve, and under pressure spherical pellets will become rhomboid. This arrangement will be more comparable to the

This work was supported in part by NSF Grant DMS-0801029.

Department of Mathematics, University of Michigan, Ann Arbor, MI 48109-1043, USA
e-mail: lagarias@umich.edu

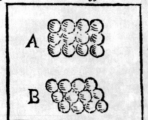

'Iam fi ad ſtructuram ſolidorum quam poteſt fieri arctiſſimam progrediaris, ordinesq, ordinibus ſuperponas, in plano prius coaptatos,
aut ij erunt quadrati A aut trigonici:
n ſi quadrati . aut ſinguli globi ordinis
ſuperioris ſingulis ſuperſtabunt ordinis inferioris aut contra ſinguli ordinis ſuperioris ſedebunt inter quaternos ordinis inferioris. Priori modo
tangitur quilibet globus à quatuor
circumſtantibus in eodem plano, ab
vno ſupra ſe, & ab vno infra ſe: & ſic in vniuerſum à ſex alijs, eritq̃,
ordo cubicus, & compreſſione facta fient cubi: ſed non erit arctiſſima
coaptatio. Poſteriori modo praeterquam quod quilibet globus à quatuor circumſtantibus in eodem plano tangitur, etiam à quatuor infra
ſe, & à quatuor ſupra ſe. & ſic in vniuerſum à duodecim tangetur;
fientq̃, compreſſione ex globoſis Rhombica. Ordo hic magis aſſimila-
bitur octaedro & Pyramidi. Coaptatio fiet arctiſſima: vt nullo pra-
terea ordine plures globuli in idem vas compingi queant. Rurſum ſi

Fig. 1 Kepler's 1611 assertion. [Image courtesy of the Division of Rare and Manuscript Collections, Cornell
University Libraries.]

octahedron and pyramid. The packing will be the tightest possible, so that in no
other arrangement could more pellets be stuffed into the same container.[1]

Kepler goes on to describe the FCC packing in more detail, saying (see Figure 2):

Thus, let B be a group of three balls; set one A, on it as apex; let there
be also another group, C, of six balls; another, D, of ten; and another, E, of
fifteen. Regularly superimpose the narrower on the wider to produce the shape
of a pyramid. Now, although in this construction each one in the upper layer is
seated between three in the lower, yet if you turn the figure round so that not
the apex but a whole side of the pyramid is uppermost, you will find, whenever
you peel off one ball from the top, four lying below it in square pattern. Again
as before, one ball will be touched by twelve others, to wit, by six neighbors in
the same plane, and by three above and three below. Thus in the closest pack in
three dimensions, the triangular pattern cannot exist without the square, and vice

---

[1] "Iam si ad structuram solidorum quam potest fieri arctissimam progrediaris, ordinesque ordinibus super-
ponas, in plano prius coaptatos aut ii erunt quadrati A aut trigonici B: si quadrati aut singuli globi ordinis
superioris singulis superstabunt ordinis inferioris aut contra singuli ordinis superioris sedebunt inter quaternos
ordinis inferioris. Priori modo tangitur quilibit globis a quattuour cirucmstantibus in eodem plano, ab uno supra
se, et ab uno infra se: et sic in universum a six aliis, eritque ordo cubicus, et compressione facta fient cubi: sed
non erit arctissima coaptatio. Posteriori modo praeterquam quod quilibet globus a quattuor circumstantibus in
eodem plano tangitur etiam a quattuor infra se, et a quattuor supra se, et sic in universum a duodecim tangetur;
fientque compressione ex globosis rhombica. Ordo hic magis assimilabitur octaedro et pyramidi. Coaptatio
fiet arctissima, ut nullo praetera ordine plures globuli in idem vas compingi queant." [English translation:
Colin Hardie [43, p. 15]]

quadrilateris.Esto enim B *copula trium globorum. Ei superpone* A *v-num pro apice, esto & alia copula senū globorū* , *& alia demū* D, *& alia quindenum* E *Impone semper angustiorem latiori, vt fiat figura Pyramidis. Etsi igitur per hanc impositionem singuli superiores sederunt inter trinos inferiores: tamen iam versa figura, vt non apex sed integrum latus pyramidis sit loco superiori, quoties vnum globulum degluberis è summis, infra stabunt quatuor ordine quadrata. Et rursum tangetur vnus globus vt prius, à duodecim alijs, à sex nempe circumstantibus in eadem plano tribus supra & tribus infra. Ita in solida coaptatione arctissima non potest esse ordo triangularis sine quadrangulari, nec vicissim. Patet igitur, acinos Punici mali, materiali necessitate concurrente cum rationibus incrementi acinorum, exprimi in figuram Rhōbici corporis: cum non infestis frontibus pertinaciter nitantur rotundi ex aduerso acini, sed cedant expulsi, in spacia inter ternos vel quaternos oppositos interiecta.*

Fig. 2 Kepler's description of the FCC "cannonball" packing. [Image courtesy of the Division of Rare and Manuscript Collections, Cornell University Libraries.]

versa. It is therefore obvious that the loculi of the pomegranate are squeezed into the shape of a solid rhomboid.[2]

The FCC packing had been noted earlier by the English mathematician Thomas Hariot [Harriot] (1560–1621). Hariot was mathematics tutor to Sir Walter Raleigh, designed some of his ships, wrote a treatise on navigation, and went on an expedition to Virginia in 1585–1587. He computed a chart in 1591 on how to most efficiently stack cannonballs using the FCC packing, and computed a table of the number of cannonballs in such stacks (cf. Shirley [55, pp. 242–243]). Hariot supported the atomic theory of matter, in which case macroscopic objects may be packed arrangements of very tiny spherical objects, i.e., atoms [42, Chap. III]. He corresponded with Kepler in 1606–1608 on optics, and mentioned the atomic theory in a Dec. 1606 letter as a possible way of explaining why some light is reflected, and some refracted, when hitting a liquid. Kepler replied in 1607,

---

[2] "Esto enim *B* copula trium globorum. Ei superpone *A* unum pro apice; esto et alia copula senum globorum *C*, et alia denum *D* et alia quindenum *E*. Impone semper angstiorem latiori, ut fiat figura pyramidis. Etsi igitur per hanc impositionem singuli superiores sederunt into trinos inferiores: tamen iam versa figura, ut non apex sed integrum latus pyramidis sit loc superiori, quoties unum globulum deglberis e summis, infra stabunt quattuor ordine quadrato. Et rursum tangetur unus globus ut prius, et duodecim aliis, a sex nempe circumstantibus in eodem plano tribus supra et tribus infra. Ita in solida coaptatione arctissima non potest ess ordo triangularis sine quadrangulari, nec vicissim. Patet igitur, acinos punici mali, materiali necessitate concurrente cum rationaibus incrementi acinorum, exprimi in figuri rhombici corporis" [English translation by Colin Hardie [43, p. 17]]

not supporting the atomic theory. The known correspondence of Hariot with Kepler does not deal directly with sphere packing.

Questions on sphere packing attracted the attention of Isaac Newton. A discussion with the mathematician David Gregory on 4 May 1694 concerning the brightest stars was summarized in a memorandum of Gregory [50, Vol III, p. 317] as:

> To discover how many stars there are of a given magnitude, he [Newton] considers how many spheres, nearest, second from them, third etc. surround a sphere in a space of three dimensions, there will be 13 of first magnitude, 4×13 of second, $9 \times 4 \times 13$ of third.[3]

Newton's own star table "A Table of ye fixed Starrs for ye yeare 1671" records 13 first magnitude stars, 43 of the second magnitude, 174 of third magnitude; cf. [50, Vol. II, p. 394]. In an (unpublished) notebook Gregory considered the packing problem in 2-dimensions and 3-dimensions and recorded that 13 spheres might touch a given equal sphere [50, Vol. III, p. 321]. It has been asserted that Newton believed only 12 spheres of fixed radius could touch a single sphere of the same radius, but I do not know of a primary reference for this assertion. Certainly every sphere in the FCC packing touches exactly 12 neighbors. It is now known that the maximum number of disjoint equal spheres that can touch a given equal spheres (the "kissing number") is 12, as observed by Hoppe [36] in 1874 and shown rigorously in 1954 by Schütte and van der Waerden [54]. See Conway and Sloane [10, Sec 1.2] and Casselman [7] for further discussion and references.

In the 19th century it was discovered that besides the FCC packing there exists another equally dense sphere packing possessing a full translation group of symmetries, the hexagonal close packing (HCP). This packing was described by William Barlow [3] in 1883, in connection with the possible internal symmetries of crystals.

In 1900 Hilbert [34] included the sphere packing problem in his famous problem list, as part of his 18th problem. He wrote:

> I point out the following question, related to the preceding one, and important to number theory and perhaps sometimes useful in physics and chemistry: How can one arrange most densely in space an infinite number of equal solids of given form, e.g., spheres with given radii or regular tetrahedra with given edges (or in prescribed positions), that is, how can one so fit them together that the ratio of the filled to the unfilled space may be as great as possible?[4]

See Milnor [48] for a further discussion and history of work on the various parts of Hilbert's 18th problem.

The topic of sphere packing is of interest in much wider situations in mathematics, physics and materials science; in the higher dimensional case it is also important in

---

[3] "Ut noscatur quot sunt stellae magnitudinis 1 ae, 2 dae, 3 ae & tc. considerando quot spherae proximae, seundae ab his 3 ae &tc spheram in spatio trium dimensionis circumstent: erunt 13 primae, 4×13 2-dae, $9 \times 4 \times 13$ 3 ae."

[4] "Ich weise auf die heirmit in Zusammenhang stehende, für die Zahlentheorie wichtige und vielleicht auch der Physik und Chemie einmal Nutzen bringende Frage hin, wie man undendlich viele Körper von das gleichen vorgeschriebenen Gestalt, etwa Kugeln mit gegebenem Radius oder reguläre Tetraeter mit gegebener Kante (bez. in vorgeschriebener Stellung) im Raume am dictesten einbetten, d. h. so lagern kann, daß des Verhältnis des erfülten Raumes zum nichterfüllten Raume möglichst groß ausfällt." [English translation: Dr. Mary Winston Newson]

communications and coding theory. For a physics viewpoint see Aste and Weaire [2] and for both mathematical and communications aspects see Conway and Sloane [10]. For purely mathematical aspects, see Böröczky [5], Fejes-Toth [15], Rogers [53], and Zong [61].

This volume includes the complete proof of the Kepler conjecture by Thomas C. Hales with Samuel P. Ferguson, first presented in six preprints posted on the math arXiv in 1998, and then after extensive revisions, published in six papers in 2006 in *Discrete & Computational Geometry*. (An abridged version of this proof was published earlier by Hales in *Annals of Mathematics* in 2005.) The proof consists of a large body of mathematical arguments and a massive computer verification of many inequalities. There is also a follow-up paper written in 2009 ("A revision of the proof of the Kepler Conjecture") and published in 2010, supplying more details about one point in the 2006 proof, and describing progress towards constructing a formal proof of the Kepler conjecture, in a formal language, completely checkable and certifiable by computer. When completed, this will be a second-generation proof.

We view these papers not just as mathematics but also as historical documents. Below we describe some aspects of the history of work on the problem, and how the work of Hales and Ferguson came about. Then we describe the peer-review process, the reliability of this proof, and the ongoing work of Hales to obtain a formal proof of the Kepler conjecture.

As background, we first explain why the problem is hard.

2 Why the Kepler Problem Is Difficult

It is not immediately apparent that Kepler's Conjecture is a problem of extraordinary difficulty.

A first difficulty is to define a rigorous notion of density of packing of spheres. This is resolved using limiting notions of packing a large region and letting its diameter increase to infinity. This notion was not clarified until the 20th century. One can then prove a result asserting that a packing exists that attains a maximal limiting density; cf. [53, Chap. 1].

A second difficulty is that notions of limiting density are very crude in the sense that one can remove spheres from an arbitrarily large finite region without affecting the limiting density. One may therefore wish to impose further local restrictions on the notion of a "densest" packing, for example; that it contain no large "holes." We define a *saturated packing* of spheres to be one into which no new sphere can be inserted, i.e., there is no "hole" of diameter 2 or larger in the packing. Besides this, one might be able to increase density locally by removing a finite collection of spheres in a region and repacking that region to squeeze in one more sphere. This sort of condition seems difficult to analyze, but it already shows that one may wish to take account of "local" conditions specifying density of a packing, compare Bezdek et al [4].

A third difficulty, peculiar to three dimensions, is that there exist uncountably many essentially different "optimally dense" packings. Here we consider packings as essentially different if they are not congruent under a Euclidean motion of space. Consider the packing that starts with a planar layer of hexagonally close packed spheres. That is, there is a planar slice through this layer that intersects all the sphere centers, giving a circle packing of the plane, and this packing is the optimal two-dimensional hexagonal

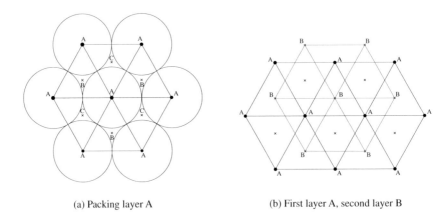

(a) Packing layer A   (b) First layer A, second layer B

Fig. 3 Hexagonal packing layers viewed from above

circle packing. Now one can fit a second identical layer of spheres on top of this layer, so that the spheres fit as deep as possible in indentations between the spheres in the first layer. If we mark one sphere as the center of the first layer, there are two possible ways to do this: each such packing occupies 3 of the 6 holes formed by a hexagon and there are two choices. One can repeat this in the layer below the first layer, and continue adding layers in this fashion, making a choice at each layer. All such packings have the same limiting density. If the stacked layers are viewed vertically from above, then spheres in all the layers can be seen to line up in three possible positions of the hexagonal lattice, which we can label A, B, C, with the packing layer with the marked sphere labelled A. These are pictured in Figure 3. Note that the centers in B and C layers are "deep holes" in layer A.

In effect the choices at each layer of the packing can be labelled using letters $\{A, B, C\}$, with no two consecutive letters the same, to uniquely label a packing (with a single marked sphere serving as origin) with a doubly infinite string of such letters. It can be shown that two packings are essentially the same (up to a Euclidean motion) if their doubly infinite strings of letters can be lined up to agree. Two of these packings are especially nice, the *face-centered-cubic lattice packing (FCC packing)* and the *hexagonal close packing (HCP packing)*, described further below. The FCC packing corresponds to a repeating pattern ABCABCABC... while the HCP packing corresponds to a repeating pattern ABABAB... (or ACACAC... or BCBCBC...). The FCC packing was found by Kepler, while the HCP packing was first described by Barlow [3] in 1883; cf. Coxeter [11, Sec. 22.4]. These packings are described at the beginning of Hales [24] (in this volume).

The collection of "optimally dense" packings just described are locally optimal in a very strong sense. A *Voronoi domain* or *Voronoi cell* around a given sphere center is the set[5] of all points in space closer to that sphere center than to any other sphere center. In saturated packings all Voronoi cells are polyhedrons. For the packings just described, all

---

[5] This defines the interior of the Voronoi domain. The Voronoi domain itself is the closure of this set, adding boundary points which are certain points equidistant to two sphere centers.

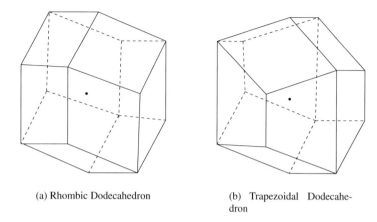

(a) Rhombic Dodecahedron   (b) Trapezoidal Dodecahe-
                            dron

Fig. 4 Voronoi Domains of (a) Type 1 (FCC) (b) Type 2 (HCP)

Voronoi domains consist of one of two shapes, each having 12 faces and 14 vertices [15, p. 173]. (See Figure 4.)
These are:

  Type 1. Rhombic Dodecahedron (all 12 faces are rhombi).
  Type 2. Trapezoidal Dodecahedron (6 faces trapezoids, 6 faces rhombi)

Here Type 1 occurs for all Voronoi cells in the middle layer of consecutive layers labelled $ABC, ACB, BAC, BCA, CAB, CBA$, while type 2 occurs for all Voronoi cells in consecutive layers labelled $ABA, ACA, BAB, BCB, CAC, CBC$. Voronoi cells of types 1 and 2 are known to have the same volume and surface area. For each of these cells, the ratio of the volume of each sphere to the volume of the Voronoi cell containing it is exactly $\frac{\pi}{\sqrt{18}}$. Thus in these packings optimality holds locally; that is, it is attained simultaneously in each Voronoi cell. In the FCC (Face-Centered Cubic) lattice packing, all Voronoi cells are of type 1. In the HCP (Hexagonal Close Packing) packing, all Voronoi cells are of type 2. (See [58].) In all remaining packings above, Voronoi cells of both types occur in the packing. The packings can be told apart by their different arrangements of types 1 and 2 Voronoi cells. All these packings have at least a two-dimensional lattice of translational symmetries; the FCC and HCP packings, and countably many others, have a full three-dimensional lattice of translational symmetries. From the mathematical viewpoint, as an optimization problem, this means that there are (at least) two different local optima, represented by the two types of Voronoi cells above.

A fourth difficulty is that the optimization problem to maximize density is a priori an infinite-dimensional problem: one has infinitely many spheres to pack. Describing a packing requires infinitely many variables: three variables each for the coordinates of each sphere center. Approaches to make headway with this problem seek to prove stronger results which only involve finite-dimensional optimizations that encode local conditions. This is the "local density inequality" approach described below and in the next section. These approaches in effect assign, by some recipe, to each sphere in a sphere packing a (weighted) sum of the covered and uncovered volume near that sphere center. This recipe is "local" in that the weighted sum for a given sphere center is completely determined by the locations of all spheres in the sphere packing nearby, within a fixed distance C of the given sphere center. When the recipe quantities are added up over all spheres, it should count all volume with weight 1. If so, then an upper bound on the weighted area will give an upper bound on global sphere packing density. We say such a local density inequality is "optimal" if it will produce the upper bound $\frac{\pi}{\sqrt{18}}$ for the sphere packing density.

A fifth difficulty is that it is not clear that "optimal" local density inequalities exist, and if they do exist they (apparently) cannot be very simple. In support of this view, the two most natural ways to locally partition space attached to a sphere packing yield local density inequalities that are not "optimal," as we now explain.

The first natural way to partition space corresponding to sphere centers is to divide it into *Voronoi cells* around these points, as described above. The region assigned to each sphere center is its Voronoi cell. However, it is known that an arrangement of 12 unit spheres touching a given unit sphere with their sphere centers being the vertices of a regular dodecahedron yields a Voronoi cell that is a regular dodecaheron of inradius 1, having a ratio of covered to uncovered volume approximately 0.754697, which exceeds $\frac{\pi}{\sqrt{18}} \approx 0.74048$.

A second natural partition of space is the *Delaunay tessellation*, which is a partition of space $\mathbb{R}^3$ into simplices (tetrahedra) having vertices among the sphere center points. We describe it further below. Such Delaunay simplices have their vertices at four sphere centers, so that their edge lengths are necessarily 2 or greater. Now we can cut up these tetrahedra, for example by barycentric subdivision, and assign parts of the uncovered area to each of the four spheres at the corners of the tetrahedron. Each sphere is now assigned certain regions associated to each of the Delaunay simplices for which its center is one vertex. Hales notes ([24, p. 13] in this volume) that there is a Delaunay tessellation having an individual tetrahedron with volume of covered to uncovered volume of 0.78469. Even if one sums over all Delaunay tetrahedra associated to a given sphere center, there are examples of local configurations with density exceeding the Kepler bound. An original approach of Hales in the early 1990s, described in [17], [18], based on the Delaunay tessellation, ground to a halt due to combinatorial difficulty (cf. Hales [24, Sec. 2.1]).

The failure of these two natural decompositions to be "optimal" indicate the necessity to consider more complicated inequalities in which the region assigned to a sphere center will need in certain cases to "borrow" an excess of uncovered volume from some spheres nearby to this sphere. This raises the spectre that perhaps the distance over which volume has to be "borrowed" is in fact unbounded. If so, no "optimal" local density inequality exists. One would have instead an infinite sequence of such local inequalities, taken over larger regions, each giving a better upper bound, tending in the limit to $\frac{\pi}{\sqrt{18}}$.

A sixth difficulty, then, is that of designing candidate local density inequalities that may be "optimal." If an "optimal" local inequality does exist, then it can in principle be verified by a finite computation, i.e., it comprises a finite dimensional non-linear optimization problem over a large number of variables, specifying the possible locations of sphere centers in a ball of radius C around a given sphere center. The size of the problem rapidly goes up as the distance from the initial sphere increases. There can be 12 spheres touching a given sphere, and the next layer of spheres can contain more than 30 spheres. There now arises a psychological difficulty, which is that the "optimality" of the local density inequality is *only certified after the fact*, when a proof is found. This means that one must first do a very large amount of work, with the downside risk of eventually determining that the inequality is not optimal. Therefore, one would like to increase one's confidence in advance that the inequality considered is likely to be "optimal," by making various preliminary experimentation and computer checks, before attempting a full scale proof.

There are now several reasonable candidates that have been put forward to be "optimal" local density inequalities. These include one proposed by László Fejes-Tóth in 1953 [15] and again in 1964 [16], one proposed by W.-Y. Hsiang [37] in 1993, and several different such inequalites proposed by T. C. Hales, both alone and with S. P. Ferguson. The original local density inequality of Fejes-Tóth considers weighted averages of Voronoi cells of nearby sphere centers, and that of Hsiang is similar in spirit. In 1991 Hales [17], [18] gave an approach proposing "optimal" local inequalities based on the Delaunay tessellation, with some Voronoi cell correction terms. In 1994 and later, Hales [20], [21], and Hales and Ferguson [25] formulated various candidate "optimal" inequalities using a hybrid decomposition of space employing elements of both the Delaunay tessellation and the Voronoi domain tessellation. The analysis of Hales and Ferguson suggests that there is a large class of such inequalities that will be "optimal," and there is some flexibility in formulating them. These inequalities will however have quite complicated "scoring" functions.

A seventh, and most crucial difficulty for "optimal" local density inequalities is the computational size of the problem. Each separate local density inequality asserts an upper bound to an enormously complicated finite-dimensional nonlinear optimization problem. This problem may simply be too large to be computationally feasible. It involves finding the global maximum of a highly non-linear function, over a high dimensional space of possible local configurations, enumerated by the location of the centers of all the nearby spheres. Further, this seems to involve on the order of 40 spheres, so that the problem has approximately 120 dimensions. General nonlinear problems of this dimensionality, with no additional structure, are too large to solve at present. In addition, the partitioning of space typically results in many combinatorially distinct configurations, so that the space of all configurations has a huge number of connected components. The complexity of the problem will also depend on the landscape of local optima, i.e., how many there are and how close they are to the global optimum. Therefore the local density inequality to prove needs to be carefully engineered to have extra properties that will simplify the computations. One of the key features of the Hales and Ferguson approach is to design such a "local density" inequality, allowing it to be complicated and inelegant, in order to make it have properties that reduce the needed computations. Their "Formulation" paper fills 49 pages to describe the partition of space and "scoring" function comprising the local density measure (see [25] in this volume). The final proof is an intricate blend of theory, in part needed to simplify the problem, together with large-scale computations.

In the Hales-Ferguson proof, in this volume, the local density inequality is designed so that the regions to be analyzed split up into (non-interacting) cones whose contributions to the "score" can be evaluated separately and then added. It is also necessary for the inequality to be proved to have a finite number of (sets of) global maxima. Then the proof will logically split into two parts: an exact analytic treatment in the neighborhood of each global maximum, proving this property, and then a cruder analysis of the remaining (very large) part of the space of possible configurations, establishing inequalities showing a bound strictly below the optimum. The second part is the vast bulk of the proof. In the actual proof there are two types of global maxima, corresponding to the two types of Voronoi domains above. (In the proof these comprise a finite number of different local configurations, called decomposition stars, that attain the maximum.) One might think that the difficulty in the proof is in checking near the global maxima, but in fact the greatest part of the difficulty is in dealing with the inequalities on the huge remaining portion of the configuration space. In fact, it is extremely hard to describe this configuration space in a way that is suitable for analysis and computation. The problem is reduced to many smaller nonlinear optimization problems, which are relaxed to convex optimization problems having a single maximum, that can be bounded above. The devil is in the details.

3 Local Density Approach: History

The first person to formulate a local density approach was László Fejes-Tóth [15], who made many deep insights into the sphere packing problem. In 1953 [15, pp. 174–181] he indicated the possibility of proving the Kepler conjecture by relating it to a nonlinear optimization problem over a compact set. More precisely, he proposed a specific inequality that might hold. This inequality involved averages of Voronoi cell volumes of close neighbors, and would imply Kepler's conjecture. In 1964 he restated this inequality [16, pp. 295–299] and added:

> Thus it seems that the problem can be reduced to the determination of a minimum of a function of a finite number of variables, providing a programme realizable in principle. In view of the intricacy of this function we are far from attempting to determine the exact minimum. But, mindful of the rapid develop ment of our computers, it is imaginable that the minimum may be approximated with great exactitude.

Such optimization problems can generically be termed "local density inequalities" as they involve some weighted measure of density associated to a neighborhood of each sphere center in a packing. Establishing a local density inequality yields a result which is stronger than the Kepler conjecture, since for optimality it requires the existence of a packing maximizing the local inequality simultaneously at every sphere center.

In particular, each local density inequality asserts a different mathematical result. This is so, although each may imply Kepler's conjecture as a corollary. Assuming that the associated optimization problem has a finite set of isolated global maxima, and that the local minimality can be verified analytically in an open neighborhood of each of these maxima, then in principle the inequality can be proved by computer elsewhere using interval arithmetic. However the size of the resulting optimization problems seemed far beyond the range of what could be solved by a computer at that time, or even now, without new ideas to reduce the complexity of the problem.

A precise framework for local density inequalities is given in the next paper in this volume [45]. Local density is measured at a sphere center by a "score function" which assigns to it the volume nearby counted with certain penalties and credits. These penalties and credits must have the property that when summed over all sphere centers they cancel out, so that volume is then counted with weight at most 1 everywhere.

The two major starting points in designing such local density "score functions" are the Voronoi tessellation (also called the Dirichlet tessellation (and, in physics, Wigner-Seitz cells) and Delaunay tessellation. Given centers of spheres in a saturated sphere packing, the *Voronoi tessellation* partitions space into Voronoi cells, which for each sphere center is the (closure of) the set of points in space closer to that sphere center than any other sphere center; this region contains the unit sphere. For saturated sphere packings with unit radius spheres, a Voronoi cell is a finite polyhedron having no point at distance exceeding 2 from the sphere center.

A *Delaunay tessellation* for a set of points in space $\mathbb{R}^3$ is a subdivision of space into simplices (tetrahedra) such that the sphere circumscribing each tetrahedron has no points of the set in its interior. Delaunay [12] showed such tessellations exist. They are efficiently constructible; cf. Watson [59]. The resulting tessellation is topologically dual to the Voronoi tessellation, i.e., vertices of the Delaunay tessellation correspond to regions in the Voronoi tessellation, edges in the Delaunay tessellation correspond to faces of Voronoi cells, faces in the Delaunay tessellation correspond to edges in Voronoi cells, and the interior of a Delaunay cell corresponds to a vertex of a Voronoi cell lying in its interior. The Delaunay tessellation is unique for points in general position (no 5 points at equal distance from one point) but may be non-unique otherwise. In a finite region there are only a finite number of choices for the Delaunay simplices.

As noted above, the Voronoi and Delaunay partitions of space do not directly yield "optimal" local density inequalities. Further averagings of space are required. The approaches to local density inequalities proposed by Fejes-Tóth and by Hsiang [38], [41] are based on averaging of Voronoi cell densities of regions near a given sphere. Hales proposed a local density inequality based on Delaunay tessellation in 1991 [17, Sec. 4], which involves a functional defined on a compact space of (abstract) Delaunay stars. His original study indicated that this inequality is near optimal, and that to get an optimal inequality some correction terms had to be added; he proposed terms that add some information associated to Voronoi cells. He then formulated a candidate "optimal" inequality based on a superposition of Delaunay and Voronoi tessellations [17]. In work after 1994 Hales [20], [21] (papers in part IV of this volume) formulated "hybrid" local inequalities which decompose space starting from a Delaunay-like decomposition but use a Voronoi-type decomposition on that part of space on which the Delaunay simplices do not have prescribed shapes. The local density inequality of Hales and Ferguson [25] uses a decomposition of space starting from a set of simplices with corners at four sphere centers, having allowed shapes, without imposing the "empty sphere" condition of the Delaunay tessellation. (They did this to make their local conditions not interact across long distances.)

In 1990 Wu-Yi Hsiang proposed a local density inequality using averages of Voronoi cell volumes and announced a proof of it; this would prove the Kepler Conjecture. This announcement received much publicity, see Szpiro [56, pp. 144–152]. One idea of Hsiang was to use spherical geometry rather than Euclidean geometry to obtain the sphere packing bounds: some of the uncovered regions could be treated as spherical triangles, with volumes computed using spherical trigonometry. A paper was subsequently submitted to *Annals of Mathematics*. The reviewers there uncovered some difficulties in the proof. In response to the reviewer's criticisms, Hsiang did not resubmit his paper to the *Annals of Mathematics*, and instead submitted it elsewhere. He eventually was able to publish a revised 93 page paper in 1993 in the *International Journal of Mathematics* (Hsiang [38]). This journal records that the paper was received 17 November 1992, revised 9 March 1993, and was in print by the end of 1993, a rapid review for a paper of its length and complexity.

Objections were voiced to the published Hsiang proof by Conway, Hales, Muder and Sloane [9], and were detailed by Hales [19] in 1994 in the *Mathematical Intelligencer*. These objections applied not only to specific results claimed in the proof, but also asserted that some methods of argument used were invalid. To this critique Hsiang [40] gave a rejoinder. Hsiang [41] later supplied in 2002 a book-length treatment of this proof, along with other topics.

The consensus of the mathematical community, as represented in *Mathematical Reviews* in the reviews of Hsiang's 1992 paper and 2002 book, is that the Hsiang proof is incomplete. The review of Hsiang's 1993 paper (by G. Fejes-Tóth) says:

> I think there is hope that Hsiang's strategy will work: at least the main inequalities seem to hold. As far as the details are concerned, my opinion is that many of the key statements have no acceptable proofs. Typically we are given arguments such as "the most critical case is…" followed by the statement that "the same method will imply the general case." The problem with arguments of this kind is they require the reader to redo some pages of calculations but, notoriously, that they occur at places where we expect difficulties and most frequently it is impossible to see how the same method works in the general case.

In particular, the critique of this proof asserts that the large mass of possible local configurations of spheres has not been shown by Hsiang to have his local averaged Voronoi density strictly below the optimal bound. Specifically, a major objection is that an essential part of Hsiang's proof approach is flawed, that of using deformation arguments to simplify configurations on which the density is nondecreasing. There are notorious examples where such monotonicity of density under deformation does not hold, as indicated in the objections of Hales [19] above.

These issues are not considered resolved by Hsiang's book. Concerning Hsiang's book [41], the reviewer for *Mathematical Reviews* (U. Hertrich-Jeromin) states:

> The present reviewer is not in a position to judge whether the proof of Main Theorem 1 or of the Kepler conjecture presented in this book is correct, complete, or adequate.

It may be that the Hsiang local density inequality is in fact an "optimal" inequality. If so, in principle this might be established by a computer-aided proof. To accomplish this there appear to remain extensive issues needing resolution to reduce the problem to computation, and it is not clear that the resulting computational problem would be of feasible size to attempt.

In any event, the work of Hales, with Ferguson, described below, successfully establishes a different local density inequality implying Kepler's conjecture.

It remains possible that in the future new local density inequalities will be discovered which will permit alternative proofs of the Kepler Conjecture which are simpler than the current proof.

4 Hales Program and Hales-Ferguson Papers

An approach to prove the Kepler conjecture was developed by T. Hales starting about 1990 (see [17], [18]), based on constructing a local density inequality starting from the Delaunay tessellation. A significant problem lies in formulating a local density inequality of this sort that is:

(1) expected to be "optimal," and

(2) which represents a non-linear optimization problem small enough to be solvable in a reasonable time.

One of Hales's contributions was the formulation of a larger, more flexible class of local density measures than had been formulated previously; cf. [45, Sec. 2] (in this volume).

As explained by Hales [24] (in this volume), he eventually found it necessary to modify this pure Delaunay tessellation approach to Kepler's conjecture. He then considered an approach that used hybrid inequalities that mixed Voronoi and Delaunay domains. A definite program to prove Kepler's Conjecture based on such a decomposition was formulated by Hales in 1994, described in [20], [21] (papers in this volume). Additional modifications of the local density inequality turned out to be required to obtain a tractable problem. In the (re)formulation process, important contributions were made by S. P. Ferguson, a doctoral student of Hales whose published version of his Ph.D. thesis work forms the paper *Sphere Packing V*. Significant innovations of Hales were needed to reduce the optimization problem to a tractable size.

A first version of a complete proof was presented in six preprints, one joint with S. P. Ferguson, and one by Ferguson alone, and was posted on the mathematics arXiv in 1998. The proofs then underwent intensive peer-reviewing and revision, culminating in the publication of the proof in 2005 and 2006.

The initial 1998 papers were written while Hales was a faculty member at the University of Michigan. At that time Ferguson was a graduate student at the University of Michigan, and he subsequently received his Michigan Ph.D. for work on the Kepler problem. Some of the revisions on the paper were done after Hales became a faculty member at the University of Pittsburgh.

5 Peer-Reviewing of the Hales-Ferguson Papers

A week-long workshop on the Hales-Ferguson proof was given at the Institute for Advanced Study in January 1999. At that time this editor wrote a summary of the top level of the proof tree, which was later included in [45] (paper in this volume). Also in 1999 Joseph Oesterlé gave a Séminaire Bourbaki exposé on the Hales-Ferguson proof [51].

The six papers were submitted to the *Annals of Mathematics*, to Robert McPherson as editor, and a team of at least 13 reviewers was assembled by Gabor Fejes-Tóth. It is very unusual to have such a large set of reviewers. The main portion of the reviewing took place in a seminar at Eötvos University, Budapest, over a three year period. Some reviewers made computer experiments, in a detailed check of specific parts of the proof. The nature of this proof, consisting in part of a large number of inequalities having little internal structure, and a complicated proof tree, makes it hard for humans to check every step reliably. In this process detailed checking of many specific assertions found them to be essentially correct in every case. This result of the reviewing process produced in these reviewers a strong degree of conviction of the essential correctness of this proof approach, and that the reduction method led to nonlinear programming problems of tractable size.

The popular book of George Szpiro [56], and a book review by Frank Morgan [49], give a status report on how the peer-review process of the proof was viewed as of 2003.

This editor became involved in the reviewing process in January 2003. At that time Hales undertook a substantial revision and reorganization of the papers, making clearer the top portion of the proof tree. This revison also led to a simplification of some definitions and lemmas. Ferguson's paper was also revised. The six final published papers include added remarks motivating the choice of the given local density inequality, which is very complicated. They also provide some reasons indicating why one should expect this particular local density inequality to hold. The revised version of the Ferguson paper "Sphere Packing V" now includes an explicit error bound away from the optimal density.

An abridged version of the proof, authored by T. Hales, was published in *Annals of Mathematics*, **162** (2005), 1065–1185. This paper presents the upper level details of the proof tree but excludes the paper of S. Ferguson handling one important case. This paper, presenting the upper level details of the proof tree but excluding the paper of S. P. Ferguson handling one important case, is supplemented with computer files and programs which are available on the *Annals of Mathematics* website.

In response to the issues raised by the reviewing of the Hales and Ferguson papers, the *Annals of Mathematics* added, at the front of the issue of the journal in which the Hales paper appears, the following "Statement by the Editors" [1].

> Computer-assisted proofs of exceptionally important mathematical theorems will be considered by the *Annals*.
>
> The human part of the proof, which reduces the original mathematical problem to one tractable by a computer, will be refereed for correctness in the traditional manner. The computer part may not be checked line-by-line, but will be examined for the methods by which the authors have eliminated or minimized possible sources of error: (e.g., round-off error eliminated by interval arithmetic, programming error minimized by transparent surveyable code and consistency checks, computer error minimized by redundant calculations, etc. [Surveyable means that the interested person can readily check that the code is essentially operating as claimed]).
>
> We will print the human part of the paper in an issue of the *Annals*. The authors will provide the computer code, documentation necessary to understand it, and the computer output, all of which will be maintained on the *Annals of Mathematics* website online.

This statement continues to reflect the views of the Editors of *Annals of Mathematics*, and currently appears on the journal's website.

In 2006 a detailed version of the proof was published in *Discrete & Computational Geometry*, Vol. 36, No. 1 (2006), 5–265. These papers appear in this book. They are revised and rearranged versions of the six 1998 preprints; the 2005 *Annals* paper represented a subset of these papers. These papers appeared as a complete issue of *Discrete & Computational Geometry*, edited by Gabor Fejes-Tóth and myself as Guest Editors. In an unusual feature, the Foreword to this issue lists by name (some of) the peer reviewers: Andras Bezdek, Michael Bleicher, Karoly Böröczky, Karoly Böröczky, Jr., Gabor Fejes-Tóth, Aladar Heppes, Wlodek Kuperberg, Endre Makai, Attila Por, Günter Rote, Istvan Talata, Bela Uhrin, and Zoltan Ujvary-Menyhard. This set of reviewers includes many well-known experts in discrete geometry.

6 Reliability of the Hales-Ferguson Proof

To what extent have Hales and Ferguson given a complete proof of the Kepler conjecture? The high level part of the proof tree is given as a standard mathematical proof. It has a definite overall strategy that subdivides the problem into manageable parts. It also supplies a rationale for the choice of local density inequality considered. At the bottom level of the proof tree are a large collection of inequalities verified by computer calculation. In some cases, where mathematical programming is involved, the computer made decisions as to how to break the problem into smaller pieces to achieve a proof. There has been considerable controversy over the use of such proofs relying on large amounts of computation, of a non-routine sort, which are hard for humans to check. Is such a proof absolutely reliable? We have reached a borderline between mathematics and science, where the computations can be viewed as an experiment that is in principle reproducible by others.

The topic of the nature and reliability of proofs involves both mathematical and philosophical issues. A recent discussion of this topic appears in the *Philosophical Transactions of the Royal Society A* [6] in October 2005, which mentions in particular the Hales-Ferguson computer-aided proof of the Kepler Conjecture.

A Fields Medal winner, W. P. Thurston [57, Sec. 4], argues that standard (pre-formal) mathematical proofs are not absolutely reliable, but "there is a strong social standard of validity and truth." He writes [57, p. 170]:

> Mathematics as we practice it is much more formally complete and precise than other sciences, but it is much less formally complete and precise for its content than computer programs.

Supporting the view of human checkability above, Thurston writes [57, p. 170]:

> People are usually not very good in checking *formal correctness* of proofs, but they are quite good at detecting potential weaknesses or flaws in proofs.

He adds, on revision of proofs:

> Mathematicians can and do fill in gaps, correct errors, and supply more detail and more careful scholarship when they are called on or motivated to do so. Our system is quite good at producing reliable theorems that can be backed up. It's just that the reliability does not primarily come from mathematicians checking formal arguments; it come from mathematicians thinking carefully and critically about mathematical ideas.

What is clear is that Hales and Ferguson have presented a proof schema for formulating "local density inequalities" of a kind believed to be optimal inequalities. They have presented an effective methodology for formulating a class of "local density" inequalities of a type expected to contain many "optimal" inequalities. These involve partitioning Euclidean space associated into sphere centers, first into Delaunay-like simplices, provided their shapes satisfy suitable restrictions, then partitioning the remaining space into Voronoi-like cells. They also assign weights to (subdivisions of) the Delaunay-like simplices, to define a "decomposition star" of quantities contributing to the score at a given sphere center. They also provided effective reduction methods for the nonlinear optimization problem, reducing it to tractable size for computation of the bottom levels of the proof tree. They have given very precise details of the proof tree to carry this out for a specific "optimal" local density inequalities. These details have been extensively peer-reviewed without significant error found. This holds whether or not one believes the software has been coded reliably, or the calculations entirely free of errors. Hales and Ferguson have presented details sufficient for reconstruction along these lines of proofs of a similar kind.

To consider the reliability question further, a proof published in a mathematical journal is written in a high level language, not in a logical language. To quote the philosopher of mathematics Imre Lakatos [47] in the article "What does a mathematical proof prove":

> Proofs can be divided into (1) *pre-formal,* (2) *formal,* and (3) *post-formal* proofs. Of these, (1) and (3) are informal proofs.

The pre-formal proofs in a mathematical journal typically only form an outline of a schema of formal proof, which is intended to be convincing by supplying details concerning the key ideas in the proof, and precise details on the most difficult parts of the proof. A well-written pre-formal proof often satisfies the dictum [44, p. 315]:

> The proof tells us where to concentrate our doubts.

A reader must have a certain mathematical background and training to read a pre-formal proof. Such a proof can be regarded as a schema intended to be convertible to a proof in a formal logical language. The latter would then constitute a formal proof. The resulting formal proof would not be easy to read by a human, but it could be reliably checked by a computer. In the realm of formal proofs, a digital computer is a more reliable proof-checker than a human.

The proof of the Kepler Conjecture that Hales and Ferguson presented in this volume is a pre-formal proof in the Lakatos sense. It provided a strong measure of conviction in the judgment of the reviewers, and is reliable in the Thurston sense.

However, here we may recall that Thurston also says [57, pp. 170–171]:

> When one considers how hard it is to write a computer program even approaching the intellectual scope of a good paper, and how much greater time and effort have to be put in to make it "almost" formally correct, it is preposterous to claim that mathematics as we practice it is anywhere near formally correct.

This observation motivates the pursuit of a formal proof of the Kepler conjecture. Constructing such a proof is a current project of Hales, discussed below.

We remark that the final category of proofs considered by Lakatos, that of post-formal proofs, consists of metamathematical arguments drawing conclusions about the meaning of what has been proved in a formal proof. We do not consider post-formal proof here; the meaning of the Kepler Conjecture seems intuitively clear.

7 Formal Proof of the Kepler Conjecture

Subsequent to the Hales-Ferguson proof, Hales embarked on a process of constructing a formal proof of the Kepler conjecture. This project is termed by Hales the FLYSPECK Project (FlysPecK is an acronym for Formal Proof of the Kepler conjecture). It is an ongoing project described in Hales et al. [31] (in this volume).

It is a much more difficult problem to produce and verify a formally checkable proof than it is to provide a pre-formal proof. In practice the proofs provided in mathematical journals do not supply sufficient details to make the conversion to a formal proof a routine procedure. That is, a "pre-formal proof" in a mathematics journal can be regarded as somewhat analogous to a software specification in computer science that prescribes a recipe for writing software code, but which could conceivably lead to some surprises in the actual process of writing and testing code. Such conversion to a formal proof requires a choice of axiom scheme and the axioms that are (often implicitly) assumed in the basic framework of the subject. The process of reducing to axiomatization often exposes hidden assumptions that are built into the proof. For a viewpoint firmly on the side of computers, see Zeilberger [60].

The Kepler Conjecture uses only elementary Euclidean geometry in its formulation and seems well suited to formalization of the proof. A detailed study of axiomatic foundations of geometry was undertaken by Hilbert [35], in a book first published in 1899. He noted that Euclid's list of axioms was incomplete, proposed formal axioms for Euclidean geometry and related geometries, and studied the interdependence of various axiom schemes. This work led to some surprises. Some of Hilbert's original axioms were found to be implied by others of his axioms; his book went through many editions and corrections; many appendices were added in later editions. In fact, a formal proof of the Kepler Conjecture necessarily requires going beyond the framework of Hilbert's axioms because these do not cover the discrete nature of packings. Further work on axiomatization is needed.

Constructing a formal proof of the Kepler Conjecture involves not only logic, it is also a giant software project. When completed, such a proof will represent the most complicated proof to date that is reduced to formalization. Details on this project are presented in the paper Hales et al. [31] (in this volume).

As mentioned, formalization of a proof can also uncover gaps in a pre-formal proof. This was the case here. The initial process of formalizing the proof tree of the published proof in the six papers of the Kepler Conjecture did uncover one logical gap in the published proof. This gap is described, and filled, in the second part of paper [31] (in this volume).

8 Applications of the Hales-Ferguson Proof Methodology

The ideas behind the "scoring" method developed for the proof of the Kepler Conjecture have been adapted by T. C. Hales, with others, to solve other geometric optimization problems.

Regarding the Voronoi tessellation, in 1943 Laszlo Fejes-Tóth[14] proposed the following conjecture:

Dodecahedral Conjecture *The volume of any Voronoi cell of a unit sphere packing in* $\mathbb{R}^3$ *is not smaller than the volume of a regular dodecahedron circumscribed around the unit sphere.*

In 1998 Hales and S. McLaughlin [32] announced a computer-aided proof of the Dodecahedral conjecture, based on similar methods to those in the Hales-Ferguson proof of the Kepler conjecture. After a very long peer review process, and substantial rewriting, this paper recently appeared in the *Journal of the American Mathematical Society* [33].

Another very old problem concerns partitioning the plane into cells of equal area. How should the cells be designed to minimize the average perimeter of the cells? The conjectured answer is that each cell should be a regular hexagon, as in a honeybee comb. This conjecture was proved by Hales [22] in 2001, by establishing various new two-dimensional "local density" inequalities appropriate to the problem.

9 Contents of This Volume

The remainder of this volume consists of a collection of papers reprinted from *Discrete & Computational Geometry*. These are grouped in four parts, as follows.

PART I. *Survey: Local Density Inequalities for Sphere Packing*

The next chapter of Part I reprints a paper of J. C. Lagarias [45] describing local density inequalities for giving upper bounds on the density of sphere packings in any dimension. This paper can serve as an introduction to the ideas of Hales and Ferguson; that is, it describes the particular local density inequality studied by Hales and Ferguson, as given in their original 1998 preprints. This paper was written in the middle of the review process of the Hales-Ferguson papers, and influenced reorganization of the the the six papers presenting the proof that appeared in print after peer review (Part II). As it turned out, the proof given in the six papers in Part II made some further small modifications and simplifications to the formulation of the local density inequality used in the 1998 preprints, as explained by T. C. Hales in the first of the six papers presenting the published proof.

PART II. *Proof: Hales-Ferguson Proof of the Kepler Conjecture*

This part of the book reprints the detailed proof of the Kepler Conjecture given in the six papers by T. C. Hales with S. P. Ferguson that appeared in *Discrete & Computational Geometry* in 2006, and form the main body of this book. It also reproduces the editorial foreword of Gábor Fejes-Tóth and Jeffrey C. Lagarias concerning the editing of this proof.

The first of the six papers, "Historical overview of the Kepler Conjecture," authored by T. C. Hales, provides a historical overview of the problem and some remarks on the main features of the proof, and gives motivation for the decomposition of space and choice of "score function" used in the proof.

The second of the six papers, "A formulation of the Kepler Conjecture," co-authored by T. C. Hales and S. P. Ferguson, formulates the partition of space and the weights giving the local density "score function." A primary feature of the partition of space is its use of Delaunay-like tetrahedra for the main part of the partition when these are of suitable shape (in the two optimal configurations only tetrahedra of this type appear); and then the use of Voronoi-like regions for the remaining part of the decomposition. The partition of space assigns a "'decomposition star" to a sphere center, which comprises (weighted) volumes of regions in a series of polyhedral cones with apex at the sphere center. There are an enormous variety of such decomposition stars. The weighting scheme introduced for the tetrahedra is a very important part of the "score function." This paper formulates the precise local density inequality that if proved, will establish Kepler's conjecture as a corollary.

The third of the six papers, "Sphere Packings III. Extremal cases," authored by T. C. Hales, formulates preliminary geometric inequalities to the proof. It treats the two extremal cases where equality will hold in the inequality, corresponding to the two types of Voronoi cells 1 and 2 above. It shows that they are local maxima of the "score function."

The fourth of the six papers, "Sphere packings IV. Detailed bounds," authored by T. C. Hales, is the technical heart of the proof. The proof studies the structure of possible decomposition stars that might produce a counterexample, termed contravening decomposition stars. These would be decomposition stars with score exceeding a quantity 8 pt, the "optimal" constant. (Here a "point" $pt := 4\tan^{-1}(\frac{\sqrt{2}}{5}) - \frac{\pi}{3} \approx 0.05537$.) Associated to a decomposition star is a graph corresponding to intersecting the cones in the decomposition with a small sphere around the sphere center. A priori the graph of a decomposition star could have disconnected pieces, as well as have regions that are not convex, etc. The proof aims to show the set of graphs of contravening decomposition stars is empty. As a step towards this, this paper shows the graphs associated to a contravening decomposition star must be very restricted in geometry. The regions in the associated graph are connected and polygonal with at most 8 sides, with various further restrictions concerning weights associated with the polygonal regions. It relies on many detailed computer calculations.

The fifth of the sixth papers, "Sphere packings V. Pentahedral prisms," authored by S. P. Ferguson, treats the case of a decomposition star of a particular shape, having a particular associated graph, called a "pentrahedral prism." It proves that the score is strictly below the optimal value 8 pt in these cases.

The sixth paper, "Sphere packings VI. Tame graphs and linear programs," authored by T. C. Hales, deals with the large mass of possible contravening decomposition stars remaining from the fourth paper. The first part of this paper describes a family of abstract planar graphs called *tame*, having no faces of degree 9 or greater, and other properties. It classifies all such graphs, showing there are a few thousand of them, and then uses the results of the fourth paper, with minor modifications, to show that any contravening decomposition star must have associated graph a tame graph, according to this definition. The second part of this paper uses linear programming to obtain upper bounds on the score of decomposition stars having associated tame graphs. These computations rule out all tame graphs as contravening, with three exceptions. The exceptions are all tame graphs corresponding to the FCC Voronoi cell, the HCP Voronoi cell, and the pentahedral prism. These three cases were already treated in the third and fifth papers. The local density inequality follows.

Strictly speaking, these six papers describe only the upper levels of the proof tree, since at the bottom lies a massive computer verification of many inequalities. References to computer verifications is provided in the papers.

As remarked above, an abridged version of this proof, omitting details of proof for a result of S. Ferguson (given in the fifth paper above, [28]). was published by T. C. Hales [23] in 2005 in the *Annals of Mathematics*, the top mathematical journal.

PART III. *Revision: A revision of the proof of the Kepler Conjecture*

Work on the proof did not cease with its publication. This part contains the paper "A revision of the Proof of the Kepler Conjecture," authored by T. C. Hales, J. Harrison, S. McLaughlin, T. Nipkow, S. Obua, and R. Zumkeller [31], which appeared in 2010.

The first part of this paper describes ongoing work to produce a formal proof of the Kepler Conjecture. This project is termed by Hales the FLYSPECK project. The goal is to provide a proof of the Kepler conjecture within a formal logical language, entirely checkable by computer. The second part of the paper deals with the discovery of one gap in logic in the original proof, uncovered in the process of formalization. This part of the paper, written by T. C. Hales alone, fills in that gap in the original proof. At the end, this paper supplies a list of errata in the original proof in *Discrete & Computational Geometry*. The locations of these errata in the original proof are indicated by asterisks added in the margins of the papers in Part II of this volume.

PART IV. *Prehistory: Initial Hales papers on the Kepler Conjecture*

For historical purposes, this volume also includes two of T. C. Hales's earlier papers in 1997 on the Kepler conjecture. These papers, "Sphere Packings I" [20] and and "Sphere Packings II" [21], give his original formulation of this research program to prove the Kepler conjecture. Later work required a modification of Hales's original approach. In particular in the published proof the scoring function was given a new definition, slightly different from the one used in "Sphere Packings I" and "Sphere Packings II." The final version of the six papers giving the proof above were written to be logically independent of the first two papers. However certain auxiliary results were first proved in "Sphere Packings I" and "Sphere Packings II."

Acknowledgements I thank Gabor Fejes-Toth for editorial work on the DCG Special Issue in July 2006. I thank him, Thomas C. Hales, Steven G. Krantz, Richard Pollack, and K. Soundararajan for helpful comments on this paper.

References

1. Annals of Mathematics Editors, *Statement by the Editors*, Annals of Mathematics **162** (2005), no. 3, un-numbered page preceding page 1165.
2. T. Aste and D. Weaire, *The Pursuit of Perfect Packing*, Institute of Physics Publishing, Bristol and Philadelphia 2000.
3. William Barlow, Probable nature of the internal symmetry of crystals, Nature **29** (1883), No. 738, 186–188.
4. A. Bezdek, K. Bezdek and R. Connelly, Finite and uniform stability of sphere packings, Disc. & Comput. Geom. **20** (1998), no, 1, 111–130.
5. K. Böröczky, Jr, *Finite Packing and Covering*, Cambridge Math. Tracts No. 154, Cambridge Univ. Press: Cambridge 2004.
6. A. Bundy (Editor), Discussion Meeting Issue 'The nature of mathematical proof' organized by A. Bundy, M. Atiyah, A. Macintyre and D. Mackenzie, Phil. Trans. of the Royal Society A Mathematical, Physical & Engineering Sciences **363** (2005), Issue 1835, October 15, 2005, pp. 2331–2461.
7. W. Casselman, *The Difficulties of Kissing int Three Dimensions,* Notices of the Amer. Math. Soc. **51**, No. 8 (2004), 884–885.
8. J. H. Conway, C. Goodman-Strauss and N. J. A. Sloane, Recent progress in sphere packing, in: *Current developments in mathematics, 1999 (Cambridge, MA)*, pp. 37–76, International Press, Somerville, MA1999.
9. J. H. Conway, T. C. Hales, D. J. Muder and N. J. A. Sloane, On the Kepler Conjecture, Letter to the Editor, Math. Intelligencer **16** (1994) No. 2, 5.
10. J. H. Conway and N. J. A. Sloane, *Sphere Packings, Lattices and Groups*, (Third Edition) Springer-Verlag: New York 1999.
11. H. S. M. Coxeter, *Introduction to Geometry*, Reprint of the 1969 Edition. John Wiley and Sons: New York 1989.
12. B. Delaunay, Sur la sphère vide, Izv. Akad. Nauk. SSSR, **7** (1934), 793–800.
13. G. Fejes Tóth and J. C. Lagarias, Guest editor's foreword, Discrete Comput. Geom. **36** (2006), 1–3.
14. L. Fejes Tóth, Über die dichteste Kugellagerung, Math. Zeitschrift **48** (1943), 676–684.
15. L. Fejes Tóth, *Lagerungen in der Ebene auf der Kugel und im Raum*, Springer-Verlag: Berlin 1953. (Second Edition 1972) [see pp. 171–181, both editions].

16. L. Fejes Tóth, *Regular Figures*, MacMillan: New York 1964.
17. T. C. Hales, The Sphere Packing Problem, J. Comp. App. Math. **44** (1992), 41–76.
18. T. C. Hales, Remarks on the density of sphere packings in three dimensions, Combinatorica, **13** (2), (1993), 181–197.
19. T. C. Hales, The status of the Kepler conjecture, Math. Intelligencer **16** (1994), no. 3, 47–58.
20. T. C. Hales, Sphere Packings I, Discrete Comput. Geom. **17** (1997), 1–51, eprint: math.MG/9811073.
21. T. C. Hales, Sphere Packings II, Discrete Comput. Geom. **18** (1997), 135–149, eprint: math.MG/9811074.
22. T. C. Hales, The honeycomb conjecture, Disc. Comput. Geom. **25** (2001), 1–22.
23. T. C. Hales, A proof of the Kepler conjecture, Annals of Math. **162** (2005), no. 3, 1065–1185.
24. T. C. Hales, Historical Overview of the Kepler Conjecture, Discrete Comput. Geom. **36** (2006), 5–20.
25. T. C. Hales and S. P. Ferguson, A Formulation of the Kepler Conjecture, Discrete Comput. Geom. **36** (2006), 21–69.
26. T. C. Hales, Sphere Packings III. Extreme cases, Discrete Comput. Geom. **36** (2006), 71-110.
27. T. C. Hales, Sphere Packings IV. Detailed bounds, Discrete Comput. Geom. **36** (2006), 111-166.
28. S. P. Ferguson, Sphere Packings V. Pentahedral prisms, Discrete Comput. Geom. **36** (2006), 167–204.
29. T. C. Hales, Sphere Packings VI. Tame graphs and linear programs, Discrete Comput. Geom. **36** (2006), 205–265.
30. T. C. Hales, Formal Proof, Notices of the American Math. Soc. **55**, No. 11 (2008). 1370-1380.
31. T. C. Hales, J. Harrison, S. McLaughlin, T. Nipkow, S. Obua and R. Zumkeller, A Revision of the Kepler Conjecture, Discrete Comput. Geom. **44** (2010), 1–34.
32. T. C. Hales and S. McLaughlin, A proof of the dodecahedral conjecture, preprint, Nov. 1998, arXiv:math/9811079 v1, 54 pp.
33. T. C. Hales and S. McLaughlin, A proof of the dodecahedral conjecture, J. Amer. Math. Soc. **23** (2010), 299–344.
34. D. Hilbert, Mathematische Probleme, Nachrichten Kon. Ges. d. Wiss. Göttingen, Math.-Phys. Kl., 1900, pp. 253–297.
    English translation in: D. Hilbert, Mathematical Problems, Bull. Amer. Math. Soc. **8** (1902), 437–479. Reprinted in: *Mathematical Developments Arising from Hilbert Problems*, Proc. Symp. Pure Math XXVIII, American Math. Soc.: Providence 1976.
35. D. Hilbert, *Grundlagen der Geometrie, 9th Edition, revised by Paul Bernays*, Teubner: Stuttgart 1962. English Translation: D. Hilbert, *Foundations of Geometry* Authorized Translation by E. J. Townsend (1902) Open Court Publ. Co. 1959.
36. R. Hoppe, Berkeung der Redaktion, Archiv der Mathematik und Physik **56** (1874), 307–312. [Comment on paper: C. Bender, Bestimmung der grössten Anzahl gleich grosser Kugeln, welche sich auf eine Kugel von demselben Radius, wie die übrigen, auflegen lassen, Acrhiv der Mathematik und Physik **56** (1874), 302-306.]
37. W.-Y. Hsiang, On the sphere packing problem and the proof of Kepler's conjecture, International J. Math. **4** (1993), No. 5, 739–831. (Math Reviews: 95g:52032)
38. W.-Y. Hsiang, On the sphere packing problem and the proof of Kepler's conjecture, in *Differential Geometry and Topology (Alghero, 1992)*, World Scientific: River Edge, NJ 1993, pp. 117–127. (Math Reviews: 96f:52028)
39. W.-Y. Hsiang, The geometry of spheres, in: *Differential Geometry (Shanghai 1991)*, World Scientific : River Edge, NJ 1993, pp. 92–107.
40. W.-Y. Hsiang, A rejoinder to T. C. Hales's article "The status of the Kepler conjecture," Math. Intelligenger **17**, No. 1 91995), 35–42. (Math Reviews: 97f:52029)
41. W.-Y. Hsiang, *Least Action Principle of Crystal Formation of Dense Packing Type and the Proof of Kepler's Conjecture,* World Scientific: River Edge, NJ 2002. (Math Reviews: 2004g:52033)
42. R. H. Kargon, *Atomism in England from Hariot to Newton,* Oxford: Clarendon Press 1966.
43. Johannes Kepler, *Strena Seu de Niue Sexangula*, Published by Godfrey Tampach at Frankfort on Main, 1611. English Translation in: J. Kepler, *The Six-Cornered Snowflake* (Colin Hardie, Translator) Oxford University Press: Oxford 1966.
44. M. Kline, *Mathematics: The Loss of Certainty*, Oxford University Press 1982.
45. J. C. Lagarias, Local density bounds for sphere packings and the Kepler Conjecture, Disc. Comp. Geom. **27** (2002), 165–193,
46. I. Lakatos, *Proofs and Refutations, The Logic of Mathematical Discovery,* (ed. J. Worrall and E. Zahar), Cambridge University Press: Cambridge 1976.
47. I. Lakatos, What does a mathematical proof prove?, pp. 61–69 in: I. Lakatos, *Mathematics, Science and Epistemology, Philosphical Papers, Volume 2,* (J.Worrell and G. Currie, Eds), Cambridge Univ. Press; Cambridge 1978.
48. J. Milnor, Hilbert's problem 18: On Crystalographic groups, fundamental domains, and on sphere packing, pp. 491–506 in: F. W. Browder, Ed., *Mathematical Developments Arising from the Hilbert Problems*, Proc. Symp. Pure Math. **28**, American Math. Soc.: Providence, 1976.

49. F. Morgan, Kepler's Conjecture and Hales's Proof. A Book Review, Notices of the American Math. Society **52** No. 1 (2005), 44–47.
50. I. Newton, *The Correspondence of Isaac Newton*, (9 volumes) H. W. Turnbull, F. R. S. (Ed.), Cambridge University Press 1961.
51. J. Oesterlé, Densité maximale des empilements de sphères en dimension 3 [d'après Thomas C. Hales et Samuel P. Ferguson], Séminaire Bourbaki, Exp. No. 863, Juin 1999.
52. C. A. Rogers, The Packing of equal spheres, Proc. London Math. Soc. **8** (1958), 609–620.
53. C. A. Rogers, *Packing and Covering*, Cambridge Tracts on Mathematics and Physical Science No. 54, Cambridge Univ. Press, New York 1964.
54. K. Schütte and B. L. van der Waerden, Das Problem der dreizehn Kugeln, Math. Annalen **125** (1953), 325–334.
55. John W. Shirley, *Thomas Hariot: A Biography*, Clarendon Press: Oxford 1983.
56. George G. Szpiro, *Kepler's Conjecture: How Some of the Greatest Minds in History Helped Solve One of the Oldest Math Problems in the World,* Wiley: New York 2003.
57. W. P. Thurston, On proof and progress in mathematics, Bull. Amer. Math. Soc. **30** (1994), no. 2, 161–177.
58. J. P. Troadec, A. Gervois and L. Oger, Statistics of Voronoi cells of slightly perturbed face-centered cubic and hexagonal close-packed lattices, Europhysics Letters **42** no. 2 (1998), 167–172.
59. D. F. Watson, Computing the n-dimensional Delaunay tesselation with applications to Voronoi polytopes, Computer J. **24** (1981), 167–172.
60. D. Zeilberger, Theorems for a price: tomorrow's semi-rigorous mathematical culture, Notices of the AMS **40** (1993), no. 8, 978–981.
61. C. Zong, *Sphere Packings*, Springer-Verlag: New York 1999.

2

Bounds for Local Density of Sphere Packings and the Kepler Conjecture, by J. C. Lagarias

This paper presents a framework for local density inequalities giving upper bounds for sphere packing density in arbitrary dimensions. It discusses in detail such inequalities in the three-dimensional case, and the Kepler conjecture.

Sections 4 and 5 of this paper outline the Hales-Ferguson proof as presented in their six 1998 preprints. The final version of the proof given in their 2006 papers, given in Part II, made further small changes from the outline here. In particular, their treatment of V-cells is modified, as described in Section 5.1 of the paper T. C. Hales and S. P. Ferguson, "A Formulation of the Kepler Conjecture"; see especially Remarks 5.10 and 5.11. The modified treatment eliminates the need to handle "tips" as described in Definition 4.5 of this paper.

Contents

Discrete Comput Geom 27:165–193 (2002)
DOI: 10.1007/s00454-001-0060-9

Bounds for Local Density of Sphere Packings and the Kepler Conjecture

J. C. Lagarias

AT&T Labs - Research,
Florham Park, NJ 07932-0971, USA
jcl@research.att.com

Abstract. This paper formalizes the local density inequality approach to getting upper bounds for sphere packing densities in $\mathbb{R}^n$. This approach was first suggested by L. Fejes Tóth in 1953 as a method to prove the Kepler conjecture that the densest packing of unit spheres in $\mathbb{R}^3$ has density $\pi/\sqrt{18}$, which is attained by the "cannonball packing." Local density inequalities give upper bounds for the sphere packing density formulated as an optimization problem of a nonlinear function over a compact set in a finite-dimensional Euclidean space. The approaches of Fejes Tóth, of Hsiang, and of Hales to the Kepler conjecture are each based on (different) local density inequalities. Recently Hales, together with Ferguson, has presented extensive details carrying out a modified version of the Hales approach to prove the Kepler conjecture. We describe the particular local density inequality underlying the Hales and Ferguson approach to prove Kepler's conjecture and sketch some features of their proof.

1. Introduction

The Kepler conjecture was stated by Kepler in 1611. It asserts that the face-centered cubic lattice gives the tightest possible packing of unit spheres in $\mathbb{R}^3$.

Kepler Conjecture. *Any packing Ω of unit spheres in $\mathbb{R}^3$ has upper packing density*

$$\bar{\rho}(\Omega) \leq \frac{\pi}{\sqrt{18}} \simeq 0.740480. \qquad (1.1)$$

The definition of upper packing density is given in Section 2. The problem of proving the Kepler conjecture appears as part of Hilbert's 18th problem, see [Hi].

There have been many attempts to prove Kepler's conjecture. Since the 1950s these have been based on finding a local density inequality that gives a (sharp) upper bound on the density. In recent years Hales has developed an approach for proving Kepler's conjecture based on such an inequality, and in 1998 he announced a proof, completed with the aid of Ferguson, presented as a set of six preprints. The proof is computer-intensive, and involves checking over 5000 subproblems. It involves several new ideas which are indicated in Sections 4 and 5.

Local density inequalities obtain upper bounds for the sphere packing constant via an auxiliary nonlinear optimization problem over a compact set of "local configurations." They measure a "local density" in the neighborhood of each sphere center separately. The general approach to the Kepler conjecture is first to find a local optimization problem that actually attains the optimal bound $\pi/\sqrt{18}$ (assuming that one exists), and then to prove it. This approach was first suggested in the early 1950s by Fejes Tóth [FT1, pp. 174–181], who presented some evidence that an optimal local density inequality might exist in three dimensions. In 1993 Hsiang [Hs1] presented another candidate for an optimal inequality.

The objects of this paper are:

(i) To formulate local density inequalities for sphere packings in arbitrary dimension $\mathbb{R}^n$, in sufficient generality to include the known candidates for optimal local inequalities in $\mathbb{R}^3$.

(ii) To review the history of local density inequalities for three-dimensional sphere packing and the Kepler conjecture.

(iii) To give a precise statement of the local density inequality considered in the Hales–Ferguson approach.

(iv) To outline some features of the Hales–Ferguson proof.

In Section 2 we present a general framework for local density inequalities, which is valid in $\mathbb{R}^n$, given as Theorem 2.1. This framework is sufficient to cover the approaches of Fejes Tóth, Hsiang, and Hales and Ferguson to Kepler's conjecture. A different framework for local density inequalities appears in [O]. In Section 3 we review the history of work on local optimization inequalities for Kepler's conjecture. In Section 4 we describe the precise local optimization problem formulated by Ferguson and Hales in [FH], which putatively attains $\pi/\sqrt{18}$. In Section 5 we remark on some details of the proof strategy taken in the papers of Hales [SP-I], [SP-II], [SP-III], [SP-IV], [KC], Ferguson and Hales [FH], and Ferguson [SP-V]. In Section 6 we make some concluding remarks.

The current status of the Hales–Ferguson proof is that it appears to be sound. The proof has been examined in fairly careful detail by a team of reviewers, but it is so long and complicated that it seems difficult for any one person to check it. This paper is intended as an aid in understanding the overall structure of the Hales–Ferguson proof approach, as presented in the preprints. For another account of the Hales and Ferguson work, see [O]. For Hales' own perspective, see [Ha4]. It may yet be some time until a version of the Hales–Ferguson proof is published; significant simplification may yet occur, and the local density inequality to be proved in the final version, which we may call "Version 2.0," may differ in some details from the one described in Section 4.

Two appendices are included which contain some information relevant to the Hales–Ferguson proof given in the preprints. Appendix A describes some of the Hales–Ferguson scoring functions. Appendix B lists references in the Hales and Ferguson preprints for proofs of lemmas and theorems stated without proof in Sections 4 and 5.

This paper is a slightly revised version of the manuscript [L].

Notation. $\mathbf{B}_n := B_n(\mathbf{0}; 1) = \{\mathbf{x} \in \mathbb{R}^n : \|\mathbf{x}\| \le 1\}$ is the unit n-sphere. It has volume $\kappa_n := \pi^{n/2} / \Gamma(n/2 + 1)$, with $\kappa_2 = \pi$ and $\kappa_3 = 4\pi/3$. We let $\mathbf{C}_n(\mathbf{x}, T) := \mathbf{x} + [0, T]^n$ denote an n-cube of sidelength T, with sides parallel to the coordinate axes, and lowest corner at $\mathbf{x} \in \mathbb{R}^n$.

2. Local Density Inequalities

In this section we present a general formulation of local density inequalities. We first recall the standard definition of sphere packing densities, following Rogers [R2]. Let Ω denote a set of unit sphere centers, so that $\|\mathbf{v} - \mathbf{v}'\| \ge 2$ for distinct $\mathbf{v}, \mathbf{v}' \in \Omega$. The associated sphere packing is $\Omega + \mathbf{B}_n$.

Definition 2.1.

(i) For a bounded region S in $\mathbb{R}^n$, and a sphere packing $\Omega + \mathbf{B}_n$ specified by the sphere centers Ω, the *density* $\rho(S) = \rho(\Omega, S)$ of the packing in the region S is

$$\rho(S) := \frac{vol(S \cap (\Omega + \mathbf{B}_n))}{vol(S)}. \tag{2.1}$$

(ii) For $T > 0$ the *upper density* $\rho(\Omega, T)$ is the maximum density of the packing Ω over all cubes of size T, i.e.,

$$\bar{\rho}(\Omega, T) := \sup_{\mathbf{x} \in \mathbb{R}^3} \rho([0, T]^n + \mathbf{x}). \tag{2.2}$$

Then the *upper packing density* of Ω is

$$\bar{\rho}(\Omega) := \limsup_{T \to \infty} \bar{\rho}(\Omega, T). \tag{2.3}$$

(iii) The *sphere packing density* $\delta(\mathbf{B}_n)$ of the ball $\mathbf{B}_n$ of unit radius is

$$\delta(\mathbf{B}_n) := \sup_{\Omega} \bar{\rho}(\Omega). \tag{2.4}$$

Definition 2.2. A sphere packing Ω is *saturated* if no new sphere centers can be added to it.

To obtain sphere packing bounds it obviously suffices to study saturated sphere packings, and in what follows we assume that all packings are saturated unless otherwise stated.

Definition 2.3. An *admissible partition rule* is a rule assigning to each saturated packing Ω in $\mathbb{R}^n$ a collection of closed sets $\mathcal{P}(\Omega) := \{R_\alpha = R_\alpha(\Omega)\}$ with the following properties:

(i) *Partition.* Each set R_α is a finite union of bounded convex polyhedra. The sets R_α cover $\mathbb{R}^3$ and have pairwise disjoint interiors.

(ii) *Locality.* There is a positive constant C (independent of Ω) such that each region R_α has

$$diameter(R_\alpha) \le C. \tag{2.5}$$

Each R_α is completely determined by the set of sphere centers $\mathbf{w} \in \Omega$ with

$$distance(\mathbf{w}, R_\alpha) \le C. \tag{2.6}$$

There are at most C regions intersecting any cube of side 1.

(iii) *Translation-Invariance.* The partition assigned to the translated packing $\Omega' = \Omega + \mathbf{x}$ consists of the sets $\{R_\alpha(\Omega) + \mathbf{x}\}$.

Definition 2.4. An *admissible weighting rule* or *admissible scoring rule* σ for an admissible partition rule in $\mathbb{R}^n$ is a rule which for each Ω assigns to each pair $(R_\alpha, \mathbf{v})$ consisting of a region $R_\alpha \in \mathcal{P}(\Omega)$ and a center $\mathbf{v} \in \Omega$, a real weight $\sigma(R_\alpha, \mathbf{v})$ which satisfies $|\sigma(R_\alpha, \mathbf{v})| < C^*$ for an absolute constant C^*, and which has the following properties:

(i) *Weighted Density Average.* There are positive constants A and B (independent of Ω) such that for each set R_α,

$$\sum_{\mathbf{v} \in \Omega} \sigma(R_\alpha, \mathbf{v}) = (A\rho(R_\alpha) - B)\, vol(R_\alpha), \tag{2.7}$$

where

$$\rho(R_\alpha)\, vol(R_\alpha) = vol(R_\alpha \cap (\Omega + \mathbf{B}_n)) \tag{2.8}$$

measures the volume covered in R_α by the sphere packing Ω with unit spheres.

(ii) *Locality.* There is an absolute constant C (independent of Ω) such that each value $\sigma(R_\alpha, \mathbf{v})$ is completely determined by the set of sphere centers $\mathbf{w} \in \Omega$ with $\|\mathbf{w} - \mathbf{v}\| \le C$. Furthermore,

$$\sigma(R_\alpha, \mathbf{v}) = 0 \quad \text{if} \quad dist(\mathbf{v}, R_\alpha) > C. \tag{2.9}$$

(iii) *Translation-Invariance.* The weight function σ' assigned to the translated packing $\Omega' = \Omega + \mathbf{x}$ satisfies

$$\sigma'(R_\alpha + \mathbf{x}, \mathbf{v} + \mathbf{x}) = \sigma(R_\alpha, \mathbf{v}). \tag{2.10}$$

Note that this definition specifically allows negative weights.

The "local density" is measured by the sum of the weights associated to a given vertex $\mathbf{v}$ in a saturated packing.

Definition 2.5.

(i) The *vertex D-star* (or *decomposition star*) $\mathcal{D}(\mathbf{v})$ at a vertex $\mathbf{v} \in \Omega$ consists of all sets $R_\alpha \in P(\Omega)$ such that $\sigma(R_\alpha, \mathbf{v}) \neq 0$.

(ii) The *total score* assigned to a vertex D-star $\mathcal{D}(\mathbf{v})$ at $\mathbf{v} \in \Omega$ is

$$Score(\mathcal{D}(\mathbf{v})) := \sum_{R_\alpha \in \mathcal{D}(\mathbf{v})} \sigma(R_\alpha, \mathbf{v}). \tag{2.11}$$

The total score at $\mathbf{v}$ depends only on regions entirely contained within distance C of $\mathbf{v}$. Any admissible partition and weight function (P, σ) together yield a local inequality for the density of sphere packings, as follows.

Theorem 2.1. *Given an admissible partition rule and an admissible weighting rule (P, σ) for saturated packings in $\mathbb{R}^n$, set*

$$\theta = \theta_{P,\sigma}(A, B) := \sup_{\Omega \ saturated} (\sup_{\mathbf{v} \in \Omega} Score(\mathcal{D}(\mathbf{v}))), \tag{2.12}$$

and suppose that $\theta < \kappa_n A$, where κ_n is the volume of the unit n-sphere. Then the maximum sphere packing density satisfies

$$\delta(\mathbf{B}_n) \leq \frac{\kappa_n B}{\kappa_n A - \theta}. \tag{2.13}$$

Remarks. (1) We let $f(A, B, \theta) := \kappa_n B / (\kappa_n A - \theta)$ denote the packing density bound as a function of the score constants A and B. The sphere packing density bound actually depends only on the *score constant ratio* B/A, rather than on B and A separately, since θ is a homogeneous linear function of A and B. This ratio determines the relative weighting of covered and uncovered volume used in the inequality.

(2) A natural approach to sphere packing bounds, used in many previous upper bounds, is to partition space into pieces $R(\mathbf{v})$ corresponding to each sphere center $\mathbf{v}$, with each piece containing the unit sphere around $\mathbf{v}$. Then one obtains a bound $\bar{\rho}(\Omega) \leq \theta$ by establishing an upper bound

$$\rho(R(\mathbf{v})) = \frac{\kappa_n}{vol(R(\mathbf{v}))} \leq \theta, \quad \text{all} \quad \mathbf{v} \in \Omega. \tag{2.14}$$

Any optimal sphere packing bound of this type must necessarily be *volume-independent*, in the sense that if equality is to be attained at all local cells $R(\mathbf{v})$ simultaneously, then they must all have the same volume. In contrast the inequality of Theorem 2.1 does take into account the volumes of the individual pieces in the vertex D-star. This flexibility results in a larger class of local density inequalities. The idea of using a score constant ratio B/A not equal to the optimal packing density is due to Hales [Ha1], [Ha2], cf. Appendix B.

Proof of Theorem 2.1. We may assume that Ω is saturated. Given $T > 0$ and any $\varepsilon > 0$, we choose a point $\mathbf{x} \in \mathbb{R}^n$ which attains the density bound $\bar{\rho}(\Omega, T)$ on the cube $\mathbf{C}_n(\mathbf{x}, T)$

to within ε. We evaluate the scores of all vertex D-stars of vertices $\mathbf{v} \in \Omega \cap \mathbf{C}_n(\mathbf{x}, T)$ in two ways. First, by definition of θ,

$$\sum_{\mathbf{v} \in \Omega \cap \mathbf{C}_n(\mathbf{x}, T)} Score(\mathcal{D}(\mathbf{v})) \le \theta \# |\Omega \cap \mathbf{C}_n(\mathbf{x}, T)|. \tag{2.15}$$

However, we also have

$$\sum_{\mathbf{v} \in \Omega \cap \mathbf{C}_n(\mathbf{x}, T)} Score(\mathcal{D}(\mathbf{v})) = \sum_{\mathbf{v} \in \Omega \cap \mathbf{C}_n(\mathbf{x}, T)} \left(\sum_\alpha \sigma(R_\alpha, \mathbf{v}) \right)$$

$$= \sum_{R_\alpha \subseteq \mathbf{C}_n(\mathbf{x}, T)} \left(\sum_{\mathbf{v} \in \Omega \cap \mathbf{C}_n(\mathbf{x}, T)} \sigma(R_\alpha, \mathbf{v}) \right) + O(T^{n-1})$$

$$= \sum_{R_\alpha \subseteq \mathbf{C}_n(\mathbf{x}; T)} (A\rho(R_\alpha) - B) \, vol(R_\alpha) + O(T^{n-1})$$

$$= \kappa_n A \# |\Omega \cap \mathbf{C}_n(\mathbf{x}, T)| - B \, vol(\mathbf{C}_n(\mathbf{x}, T)) + O(T^{n-1})$$

$$= \kappa_n A \# |\Omega \cap \mathbf{C}_n(\mathbf{x}, T)| - BT^n + O(T^{n-1}). \tag{2.16}$$

Here we use that fact that $\{R_\alpha\}$ partitions $\mathbb{R}^n$, so covers the cube, and the $O(T^{n-1})$ error terms above occur because the counting is not perfect within a constant distance C of the boundary of the cube. Combining these evaluations yields

$$(\kappa_n A - \theta) \# |\Omega \cap \mathbf{C}_n(\mathbf{x}, T)| \le BT^n + O(T^{n-1}).$$

If $\theta < \kappa_n A$, then we can rewrite this as

$$\frac{\# |\Omega \cap \mathbf{C}_n(\mathbf{x}, T)|}{T^n} \le \frac{B}{\kappa_n A - \theta} + O\left(\frac{1}{T}\right). \tag{2.17}$$

By assumption

$$\bar{\rho}(\Omega, T) - \varepsilon \le \frac{vol(\mathbf{C}(\mathbf{x}, T) \cap (\Omega + \mathbf{B}))}{T^3}$$

$$= \kappa_n \frac{\# |\Omega \cap \mathbf{C}(\mathbf{x}, T)|}{T^n} + O\left(\frac{1}{T}\right).$$

Together with (2.17), this yields

$$\bar{\rho}(\Omega, T) - \varepsilon \le \frac{\kappa_n B}{\kappa_n A - \theta} + O\left(\frac{1}{T}\right), \tag{2.18}$$

with an O-symbol constant independent of ε. Letting $\varepsilon \to 0$ and then $T \to \infty$ gives the inequality for $\bar{\rho}(\Omega)$. Since this holds for all saturated packings the result follows. $\square$

Determining the quantity $\theta_{\mathcal{P}, \sigma}(A, B)$ for fixed A, B can be viewed as a nonlinear optimization problem over a compact set. The translation-invariance property of $(\mathcal{P}, \sigma)$

allows the supremum (2.12) to be taken over the smaller set with $\mathbf{v} = \mathbf{0}$ and admissible Ω containing $\mathbf{0}$. The locality property shows that the vertex D-star at $\mathbf{0}$ is completely determined by $\mathbf{w} \in \Omega$ with $\|\mathbf{w}\| \leq C$. The set of such configurations of nearby sphere centers forms a compact set in the Euclidean topology. Actually the partition and weight functions may be discontinuous functions of the locations of sphere centers, so the optimization problem above is not genuinely over a compact set. One must compactify the space of allowable vertex D-stars by allowing some sets of sphere centers to be assigned more than one possible vertex D-star. In practical cases there will be a finite set of types of vertex D-stars, hence a finite upper bound on the number of possibilities.

Definition 2.6. A local density inequality in $\mathbb{R}^n$ is *optimal* if

$$f(A, B, \theta_{P,\sigma}) = \delta(\mathbf{B}_n). \tag{2.19}$$

Optimal local density inequalities exist in one and two dimensions. In discussing the three-dimensional case, we presume that $\delta(\mathbf{B}_3) = \pi/\sqrt{18}$, so that an optimal density inequality in $\mathbb{R}^3$ will refer to one achieving this value. The evidence indicates that there are many different possible optimal local density inequalities in three dimensions, including that of the Hales and Ferguson proof.

There are currently four candidates for local density inequalities that may be optimal in three dimensions. The first is that of Fejes Toth, described in Section 3, which uses averages over Voronoi domains, in which the score constant ratio $B/A = \pi/\sqrt{18}$. The second is that of Hsiang [Hs1], which is a modification of the Fejes Tóth averaging, and uses the same score constant ratio. The third is due to Hales [SP-I], and is based on the Delaunay triangulation, using a modified scoring rule described in Section 3. The fourth is that given by Ferguson and Hales [FH], and uses a combination of Voronoi-type domains and Delaunay-type simplices, with a complicated scoring rule, described in Section 4. In the latter two cases the score constant ratio is

$$\frac{B}{A} = \delta_{\text{oct}} = \frac{-3\pi + 12 \arccos(1/\sqrt{3})}{\sqrt{8}} \approx 0.720903.$$

In all four cases the compact set of local configurations to be searched has very high dimension. Each sphere center has three degrees of freedom, and the number of sphere centers involved in these methods to determine a vertex D-star seems to be around 50, so the search space consists of components of dimension up to roughly 150.

It is unknown whether *optimal* local density inequalities exist for the sphere packing problem in $\mathbb{R}^n$ in any dimension $n \geq 4$. In dimensions 4, 8, and 24 it seems plausible that the minimal volume Voronoi cell in any sphere packing actually occurs in the densest lattice packing. If so, the Voronoi cell decomposition would yield an optimal local inequality in these dimensions, and the densest packing would be a lattice packing in these dimensions. Another question asks: in which dimensions is the maximal sphere packing density attained by a sphere packing whose centers form a finite number of cosets of an n-dimensional lattice? Perhaps in such dimensions an optimal local density inequality exists.

The state of the art in sphere packings in dimensions 4 and above is given in [CS] and [Z]. Apart from this, a very interesting new approach giving upper bounds for the density of sphere packings appears in the thesis of Henry Cohn [C]. His bounds are nearly sharp in dimensions 8 and 24, and strongly suggest that the E_8 and Leech lattice Λ_{24} packings are optimal in these dimensions, respectively.

3. History

We survey results on local density inequalities in three dimensions. The work on local density bounds was originally based on two partitions of $\mathbb{R}^3$ associated to a set Ω of sphere centers: the Voronoi tesselation and the Delaunay triangulation. Since they will play an important role, we recall their definitions.

Definition 3.1. The *Voronoi domain* (or *Voronoi cell*) of $\mathbf{v} \in \Omega$ is

$$V_{\text{vor}}(\mathbf{v}) = V_{\text{vor}}(\mathbf{v}, \Omega) := \{\mathbf{x} \in \mathbb{R}^3 \colon \|\mathbf{x} - \mathbf{v}\| \leq \|\mathbf{x} - \mathbf{w}\| \text{ for all } \mathbf{w} \in \Omega\}. \qquad (3.1)$$

The *Voronoi tesselation* for Ω is the set of Voronoi domains $\{V_{\text{vor}}(\mathbf{v}) \colon \mathbf{v} \in \Omega\}$.

The Voronoi tesselation is a partition of space up to boundaries of measure zero. If Ω is a saturated sphere packing, then all Voronoi domains are compact sets with diameter bounded by 4. Indeed, for a saturated packing the Voronoi domain is contained in the sphere of radius 2 centered at $\mathbf{v}$, as these spheres form a covering.

Definition 3.2. The *Delaunay triangulation* associated to a set Ω is dual to the Voronoi tesselation. It contains an edge between every pair of vertices that have Voronoi domains that share a common face. Suppose now that the points of Ω are in general position, which means that each corner of a Voronoi domain has exactly four incident Voronoi domains. In this case these Voronoi domains have between them four faces that touch this corner, and these faces in turn determine (four edges of) a Delaunay simplex. The resulting Delaunay simplices partition $\mathbb{R}^3$ and make up the Delaunay triangulation. In the case of non-general position Ω the Delaunay triangulation is not unique. The possible Delaunay triangulations are determined locally as limiting cases of general position points. (There are only finitely many triangulations possible in any bounded region of space.)

All the simplices in a Delaunay triangulation have vertices $\mathbf{v}_i \in \Omega$ and contain no other point $\mathbf{v} \in \Omega$. We define, more generally:

Definition 3.3. A *D-simplex* (or *Delaunay-type simplex*) for Ω is any tetrahedron T with vertices $\mathbf{v}_1, \mathbf{v}_2, \mathbf{v}_3, \mathbf{v}_4 \in \Omega$ such that no other $\mathbf{v} \in \Omega$ is in the closure of T. We denote it $D(\mathbf{v}_1, \mathbf{v}_2, \mathbf{v}_3, \mathbf{v}_4)$.

In the literature a *Delaunay simplex* associated to a point set Ω is any simplex with vertices in Ω whose circumscribing sphere contains no other vertex of Ω in its interior. All simplices in a Delaunay triangulation of Ω are Delaunay simplices in this sense, so are necessarily D-simplices, but the converse need not hold.

The admissible partitions that have been seriously studied all consist of a domain $V(\mathbf{v})$ associated to each vertex $\mathbf{v} \in V$, which we call a *V-cell*, together with a collection of certain D-simplices $D(\mathbf{v}_1, \mathbf{v}_2, \mathbf{v}_3, \mathbf{v}_4)$ which we call the D-*system* of the partition. We use the term D-*set* to refer to a D-simplex included in the D-system. We note that a V-cell may consist of several polyhedral pieces, and may even be disconnected.

The original approach of Fejes Tóth to getting local upper bounds for sphere packing in $\mathbb{R}^3$ used the Voronoi tesselation associated to Ω. If Ω is a saturated packing, then each Voronoi domain $V_{\text{vor}}(\mathbf{v})$ is a bounded polyhedron consisting of points within distance at most 4 of $\mathbf{v}$. Examples are known of Voronoi domains in a saturated packing that have 44 faces; an upper bound for the number of faces of a Voronoi domain of a saturated packing is 49. The *Voronoi partition* takes the V-sets $V(\mathbf{v})$ to be the Voronoi domains of $\mathbf{v}$, with no D-sets, and the vertex D-star $\mathcal{D}_{\text{vor}}(\mathbf{v})$ is just $V_{\text{vor}}(\mathbf{v})$. A *Voronoi scoring rule* is

$$Score(\mathcal{D}_{\text{vor}}(\mathbf{v})) := (\rho(V(\mathbf{v})) - B)\, vol(V(\mathbf{v})), \qquad (3.2)$$

with the score constants $A = 1$ and B is to be chosen optimally. Such scoring rules are admissible. However, it has long been known that no Voronoi scoring rule gives an optimal inequality (2.19). The dodecahedral conjecture states that the maximum packing density of a Voronoi domain is attained for a local configuration of 12 spheres touching at the center of faces of a circumscribed regular dodecahedron.

Dodecahedral Conjecture. *For a Voronoi domain $V_{\text{vor}}(\mathbf{v})$ of a unit sphere packing,*

$$\rho(V_{\text{vor}}(\mathbf{v})) \leq \frac{\pi}{15(1 - \cos(\pi/5)\tan(\pi/3))} \simeq 0.754697 \qquad (3.3)$$

and equality is attained for the dodecahedral configuration.

A proof of the dodecahedral conjecture has been announced by Hales and McLaughlin [Dod], based on similar ideas to the Hales' approach to the Kepler conjecture.

In 1953 Fejes Tóth [FT1] proposed that an optimal inequality might exist based on a weighted averaging over Voronoi domains near a given sphere center, and in 1964 he made a specific proposal for such an optimal inequality. In the notation of this paper he used the Voronoi partition and an admissible scoring function of the form

$$\sigma(V_{\text{vor}}(\mathbf{w}), \mathbf{v}) := \omega(\mathbf{w}, \mathbf{v})\{(A\rho(V_{\text{vor}}(\mathbf{w})) - B)\, vol(V_{\text{vor}}(\mathbf{w}))\}, \qquad (3.4)$$

in which $A = 1$, $B = \pi/\sqrt{18}$, and the *weights* $\omega(\mathbf{w}, \mathbf{v})$ are given by

$$\omega(\mathbf{w}, \mathbf{v}) := \begin{cases} \frac{1}{12} & \text{if} \quad 2 \leq \|\mathbf{w} - \mathbf{v}\| \leq 2 + t, \\ 0 & \text{if} \quad \|\mathbf{w} - \mathbf{v}\| > 2 + t, \end{cases} \qquad (3.5)$$

and

$$\omega(\mathbf{v}, \mathbf{v}) := 1 - \sum_{\mathbf{w} \neq \mathbf{v}} \omega(\mathbf{w}, \mathbf{v}). \qquad (3.6)$$

Here $t \geq 0$ is a fixed constant. The resulting inequality in Theorem 2.1 is optimal if the associated constant $\theta_{\mathcal{P}, \omega}(A, B) = 0$. The Fejes-Tóth scoring function corresponds to a weighted averaging over the spheres touching a central sphere at $\mathbf{v}$, where the weight assigned to the central sphere depends on how many spheres touch it. Fejes Tóth considered choosing t as large as possible consistent with requiring that $\omega(\mathbf{v}, \mathbf{v}) \geq 0$, which is equivalent to requiring that it is impossible to pack 13 spheres around a given sphere, with all 13 sphere centers within distance $2 + t$ of the center of the given sphere. In 1964 he suggested [FT2, p. 299] that one could take the value $t = 0.0534$, and we consider this to be Fejes Tóth's candidate for an optimal inequality. It has not been demonstrated that $\omega(\mathbf{v}, \mathbf{v}) \geq 0$ holds for this value of t, but note that the argument of Theorem 2.1 is valid for any value of t, even if some $\omega(\mathbf{v}, \mathbf{v})$ are negative. The only issue is whether the resulting sphere packing bound is optimal. Fejes Tóth [FT2] explicitly noted that establishing an optimal inequality, if it is true, reduces the problem in principle to one in a finite number of variables, possibly amenable to solution by computer.

In 1993 Hsiang [Hs1] studied a variant of the Fejes Tóth approach. He used the Voronoi partition and an admissible scoring function[1] of the form (3.4), with the same $A = 1$, $B = \pi/\sqrt{18}$, and with the *weights* $\omega(\mathbf{w}, \mathbf{v}) \geq 0$ given by

$$\omega(\mathbf{w}, \mathbf{v}) := \begin{cases} \dfrac{1}{1 + N(\mathbf{v})} & \text{if} \quad \|\mathbf{w} - \mathbf{v}\| \leq \frac{218}{100}, \\ 0 & \text{if} \quad \|\mathbf{w} - \mathbf{v}\| > \frac{218}{100}, \end{cases} \tag{3.7}$$

where

$$N(\mathbf{v}) := \#\{\mathbf{w} \in \Omega : 0 < \|\mathbf{w} - \mathbf{v}\| \leq \tfrac{218}{100}\} \tag{3.8}$$

counts the number of "near neighbors" of $\mathbf{v}$. Hsiang announced that his local inequality is optimal (with $\theta_{\mathcal{P}, \omega}(A, B) = 0$), and that he had proved it, which would then constitute a proof of Kepler's conjecture. However, the proof of optimality presented in [Hs1] is regarded as incomplete by the mathematical community, see G. Fejes Tóth's review of Hsiang's paper in *Mathematical Reviews*, the critique in [Ha3], and Hsiang's rejoinder [Hs2].

In 1992 Hales [Ha1], [Ha2] studied the Delaunay triangulation, which partitions $\mathbb{R}^3$ into D-sets. There are a finite number of (local) choices for Delaunay triangulations of a neighborhood of a fixed $\mathbf{v} \in \Omega$. Hales used the following function in defining his associated weight function.

Definition 3.4. The *compression* of a finite region R in $\mathbb{R}^3$ with respect to a sphere packing Ω is

$$\Gamma(R) := (\rho(R) - \delta_{\text{oct}}) \, vol(R) \tag{3.9}$$

in which

$$\delta_{\text{oct}} := \frac{-3\pi + 12 \arccos(1/\sqrt{3})}{\sqrt{8}} \approx 0.720903 \tag{3.10}$$

is the fraction of the volume of the regular octahedron of sidelength 2 covered by unit spheres centered at its vertices.

---

[1] More accurately, the locally averaged density in Section 3 of [Hs1] is converted to the form given in Section 2 by clearing denominators, and cancelling out Hsiang's factor of 13. Note also that each Voronoi domain contains exactly one sphere, so that $\rho(V(\mathbf{v})) \, vol(V(\mathbf{v})) = 4\pi/3$.

Hales initially considered the admissible weight function

$$\sigma(D(\mathbf{v}_1, \mathbf{v}_2, \mathbf{v}_3, \mathbf{v}_4), \mathbf{v}_i) := \Gamma(D(\mathbf{v}_1, \mathbf{v}_2, \mathbf{v}_3, \mathbf{v}_4)). \tag{3.11}$$

The vertex D-star of $\mathbf{v}$ consists of all the simplices in the Delaunay triangulation that have $\mathbf{v}$ as a vertex; we call this set of simplices the *Delaunay D-star* $\mathcal{D}_{\mathrm{Del}}(\mathbf{v})$ at $\mathbf{v}$. Hales used score constants $A = 4$ and $B = 4\delta_{\mathrm{oct}}$, with $A = 4$ used since each simplex is counted four times. However, he discovered that the *pentagonal prism* attained a score value exceeding what is needed to prove Kepler's conjecture. The pentagonal prism is conjectured to be extremal for this score function.

The fact that the (conjectured) extremal configurations for the Voronoi tesselation and Delaunay triangulation do not coincide suggested to Hales that a hybrid scoring rule be considered that combines the best features of the Voronoi and Delaunay scoring function. In 1997 Hales again considered a Delaunay triangulation, but modified the scoring rule to depend on the shape of the D-simplex $D(\mathbf{v}_1, \mathbf{v}_2, \mathbf{v}_3, \mathbf{v}_4)$. For some simplices he used the weight function above, while for others he cut the simplex into four pieces, one for each vertex, called the pieces $V(D, \mathbf{v}_i)$, and assigned the weights[2]

$$\sigma(D(\mathbf{v}_1, \mathbf{v}_2, \mathbf{v}_3, \mathbf{v}_4), \mathbf{v}_i) = 4\Gamma(V(D, \mathbf{v}_i)),$$

for $1 \leq i \leq 4$. He also partitioned a vertex D-star into pieces called "clusters" whose score functions could be evaluated separately and added up to get the total score. Each "cluster" is a finite union of Delaunay simplices filling up that part of the vertex D-star at $\mathbf{v}$ lying in a pointed cone with vertex $\mathbf{v}$. This vertex cone subdivision facilitates computer-aided proofs by decomposing the problem into smaller subproblems. Hales [SP-I], [SP-II] presented evidence that this modified scoring function satisfies an optimal local inequality. He showed that the two known local extremal configurations[3] gave local maxima of the score of the Delaunay D-star in the configuration space. The associated scoring function is

$$Score(\mathcal{D}_{\mathrm{Del}}(\mathbf{v})) = 8pt, \tag{3.12}$$

where

$$pt := \frac{11\pi}{3} - 12 \arccos\left(\frac{1}{\sqrt{3}}\right) \simeq 0.0553736, \tag{3.13}$$

which is the desired optimal value. He also showed that this was a global upper bound over the subset of configurations described by a vertex map[4] $\mathcal{G}(\mathbf{v})$ that is triangulated. Hales [SP-I, Conjecture 2.2] conjectured that the modified score function achieved the optimal inequality

$$Score(\mathcal{D}_{\mathrm{Del}}(\mathbf{v})) \leq 8pt, \tag{3.14}$$

---

[2] More precisely, he used here the "analytic continuation" of this scoring function that is described in Appendix A.

[3] These correspond to Voronoi cells being the rhombic dodecahedron or trapezo-rhombic dodecahedron on p. 295 of [FT2].

[4] See Section 5 for a definition of vertex map $\mathcal{G}(\mathbf{v})$.

for all Delaunay D-stars $\mathcal{D}_{\mathrm{Del}}(\mathbf{v})$. However, he and Ferguson (his student) [FH] discovered that a pentagonal prism configuration comes very close to violating the inequality (3.14). Furthermore, there turned out to be many similar difficult configurations which might possibly violate the inequality. These and other difficulties indicated that it was not numerically feasible to prove (3.14) by a computer proof, assuming that (3.14) is actually true.

Hales and Ferguson together then further modified both the partition rule $\mathcal{P}$ and the scoring rule σ to obtain a rule with the following properties:

(i) It makes the score inequality stronger on the known bad cases related to the pentagonal prism configuration.

(ii) It uses a more complicated notion of "cluster," which includes Voronoi pieces as well as D-sets, and which retains the "decoupling" property that it is completely determined by vertices of Ω in the cone above it.

(iii) It chooses a scoring rule which when combined with "truncation" on clusters is still strong enough to rule out most configurations. The "truncation" operation greatly reduces the number of configurations to be checked, at the cost of weakening the inequality to be proved.

In Section 4 we give a precise description of the Hales–Ferguson rules $(\mathcal{P}, \sigma)$.

4. Hales–Ferguson Partition Rule and Score Function

Ferguson and Hales [FH] use the following partition rule and scoring rule. The partition uses two types of D-simplices, with a complicated rule for picking which ones to include as D-sets in the partition. Modified Voronoi domains $V(\mathbf{v})$ are used as V-sets. These differ from the usual Voronoi domain (with the D-sets removed) by mutually exchanging some regions called "tips." The scoring rule is also complicated: the weight function used on a D-simplex no longer depends on just its shape, but depends on the structure of nearby D-sets.

We begin by defining the two types of D-simplices.

Definition 4.1. *A QR-tetrahedron (or quasi-regular tetrahedron) is any tetrahedron with all vertices in Ω and all edges of length $\leq \frac{251}{100}$.*

Definition 4.2. *A QL-tetrahedron (or quarter) is any tetrahedron with all vertices in Ω and five edges of length $\leq \frac{251}{100}$ and one edge with length $\frac{251}{100} < l \leq 2\sqrt{2}$. The long edge is called the spine (or diagonal) of the QL-tetrahedron.*

For some purposes[5] the case of a spine of length exactly $\frac{251}{100}$ should be considered as either a QL-tetrahedron or a QR-tetrahedron. Here we treat it exclusively as a QR-tetrahedron.

---

[5] In proving inequalities one wants to work on a compact set. In compactifying the space of configurations, this requires allowing the lower inequality in the definition of the QL-tetrahedron to be an equality.

Neither kind of tetrahedron is guaranteed to be included in the Delaunay triangulation of Ω, but we do have:

Lemma 4.1. *All QR-tetrahedra and QL-tetrahedra are D-simplices.*

The Hales–Ferguson partition rule starts by selecting which D-simplices to include in the D-system. These consist of:

(i) All QR-tetrahedra.
(ii) Some QL-tetrahedra. The QL-tetrahedra included in the partition satisfy the *common spine condition* which states that for a given spine, either all QL-tetrahedra having that spine are included or none are.

This collection of tetrahedra is required (by the definition of an admissible partition) to form a non-overlapping set, where we say that two sets S_1 and S_2 *overlap* if $\bar{S}_1 \cap \bar{S}_2$ has positive Lebesgue measure in $\mathbb{R}^3$. To show this is possible with (i) holding we have:

Lemma 4.2. *No two QR-tetrahedra overlap.*

QL-tetrahedra may overlap QR-tetrahedra or other QL-tetrahedra, hence one needs a rule for deciding which QL-tetrahedra to include. To begin with, QL-tetrahedra can overlap QR-tetrahedra in essentially one way.

Lemma 4.3. *If a QL-tetrahedron and QR-tetrahedron overlap, then the QR-tetrahedron has a common face with an adjacent QR-tetrahedron, and the two unshared vertices of these QR-tetrahedra are the endpoints of the spine of the QL-tetrahedron. The union of these two QR-tetrahedra can be partitioned into three QL-tetrahedra having the given spine, which includes the given QL-tetrahedron. Aside from these QL-tetrahedra, no other QL-tetrahedron overlaps either of these two QR-tetrahedra.*

This lemma shows that the QL-tetrahedra having a given spine have the property that either all of them or none of them overlap the set of QR-tetrahedra. We next consider how QL-tetrahedra can overlap other QL-tetrahedra. The following configuration plays an important role.

Definition 4.3. A Q-octahedron is an octahedron whose six vertices $\mathbf{v}_i \in \Omega$ and whose twelve edges each have lengths $2 \leq l \leq \frac{251}{100}$.

A Q-octahedron has three interior diagonals. If a diagonal has length $2 < l \leq \frac{251}{100}$, then it partitions the Q-octahedron into four QR-tetrahedra. If a diagonal has length $\frac{251}{100} < l \leq 2\sqrt{2}$, then it partitions the Q-octahedron into four QL-tetrahedra of which it is the common spine. If a diagonal has length $l > 2\sqrt{2}$ it yields no partition. A Q-octahedron thus gives between zero and three different partitions into four QR-tetrahedra or QL-tetrahedra. We call it a *live Q-octahedron* if it has at least one such partition. Lemma 4.3 implies that if it has a partition into QR-tetrahedra, then it has no other partition into QR-tetrahedra or QL-tetrahedra.

Lemma 4.4. *A QL-tetrahedron having a spine which is a diagonal of a Q-octahedron does not overlap any QL-tetrahedron whose spine is not a diagonal of the same Q-octahedron.*

The selection rules for choosing QL-tetrahedra to include in the D-system require that one either includes all QL-tetrahedra having the same spine or else none of them. Thus the selection rule really specifies which spines to include.

Definition 4.4. Consider an edge $[\mathbf{v}_1, \mathbf{v}_2]$ with $\mathbf{v}_1, \mathbf{v}_2 \in \Omega$ and $\frac{251}{100} < \|\mathbf{v}_1 - \mathbf{v}_2\| \le 2\sqrt{2}$. A vertex $\mathbf{w} \in \Omega$ is called an *anchor* of the edge $[\mathbf{v}_1, \mathbf{v}_2]$ if

$$\|\mathbf{w} - \mathbf{v}_i\| \le \tfrac{251}{100} \qquad \text{for} \quad i = 1, 2.$$

Ferguson and Hales use the number of QL-tetrahedra having a spine $[\mathbf{v}_1, \mathbf{v}_2]$ and the number of anchors of that spine in deciding which QL-tetrahedra to include in the D-system. Call a QL-tetrahedron *isolated* if it is the only QL-tetrahedron on its spine $[\mathbf{v}_1, \mathbf{v}_2]$. The inclusion rule for an isolated QL-tetrahedron is:

(QL0) An isolated QL-tetrahedron is included in the D-system if and only if it overlaps[6] no other QL-tetrahedron or QR-tetrahedron.

Next consider spines $[\mathbf{v}_1, \mathbf{v}_2]$ which have two or more associated QL-tetrahedra. Such spines have at least three anchors, and the inclusion rules are:

(QL1) Each non-isolated QL-tetrahedron on a spine with five or more anchors is included in the D-system.

(QL2) Each non-isolated QL-tetrahedron on a spine with four anchors is included in the D-system, if the spine is not a diagonal of some Q-octahedron. In the case of a live Q-octahedron, we include all QL-tetrahedra having one particular diagonal, and exclude all QL-tetrahedra on other diagonals. For definiteness, we choose the spine to be the shortest diagonal. In the case of a tie for the shortest diagonal, a suitable tie-breaking rule is used.

(QL3) Each non-isolated QL-tetrahedron on a spine with three anchors is included in the D-system if each QL-tetrahedron on the spine does not overlap any other QL-tetrahedron or QR-tetrahedron, or overlaps only isolated QL-tetrahedra. It is excluded from the D-system if some tetrahedron on the spine overlaps either a QR-tetrahedron or a non-isolated QL-tetrahedron having four or more anchors. Finally, if some tetrahedron on the spine overlaps a non-isolated QL-tetrahedron having exactly three anchors, then the spine of the overlapped set is unique, and exactly one of these two sets of non-isolated QL-tetrahedra with three anchors is to be included in the D-system, according to a tie-breaking rule.[7]

---

[6] It cannot overlap a QR-tetrahedron by Lemma 4.3.

[7] The tie-breaking rule could be to include the spine with the lowest endpoint using a lexicographic ordering of points in $\mathbb{R}^3$. It appears to me that Hales would permit an arbitrary choice of which one to include, see Lemma 4.5(iii).

The set of *QL*-tetrahedra selected above are pairwise disjoint and are disjoint from all *QR*-tetrahedra. This is justified by the following lemma.

Lemma 4.5.

(i) *If two QL-tetrahedra overlap, then at most one of them has five or more anchors.*
(ii) *If two overlapping QL-tetrahedra each have four anchors, then their spines are (distinct) diagonals of some Q-octahedron.*
(iii) *If a non-isolated QL-tetrahedron with three anchors overlaps another QL-tetrahedron having three anchors, then each of their spines contains exactly two non-isolated QL-tetrahedra, and these four QL-tetrahedra overlap no other QL-tetrahedron or QR-tetrahedron.*

We call the set of *QR*-tetrahedra and *QL*-tetrahedra selected as above the Hales–Ferguson *D*-system. (Hales and Ferguson call this a *Q*-system.)

We now define the *V*-cells of the Hales–Ferguson partition. To begin, we take the Voronoi domain $V_{vor}(\mathbf{v})$ at vertex $\mathbf{v}$ and remove from it the union of all *D*-simplices in the *D*-system to obtain a reduced Voronoi region $V_{red}(\mathbf{v})$. Next we move certain regions of $V_{red}(\mathbf{v})$ called "tips" to neighboring reduced Voronoi regions to obtain modified regions $V_{mod}(\mathbf{v})$ and finally we define the *V*-cell $V(\mathbf{v})$ at $\mathbf{v}$ to be the closure of $V_{mod}(\mathbf{v})$.

Definition 4.5. Let *T* be any tetrahedron such that the center $\mathbf{x} = \mathbf{x}(T)$ of its circumscribing sphere lies outside *T*. A vertex $\mathbf{v}$ of *T* is *negative* if the plane *H* determined by the face *F* of *T* opposite $\mathbf{v}$ separates $\mathbf{v}$ from $\mathbf{x}$. The "*tip*" $\Delta(T, \mathbf{v})$ of *T* associated to a negative vertex $\mathbf{v}$ is that part of the Voronoi region of $\mathbf{v}$ with respect to the points $\{\mathbf{v}_1, \mathbf{v}_2, \mathbf{v}_3, \mathbf{v}_4\}$ that lies in the closed half-plane H^+ determined by *H* that contains $\mathbf{x}$. The "tip" region $\Delta(T, \mathbf{v})$ does not overlap *T*, and is a tetrahedron having $\mathbf{x}$ as a vertex, and has three other vertices lying on *H*.

Lemma 4.6.

(i) *A QR-tetrahedron or QL-tetrahedron T has at most one negative vertex.*
(ii) *If a negative vertex is present, then the three vertices of the associated "tip" that lie on H actually lie in the face of T opposite to the negative vertex.*
(iii) *The "tip" of any tetrahedron in the Hales–Ferguson D-system either does not overlap any D-simplex in the Hales–Ferguson D-system, or else is entirely contained in the union of the D-simplices in the D-system.*

We say that a "tip" that does not overlap any *D*-set is *uncovered*. The lemma shows that uncovered "tips" lie in the union of the Voronoi regions $\{V_{red}(\mathbf{w}): \mathbf{w} \in \Omega\}$, so that rearrangement of uncovered "tips" is legal. There is an a priori possibility that two "tips" may overlap[8] each other.

---

[8] I do not know if this possibility can occur.

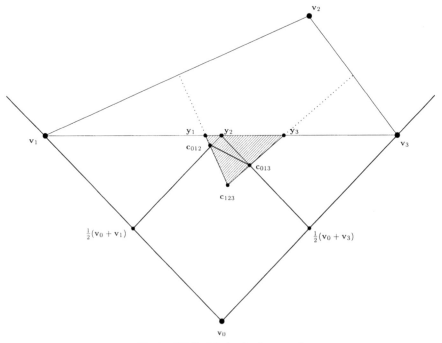

Fig. 1.   "Tip" of D-simplex $[\mathbf{v}_1, \mathbf{v}_2, \mathbf{v}_3]$.

Uncovered Tip Rearrangement Rule.   Each $\mathbf{y} \in \mathbb{R}^3$ that belongs to an uncovered "tip" is reassigned to the nearest *vertex* $\mathbf{w} \in \Omega$ such that $\mathbf{y}$ is not in an uncovered "tip" of any pair $(T, \mathbf{w})$ where T is in the Hales–Ferguson D-system and $\mathbf{w}$ is a negative vertex of T. (A tie-breaking rule is used if the two nearest vertices $\mathbf{w}$ are equidistant.)

This rule cuts an uncovered "tip" into a finite number of polyhedral pieces and reassigns the pieces to different reduced Voronoi regions. This prescribes how $V_{\text{mod}}(\mathbf{v})$ is constructed, and thus defines the Hales–Ferguson V-cells $V(\mathbf{v})$.

A two-dimensional analogue of a "tip" is pictured[9] in Fig. 1. In this figure the triangle $T = [\mathbf{v}_1, \mathbf{v}_2, \mathbf{v}_3]$ plays the role of a D-simplex, with $\mathbf{v}_2$ as a negative vertex and the "tip" is the shaded region. The points $\mathbf{c}_{012}, \mathbf{c}_{013}, \mathbf{c}_{123}$ are centroids of the triangles determined by the corresponding $\mathbf{v}_i$'s. The shaded triangle $[\mathbf{c}_{012}, \mathbf{c}_{013}, \mathbf{c}_{123}]$ is in the Voronoi cell $V_{\text{vor}}(\mathbf{v}_0)$ while the remainder of the "tip" is in the Voronoi cell $V_{\text{vor}}(\mathbf{v}_2)$. The uncovered tip rearrangement rule partitions the part in $V_{\text{vor}}(\mathbf{v}_2)$ into three triangles which are reassigned to the V-cells $V(\mathbf{v}_0)$, $V(\mathbf{v}_1)$, and $V(\mathbf{v}_3)$, e.g., $[\mathbf{y}_1, \mathbf{y}_2, \mathbf{c}_{012}]$ is reassigned to $V(\mathbf{v}_1)$. The reassignment of the "tip" ensures that the pointed cone over $\mathbf{v}_2$ generated by the D-simplex $[\mathbf{v}_1, \mathbf{v}_2, \mathbf{v}_3]$ does not contain any part of the V-cell at $\mathbf{v}_2$. In this example the V-cell at $\mathbf{v}_0$ does not feel the effect of the vertex $\mathbf{v}_2$, due to the rearrangement.

We now turn to the Hales–Ferguson scoring rules. These use the compression function $\Gamma(S)$ given in (3.9). The compression function is additive: If $S = S_1 \cup S_2$ is a partition,

---

[9] See Figure 2.1 of Hales [SP-II] for another example.

then

$$\Gamma(S) = \Gamma(S_1) + \Gamma(S_2).\tag{4.1}$$

For a D-simplex T,

$$vol(T)\rho(T) = \sum_{i=1}^{4} \frac{(\text{solid angle})_i}{3},\tag{4.2}$$

where a full solid angle is 4π.

The Hales–Ferguson weight function for a V-cell is as follows:

(S1) For a V-cell $V(\mathbf{v})$,

$$\sigma_{\text{HF}}(V(\mathbf{v}), \mathbf{w}) = \begin{cases} 4\Gamma(V(\mathbf{v})) & \text{if} \quad \mathbf{v} = \mathbf{w}, \\ 0 & \text{if} \quad \mathbf{v} \neq \mathbf{w}. \end{cases}\tag{4.3}$$

We next consider the weight function for D-sets. Let $(T, \mathbf{v})$ denote a D-simplex together with a vertex $\mathbf{v}$ of it.

Definition 4.6. The *Voronoi measure* $vor(T, \mathbf{v})$ is defined as follows. If the center of the circumscribing sphere of T lies inside T, then T is partitioned into four pieces:

$$V_{\text{vor}}^{+}(T, \mathbf{v}_i) := \{\mathbf{x} \in T : \|\mathbf{x} - \mathbf{v}_i\| \leq \|\mathbf{x} - \mathbf{v}_j\| \text{ for } 1 \leq j \leq 4\}$$

and then

$$vor(T, \mathbf{v}) := \Gamma(V_{\text{vor}}^{+}(T, \mathbf{v})).\tag{4.4}$$

There is an analytic formula for the right side of (4.4) given in Appendix A, and this formula is used to define $vor(T, \mathbf{v})$ in cases where the circumcenter falls outside T.

In cases where the circumcenter is outside T, and $\mathbf{v}$ is a negative vertex, then

$$vor(T, \mathbf{v}) = \Gamma(V_{\text{vor}}(T, \mathbf{v}) \cup \text{``tip''})\tag{4.5}$$

while for the other three vertices parts of the "tip" are counted with a negative weight, in such a way that

$$\sum_{i=1}^{4} vor(T, \mathbf{v}_i) = 4\Gamma(T)\tag{4.6}$$

holds in all cases. The weight function for a D-set is given as follows:

(S2) For a QR-tetrahedron $T = D(\mathbf{v}_1, \mathbf{v}_2, \mathbf{v}_3, \mathbf{v}_4)$ in the D-system,

$$\sigma_{\text{HF}}(T, \mathbf{v}) = \begin{cases} \Gamma(T) & \text{if the circumradius of } T \text{ is at most } \frac{141}{100}, \\ vor(T, \mathbf{v}) & \text{if the circumradius of } T \text{ exceeds } \frac{141}{100}. \end{cases}\tag{4.7}$$

The QL-tetrahedron scoring function is complicated. For a QL-tetrahedron T, let $\eta^+(T)$ be the maximum of the circumradii of the two triangular faces of T adjacent to the spine of T, and define the function

$$\mu(T, \mathbf{v}) := \begin{cases} \Gamma(T) & \text{if } \eta^+(T) \leq \sqrt{2}, \\ vor(T, \mathbf{v}) & \text{if } \eta^+(T) > \sqrt{2}. \end{cases} \tag{4.8}$$

Then the QL-tetrahedron scoring function is defined by:

(S3) ("Flat quarter" case) For a QL-tetrahedron T and a vertex $\mathbf{v}$ not on its spine,

$$\sigma_{\mathrm{HF}}(T, \mathbf{v}) := \mu(T, \mathbf{v}). \tag{4.9}$$

(S4) ("Upright quarter" case) For a QL-tetrahedron T with vertex $\mathbf{v}$ on its spine, let $\hat{\mathbf{v}}$ denote the opposite vertex on the spine. If T is an isolated QL-tetrahedron, set

$$\sigma_{\mathrm{HF}}(T, \mathbf{v}) := \mu(T, \mathbf{v}). \tag{4.10}$$

If T is part of a Q-octahedron, set

$$\sigma_{\mathrm{HF}}(T, \mathbf{v}) := \tfrac{1}{2}(\mu(T, \mathbf{v}) + \mu(T, \hat{\mathbf{v}})). \tag{4.11}$$

In all other cases, set

$$\sigma_{\mathrm{HF}}(T, \mathbf{v}) := \tfrac{1}{2}(\mu(T, \mathbf{v}) + \mu(T, \hat{\mathbf{v}})) + \tfrac{1}{2}(vor_0(T, \mathbf{v}) - vor_0(T, \hat{\mathbf{v}})), \tag{4.12}$$

in which $vor_0(T, \mathbf{v})$ is a "truncated Voronoi measure" that only counts volume within radius $\frac{1}{2}(\frac{251}{100})$ of vertex $\mathbf{v}$, which is defined in Appendix A, and on pp. 9–11 of [FH].

The scoring rule (S4) is the most complicated one. In it the definition (4.11) plays an important role in obtaining good bounds for the pentagonal prism case treated in [SP-V], while the definition (4.12) is important in analyzing general configurations using truncation in [SP-IV].

Theorem 4.1. *The Hales–Ferguson partition rule and scoring rule $(\mathcal{P}_{\mathrm{HF}}, \sigma_{\mathrm{HF}})$ are both admissible, with score constants $A = 4$ and $B = 48\delta_{\mathrm{oct}}$.*

Proof. It is easy to verify that the definitions for scoring QR-tetrahedra and QL-tetrahedra satisfy the weighted density average property

$$\sum_{i=1}^{4} \sigma_{\mathrm{HF}}(T, \mathbf{v}_i) = 4\Gamma(T), \tag{4.13}$$

which correspond to $A = 4$ and $B = 48\delta_{\mathrm{oct}}$, using (4.6). Most of the remaining admissibility conditions are verified by Lemmas 4.1–4.6 except for locality. For locality, a conservative estimate indicates that the rules for removing and adding "tips" to determine the V-cell $V(\mathbf{v})$ are determined by sphere centers $\mathbf{w} \in \Omega$ with $\|\mathbf{w} - \mathbf{v}\| \leq 12\sqrt{2}$. Finally the score function on the D-simplices is determined by vertices within distance $6\sqrt{2}$ of $\mathbf{v}$. $\qquad\square$

Based on this result, Theorem 2.1 associates to $(\mathcal{P}_{HF}, \sigma_{HF})$ a sphere-packing bound; the Hales program asserts this bound will be optimal. To establish the Kepler bound

$$\bar{\rho}(\Omega) \leq \frac{\pi}{\sqrt{18}}, \tag{4.14}$$

via (2.15), one must prove that

$$\theta := \theta_{\mathcal{P}_{HF}, \sigma_{HF}}(4, 4\delta_{oct}) = 8pt, \tag{4.15}$$

where

$$pt := \frac{11\pi}{3} - 12 \arccos\left(\frac{1}{\sqrt{3}}\right) \simeq 0.0553736. \tag{4.16}$$

The score function $Score(\mathcal{D}_{HF}(\mathbf{v}))$ is discontinuous as a function of the sphere centers in Ω near $\mathbf{v}$, because it is a sum of contributions of pieces which may appear and disappear as sphere centers move, and discontinuities occur when QL-tetrahedra convert to QR-tetrahedra. To deal with this, one compactifies the configuration space by allowing some sphere center configurations to have more than one legal decomposition into pieces (but at most finitely many). The optimization problem can then be split into a finite number of subproblems on each of which σ_{HF} is continuous.

The complexity of the definition of $(\mathcal{P}_{HF}, \sigma_{HF})$ is designed to yield a computationally tractable nonlinear optimization problem. The introduction of QL-tetrahedra and the complicated score function on them is designed to help get good bounds for the pentagonal prism case and similar cases. The rule for moving "tips" is intended to facilitate decomposition of the nonlinear optimization problem into more tractable pieces via Theorem 5.5 below, and the use of "truncation."

5. Kepler Conjecture

The main result to be established by the Hales program is the following.

Theorem 5.1 (Main Theorem). *For the Hales–Ferguson partition rule and scoring rule $(\mathcal{P}_{HF}, \sigma_{HF})$, and any $\mathbf{v} \in \Omega$ in a saturated sphere-packing, the vertex D-star $\mathcal{D}_{HF}(\mathbf{v})$ at $\mathbf{v}$ satisfies*

$$Score(\mathcal{D}_{HF}(\mathbf{v})) \leq 8pt, \tag{5.1}$$

where $pt := 11\pi/3 - 12 \arccos(1/\sqrt{3}) \simeq 0.0553736.$

The Kepler conjecture follows by Theorem 2.1.

To prove inequality (5.1), by translation-invariance we can reduce to the case $\mathbf{v} = \mathbf{0}$ and search the set of all possible vertex stars, which by Section 4 are determined by those points $\mathbf{w} \in \Omega$ with $\|\mathbf{w}\| \leq 12\sqrt{2}$. The space of possible sphere centers $\{\mathbf{w} \in \Omega : \|\mathbf{w}\| \leq 12\sqrt{2}\}$ is compact. It can be decomposed into a large number of pieces, on each of which the score function is continuous. To obtain *compact* pieces, we must compactify the configuration space by assigning more than one possible local D-star $\mathcal{D}(\mathbf{v})$ to certain arrangements of sphere centers. The compactification assigns at most finitely many possibilities to each arrangement, with an absolute upper bound on the number of possibilities.

In what follows we generally assume that $0 \in \Omega$ and we study the vertex star at $\mathbf{v} = 0$. However, we state definitions and lemmas to be valid for general Ω.

The definition of the score $Score(\mathcal{D}(\mathbf{v}))$ involves a sum over the V-sets and D-sets. The usefulness of the compression measure $\Gamma(S)$ is justified by the following lemma.

Lemma 5.1.

(i) *Every QR-tetrahedron T satisfies*

$$\Gamma(T) \le pt, \tag{5.2}$$

with equality occurring only when T is a regular tetrahedron of edge length 2.

(ii) *A QL-tetrahedron T has*

$$\Gamma(T) \le 0, \tag{5.3}$$

with equality occurring for those T having five edges of length 2 and a spine of length $2\sqrt{2}$.

Result (ii) illustrates a somewhat counterintuitive behavior of the local density function: when holding five edges of a tetrahedron fixed of length 2, and allowing the sixth edge to vary over $\frac{251}{100} \le l \le 2\sqrt{2}$, the local density measure is largest for a spine of *maximal* length.

The vertices $\mathbf{w} \in \Omega$ with $\|\mathbf{w}\| \le \frac{251}{100}$ play a particularly important role, for they determine all QR-simplices of Ω containing 0 as a vertex.

Definition 5.1. The *planar map* (or *graph*) $\mathcal{G}(\mathbf{v})$ associated to a vertex $\mathbf{v} \in \Omega$ consists of the radial projection onto the unit sphere $\partial B(\mathbf{v}; 1) = \{\mathbf{x} \in \mathbb{R}^3 : \|\mathbf{x} - \mathbf{v}\| = 1\}$ centered at $\mathbf{v}$ of all vertices $\mathbf{w} \in \Omega$ with $\|\mathbf{w} - \mathbf{v}\| \le \frac{251}{100}$ plus all those edges $[\mathbf{w}, \mathbf{w}']$ between two such vertices which have length $\|\mathbf{w} - \mathbf{w}'\| \le \frac{251}{100}$.

Here we regard the planar map $\mathcal{G}(\mathbf{v})$ as being given with its embedding as a set of arcs on the sphere. The following lemma asserts that no new vertices are introduced other than those coming from points of Ω with $\|\mathbf{w}\| \le \frac{251}{100}$.

Lemma 5.2. *The radial projection of two edges $[\mathbf{w}_1, \mathbf{w}_2]$, $[\mathbf{w}'_1, \mathbf{w}'_2]$ as above onto the unit sphere $\partial B(\mathbf{v}; 1)$ give two arcs in $\mathcal{G}(\mathbf{v})$ which either are disjoint or which intersect at an endpoint of both arcs.*

We study local configurations classified by the planar map $\mathcal{G}(\mathbf{v})$. The planar map $\mathcal{G}(0)$, which is determined by the vertices $\|\mathbf{w}\| \le \frac{251}{100}$, does not in general uniquely determine the vertex D-star $\mathcal{D}_{HF}(0)$, but does determine all points $\mathbf{x}$ in it with $\|\mathbf{x}\| \le \frac{251}{200}$.

Definition 5.2. The part of $\mathcal{D}_{HF}(\mathbf{v})$ that lies in the pointed cone with base point $\mathbf{v}$ determined by a face of the map $\mathcal{G}(\mathbf{v})$ is called the *cluster* over that face. Note that the face need not be convex, or even simply connected—it could be topologically an annulus, for example.

The following lemma shows that the vertex D-star $\mathcal{D}_{\mathrm{HF}}(\mathbf{0})$ can be cut up into clusters in a way compatible with the scoring function.

Lemma 5.3. *Each QR-tetrahedron or QL-tetrahedron in the D-star $\mathcal{D}_{\mathrm{HF}}(\mathbf{0})$ is contained in a single cluster. Furthermore, all such tetrahedra having a common spine are contained in a single cluster.*

In effect the partition of the vertex D-star into clusters partitions the V-cell into smaller pieces, while leaving the D-sets unaffected. The scoring function is additive over any partition of a V-cell into smaller pieces, according to (4.1) and (4.3). The *score* $\sigma_{\mathrm{HF}}(F)$ of the cluster determined by a face F of $\mathcal{G}(\mathbf{v})$ is the sum of the scores of the QR-tetrahedra and QL-tetrahedra in the cluster, plus the Voronoi score $4\Gamma(R)$ of the remaining part R of the cluster. We then have

$$Score(\mathcal{D}_{\mathrm{HF}}(\mathbf{v})) = \sum_{F \in \mathcal{G}(\mathbf{v})} \sigma_{\mathrm{HF}}(F). \tag{5.4}$$

We now consider clusters associated to the simplest faces F in the graph $\mathcal{G}(\mathbf{v})$. Each triangular face corresponds to a QR-tetrahedron in $\mathcal{D}_{\mathrm{HF}}(\mathbf{v})$, and, conversely, each QR-tetrahedron in $\mathcal{D}_{\mathrm{HF}}(\mathbf{v})$ produces a triangular face. A *quad cluster* is a cluster over a quadrilateral face. A Q-octahedron with spine ending at $\mathbf{0}$ results in a quadrilateral face, but there are many other kinds of quad clusters. In the case of faces F with ≥ 5 edges, the cluster may consist of a V-cell plus some QL-tetrahedra, in many possible ways. All the possible decompositions into such pieces have to be considered as separate configurations.

Lemma 5.4.

    (i) *A cluster over a triangular face F consists of a single QR-tetrahedron, and conversely. The score of such a cluster is at most $1pt$, and equality holds if and only if it is a regular tetrahedron of edge length 2.*

    (ii) *The sum of the score functions over any quad cluster is at most zero. Equality can occur only if the four sphere centers $\mathbf{v}_i$ corresponding to the vertices of the quad cluster each lie at distance 2 from $\mathbf{v}$ and also from each other, if they share an edge of the quad cluster.*

    (iii) *The score of a cluster over any face with five or more sides is strictly negative.*

The extremal graphs where equality is known to occur in (5.1) have eight triangular faces and six quadrilateral faces. The upper bound of $8pt$ for these cases is implied by this lemma. (It appeared first in [SP-II, Theorem 4.1].)

The following result rules out graphs $\mathcal{G}(\mathbf{v})$ with faces of high degree.

Theorem 5.2. *All decomposition stars $\mathcal{D}_{\mathrm{HF}}(\mathbf{0})$ with planar maps $\mathcal{G}(\mathbf{0})$ satisfy*

$$Score(\mathcal{D}_{\mathrm{HF}}(\mathbf{0})) \leq 8pt \tag{5.5}$$

unless the planar map $\mathcal{G}(0)$ consists entirely of (not necessarily convex) faces of the following kinds: polygons having at most eight sides, in which pentagons and hexagons may contain an isolated interior vertex or a single edge from an interior vertex to an outside vertex, and a pentagon may exclude from its interior a triangle with two interior vertices.

There remain a finite set of possible map structures that satisfy the conditions of Theorem 5.2. Here we use the fact that there can be at most 50 vertices $\mathbf{v}$ with $\|\mathbf{v}\| \leq \frac{251}{100}$. The list is further pruned by various methods, and reduced to about 5000 cases. Since the (putative) extremal cases are already covered by Lemma 5.4, in the remaining cases one wishes to prove a strict inequality in (5.1), and such bounds can be obtained in principle by computer.

Most of the remaining cases are eliminated by linear programming bounds. The linear programs obtain upper bounds for the score function $Score(\mathcal{D}_{\mathrm{HF}}(\mathbf{0}))$ for a planar map $\mathcal{G}$ of a particular configuration type, using the score function as the objective function, in the form

$$\text{Maximize } Score(\mathcal{D}_{\mathrm{HF}}(0)) := \sum_{\substack{\text{faces} \\ F}} \sigma(F), \qquad (5.6)$$

where the variable $\sigma(F)$ is the sum of weights associated to the cluster over the face F. The use of linear programming relaxations of the nonlinear program seems to be a necessity in bounding the score function. For example, the compression function $\Gamma(R)$ for different regions R is badly behaved: it is neither convex nor concave in general. The linear constraints include hyperplanes bounding the convex hull of the score function over the variable space.

One can decouple the contributions of the separate faces F of $\mathcal{G}(\mathbf{v})$ using the following result.

Lemma 5.5 (Decoupling Lemma). *Let $\mathbf{v} \in \Omega$ be a vertex of a saturated packing and let F be a face of the associated planar map $\mathcal{G}(\mathbf{v})$, and let C_F denote the (closed) pointed cone over F with vertex $\mathbf{v}$, and let $C_{F,\mathrm{red}}$ denote the closure of the cone over F obtained by removing from C_F all cones over D-sets with a corner at $\mathbf{v}$. Then the portion of the V-cell $V(\mathbf{v})$ that lies in C_F is completely determined by the vertices of Ω that fall in the smallest closed convex cone $\bar{C}_F$ containing C_F. In particular,*

$$V_F := V(\mathbf{v}) \cap C_F = V(\Omega \cap \bar{C}_F, \mathbf{v}) \cap C_{F,\mathrm{red}}. \qquad (5.7)$$

To obtain such a decoupling lemma requires the exchange of "tips" between Voronoi domains, as described in Section 4.

The decoupling lemma permits the score function $\sigma(V(\mathbf{v}), \mathbf{v})$ to be decomposed into polyhedral pieces that depend on only a few of the nearby vertices. This decomposes the problem into a sum of smaller problems, to bound the scores of the pieces $\sigma(V(\mathbf{v}) \cap C_F, \mathbf{v})$ in terms of these vertices. It will often be applied when the face F is convex, in which case $\bar{C}_F = C_F$.

A futher very important relaxation of the linear programs involves "truncation." The *truncated V-cell* is

$$V_{\text{trunc}}(\mathbf{v}) := V(\mathbf{v}) \cap \mathbf{B}(\mathbf{v}: \tfrac{251}{200}). \tag{5.8}$$

The following lemma says that we may choose whether to apply truncation or not separately to the parts of the *V-cell* which lies over each face of $\mathcal{G}$.

Lemma 5.6. *Let F be a face of $\mathcal{G}(\mathbf{v})$ and let C_F be the cone over that face. The region $V_{\text{trunc}}(\mathbf{v}) \cap C_F$ is entirely determined by the vertices of $\mathcal{G}(\mathbf{v})$ in C_F. If $\mathcal{V}_F$ denotes this set of vertices, together with $\mathbf{v}$, then this region is the closure of $(V_{\text{vor}}(\mathcal{V}, \mathbf{v}) \cap C_F) - \{D\text{-sets}\}$. The compression function satisfies the bound*

$$\Gamma(V_{\text{trunc}}(\mathbf{v}) \cap C_F) \geq \Gamma(V(\mathbf{v}) \cap C_F). \tag{5.9}$$

Inequality (5.9) implies that replacing a Voronoi-type region by a truncated region can only increase the score, hence one can relax the linear program by using the score of truncated regions. If one is lucky the linear programming bounds using truncated regions will still be strong enough to give the desired inequality. The use of truncation greatly reduces the number of configurations that must be examined. Truncation bounds were also used in proving Theorem 5.2 above.

We add the following remarks about the construction of the linear programming problems:

(1) For each face F of a given graph type $\mathcal{G}$, Hales and Ferguson construct a large number of linear programming constraints in terms of the edge lengths, dihedral angles, and solid angles of the polyhedral pieces making up the cluster of $\mathcal{D}_{\text{HF}}(\mathbf{v})$ over face F of the graph $\mathcal{G}$. The edge lengths, dihedral angles, and solid angles are variables in the linear program. Some of the constraints embody geometric restrictions that a polyhedron of the given type must satisfy. Others of them are inequalities relating the weight function of the polyhedron, which is also a variable in the linear program, to the geometric quantities. The inequalities bound the score function on the cluster (either as a V-cell or as D-sets) in terms of these variables. There are also some global constraints in the linear program, for example, that the solid angles of the faces around $\mathbf{v}$ add up to 4π.

(2) The weight function for D-sets does not permit subdivision of the simplex, but the weight function on the V-cell is additive under subdivision, so one can cut up such regions into smaller pieces if necessary, to get improved linear programming bounds, by including more stringent constraints.

(3) In the linear programming relaxation, a feasible solution to the constraints need not correspond to any geometrically constructible vertex D-star. All that is required is that every vertex D-star of the particular configuration type corresponds to some feasible point of the linear program.

In this fashion one obtains a long list of linear programs, one for each configuration type, and to rule out a map type $\mathcal{G}$ one needs an upper bound for the linear program's objective function strictly below $8pt$. To obtain such an upper bound rigorously, it suffices to find a feasible solution to the dual linear program, and to obtain a good upper bound one wants the dual feasible solution close to a dual optimal solution. The value of the dual linear program's objective function is then a certified upper bound to the primal linear program. To obtain such a certification, it is useful to formulate the linear programs so that the dual linear program has only inequality constraints, with no equality constraints, so that the feasible region for it is full-dimensional. This way, one can guarantee that the dual feasible solution is *strictly* inside the dual feasible region which facilitates checking feasibility. This is necessary because the linear program put on the computer is only an approximation to the true linear program. For example, certain constraints of the true linear program involve transcendental numbers like π, and one considers an approximation. The effect of these errors is to perturb the *objective function* of the dual linear program. Thus a rigorous bound on the effect of these perturbations on the upper bound can be obtained in terms of the dual feasible solution. In this way one can (in principle) get a certified upper bound[10] on the score for a map type $\mathcal{G}$, using a computer.

The linear programming bounds in the Ferguson–Hales approach above suffice to eliminate all map types $\mathcal{G}$ not ruled out by Theorem 5.2 except for about 100 "bad" cases. These are then handled by ad hoc methods. I have not studied the details about how these remaining "bad" cases are handled. Presumably they are split into smaller pieces, extra inequalities are generated somehow, and perhaps specific information on the location of vertices of more than $\frac{251}{100}$ is incorporated into the linear programs.

6. Concluding Remarks

The Kepler conjecture appears to be an extraordinarily difficult nonlinear optimization problem. The "configuration space" to be optimized over has an extremely complicated structure, of high dimensionality, and the function being optimized is highly nonlinear and nonconvex, and lacks good monotonicity properties. The crux of the Hales approach is to select a formulation of an optimization problem that can be carried out (mostly by computer) in a reasonable length of time. This led to the Hales–Ferguson choice of a very complicated partition and score function, giving an inelegant local inequality, which however has good decomposition properties in terms of the nonlinear program. Much of the work in the proof lies in the reductions to reasonable sized cases and the use of linear programming relaxations. The elimination of the most complicated cases in Theorem 5.2 was a major accomplishment of this approach. The use of Delaunay simplices to cover most of the volume where density is high seems important to the proof and to the choice of score functions, since simple analytic formulas are available for tetrahedra. The Hales–Ferguson proof, assumed correct, is a tour de force of nonlinear optimization.

In contrast, the Hsiang approach formulates a relatively elegant local inequality, involving only Voronoi domains and a fairly simple weight function: only nearest neighbor regions are counted. It is conceivable that a rigorous proof of the Hsiang inequality can be established, but it very likely will require an enormous computer-aided proof of a sort very similar to the Hales approach. Voronoi domains do not seem well suited to

---

[10] The Hales proof in the preprints used a linear programming package CPLEX that does not supply such certificates. Therefore the linear programming part of the Ferguson–Hales proof needs to be re-done to obtain *guaranteed* certificates.

computer proof: they may have 40 or more faces each, and the Hsiang approach requires considering up to 20 of them at a time. A computer-aided proof would likely have to dissect the Voronoi domains into pieces, further increasing the size of the problem.

Acknowledgments

I am indebted to T. Hales for critical reading of a preliminary version, with many suggestions and corrections. I am indebted to G. Ziegler and the referee for helpful comments and corrections.

Appendix A. Hales Score Function Formulas

These definitions are taken from Section 8 of [SP-I] and pp. 8–11 of [FH]. A tetrahedron $T(l_1, l_2, l_3, \ldots, l_6)$ is uniquely determined by its six edge lengths l_i. Let the vertices of T be $\mathbf{v}_0, \mathbf{v}_1, \mathbf{v}_2, \mathbf{v}_3$ and number the edges as

$$l_i = \|\mathbf{v}_0 - \mathbf{v}_i\| \quad \text{for} \quad 1 \leq i \leq 3, \qquad l_4 = \|\mathbf{v}_2 - \mathbf{v}_3\|, \qquad l_5 = \|\mathbf{v}_1 - \mathbf{v}_3\|,$$

$$\text{and} \quad l_6 = \|\mathbf{v}_1 - \mathbf{v}_2\|. \tag{A.1}$$

We take $\mathbf{v}_0 = \mathbf{0}$ for convenience.

Suppose that the circumcenter $\mathbf{w}_c = \mathbf{w}_c(T)$ of T is contained in the pointed cone over vertex $\mathbf{v}_0$, determined by T. Let $\hat{T}_0$ denote the part of the Voronoi cell of $\mathbf{v}_0$ with respect to the set $\Omega = \{\mathbf{v}_0, \mathbf{v}_1, \mathbf{v}_2, \mathbf{v}_3\}$ of vertices of T that lies in T. Suppose in addition that the three faces of T containing $\mathbf{v}_0$ are each *non-obtuse* triangles. Then the set $\hat{T}_0$ subdivides into six pieces, called Rogers simplices by Hales [SP-I, p. 31]. A *Rogers simplex* in T is the convex hull of $\mathbf{v}_0$, the midpoint of an edge emanating from $\mathbf{v}_0$, the circumcenter of one face of T containing that edge, and the circumcenter $\mathbf{w}_c = \mathbf{w}_c(T)$. If a denotes the half-length of an edge, b the circumradius of a face, and $c = \|\mathbf{y}_c\|$ is the circumradius of T, then the associated Rogers simplex has shape

$$R(a, b, c) := T(a, b, c, (c^2 - b^2)^{1/2}, (c^2 - a^2)^{1/2}, (b^2 - a^2)^{1/2}), \tag{A.2}$$

with the positive square root taken. The intersection of a unit sphere centered at $\mathbf{0}$ with $R(a, b, c)$ has volume $\frac{1}{3} Sol(\mathbf{v}_0; R(a, b, c))$, where $Sol(\mathbf{v}_0; R(a, b, c))$ denotes the solid angle of $R(a, b, c)$ at $\mathbf{v}_0$, normalized so that a total solid angle is 4π. The values

$$x_i := l_i^2 \tag{A.3}$$

are the squares of the edge lengths.

Lemma A.1. *The solid angle $Sol(\mathbf{v}_0, T)$ of a tetrahedron $T(l_1, l_2, l_3, l_4, l_5, l_6)$ is given by*

$$Sol(\mathbf{v}_0, T) := 2 \operatorname{arccot}\left(\frac{2A}{\Delta^{1/2}}\right) \tag{A.4}$$

in which the positive square root of Δ *is taken, the value of* arccot *lies in* $[0, \pi]$,

$$A(l_1, l_2, l_3, l_4, l_5, l_6) := l_1 l_2 l_3 + \tfrac{1}{2} l_1 (l_2^2 + l_3^2 - l_4^2) + \tfrac{1}{2} l_2 (l_1^2 + l_3^2 - l_5^2) + \tfrac{1}{2} l_3 (l_1^2 + l_2^2 - l_6^2) \quad \text{(A.5)}$$

and

$$\Delta(l_1, l_2, l_3, l_4, l_5, l_6)$$
$$:= l_1^2 l_4^2 (-l_1^2 + l_2^2 + l_3^2 - l_4^2 + l_5^2 - l_6^2) + l_2^2 l_5^2 (l_1^2 - l_2^2 + l_3^2 + l_4^2 - l_5^2 + l_6^2)$$
$$+ l_3^2 l_6^2 (l_1^2 + l_2^2 - l_3^2 + l_4^2 + l_5^2 - l_6^2)$$
$$- l_2^2 l_3^2 l_4^2 - l_1^2 l_3^2 l_5^2 - l_1^2 l_2^2 l_6^2 - l_4^2 l_5^2 l_6^2. \quad \text{(A.6)}$$

Definition A.1.

(i) For a tetrahedron $T(l_1, l_2, \ldots, l_6)$ with vertex $\mathbf{v}_0$, if the circumcenter $\mathbf{w}_c$ of T falls inside the cone determined by T at $\mathbf{v}_0$, then we set

$$vor(T, \mathbf{v}_0) := 4 \sum_{i=1}^{6} \{ vol(R_i(a, b, c))(-\delta_{\text{oct}}) + \tfrac{1}{3} Sol(R_i, \mathbf{v}_0) \} \quad \text{(A.7)}$$

with

$$vol(R(a, b, c)) := \frac{a(b^2 - a^2)^{1/2}(c^2 - b^2)^{1/2}}{6},$$
$$\text{for} \quad 1 \le a \le b \le c. \quad \text{(A.8)}$$

This formula satisfies $vor(T, \mathbf{v}_0) = 4\Gamma(\hat{T}_0)$.

(ii) The six tetrahedra $R_i(a, b, c)$ are still defined even when the circumcenter $\mathbf{w}_c$ falls outside the cone of T at vertex $\mathbf{v}_0$, and we still take the formula (A.7) to define $vor(T, \mathbf{v}_0)$, except that both $vol(R_i(a, b, c))$ and $Sol(R_i, \mathbf{v}_0)$ are counted with a negative sign: each tetrahedron $R_i(a, b, c)$ falls outside T, and has no interior in common with it.

Hales calls the definition (ii) the "analytic continuation" of case (i). It has a geometric interpretation.

The truncated Voronoi function $vor(T, \mathbf{v}_0; t)$ of a tetrahedron T at vertex $\mathbf{v}_0$ is intended to measure the compression $\Gamma(\hat{T}_0 \cap B(\mathbf{v}_0; t))$. Here we have truncated the region $\hat{T}_0$ by removing from it all points at distance greater than t from $\mathbf{v}_0$. We set

$$vor_0(T, \mathbf{v}_0) := vor(T, \mathbf{v}_0, \tfrac{251}{200}). \quad \text{(A.9)}$$

The definition

$$vor(T, \mathbf{v}_0; t) := \Gamma(\hat{T}_0 \cap B(\mathbf{v}_0; t)) \quad \text{(A.10)}$$

is valid only when the circumcenter $\mathbf{w}_c$ of T lies in the cone generated from T at vertex $\mathbf{v}_0$. In the remaining case one must construct an analytic representation analogous to (A.7) for $vor(T, \mathbf{v}_0; t)$. This is done on pp. 9–10 of [FH].

Appendix B.   References to the Hales Program Results

This paper was written to state the Hales–Ferguson local inequality in as simple a way as I could find, and does not match the order in which things are done in the preprints of Hales and Ferguson. Also, the lemmas and theorems stated here are not all stated in the Hales and Ferguson preprints; some of them are based on talks that Hales gave at IAS in January 1999. The pointers below indicate where to look in the preprints for the results I formulate as lemmas and theorems. *Warning*: The Hales–Ferguson partition and scoring function given in [FH], which are the ones actually used for the proof of the Kepler conjecture, differ from those used earlier by Hales in [SP-I] and [SP-II].

(0) The idea of considering local inequalities that weight total area and covered area by spheres in a ratio B/A that is not equal to the optimal density occurs in Hales' original approach based on Delaunay triangulations, see [Ha1] and [Ha2]. It also appears in [SP-I, Lemma 2.1] and in [FH, Proposition 3.14]. I have inserted the parameters A and B in order to include the density inequality of Hsiang [Hs1] in the same framework.

(1) *Definitions* 4.1 and 4.2 appear on p. 2 of [FH].

(2) *Lemma* 4.1 is Lemma 1.2 of [FH], proved in Lemma 3.5 of [SP-I]. (The fact that no vertex of Ω occurs inside a face of a QL-tetrahedron or a QR-tetrahedron requires additional argument.)

(3) *Lemma* 4.2 follows from Lemma 1.3 of [FH].

(4) *Lemma* 4.3 is proved at the bottom of p. 3 of [FH].

(5) *Lemma* 4.4 is covered in the discussion on pp. 5–6 of [FH].

(6) *Lemma* 4.5(i)–(iii) is covered in the discussion on p. 5 of [FH], including Lemma 1.8 of that paper.

(7) The notion of "tip" is discussed at length in Section 2 of [SP-II]. In part II "tips" are not actually reassigned, although this is mentioned. Instead their existence affects (i.e., is encoded in) the scoring rule used for the associated Delaunay simplex which the "tip" is associated to. The rules for moving "tips" around to make V-cells in the Hales–Ferguson approach are discussed on p. 8 of [FH]. *Warning*: The way that "tips" are handled in part II and in [FH] may not be the same: [FH] takes priority.

(8) *Lemma* 4.6(i) is Lemma 2.2 of [SP-II] and Lemma 4.17 of [FH]. Facts related to (ii) are discussed in Section 8.6.7 of [SP-I]. (For the second part I do not have a reference.) (iii) Hales mentioned this in IAS lectures, and sent me a proof sketch, which I expanded into the following: Let S be a simplex in the D-system that overlaps a "tip" protuding from $\mathbf{v}$. Say that the "tip" overlaps by pointing to S along a face F of S. Thus F is a negatively oriented face of $S' = (F, \mathbf{v})$, which means that the simplex S' is a QR-tetrahedron or else a QL-tetrahedron with spine on F. Suppose first that S' is a QL-tetrahedron. It now follows that S must be a QL-tetrahedron with its spine on F by Lemma 2.2 of [FH]. So S' and S are adjacent QL-tetrahedra with spines on their common face F. Now S is in the D-system since S' is in the D-system. Thus the distance of $\mathbf{v}$ to the vertices in F is at most $\frac{251}{100}$, since $\mathbf{v}$ is not on the spine. We now suppose that the "tip" is not entirely contained in S, and derive a contradiction. If it is not contained in S, then

it crosses out through a face F' of S. By the same argument, the distance from $\mathbf{v}$ to the vertices of F' is at most $\frac{251}{100}$. Thus $\mathbf{v}$ has distances at most $\frac{251}{100}$ from all vertices of S, which is impossible by Lemmas 1.2 and 1.3 of [FH]. Suppose secondly that S' is a QR-tetrahedron. Then one shows that S is also a QR-tetrahedron, hence is in the D-system. The rest of the argument goes as before, to the same contradiction.

(9) *Theorem* 5.1. The main theorem is first stated as Conjecture 3.15 on p. 13 of [FH]. It is the theorem asserted to be proved in [KC].

(10) *Lemma* 5.1(i) appears as Lemma 3.13 in [FH]. Lemma 5.1(ii) is a special case of Lemma 3.13 of [FH] for a quad cluster, which can consist of four congruent QL-tetrahedra.

(11) The *standard regions* corresponding to the graph $\mathcal{G}(\mathbf{v})$ are defined on p. 4 of [FH]. ("Planar map that breaks unit sphere into regions.")

(12) *Lemma* 5.2 follows from Lemma 1.6 of [FH], which implies that crossing lines come from QL-tetrahedra only.

(13) *Lemma* 5.3 is an immediate consequence of my Lemma 5.2 and Lemma 1.3 of [FH].

(14) *Lemma* 5.4 appears as Lemma 3.13 in [FH].

(15) *Theorem* 5.2 follows from the corollary to Theorem 4.4 of [SP-IV]. See also Proposition 7.1 of [SP-III].

(16) *Lemmas* 5.5 and 5.6. These results are briefly stated at the bottom of p. 8 of [FH]. There are also some relevant details in Section 2.2 of [SP-II]. (I do not know an exact reference for a detailed proof.)

Hales–Ferguson terminology	Terminology in this paper
(1) Decomposition star	Vertex D-star
(2) Quasiregular tetrahedron	QR-tetrahedron
(3) Quarter	QL-tetrahedron
(4) Diagonal (of quarter)	Spine (of QL-tetrahedron)
(5) Q-system	D-system
(6) Score $\sigma(R, \mathbf{v})$	Weight function $\sigma(R, \mathbf{v})$
(7) Standard cluster	Cluster

References

[C] H. Cohn, New Bounds on Sphere Packings, Thesis, Harvard University, April 2000.

[CS] J. H. Conway and N. J. A. Sloane, *Sphere Packings, Lattices and Codes*, Third edition, Springer-Verlag: New York, 1999.

[Dod] T. C. Hales and S. McLaughlin, A proof of the dodecahedral conjecture, eprint: arXiv math.MG/9811079.

[FT1] L. Fejes Tóth, *Lagerungen in der Ebene auf der Kugel und im Raum*, Springer-Verlag: Berlin, 1953. (Second edition, 1972.)

[FT2] L. Fejes Tóth, *Regular Figures*, MacMillan: New York, 1964.

[FH] S. P. Ferguson and T. C. Hales, A formulation of the Kepler conjecture, eprint: arXiv math.MG/9811072.

[Ha1] T. C. Hales, The sphere packing problem, *J. Comput. Appl. Math.* **44** (1992), 41–76.

[Ha2] T. C. Hales, Remarks on the density of sphere packings in three dimensions, *Combinatorica* **13**(2) (1993), 181–197.

[Ha3] T. C. Hales, The status of the Kepler conjecture, *Math. Intelligencer* **16**(3) (1994), 47–58.

[Ha4] T. C. Hales, Cannonballs and honeycombs, *Notices Amer. Math. Soc.* **47**(4) (2000), 440–449.

[Hi] D. Hilbert, Mathematical problems, *Bull. Amer. Math. Soc.* **8** (1902), 437–479. Reprinted in *Mathematical Developments Arising from Hilbert Problems*, Proceedings of Symposia in Pure Mathematics, XXVIII, American Mathematical Society: Providence, RI, 1976.

[Hs1] W.-Y. Hsiang, On the sphere problem and Kepler's conjecture, *Internat. J. Math.* **4**(5) (1993), 739–831. (*MR* 95g: 52032.)

[Hs2] W.-Y. Hsiang, A rejoinder to Hales' article, *Math. Intelligencer* **17**(1) (1995), 35–42.

[KC] T. C. Hales, The Kepler conjecture, eprint: arXiv math.MG/9811078.

[KC0] T. C. Hales, An overview of the Kepler conjecture, eprint: arXiv math.MG/9811071.

[L] J. C. Lagarias, Notes on the Hales approach to the Kepler conjecture, manuscript, May 1999.

[O] J. Oesterlé, Densité maximale des empilements de sphères en dimension 3 [d'après Thomas C. Hales et Samuel P. Ferguson], *Séminaire Bourbaki*, Vol. 1998/99. *Astérisque* **266** (2000), Exp. No. 863, 405–413.

[R1] C. A. Rogers, The packing of equal spheres, *Proc. London Math. Soc.* **8** (1958), 609–620.

[R2] C. A. Rogers, *Packing and Covering*, Cambridge University Press: Cambridge, 1964.

[SP-I] T. C. Hales, Sphere packings, I, *Discrete Comput. Geom.* **17** (1997), 1–51, eprint: arXiv math.MG/9811073.

[SP-II] T. C. Hales, Sphere packings, II, *Discrete Comput. Geom.* **18** (1997), 135–149, eprint: arXiv math.MG/9811074.

[SP-III] T. C. Hales, Sphere packings, III, eprint: arXiv math.MG/9811075.

[SP-IV] T. C. Hales, Sphere packings, IV, eprint: arXiv math.MG/9811076.

[SP-V] S. P. Ferguson, Sphere packings, V, Thesis, University of Michigan, 1997, eprint: arXivmath.MG/9811077.

[Z] C. Zong, *Sphere Packings*, Springer-Verlag: New York, 1999.

Received November 19, 1999, *and in revised form April* 17, 2001. *Online publication December* 17, 2001.

Proof of the Kepler Conjecture

Introduction to Part II

Part II contains the published issue of *Discrete & Computational Geometry* containing the Hales-Ferguson proof of the Kepler Conjecture (*Discrete Comput. Geom.*, **36** (2006), 1–269). This consists of six papers together with a Guest Editors' Foreword.

The Guest Editor's Foreword include remarks on the proof and on the seven-year reviewing process.

These six papers published in this issue represent the top of the proof tree. At the bottom layers of the proof are a large number of inequalities checked by computer. Collections of these inequalities appear in a database maintained by *Annals of Mathematics*, and also in another database maintained by T. C. Hales.

Discrete & Computational Geometry

Volume 36   No. 1   July 2006

THE KEPLER CONJECTURE

by

Thomas C. Hales, with Samuel P. Ferguson

Special Issue

Edited by G. Fejes Tóth and J. C. Lagarias

 Springer

454 DISCRETE COMPUT. GEOM.
ISSN 0179-5376
36(1) 1–272 (2006)

Online First
Immediately Online
www.springerlink.com

Faster publication!

The cover of *Discrete & Computational Geometry*, **36**-1 (2006)—the special issue on the Kepler Conjecture.

Discrete Comput Geom 36:1–3 (2006)
DOI: 10.1007) s00454-005-1209-8

Discrete & Computational
© 2006 Springer Science+Business Media, Inc.

Guest Editors' Foreword

This issue of *Discrete & Computational Geometry* contains the detailed proof by T. Hales and S. P. Ferguson of the Kepler conjecture that the densest packing of three-dimensional Euclidean space by equal spheres is attained by "cannonball" packing. This is a landmark result. This conjecture, formulated by Kepler in 1611, was stated in Hilbert's formulation of his 18th problem [8]. The proof consists of mathematical arguments and a massive computer veri cation of many inequalities.

In 1953 László Fejes Tóth [1] indicated the possibility of proving the Kepler conjecture by relating it to a nonlinear optimization problem over a compact set, and in 1964 he proposed a speci c such inequality [2, pp. 295–299]. Such optimization problems can generically be termed "local density inequalities" as they involve some weighted measure of density associated to a neighborhood of each sphere center in a packing. Establishing a local density inequality yields a result stronger than the Kepler conjecture, since it requires the existence of a packing maximizing the local inequality simultaneously at every sphere center. Thus different local density inequalities are different mathematical results, although each may imply Kepler's conjecture as a corollary. Assuming that the optimization problem has a nite set of isolated global minima, and that the local minimality can be veri ed analytically in an open neighborhood of each of these minima, in principle the inequality can be proved by computer elsewhere using interval arithmetic. However, the size of the resulting optimization problems seemed far beyond the range of what could be solved by computer at that time, or even now, without new ideas to reduce the complexity of the problem.

An approach to prove the Kepler conjecture in this fashion was developed by Hales starting about 1992 (see [3] and [4]). A signi cant problem lies in formulating a local density inequality of this kind that is expected to be valid, and which represents an optimization problem small enough to be solvable in a reasonable time. A de nite program to prove it was then initiated by Hales after 1995 in [5] and [6]. Further modi cations of the local density inequality were required to obtain a tractable problem, and in this formulation process, important contributions were made by S. P. Ferguson, a student of Hales whose doctoral thesis work forms the paper Sphere Packing V in this issue. Signif-icant innovations of Hales were needed to reduce the optimization problem to a tractable size. A rst version of a complete proof was presented in six preprints, one joint with his

Ph.D. student Ferguson, and one by Ferguson alone, posted on the mathematics arXiv in 1998.

A week-long workshop on the proof was given at the Institute for Advanced Study in January 1999. At that time the second editor wrote a summary of the top level of the proof tree, which was later included in [9]. The six papers were submitted to *Annals of Mathematics* and a team of reviewers was assembled by the rst editor. The main portion of the reviewing took place in a seminar run at Eötvos University over a 3 year period. Some computer experiments were done in a detailed check. The nature of this proof, consisting in part of a large number of inequalities having little internal structure, and a complicated proof tree, makes it hard for humans to check every step reliably. Detailed checking of speci c assertions found them to be essentially correct in every case tested. The reviewing process produced in the reviewers a strong degree of conviction of the essential correctness of this proof approach, and that the reduction method led to nonlinear programming problems of tractable size.

The second editor became involved in the reviewing process in January 2003. This resulted in a substantial revision and reorganization of the papers making clearer the top portion of the proof tree, and leading to a simpli cation of some de nitions. The papers now include extra remarks motivating the choice of the given local density inequality, which is very complicated, and provide some reasons indicating why one should expect this particular local density inequality to hold. The revised version of the Ferguson paper now includes an explicit error bound away from the optimal density. An abridged version of the proof (excluding the paper of Ferguson) appears in *Annals of Mathematics* by Hales [7], with computer les and programs available on the *Annals of Mathematics* website. The detailed version of the proof, representing revised versions of the six 1998 preprints, appears in this special issue of *Discrete & Computational Geometry*.

Summaries of some aspects of this proof and other local density approaches to the Kepler conjecture are discussed in the Seminar Bourbaki report of Oesterlé [10] and in [9].

The reviewing of these papers was a particularly enormous and daunting task. The rst editor wishes to thank the following reviewers who substantially contributed to the process: Andras Bezdek, Michael Bleicher, Karoly Böröczky, Karoly Böröczky, Jr., Aladar Heppes, Wlodek Kuperberg, Endre Makai, Attila Por, Günter Rote, Istvan Talata, Bela Uhrin, and Zoltan Ujvary-Menyhard.

References

1. L. Fejes Tóth, *Lagerungen in der Ebene auf der Kugel und im Raum*, Springer-Verlag: Berlin, 1953. (Second edition 1972.) [See pp. 174–181.]
2. L. Fejes Tóth, *Regular Figures*, MacMillan: New York, 1964.
3. T. C. Hales, The sphere packing problem, *J. Comput. Appl. Math.* **44** (1992), 41–76.
4. T. C. Hales, Remarks on the density of sphere packings in three dimensions, *Combinatorica* **13**(2) (1993), 181–197.
5. T. C. Hales, Sphere packings, I, *Discrete Comput. Geom.* **17** (1997), 1–51, eprint: math.MG/9811073.
6. T. C. Hales, Sphere packings, II, *Discrete Comput. Geom.* **18** (1997), 135–149, eprint: math.MG/9811074.
7. T. C. Hales, A proof of the Kepler conjecture, *Ann. of Math.* **162**(3) (2005), 1065–1185.

8. D. Hilbert, Mathematical problems, *Bull. Amer. Math. Soc.* **8** (1902), 437–479. Reprinted in: *Mathematical Developments Arising from Hilbert Problems*, Proc. Symp. Pure Math. XXVIII, American Mathematical Society: Providence, RI, 1976.

9. J. C. Lagarias, Local density bounds for sphere packings and the Kepler Conjecture, *Discrete Comput. Geom.* **27** (2002), 165–193.

10. J. Oesterlé, Densité maximale des empilements de sphères en dimension 3 [d'après Thomas C. Hales et Samuel P. Ferguson], Séminaire Bourbaki, Exp. No. 863, Juin 1999.

Gábor Fejes Tóth
Hungarian Academy of Sciences

Jeffrey C. Lagarias
University of Michigan

3

Historical Overview of the Kepler Conjecture, by T. C. Hales

The paper is the first of the six papers in the Kepler Conjecture proof. It comprises Chapters 1 and 2 of the proof, giving history and an overview.

Contents

Historical Overview of the Kepler Conjecture, by T. C. Hales (Discrete Comput. Geom., **36** (2006), 5–20).

The original version of this chapter was revised. An erratum to this chapter can be found at
http://dx.doi.org/10.1007/978-1-4614-1129-1_12

Discrete Comput Geom 36:5–20 (2006)
DOI: 10.1007/s00454-005-1210-2

Historical Overview of the Kepler Conjecture

Thomas C. Hales

Department of Mathematics, University of Pittsburgh,
Pittsburgh, PA 15217, USA
hales@pitt.edu

Abstract.   This paper is the first in a series of six papers devoted to the proof of the Kepler conjecture, which asserts that no packing of congruent balls in three dimensions has density greater than the face-centered cubic packing. After some preliminary comments about the face-centered cubic and hexagonal close packings, the history of the Kepler problem is described, including a discussion of various published bounds on the density of sphere packings. There is also a general historical discussion of various proof strategies that have been tried with this problem.

1.   Introduction

1.1.   *The Face-Centered Cubic Packing*

A packing of spheres is an arrangement of nonoverlapping spheres of radius 1 in Euclidean space. Each sphere is determined by its center, so equivalently it is a collection of points in Euclidean space separated by distances of at least 2. The density of a packing is defined as the lim sup of the densities of the partial packings formed by spheres inside a ball with fixed center of radius R. (By taking the lim sup, rather than lim inf, as the density, we prove the Kepler conjecture in the strongest possible sense.) Defined as a limit, the density is insensitive to changes in the packing in any bounded region. For example, a finite number of spheres can be removed from the face-centered cubic packing without affecting its density.

Consequently, it is not possible to hope for any strong uniqueness results for packings of optimal density. The uniqueness established by this work is as strong as can be hoped for. It shows that certain local structures (decomposition stars) attached to the face-centered cubic (fcc) and hexagonal-close packings (hcp) are the only structures that maximize a local density function.

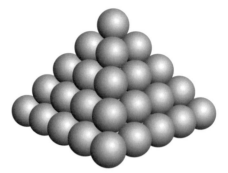

Fig. 1.1.   The fcc packing.

Although we do not pursue this point, Conway and Sloane develop a theory of tight packings that is more restrictive than having the greatest possible density [CS1]. An open problem is to prove that their list of tight packings in three dimensions is complete.

The fcc packing appears in Fig. 1.1.

The following facts about packings are well known. However, there is a popular and persistent misconception in the popular press that the fcc packing is the only packing with density $\pi/\sqrt{18}$. The comments that follow correct this misconception.

In the fcc packing, each ball is tangent to twelve others. For each ball in the packing, this arrangement of twelve tangent balls is the same. We call it the fcc pattern. In the hcp, each ball is tangent to twelve others. For each ball in the packing, the arrangement of twelve tangent balls is again the same. We call it the hcp pattern. The fcc pattern is different from the hcp pattern. In the fcc pattern there are four different planes through the center of the central ball that contain the centers of six other balls at the vertices of a regular hexagon. In the hcp pattern there is only one such plane. We call the arrangement of balls tangent to a given ball the *local tangent arrangement* of the ball.

There are uncountably many packings of density $\pi/\sqrt{18}$ that have the property that every ball is tangent to twelve others and such that the tangent arrangement around each ball is either the fcc pattern or the hcp pattern.

By *hexagonal layer*, we mean a translate of the two-dimensional lattice of points M in the A_2 arrangement. That is, M is a translate of the planar lattice generated by two vectors of length 2 and angle $2\pi/3$. The fcc packing is an example of a packing built from hexagonal layers.

If M is a hexagonal layer, a second hexagonal layer M' can be placed parallel to the first so that each lattice point of M' has distance 2 from three different vertices of M. When the second layer is placed in this manner, it is as close to the first layer as possible. Fix M and a unit normal to the plane of M. The normal allows us to speak of the second layer M' as being "above" or "below" layer M. There are two different positions in which M' can be placed closely above M and two different positions in which M' can be placed closely below M. As we build a packing, layer by layer (M, M', M'', and so forth), there are two choices at each stage of the close placement of the layer above the previous layer. Running through different sequences of choices gives uncountably many packings. In each of these packings the tangent arrangement around each ball is that of the twelve spheres in the fcc or the twelve spheres in the hcp.

Let Λ be a packing built as a sequence of close-packed hexagonal layers in this fashion. If P is any plane parallel to the hexagonal layers, then there are at most three different orthogonal projections of layers M to P. Call these projections A, B and C. Each hexagonal layer has a different projection than the layers immediately above and below it. In the fcc packing, the successive layers are A, B, C, A, B, C, ... In the hcp packing, the successive layers are A, B, A, B, ... If we represent A, B, and C as the vertices of a triangle, then the succession of hexagonal layers can be described by a walk along the vertices of the triangle. Different walks through the triangle describe different packings.

In fact, the different walks through a triangle give all packings of infinitely many equal balls in which the tangent arrangement around every ball is either the fcc pattern of twelve balls or the hcp pattern of twelve balls.

We justify the fact that different walks through a triangle give all such packings. Assume first that a packing Λ contains a ball (centered at v_0) in the hcp pattern. The hcp pattern contains a uniquely determined plane of symmetry. This plane contains v_0 and the centers of six others arranged in a regular hexagonal. If v is the center of one of the six others in the plane of symmetry, its local tangent arrangement of twelve balls must include v_0 and an additional four of the twelve balls around v_0. These five centers around v are not a subset of the fcc pattern. They can be uniquely extended to twelve centers arranged in the hcp pattern. This hcp pattern has the same plane of symmetry as the hcp pattern around v_0. In this way, as soon as there is a single center with the hcp pattern, the pattern propagates along the plane of symmetry to create a hexagonal layer M.

Once a packing Λ contains a single hexagonal layer, the condition that each ball be tangent to twelve others forces a hexagonal layer M' above M and another hexagonal layer below M. Thus, a single hexagonal layer forces a sequence of close-packed hexagonal layers in both directions.

We have justified the claim under the hypothesis that Λ contains at least one ball with the hcp pattern.

Assume that Λ does not contain any balls whose local tangent arrangement is the hcp pattern. Then every local tangent arrangement is the fcc pattern, and Λ itself is then the face-centered cubic packing. This completes the proof.

1.2. Early History, Hariot, and Kepler

The study of the mathematical properties of the fcc packing can be traced back to a Sanskrit work composed around 499 CE [P].

The modern mathematical study of spheres and their close packings can be traced to T. Hariot. Hariot's work—unpublished, unedited, and largely undated—shows a preoccupation with sphere packings. He seems to have first taken an interest in packings at the prompting of Sir Walter Raleigh. At the time, Hariot was Raleigh's mathematical assistant, and Raleigh gave him the problem of determining formulas for the number of cannonballs in regularly stacked piles. In 1591 he prepared a chart of triangular numbers for Raleigh. Shirley, Hariot's biographer, writes that the charts prepared for Raleigh led

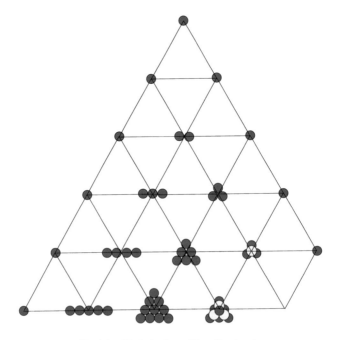

Fig. 1.2.   Hariot's view of Pascal's triangle.

to the study of the sums of squares, "a study which led inevitably to the corpuscular or atomic theory of matter orginally deriving from Lucretius and Epicurus" [Sh, p. 242].

Hariot connected sphere packings to Pascal's triangle long before Pascal introduced the triangle. See Fig. 1.2.

Hariot was the first to distinguish between the fcc and hcp [Ma, p. 52].

Kepler became involved in sphere packings through his correspondence with Hariot in the early years of the 17th century. Despite Kepler's initial reluctance to adopt an atomic theory, he was eventually swayed, and in 1611 he published an essay that explores the consequences of a theory of matter composed of small spherical particles. Kepler's essay was the "first recorded step towards a mathematical theory of the genesis of inorganic or organic form" [W, p. v].

Kepler's essay describes the fcc packing and asserts that "the packing will be the tightest possible, so that in no other arrangement could more pellets be stuffed into the same container." This assertion has come to be known as the Kepler conjecture. The purpose of this collection of papers is to give a proof of this conjecture.

1.3.   *History*

The next episode in the history of this problem is a debate between Isaac Newton and David Gregory. Newton and Gregory discussed the question of how many spheres of equal radius can be arranged to touch a given sphere. This is the three-dimensional analogue of the simple fact that in two dimensions six pennies, but no more, can be arranged to touch a central penny. This is the kissing-number problem in *n*-dimensions.

In three dimensions Newton said that the maximum was twelve spheres, but Gregory claimed that thirteen might be possible.

Newton was correct. In the 19th century, the first papers claiming a proof of the kissing-number problem appeared in [Ben], [Gu], and [Ho]. Although some writers cite these papers as a proof, they are hardly rigorous by today's standards. Another incorrect proof appears in [Bo]. The first proper proof was obtained by Schütte and van der Waerden in 1953 [SW]. An elementary proof appears in Leech [Le]. The influence of van der Waerden, Schütte, and Leech upon the papers in this collection is readily apparent. Although the connection between the Newton–Gregory problem and Kepler's problem is not obvious, L. Fejes Tóth in 1953, in the first work describing a strategy to prove the Kepler conjecture, made a quantitative version of the Gregory–Newton problem the first step [Fej6].

The two-dimensional analogue of the Kepler conjecture is to show that the honeycomb packing in two dimensions gives the highest density. This result was established in 1892 by Thue, with a second proof appearing in 1910 [T1], [T2]. Szpiro's book on the Kepler conjecture calls Thue's proofs into question [Sz]. C. Siegel said that Thue's original proof is "reasonable, but full of holes" [Sz]. A number of other proofs have appeared since then. Three are particularly notable. Rogers's proof generalizes to give a bound on the density of packings in any dimension [Ro1]. A proof by L. Fejes Tóth extends to give bounds on the density of packings of convex disks [Fej5]. A third proof, also by L. Fejes Tóth, extends to non-Euclidean geometries [Fej6]. Another early proof appears in [SM].

In 1900, Hilbert made the Kepler conjecture part of his 18th problem [Hi]. Milnor, in his review of Hilbert's 18th problem, breaks the problem into three parts [Mi]. The third part asks for the densest arrangements of congruent solids such as "spheres with given radii." Milnor writes that it is a "scandalous situation" that the sphere packing "problem in 3 dimensions remains unsolved."

1.4. The Literature

Past progress toward the Kepler conjecture can be arranged into four categories:

- bounds on the density,
- descriptions of classes of packings for which the bound of $\pi/\sqrt{18}$ is known,
- convex bodies other than spheres for which the packing density can be determined precisely,
- strategies of proof.

1.4.1. Bounds. Various upper bounds have been established on the density of packings:

0.884 (Blichfeldt) [Bl1],
0.835 (Blichfeldt) [Bl2],
0.828 (Rankin) [Ra],
0.7797 (Rogers) [Ro1],
0.77844 (Lindsey) [Li],
0.77836 (Muder)[Mu1],
0.7731 (Muder) [Mu2].

Rogers's is a particularly natural bound. As the dates indicate, it remained the best available bound for many years. His monotonicity lemma and his decomposition of Voronoi cells into simplices have become important elements in the proof of the Kepler conjecture. We give a new proof of Rogers's bound in "Sphere Packings, III." A function τ, used throughout this collection, measures the departure of various objects from Rogers's bound.

Muder's bounds, although they appear to be rather small improvements of Rogers's bound, are the first to make use of the full Voronoi cell in the determination of densities. As such, they mark a transition to a greater level of sophistication and difficulty. Muder's influence on the work in this collection is also apparent.

A sphere packing admits a Voronoi decomposition: around every sphere take the convex region consisting of points closer to that sphere center than to any other sphere center. L. Fejes Tóth's dodecahedral conjecture asserts that the Voronoi cell of smallest volume is a regular dodecahedron with inradius 1 [Fej4]. The dodecahedral conjecture implies a bound of 0.755 on sphere packings. L. Fejes Tóth actually gave a complete proof except for one estimate. A footnote in his paper documents the gap, "In the proof, we have relied to some extent solely on intuitive observation [Anschauung]." As L. Fejes Tóth pointed out, that estimate is extraordinarily difficult, and the dodecahedral conjecture has resisted all efforts until now [Mc].

The missing estimate in L. Fejes Tóth's paper is an explicit form of the Newton–Gregory problem. What is needed is an explicit bound on how close the thirteenth sphere can come to touching the central sphere, or, more generally, to minimize the sum of the distances of the thirteen spheres from the central sphere. No satisfactory bounds are known. Boerdijk has a conjecture for the arrangement that minimizes the average distance of the thirteen spheres from the central sphere. Van der Waerden has a conjecture for the closest arrangement of thirteen spheres in which all spheres have the same distance from the central sphere. Bezdek has shown that the dodecahedral conjecture would follow from weaker bounds than those originally proposed by L. Fejes Tóth [Bez2].

A proof of the dodecahedral conjecture has traditionally been viewed as the first step toward a proof of the Kepler conjecture, and if little progress has been made until now toward a complete solution of the Kepler conjecture, the difficulty of the dodecahedral conjecture is certainly responsible to a large degree.

1.4.2. *Classes of Packings.*   If the infinite-dimensional space of all packings is too unwieldy, we can ask if it is possible to establish the bound $\pi/\sqrt{18}$ for packings with special structures.

If we restrict the problem to packings whose sphere centers are the points of a lattice, the packings are described by a finite number of parameters, and the problem becomes much more accessible. Lagrange proved that the densest lattice packing in two dimensions is the familiar honeycomb arrangement [La]. Gauss proved that the densest lattice packing in three dimensions is the fcc [G]. In dimensions 4–8 the optimal lattices are described by their root systems, A_2, A_3, D_4, D_5, E_6, E_7, and E_8. Korkine and Zolotareff showed that D_4 and D_5 are the densest lattice packings in dimensions 4 and 5 [KZ1], [KZ2]. Blichfeldt determined the densest lattice packings in dimensions 6–8 [Bl3]. Cohn and Kumar solved the problem in dimension 24 [CK]. With the exception of dimension 24, beyond dimension 8, there are no proofs of optimality, and yet there are many excellent candidates for the densest lattice packings. For a proof of the existence of optimal lattices, see [O].

Although lattice packings are of particular interest because they relate to so many different branches of mathematics, Rogers has conjectured that in sufficiently high dimensions, the densest packings are not lattice packings [Ro2]. In fact, the densest known packings in various dimensions are not lattice packings. The third edition of [CS2] gives several examples of nonlattice packings that are denser than any known lattice packings (dimensions 10, 11, 13, 18, 20, 22). The densest packings of typical convex sets in the plane, in the sense of Baire categories, are not lattice packings [Fej1].

Gauss's theorem on lattice densities has been generalized by Bezdek, Kuperberg, and Makai, Jr. [BKM]. They showed that packings of parallel strings of spheres never have density greater than $\pi/\sqrt{18}$.

1.4.3. *Other Convex Bodies.*

If the optimal sphere packings are too difficult to determine, we might ask whether the problem can be solved for other convex bodies. To avoid trivialities, we restrict our attention to convex bodies whose packing density is strictly less than 1.

The first convex body in Euclidean 3-space that does not tile for which the packing density was explicitly determined is an infinite cylinder [BK]. Here Bezdek and Kuperberg prove that the optimal density is obtained by arranging the cylinders in parallel columns in the honeycomb arrangement.

In 1993 Pach exposed the humbling depth of our ignorance when he issued the challenge to determine the packing density for some bounded convex body that does not tile space [MP]. (Pach's question is more revealing than anything I can write on the subject of discrete geometry.) This question was answered by Bezdek [Bez1], who determined the packing density of a rhombic dodecahedron that has one corner clipped so that it no longer tiles. The packing density equals the ratio of the volume of the clipped rhombic dodecahedron to the volume of the unclipped rhombic dodecahedron.

1.4.4. *Strategies of Proof.*

In 1953 L. Fejes Tóth proposed a program to prove the Kepler conjecture [Fej6]. A single Voronoi cell cannot lead to a bound better than the dodecahedral conjecture. L. Fejes Tóth considered weighted averages of the volumes of collections of Voronoi cells. These weighted averages involve up to thirteen Voronoi cells. He showed that if a particular weighted average of volumes is greater than the volume of the rhombic dodecahedron, then the Kepler conjecture follows. The Kepler conjecture is an optimization problem in an infinite number of variables. L. Fejes Tóth's weighted-average argument was the first indication that it might be possible to reduce the Kepler conjecture to a problem in a finite number of variables. Needless to say, calculations involving the weighted averages of the volumes of several Voronoi cells will be significantly more difficult than those involved in establishing the dodecahedral conjecture.

To justify his approach, which limits the number of Voronoi cells to thirteen, L. Fejes Tóth needs a preliminary estimate of how close a thirteenth sphere can come to a central sphere. It is at this point in his formulation of the Kepler conjecture that an explicit version of the Newton–Gregory problem is required. How close can thirteen spheres come to a central sphere, as measured by the sum of their distances from the central sphere?

L. Fejes Tóth made another significant suggestion in [Fej7]. He was the first to suggest the use of computers in the Kepler conjecture. After describing his program, he mentions that his approach reduces the problem to an optimization problem in a finite number of variables. He suggests that computers might be used to approximate the minimum to this optimization problem.

The most widely publicized attempt to prove the Kepler conjecture was that of Hsiang [Hs1]. (See also [Hs2], [Hs3], and [Hs5].) Hsiang's approach can be viewed as a continuation and extension of L. Fejes Tóth's program. Hsiang's paper contains major gaps and errors [CHMS]. The mathematical arguments against his argument appear in my debate with him in the *Mathematical Intelligencer* [Ha3], [Hs4]. There are now many published sources that agree with the central claims of [Ha3] against Hsiang. Conway and Sloane report that the paper "contains serious flaws." G. Fejes Tóth feels that "the greater part of the work has yet to be done" [Fej1]. Bezdek concluded, after an extensive study of Hsiang's work, "his work is far from being complete and correct in all details" [Bez2]. Muder writes, "the community has reached a consensus on it: no one buys it" [Mu3].

2. Overview of the Proof

2.1. *Experiments with other Decompositions*

The following two sections (added January 2003) describe some of the motivation behind the partitions of space that have been used in the proof of the Kepler conjecture. This discussion includes various ideas that were tried, found wanting, and discarded. However, this discussion provides motivation for some of the choices that appear in the proof of the Kepler conjecture.

Let S be a regular tetrahedron of side length 2. If we place a unit ball at each of the four vertices, the fraction of the tetrahedral solid occupied by the part of the four balls within the tetrahedron is $\delta_{\text{tet}} \approx 0.7797$. Let O be a regular octahedron of side length 2. If we place a unit ball at each of the four vertices, the fraction of the octahedral solid occupied by the four balls is $\delta_{\text{oct}} \approx 0.72$. The fcc packing can be obtained by packing eight regular tetrahedra and six regular octahedra around each vertex. The density $\pi/\sqrt{18}$ of this packing is a weighted average of δ_{tet} and δ_{oct}:

$$\frac{\pi}{\sqrt{18}} = \tfrac{1}{3}\delta_{\text{tet}} + \tfrac{2}{3}\delta_{\text{oct}}.$$

My early conception (around 1989) was that for every packing of congruent balls, there should be a corresponding partition of space into regions of high density and regions of low density. Regions of high density should be defined as regions having density between δ_{oct} and δ_{tet}, and regions of low density should be defined as those regions of density at most δ_{oct}. It was my intention to prove that all regions of high density had to be confined to a set of nonoverlapping tetrahedra whose vertices are centers of the balls in the packing.

Thus, the question naturally arises of how much a regular tetrahedron of edge length 2 can be deformed before its density drops below that of a regular octahedron δ_{oct}.

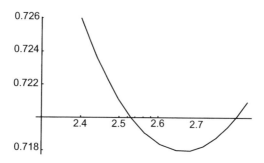

Fig. 2.1. The origin of the constant 2.51.

The graph in Fig. 2.1 shows the density of a tetrahedron with five edges of length 2 and a sixth edge of length x. Numerically, we see that the density drops below δ_{oct}, when $x = x_0 \approx 2.504$. To achieve the design goal of confining regions of high density to tetrahedra, we want a tetrahedron of edge lengths $2, 2, 2, 2, 2, x$, for $x \le x_0$, to be counted as a region of high density. Rounding upward, this example led to the cutoff parameter of 2.51 that distinguishes the tetrahedra (in the high density region) from the rest of space. This is the origin of the constant 2.51 that appears in the proof.

Since the tetrahedra are chosen to have vertices at the centers of the balls in the packing, it was quite natural to base the decomposition of space on the Delaunay decomposition. According to this early conception, space was to be partitioned into Delaunay simplices. A Delaunay simplex whose edge lengths are at most 2.51 is called a quasi-regular tetrahedron. These were the regions of presumably high density. According to the strategy in those early days, all other Delaunay simplices were to be shown to belong to regions of density at most δ_{oct}.

The following problem occupied my attention for a long period.

Problem. Fix a saturated packing. Let $X(oct)$ be the part of space of a saturated packing that is occupied by the Delaunay simplices having at least one edge of length at least 2.51. Let $X(tet)$ be the union of the complementary set of Delaunay simplices. Is it always true that the density of $X(oct)$ is at most δ_{oct}?

Early on, I viewed the positive resolution of this problem as crucial to the solution of the Kepler conjecture. Eventually, when I divided the proof of the Kepler conjecture into a five-step program, a variant of this problem became the second step of the program. See [Ha7].

To give an indication of the complexity of this problem, consider the simplex with edge lengths $2, 2, 2, 2, \ell, \ell$, where $\ell = \sqrt{2(3 + \sqrt{6})} \approx 3.301$. Assume that the two longer edges meet at a vertex. This simplex can appear as the Delaunay simplex in a saturated packing. Its density is about 0.78469. This constant is not only greater than δ_{oct}; it is even greater than δ_{tet}, so that the problem is completely misguided at the level of individual Delaunay simplices in $X(oct)$. It is only when the union of Delaunay simplices is considered that we can hope for an affirmative answer to the problem.

By the summer of 1994, I had lost hope of finding a partition of the set $X(oct)$ into small clusters of Delaunay simplices with the property that each cluster had density at most δ_{oct}. Progress had ground to a halt. The key insight came in the fall of 1994 (on Nov. 12, 1994 to be precise). On that day, I introduced a hybrid decomposition that relied on the Delaunay simplices in the regions $X(tet)$ formed by quasi-regular tetrahedra, but that switched to the Voronoi decomposition in certain regions of $X(oct)$. By April 1995 I had reformulated the problem, worked out a proof of the problem [Ha7] in its new form, and submitted it for publication. I submitted a revised version of [Ha6] that same month. The revision mentions the new strategy: "The rough idea is to let the score of a simplex in a cluster be the compression $\Gamma(S)$ [a function based on the Delaunay decomposition] if the circumradius of every face of S small, and otherwise to let the score be defined by Voronoi cells (in a way that generalizes the definition for quasi-regular tetrahedra)." See p. 6 of [Ha6].

The situation is somewhat more complicated than the previous paragraph suggests. Consider a Delaunay simplex S with edge lengths $2, 2, 2, 2, 2, 2.52$. Such a simplex belongs to the region $X(oct)$. However, if we break it into four pieces according to the Voronoi decomposition, the density of two of the pieces is about $0.696 < \delta_{oct}$ and the density of the other two is about $0.7368 > \delta_{oct}$. It is desirable not to have any separate regions in $X(oct)$ of density greater than δ_{oct}. Hence it is preferable to keep the four Voronoi regions in S together as a single Delaunay simplex. A second reason to keep S together is that the proof of the local optimality of the fcc packing and hcp seems to require it. A third reason was to treat pentahedral prisms. (This is a thorny class of counterexamples to a pure Delaunay simplex approach to the proof of the Kepler conjecture. See [Ha1], [Ha2], and [Fer].) For these reasons, we identify a class of Delaunay simplices in $X(oct)$ (such as S) that are to be treated according to a special set of rules. They are called *quarters*. As the name suggests, they often occur as the four simplices comprising an octahedron that has been "quartered."

One of the great advantages of a hybrid approach is that there is a tremendous amount of flexibility in the choice of the details of the decomposition. The details of the decomposition continued to evolve during 1995 and 1996. Finally, during a stay in Budapest following the Second European Congress in 1996, I abandoned all vestiges of the Delaunay decomposition, and adopted definitions of quasi-regular tetrahedra and quarters that rely only on the metric properties of the simplices (as opposed to the Delaunay criterion based on the position of other sphere centers in relation to the circumscribing sphere of the simplex). This decomposition of space is essentially what is used in the final proof.

The hybrid construction depends on certain choices of functions (satisfying a rather mild set of constraints). To solve the Kepler conjecture appropriate functions had to be selected, and an optimization problem based on those functions had to be solved. This function is called *the score*. Samuel Ferguson and I realized that every time we encountered difficulties in solving the minimization problem, we could adjust the scoring function σ to skirt the difficulty. The function σ became more complicated, but with each change we cut months—or even years—from our work. This incessant fiddling was unpopular with my colleagues. Every time I presented my work in progress at a conference, I was minimizing a different function. Even worse, the function was mildly incompatible with what I did in earlier papers [Ha6], [Ha7], and this required going back and patching the earlier papers.

The definition of the scoring function σ did not become fixed until it came time for Ferguson to defend his thesis, and we finally felt obligated to stop tampering with it. The final version of the scoring function σ is rather complicated. The reasons for the precise form of σ cannot be described without a long and detailed description of dozens of sphere clusters that were studied in great detail during the design of this function. However, a few general design principles can be mentioned. These comments assume a certain familiarity with the design of the proof.

(1) Simplices (with vertices at the centers of the balls in the packing) should be used whenever careful estimates of the density are required. Voronoi cells should be used whenever crude estimates suffice. For Voronoi cells, it is clear what the scoring function should be $\mathrm{vor}(R)$ (and its truncated versions $\mathrm{vor}_0(R)$, and so forth).

(2) The definition of the scoring function for quasi-regular tetrahedra was fixed in [Ha6] and this definition had to remain fixed to avoid rewriting that long paper.

Because of these first two points, most of the design effort for the function σ was focused on quarters.

(3) The decision to make the scoring for a quarter change when the circumradius of a face reaches $\sqrt{2}$ is to make the proof of the local optimality of the fcc and hcp run smoothly. From [Ha7], we see that the cutoff value $\sqrt{2}$ is important for the success of that proof. The cutoff $\sqrt{2}$ is also important for the proof that standard regions (other than quasi-regular tetrahedra) score at most 0 *pt*.

(4) The purpose of adding terms to the scoring function σ that depend on the truncated Voronoi function vor_0 is to make interval arithmetic comparisons between σ and vor_0 easier to carry out. This is useful in arguments about "erasing upright quarters."

2.2. *Contents of the Papers*

In [Ha6] a five-step program was described to prove the Kepler conjecture. It was planned that there would be five papers, each proving one step in the program. The papers [Ha6] and [Ha7] carry out the first two steps in the program. Because of the changes in the scoring function, it was necessary to issue a short paper [FH] mid-stream whose purpose was to give some adjustments to the five-step program. This paper adjusts the definitions from [Ha6] and checks that none of the results from [Ha6] and [Ha7] are affected in an essential way by these changes. Following this, the papers [Ha9] and [Fer] appeared in preprint form, completing the third and fifth steps of the program. The fourth step turned out to be particularly difficult. It occupies two separate papers [Ha10] and [Ha11].

The original series of papers suffers from the defect of being written over a span of several years. Some shifts in the conceptual framework of the research are evident. Based on comments from referees, a revision of these papers was prepared in 2002. The revisions were small, except for the paper [Ha11], which was completely rewritten. The structure of the proof remains the same, but it adds a substantial amount of introductory material that lessens the dependence on [Ha6] and [Ha7].

The papers were reorganized again in 2003. The series of papers is no longer organized along the original five steps with a mid-stream correction. Instead, the proof is now arranged according to the logical development of the subject matter. Only minor modifications have been made to the original proof. (The earlier versions are still available from [arXiv].) In the 2003 revision the exposition of the proof is entirely independent of the earlier papers [Ha6] and [Ha7].

An introduction to the ideas of the proof can be found in [Ha12]. An introduction to the algorithms can be found in [Ha14]. Speculation on a second-generation design of a proof can be found in [Ha14] and [Ha13].

2.3. *Complexity*

Why is this a difficult problem? There are many ways to answer this question.

This is an optimization problem in an infinite number of variables. In many respects the central problem has been to formulate a good finite-dimensional approximation to the density of a packing. Beyond this, there remains an extremely difficult problem in global optimization, involving nearly 150 variables. We recall that even very simple classes of nonlinear optimization problems, such as quadratic optimization problems, are NP-hard [HPT]. A general highly nonlinear program of this size is regarded by most researchers as hopeless (at least as far as rigorous methods are concerned).

There is a considerable literature on many closely related nonlinear optimization problems (the Tammes problem, circle packings, covering problems, the Lennard–Jones potential, Coulombic energy minimization of point particles, and so forth). Many of our expectations about nonlattice packings are formed by the extensive experimental data that have been published on these problems. The literature leads one to expect a rich abundance of critical points, and yet it leaves one with a certain skepticism about the possibility of establishing general results rigorously.

The extensive survey of circle packings in [Me] gives a broad overview of the progress and limits of the subject. Problems involving a few circles can be trivial to solve. Problems involving several circles in the plane can be solved with sufficient ingenuity. With the aid of computers, various problems involving a few more circles can be treated by rigorous methods. Beyond that, numerical methods give approximations but no rigorous solutions. Melissen's account of the 20-year quest for the best separated arrangement of ten points in a unit square is particularly revealing of the complexities of the subject.

Kepler's problem has a particularly rich collection of (numerical) local maxima that come uncomfortably close to the global maximum [Ha1]. These local maxima explain in part why a large number (around 5000) of planar maps are generated as part of the proof of the conjecture. Each planar map leads to a separate nonlinear optimization problem.

2.4. *Computers*

As this project has progressed, the computer has replaced conventional mathematical arguments more and more, until now nearly every aspect of the proof relies on computer verifications. Many assertions in these papers are results of computer calculations. To make the proof of Kepler's conjecture more accessible, I have posted extensive resources [arXiv].

Computers are used in various significant ways. They will be mentioned briefly here, and then developed more thoroughly elsewhere in the collection, especially in the final paper.

1. *Proof of inequalities by interval arithmetic.* "Sphere Packings, I" describes a method of proving various inequalities in a small number of variables by computer by interval arithmetic.
2. *Combinatorics.* A computer program classifies all of the planar maps that are relevant to the Kepler conjecture.
3. *Linear programming bounds.* Many of the nonlinear optimization problems for the scores of decomposition stars are replaced by linear problems that dominate the original score. They are solved by linear programming methods by computer. A typical problem has between 100 and 200 variables and 1000 and 2000 constraints. Nearly 100,000 such problems enter into the proof.
4. *Branch and bound methods.* When linear programming methods do not give sufficiently good bounds, they have been combined with branch and bound methods from global optimization.
5. *Numerical optimization.* The exploration of the problem has been substantially aided by nonlinear optimization and symbolic mathematical packages.
6. *Organization of output.* The organization of the few gigabytes of code and data that enter into the proof is in itself a nontrivial undertaking.

Acknowledgments

I am indebted to G. Fejes Tóth's survey of sphere packings in the preparation of this overview [Fej3]. For a much more comprehensive introduction to the literature on sphere packings, I refer the reader to that survey and to standard references on sphere packings such as [CS2], [PA], [GO], [Ro2], [Fej7], and [Fej8].

A detailed strategy of the proof was explained in lectures I gave at Mount Holyoke and Budapest during the summer of 1996 [Ha4]. See also the 1996 preprint, "Recent Progress on the Kepler Conjecture" [Ha4].

I owe the success of this project to a significant degree to S. Ferguson. His thesis solves a major step of the program. He has been highly involved in various other steps of the solution as well. He returned to Ann Arbor during the final 3 months of the project to verify many of the interval-based inequalities appearing in the appendices of "Sphere Packings, IV" and "The Kepler Conjecture." It is a pleasure to express my debt to him.

Sean McLaughlin has been involved in this project through his fundamental work on the dodecahedral conjecture. By detecting many of my mistakes, by clarifying my arguments, and in many other ways, he has made an important contribution.

I thank S. Karni, J. Mikhail, J. Song, D. J. Muder, N. J. A. Sloane, W. Casselman, T. Jarvis, P. Sally, E. Carlson, G. Bauer, and S. Chang for their contributions to this project. I express particular thanks to L. Fejes Tóth for the inspiration he provided during the course of this research. I thank J. Lagarias, G. Fejes Tóth, R. MacPherson, and G. Rote for their efforts as editors and referees. More broadly, I thank all the participants at the 1999 workshop on Discrete Geometry and the Kepler Problem.

This project received generous institutional support from the University of Chicago mathematics department, the Institute for Advanced Study, the journal *Discrete and Computational Geometry*, the School of Engineering at the University of Michigan (CAEN), and the National Science Foundation. Software (*cfsqp*)[1] for testing nonlinear inequalities was provided by the Institute for Systems Research at the University of Maryland.

Finally, I wish to give my special thanks to Kerri Smith, who has been my greatest source of support and encouragement through it all.

References

[arXiv] http://xxx.lanl.gov.

[Ben] C. Bender, Bestimmung der grössten Anzahl gleich grosser Kugeln, welche sich auf eine Kugel von demselben Radius, wie die übrigen, auflegen lassen, *Arch. Math. Phys.* **56** (1874), 302–306.

[Bez1] A. Bezdek, A remark on the packing density in the 3-space, in *Intuitive Geometry*, eds. K. Böröczky and G. Fejes Tóth, Colloquia Mathematica Societati János Bolyai, vol. 63, North-Holland, Amsterdam, 1994, pp. 17–22.

[Bez2] K. Bezdek, Isoperimetric inequalities and the dodecahedral conjecture, *Internat. J. Math.* **8**(6) (1997), 759–780.

[BK] A. Bezdek and W. Kuperberg, Maximum density space packing with congruent circular cylinders of infinite length, *Mathematica* **37** (1990), 74–80.

[BKM] A. Bezdek, W. Kuperberg, and E. Makai Jr., Maximum density space packing with parallel strings of balls, *Discrete Comput. Geom.* **6** (1991), 227–283.

[Bl1] H. F. Blichfeldt, Report on the theory of the geometry of numbers, *Bull. Amer. Math. Soc.* **25** (1919), 449–453.

[Bl2] H. F. Blichfeldt, The minimum value of quadratic forms and the closest packing of spheres, *Math. Ann.* **101** (1929), 605–608.

[Bl3] H. F. Blichfeldt, The minimum values of positive quadratic forms in six, seven and eight variables, *Math. Z.* **39** (1935), 1–15.

[Bo] A. H. Boerdijk, Some remarks concerning close-packing of equal spheres, *Philips Res. Rep.* **7** (1952), 303–313.

[CHMS] J. H. Conway, T. C. Hales, D. J. Muder, and N. J. A. Sloane, On the Kepler conjecture, *Math. Intelligencer* **16**(2) (1994), 5.

[CK] H. Cohn and A. Kumar, The densest lattice in twenty-four dimensions, math.MG/0408174 (2004).

[CS1] J. H. Conway and N. J. A. Sloane, What are all the best sphere packings in low dimensions? *Discrete Comput. Geom.* **13** (1995), 383–403.

[CS2] J. H. Conway and N. J. A. Sloane, *Sphere Packings, Lattices and Groups*, third edition, Springer-Verlag, New York, 1998.

[Fej1] G. Fejes Tóth, Review of [Hsi1], *MR* 95g#52032, 1995.

[Fej3] G. Fejes Tóth, Recent progress on packing and covering, in *Advances in Discrete and Computational Geometry* (*South Hadley, MA, 1996*), Contemporary Mathematics, vol. 223, AMS, Providence, RI, 1999, pp. 145–162. *MR* 99g:52036.

[Fej4] L. Fejes Tóth, Über die dichteste Kugellagerung, *Math. Z.* **48** (1942–1943), 676–684.

[Fej5] L. Fejes Tóth, Some packing and covering theorems, *Acta Sci. Math.* (*Szeded*) **12**/A (1950), 62–67.

[Fej6] L. Fejes Tóth, *Lagerungen in der Ebene auf der Kugel und im Raum*, first edition, Springer-Verlag, Berlin, 1953.

[Fej7] L. Fejes Tóth, *Regular Figures*, Pergamon Press, Oxford, 1964.

---

[1] www.isr.umd.edu/Labs/CACSE/FSQP/fsqp.html.

[Fej8] L. Fejes Tóth, *Lagerungen in der Ebene auf der Kugel und im Raum*, second edition, Springer-Verlag, Berlin, 1972.

[Fer] S. P. Ferguson, Sphere Packings, V, Thesis, University of Michigan, 1997.

[FH] S. P. Ferguson and T. C. Hales, A formulation of the Kepler conjecture, Preprint, 1998.

[G] C. F. Gauss, Untersuchungen über die Eigenscahften der positiven ternären quadratischen Formen von Ludwig August Seber, *Göttingische gelehrte Anzeigen*, 1831 Juli 9, also published in *J. Reine Angew. Math.* **20** (1840), 312–320, and *Werke*, vol. 2, Königliche Gesellschaft der Wissenschaften, Göttingen, 1876, pp. 188–196.

[GO] J. E. Goodman and J. O'Rourke, *Handbook of Discrete and Computational Geometry*, CRC Press, Boca Raton, FL, 1997.

[Gu] S. Günther, Ein stereometrisches Problem, *Arch. Math. Phys.* **57** (1875), 209–215.

[Ha1] T. C. Hales, The sphere packing problem, *J. Comput. Appl. Math.* **44** (1992), 41–76.

[Ha2] T. C. Hales, Remarks on the density of sphere packings in three dimensions, *Combinatorica* **13** (1993), 181–187.

[Ha3] T. C. Hales, The status of the Kepler conjecture, *Math. Intelligencer* **16**(3) (1994), 47–58.

[Ha4] T. C. Hales, http://www.pitt.edu/~thales/kepler98/holyoke.html.

[Ha6] T. C. Hales, Sphere packings, I, *Discrete Comput. Geom.* **17** (1977), 1–51.

[Ha7] T. C. Hales, Sphere packings, II, *Discrete Comput. Geom.* **18** (1997), 135–149.

[Ha9] T. C. Hales, Sphere packings, III, math.MG/9811075.

[Ha10] T. C. Hales, Sphere packings, IV, math.MG/9811076.

[Ha11] T. C. Hales, The Kepler conjecture, math.MG/9811078.

[Ha12] T. C. Hales, Cannonballs and honeycombs, *Notices Amer. Math. Soc.* **47**(4), 440–449.

[Ha13] T. C. Hales, Sphere Packings in 3 Dimensions, Arbeitstagung, 2001, math.MG/0205208.

[Ha14] T. C. Hales, Some algorithms arising in the proof of the Kepler conjecture, in *Discrete and Computational Geometry*, Algorithms and Combinatorics, vol. 25, Springer-Verlag, Berlin, July 2003, pp. 489–507.

[Hi] D. Hilbert, Mathematische Probleme, *Archiv Math. Phys.* **1** (1901), 44–63, also in *Mathematical Developments Arising from Hilbert Problems*, Proceedings of Symposia in Pure Mathematics, vol. 28, AMS, Providence, RI, 1976, pp. 1–34.

[Ho] R. Hoppe, Bemerkung der Redaction, *Math. Phys.* **56** (1874), 307–312.

[HPT] R. Horst, P. M. Pardalos, and N. V. Thoai, *Introduction to Global Optimization*, Kluwer, Dordrecht, 1995.

[Hs1] W.-Y. Hsiang, On the sphere packing problem and the proof of Kepler's conjecture, *Internat. J. Math.* **93** (1993), 739–831.

[Hs2] W.-Y. Hsiang, On the sphere packing problem and the proof of Kepler's conjecture, in *Differential Geometry and Topology* (Alghero, 1992), World Scientific, River Edge, NJ, 1993, pp. 117–127.

[Hs3] W.-Y. Hsiang, The geometry of spheres, in *Differential Geometry* (Shanghai, 1991), World Scientific, River Edge, NJ, 1993, pp. 92–107.

[Hs4] W.-Y. Hsiang, A rejoinder to T. C. Hales's article "The status of the Kepler conjecture," *Math. Intelligencer* **17**(1) (1995), 35–42.

[Hs5] W.-Y. Hsiang, *Least Action Principle of Crystal Formation of Dense Packing Type and the Proof of Kepler's Conjecture*, World Scientific, River Edge, NJ, 2002.

[Ka] R. Kargon, *Atomism in England from Hariot to Newton*, Oxford, 1966.

[Ke] J. Kepler, *The Six-Cornered Snowflake*, Oxford University Press, Oxford, 1966, Foreword by L. L. Whyte.

[KZ1] A. Korkine and G. Zolotareff, Sur les formes quadratiques, *Math. Ann.* **6** (1873), 366–389.

[KZ2] A. Korkine and G. Zolotareff, Sur les formes quadratiques positives, *Math. Ann.* **11** (1877), 242–292.

[La] J. L. Lagrange, Recherches d'arithmétique, *Nov. Mem. Acad. Roy. Sc. Bell Lettres Berlin* 1773, in *Œuvres*, vol. 3, Gauthier-Villars, Paris, 1867–1892, pp. 693–758.

[Le] J. Leech, The Problem of the thirteen spheres, *Math. Gaz.* **40** (Feb 1956), 22–23.

[Li] J. H. Lindsey II, Sphere packing in R^3, *Mathematika* **33** (1986), 137–147.

[Ma] B. J. Mason, On the shapes of snow crystals, in [Ke].

[Mc] S. McLaughlin, A proof of the dodecahedral conjecture, Preprint, math.MG/9811079.

[Me] J. B. M. Melissen, Packing and Covering with Circles, Ph.D. dissertation, University of Utrecht, Dec. 1997.

[Mi] J. Milnor, Hilbert's problem 18: on crystallographic groups, fundamental domains, and on sphere packings, in *Mathematical Developments Arising from Hilbert Problems*, Proceedings of Symposia in Pure Mathematics, vol. 28, AMS, Providence, RI, 1976, pp. 491–506.

[MP] W. Moser and J. Pach, Research problems in discrete geometry, DIMACS Technical Report 93032, 1993.

[Mu1] D. J. Muder, Putting the best face on a Voronoi polyhedron, *Proc. London Math. Soc.* (3) **56** (1988), 329–348.

[Mu2] D. J. Muder A new bound on the local density of sphere packings, *Discrete Comput. Geom.* **10** (1993), 351–375.

[Mu3] D. J. Muder, Letter, in *Fermat's Enigma*, by S. Singh, Walker, New York, 1997.

[O] J. Oesterlé, Empilements de sphères, Séminaire Bourbaki, vol. 1989/90, *Astérisque*, No. 189–190 exp. no. 727, (1990), 375–397.

[P] K. Plofker, Private communication, January 2000.

[PA] J. Pach and P.K. Agarwal, *Combinatorial Geometry*, Wiley, New York, 1995.

[Ra] R. A. Rankin, *Ann. of Math.* **48** (1947), 228–229.

[Ro1] C. A. Rogers, The packing of equal spheres, *Proc. London Math. Soc.* (3) **8** (1958), 609–620.

[Ro2] C. A. Rogers, *Packing and Covering*, Cambridge University Press, Cambridge, 1964.

[Sh] J. W. Shirley, *Thomas Harriot: a Biography*, Oxford, 1983.

[SHDC] N. J. A. Sloane, R. H. Hardin, T. D. S. Duff, and J. H. Conway, Minimal-energy clusters of hard spheres, *Discrete Comput. Geom.* **14**(3) (1995), 237–259.

[SM] B. Segre and K. Mahler, On the densest packing of circles, *Amer. Math Monthly* (1944), 261–270.

[SW] K. Schütte and B.L. van der Waerden, Das Problem der dreizehn Kugeln, *Math. Ann.* **125** (1953), 325–334.

[Sz] G. G. Szpiro, *Kepler's Conjecture*, Wiley, New York, 2002.

[T1] A. Thue, Om nogle geometrisk taltheoretiske Theoremer, *Forand. Skand. Natur.* **14** (1892), 352–353.

[T2] A. Thue, Über die dichteste Zusammenstellung von kongruenten Kreisen in der Ebene, *Christinia Vid. Selsk. Skr.* **1** (1910), 1–9.

[W] L. L. Whyte, Foreword to [Ke].

Received November 11, 1998, *and in revised form September* 12, 2003, *and July* 25, 2005.
Online publication February 27, 2006.

4

A Formulation of the Kepler Conjecture, by
T. C. Hales and S. P. Ferguson

This paper is the second of six papers giving the Kepler Conjecture proof. It contains Chapters 3–7 of the proof. It formulates the particular local density inequality to be proved in this approach.

The asterisks added in the margin of some pages indicate that changes are required, as listed in the Hales list of errata given on p. 372 of this volume.

Contents

The original version of this chapter was revised. An erratum to this chapter can be found at http://dx.doi.org/10.1007/978-1-4614-1129-1_12

Discrete Comput Geom 36:21–69 (2006)
DOI: 10.1007/s00454-005-1211-1

Discrete & Computational
Geometry
© 2006 Springer Science+Business Media, Inc.

A Formulation of the Kepler Conjecture

Thomas C. Hales[1] and Samuel P. Ferguson[2]

[1]Department of Mathematics, University of Pittsburgh,
Pittsburgh, PA 15217, USA
hales@pitt.edu

[2]5960 Millrace Court B-303,
Columbia, MD 21045, USA
samf2@comcast.net

Abstract. This paper is the second in a series of six papers devoted to the proof of the Kepler conjecture, which asserts that no packing of congruent balls in three dimensions has density greater than the face-centered cubic packing. The top level structure of the proof is described. A compact topological space is described. Each point of this space can be described as a finite cluster of balls with additional combinatorial markings. A continuous function on this compact space is defined. It is proved that the Kepler conjecture will follow if the value of this function is never greater than a given explicit constant.

Introduction

The following papers give a proof of the Kepler conjecture, which asserts that no packing of congruent balls in three-dimensional Euclidean space has density exceeding that of the face-centered cubic (fcc) packing.

A historical overview of the Kepler conjecture is found in the first paper in this series. Since the history of this problem is treated there, this paper does not go into the details of the extensive literature on this problem. We mention that Hilbert included the Kepler conjecture as part of his eighteenth problem [Hi]. L. Fejes Tóth was the first to formulate a plausible strategy for a proof [Fej8]. He also suggested that computers might play a role in the solution of this problem. The historical account also discusses the development of some of the key concepts of this paper.

An expository account of the proof is contained in [Ha12]. A general reference on sphere packings is [CS2]. A general discussion of the computer algorithms that are used in the proof can be found in [Ha14]. Some speculations on the structure of a second-generation proof can be found in [Ha13]. Details of computer calculations can be found on the internet at [Ha16].

The first section of this paper gives the top level structure of the proof of the Kepler conjecture. The next two sections describe the fundamental decompositions of space that are needed in the proof. The first decomposition, which is called the Q-system, is a collection of simplices that do not overlap. This decomposition was originally inspired by the Delaunay decomposition of space. The other decomposition, which is called the V-cell decomposition, is closely related to the Voronoi decomposition of space. In the following section these two decompositions of space are combined into geometrical objects called *decomposition stars*. The decomposition star is the fundamental geometrical object in the proof of the Kepler conjecture.

The final section of this paper, which was coauthored with Samuel P. Ferguson, describes a particular nonlinear function on the set of all decomposition stars, called the scoring function. The Kepler conjecture reduces to an optimization problem involving this nonlinear function on the set of all decomposition stars. This is an optimization problem in a finite number of variables. The subsequent papers (Papers III–VI) solve that optimization problem.

The choice of the particular scoring function to use was arrived at jointly with Samuel P. Ferguson. He has contributed to this project in many important ways, including the results in Section 7.

Some history of the proof and this paper is as follows. The original proof, as envisioned in 1994 and accomplished in 1998, was divided into a five-step program. As a result, the original papers were called "Sphere Packings, I," "Sphere Packings, II," and so forth. The first two papers in the series were published in an earlier volume of *DCG*. As it turned out, the fourth step "Sphere Packings, IV" is considerably more difficult than the other steps in the program. It became clear that a single paper would not suffice, and the fourth step of the proof was divided into two parts "Sphere Packings, IV" and "Kepler Conjecture (Sphere Packings, VI)." Samuel Ferguson's thesis "Sphere Packings, V" solved one of the five major steps in the proof. (Although "Sphere Packings, IV" and "Sphere Packings, VI" belonged together, because of the numbering scheme, Ferguson's theses "Sphere Packings, V" was inserted between these two papers.)

The proof that is contained in this volume is a rewritten version of the proof. For historical reasons, the papers in this volume have retained the original titles, but because of extensive revisions over the past several years, the proof is no longer arranged according to the five steps of the 1994 program.

In addition to the $5 + 1$ papers corresponding to the five steps of the original program, there is the current paper. It has the following origin. In 1996 it became clear that progress on the problem required some adjustments in the main nonlinear optimization problem of "Sphere Packings, I" and "II." As the original 1996 manuscript put it, "There are infinitely many scoring schemes that should lead to a proof of the Kepler conjecture. The problem is to formulate the scheme that makes the Kepler conjecture as accessible as possible" [Ha5]. The original purpose of this paper was to make some useful improvements in the scoring function from "Sphere Packings, I" and "II" and to make the changes in such a way that the main results of those papers would still hold true.

Over the past years, this paper has grown considerably in scope to the point that it is now lays the foundation for all of the papers in the series. In fact, all of the foundational material from "Sphere Packings, I" and "II," and the 1998 preprint series has been collected together in this article. The scoring function is no longer the same as the one presented in "Sphere Packings, I" and "II." This paper adapts the relevant material from these earlier papers to the current scoring function. This paper has expanded to the point that it is now possible to understand the entire proof of the Kepler conjecture without reading "I" and "II."

3.   The Top-Level Structure of the Proof

This section describes the structure of the proof of the Kepler conjecture.

3.1.   Statement of Theorems

Theorem 3.1 (The Kepler Conjecture).   *No packing of congruent balls in Euclidean three space has density greater than that of the face-centered cubic packing (see Fig. 3.1).*

This density is $\pi/\sqrt{18} \approx 0.74$.

The proof of this result is presented in this paper. Here, we describe the top-level outline of the proof and give references to the sources of the details of the proof.

By a *packing*, we mean an arrangement of congruent balls that are nonoverlapping in the sense that the interiors of the balls are pairwise disjoint. Consider a packing of congruent balls in Euclidean three space. There is no harm in assuming that all the balls have unit radius. The density of a packing does not decrease when balls are added to the packing. Thus, to answer a question about the greatest possible density we may add nonoverlapping balls until there is no room to add further balls. Such a packing is said to be *saturated*.

Let Λ be the set of centers of the balls in a saturated packing. Our choice of radius for the balls implies that any two points in Λ have distance at least 2 from each other. We call the points of Λ *vertices*. Let $B(x, r)$ denote the closed ball in Euclidean three space at center x and radius r. Let $\delta(x, r, \Lambda)$ be the finite density, defined as the ratio of the volume of $B(x, r, \Lambda)$ to the volume of $B(x, r)$, where $B(x, r, \Lambda)$ is defined as the intersection with $B(x, r)$ of the union of all balls in the packing. Set $\Lambda(x, r) = \Lambda \cap B(x, r)$.

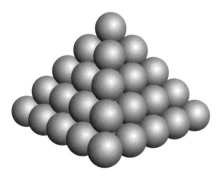

Fig. 3.1.   The fcc packing.

Recall that the Voronoi cell $\Omega(v) = \Omega(v, \Lambda)$ around a vertex $v \in \Lambda$ is the set of points closer to v than to any other ball center. The volume of each Voronoi cell in the fcc packing is $\sqrt{32}$. This is also the volume of each Voronoi cell in the hexagonal-close packing (hcp).

Definition 3.2. Let $A: \Lambda \to \mathbb{R}$ be a function. We say that A is *negligible* if there is a constant C_1 such that for all $r \geq 1$ and all $x \in \mathbb{R}^3$,

$$\sum_{v \in \Lambda(x,r)} A(v) \leq C_1 r^2.$$

We say that the function $A: \Lambda \to \mathbb{R}$ is *fcc-compatible* if for all $v \in \Lambda$ we have the inequality

$$\sqrt{32} \leq \text{vol}(\Omega(v)) + A(v).$$

The value $\text{vol}(\Omega(v)) + A(v)$ may be interpreted as a *corrected* volume of the Voronoi cell. Fcc-compatibility asserts that the corrected volume of the Voronoi cell is always at least the volume of the Voronoi cells in the fcc and hcp.

Lemma 3.3. *If there exists a negligible fcc-compatible function $A: \Lambda \to \mathbb{R}$ for a saturated packing Λ, then there exists a constant C such that for all $r \geq 1$ and all $x \in \mathbb{R}^3$,*

$$\delta(x, r, \Lambda) \leq \pi/\sqrt{18} + C/r.$$

The constant C depends on Λ only through the constant C_1.

Proof. The numerator $\text{vol } B(x, r, \Lambda)$ of $\delta(x, r, \Lambda)$ is at most the product of the volume of a ball $4\pi/3$ with the number $|\Lambda(x, r + 1)|$ of balls intersecting $B(x, r)$. Hence

$$\text{vol } B(x, r, \Lambda) \leq |\Lambda(x, r + 1)| 4\pi/3. \tag{3.1}$$

In a saturated packing each Voronoi cell is contained in a ball of radius 2 centered at the *center* of the cell. The volume of the ball $B(x, r + 3)$ is at least the combined volume of Voronoi cells whose center lies in the ball $B(x, r + 1)$. This observation, combined with fcc-compatibility and negligibility, gives

$$\sqrt{32} |\Lambda(x, r + 1)| \leq \sum_{v \in \Lambda(x,r+1)} (A(v) + \text{vol}(\Omega(v)))$$

$$\leq C_1(r + 1)^2 + \text{vol } B(x, r + 3)$$

$$\leq C_1(r + 1)^2 + (1 + 3/r)^3 \text{vol } B(x, r). \tag{3.2}$$

Recall that $\delta(x, r, \Lambda) = \text{vol } B(x, r, \Lambda)/\text{vol } B(x, r)$. Divide inequality (3.1) through by vol $B(x, r)$. Use inequality (3.2) to eliminate $|\Lambda(x, r + 1)|$ from the resulting inequality. This gives

$$\delta(x, r, \Lambda) \leq \frac{\pi}{\sqrt{18}} \left(1 + \frac{3}{r}\right)^3 + C_1 \frac{(r + 1)^2}{r^3 \sqrt{32}}.$$

The result follows for an appropriately chosen constant C. $\qquad \square$

An analysis of the preceding proof shows that fcc-compatibility leads to the particular value $\pi/\sqrt{18}$ in the statement of Lemma 3.3. If fcc-compatibility were to be dropped from the hypotheses, any negligible function A would still lead to an upper bound $4\pi/(3L)$ on the density of a packing, expressed as a function of a lower bound L on all vol $\Omega(v) + A(v)$.

Remark 3.4. We take the precise meaning of the Kepler conjecture to be a bound on the essential supremum of the function $\delta(x, r, \Lambda)$ as r tends to infinity. Lemma 3.3 implies that the essential supremum of $\delta(x, r, \Lambda)$ is bounded above by $\pi/\sqrt{18}$, provided a negligible fcc-compatible function can be found. The strategy will be to define a negligible function, and then to solve an optimization problem in finitely many variables to establish that it is fcc-compatible.

Section 6 defines a compact topological space DS (the space of decomposition stars, Definition 6.2) and a continuous function σ on that space, which is directly related to packings.

If Λ is a saturated packing, then there is a geometric object $D(v, \Lambda)$ constructed around each vertex $v \in \Lambda$. $D(v, \Lambda)$ depends on Λ only through the vertices in Λ that are at most a constant distance away from v. That constant is independent of v and Λ. The objects $D(v, \Lambda)$ are called *decomposition stars*, and the space of all decomposition stars is precisely DS. Section 6.2 shows that the data in a decomposition star are sufficient to determine a Voronoi cell $\Omega(D)$ for each $D \in$ DS. The same section shows that the Voronoi cell attached to D is related to the Voronoi cell of v in the packing by relation

$$\text{vol } \Omega(v) = \text{vol } \Omega(D(v, \Lambda)).$$

Section 7 defines a continuous real-valued function A_0: DS $\to \mathbb{R}$ that assigns a "weight" to each decomposition star. The topological space DS embeds into a finite-dimensional Euclidean space. The reduction from an infinite-dimensional to a finite-dimensional problem is accomplished by the following results.

Theorem 3.5. *For each saturated packing Λ, and each $v \in \Lambda$, there is a decomposition star $D(v, \Lambda) \in$ DS such that the function A: $\Lambda \to \mathbb{R}$ defined by*

$$A(v) = A_0(D(v, \Lambda))$$

is negligible for Λ.

This is proved as Theorem 7.11. The main object of the proof is then to show that the function A is fcc-compatible. This is implied by the inequality (in a finite number of variables)

$$\sqrt{32} \leq \text{vol } \Omega(D) + A_0(D), \tag{3.3}$$

for all $D \in$ DS.

In the proof it is convenient to reframe this optimization problem by composing it with a linear function. The resulting continuous function σ: DS $\to \mathbb{R}$ is called the *scoring function*, or *score*.

Let δ_{tet} be the packing density of a regular tetrahedron. That is, let S be a regular tetrahedron of edge length 2. Let B be the part of S that lies within distance 1 of some vertex. Then δ_{tet} is the ratio of the volume of B to the volume of S. We have $\delta_{tet} = \sqrt{8}\arctan(\sqrt{2}/5)$.

Let δ_{oct} be the packing density of a regular octahedron of edge length 2, again constructed as the ratio of the volume of points within distance 1 of a vertex to the volume of the octahedron.

The density of the fcc packing is a weighted average of these two ratios

$$\frac{\pi}{\sqrt{18}} = \frac{\delta_{tet}}{3} + \frac{2\delta_{oct}}{3}.$$

This determines the exact value of δ_{oct} in terms of δ_{tet}. We have $\delta_{oct} \approx 0.72$.

In terms of these quantities,

$$\sigma(D) = -4\delta_{oct}(\text{vol}(\Omega(D)) + A_0(D)) + \frac{16\pi}{3}. \tag{3.4}$$

Definition 3.6.   We define the constant

$$pt = 4\arctan\left(\frac{\sqrt{2}}{5}\right) - \frac{\pi}{3}.$$

Its value is approximately $pt \approx 0.05537$. Equivalent expressions for pt are

$$pt = \sqrt{2}\delta_{tet} - \frac{\pi}{3} = -2\left(\sqrt{2}\delta_{oct} - \frac{\pi}{3}\right).$$

In terms of the scoring function σ, the optimization problem in a finite number of variables (inequality (3.3)) takes the following form. The proof of this inequality is a central concern in this paper.

Theorem 3.7 (Finite-Dimensional Reduction).   *The maximum of σ on the topological space* DS *of all decomposition stars is the constant* $8\,pt \approx 0.442989$.

Remark 3.8.   The Kepler conjecture is an optimization problem in an infinite number of variables (the coordinates of the points of Λ). The maximization of σ on DS is an optimization problem in a finite number of variables. Theorem 3.7 may be viewed as a finite-dimensional reduction of the Kepler conjecture.

Let $t_0 = 1.255$ $(2t_0 = 2.51)$. This is a parameter that is used for truncation throughout this paper.

Let $U(v, \Lambda)$ be the set of vertices in Λ at nonzero distance at most $2t_0$ from v. From v and a decomposition star $D(v, \Lambda)$ it is possible to recover $U(v, \Lambda)$, which we write as $U(D)$. We can completely characterize the decomposition stars at which the maximum of σ is attained.

Theorem 3.9. *Let D be a decomposition star at which the function* $\sigma: \text{DS} \to \mathbb{R}$ *attains its maximum. Then the set U(D) of vectors at distance at most* $2t_0$ *from the center has cardinality twelve. Up to Euclidean motion, U(D) is one of two arrangements: the kissing arrangement of the twelve balls around a central ball in the fcc packing or the kissing arrangement of twelve balls in the hcp.*

There is a complete description of all packings in which every sphere center is surrounded by twelve others in various combinations of these two patterns. All such packings are built from parallel layers of the A_2 lattice. (The A_2 lattice formed by equilateral triangles, is the optimal packing in two dimensions.) See the first paper I.

3.2. *Basic Concepts in the Proof*

To prove Theorems 3.1, 3.7, and 3.9, we wish to show that there is no counterexample. In particular, we wish to show that there is no decomposition star D with value $\sigma(D) > 8\,pt$. We reason by contradiction, assuming the existence of such a decomposition star. With this in mind, we call D a *contravening decomposition star* if

$$\sigma(D) \geq 8\,pt.$$

In much of what follows we tacitly assume that every decomposition star under discussion is a contravening one. Thus, when we say that no decomposition stars exist with a given property, it should be interpreted as saying that no such contravening decomposition stars exist.

To each contravening decomposition star D, we associate a (combinatorial) plane graph $G(D)$. A restrictive list of properties of plane graphs is described in Section 18.3. Any plane graph satisfying these properties is said to be *tame*. All tame plane graphs have been classified. There are several thousand, up to isomorphism. The list appears in [Ha16]. We refer to this list as the *archival list* of plane graphs.

A few of the tame plane graphs are of particular interest. Every decomposition star attached to the fcc packing gives the same plane graph (up to isomorphism). Call it G_{fcc}. Likewise, every decomposition star attached to the hcp gives the same plane graph G_{hcp} (see Fig. 3.2).

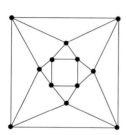 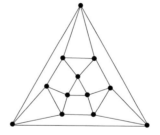

Fig. 3.2.   The plane graphs G_{fcc} and G_{hcp}.

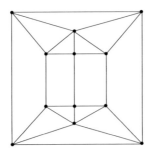

Fig. 3.3. The plane graph G_{pent} of the pentahedral prism.

There is one more tame plane graph that is particularly troublesome. It is the graph G_{pent} obtained from the pictured configuration of twelve balls tangent to a given central ball (Fig. 3.3). (Place a ball at the north pole, another at the south pole, and then form two pentagonal rings of five balls.) This case requires individualized attention. S. Ferguson proves the following theorem in Paper V.

Theorem 3.10 (Ferguson). *There are no contravening decomposition stars D whose associated plane graph is isomorphic to G_{pent}.*

3.3. *Logical Skeleton of the Proof*

Consider the following six claims. Eventually we will give a proof of all six statements. First, we draw out some of their consequences. The main results (Theorems 3.1, 3.7, and 3.9) all follow from these claims.

Claim 3.11. *If the maximum of the function σ on DS is $8\,pt$, then for every saturated packing Λ there exists a negligible fcc-compatible function A.*

Claim 3.12. *Let D be a contravening decomposition star. Then its plane graph $G(D)$ is tame.*

Claim 3.13. *If a plane graph is tame, then it is isomorphic to one of the several thousand plane graphs that appear in the archival list of plane graphs.*

Claim 3.14. *If the plane graph of a contravening decomposition star is isomorphic to one in the archival list of plane graphs, then it is isomorphic to one of the following three plane graphs: G_{pent}, G_{hcp}, or G_{fcc}.*

Claim 3.15. *There do not exist any contravening decomposition stars D whose associated graph is isomorphic to G_{pent}.*

Claim 3.16. *Contravening decomposition stars exist. If D is a contravening decomposition star, and if the plane graph of D is isomorphic to G_{fcc} or G_{hcp}, then $\sigma(D) = 8\,pt$. Moreover, up to Euclidean motion, $U(D)$ is the kissing arrangement of the twelve balls around a central ball in the fcc packing or the kissing arrangement of twelve balls in the hcp.*

Next, we state some of the consequences of these claims.

Lemma 3.17. *Assume Claims* 3.12–3.15. *If D is a contravening decomposition star, then its plane graph $G(D)$ is isomorphic to G_{hcp} or G_{fcc}.*

Proof. Assume that D is a contravening decomposition star. Then its plane graph is tame, and consequently appears on the archival list of plane graphs. Thus, it must be isomorphic to one of G_{fcc}, G_{hcp}, or G_{pent}. The final graph is ruled out by Claim 3.15. □

Lemma 3.18. *Assume Claims* 3.12–3.16. *Then Theorem* 3.7 *holds.*

Proof. By Claim 3.16 and Lemma 3.17, the value $8\,pt$ lies in the range of the function σ on DS. Assume for a contradiction that there exists a decomposition star $D \in$ DS that has $\sigma(D) > 8\,pt$. By definition, this is a contravening star. By Lemma 3.17, its plane graph is isomorphic to G_{hcp} or G_{fcc}. By Claim 3.16, $\sigma(D) = 8\,pt$, in contradiction with $\sigma(D) > 8\,pt$. □

Lemma 3.19. *Assume Claims* 3.12–3.16. *Then Theorem* 3.9 *holds.*

Proof. By Theorem 3.7, the maximum of σ on DS is $8\,pt$. Let D be a decomposition star at which the maximum $8\,pt$ is attained. Then D is a contravening star. Lemma 3.17 implies that the plane graph is isomorphic to G_{hcp} or G_{fcc}. The hypotheses of Claim 3.16 are satisfied. The conclusion of Claim 3.16 is the conclusion of Theorem 3.9. □

Lemma 3.20. *Assume Claims* 3.11–3.16. *Then the Kepler conjecture* (*Theorem* 3.1) *holds.*

Proof. As pointed out in Remark 3.4, the precise meaning of the Kepler conjecture is for every saturated packing Λ, the essential supremum of $\delta(x, r, \Lambda)$ is at most $\pi/\sqrt{18}$.

Let Λ be the set of centers of a saturated packing. Let $A: \Lambda \to \mathbb{R}$ be the negligible, fcc-compatible function provided by Claim 3.11 (and Lemma 3.18). By Lemma 3.3, the function A leads to a constant C such that for all $r \geq 1$ and all $x \in \mathbb{R}^3$, the density $\delta(x, r, \Lambda)$ satisfies

$$\delta(x, r, \Lambda) \leq \pi/\sqrt{18} + C/r.$$

This implies that the essential supremum of $\delta(x, r, \Lambda)$ is at most $\pi/\sqrt{18}$. □

Remark 3.21. One other theorem (Theorem 3.5) was stated without proof in Section 3.1. This result was placed there to motivate the other results. However, it is not an immediate consequence of Claims 3.11–3.16. Its proof appears in Theorem 7.11.

3.4. *Proofs of the Central Claims*

The previous section showed that the main results in Section 3.1 (Theorems 3.1, 3.7, and 3.9) follow from six claims. This section indicates where each of these claims is proved, and mentions a few facts about the proofs.

Claim 3.11 is proved in Theorem 7.14. Claim 3.12 is proved in Theorem 20.20. Claim 3.13, the classification of tame graphs, is proved in Theorem 19.1. By the classification of such graphs, this reduces the proof of the Kepler conjecture to the analysis of the decomposition stars attached to the finite explicit list of tame plane graphs. We will return to Claim 3.14 in a moment. Claim 3.15 is Ferguson's thesis, cited as Theorem 3.10.

Claim 3.16 is the local optimality of the fcc and hcp. In Section 8, the necessary local analysis is carried out to prove Claim 3.16 as Corollary 8.3.

Now we return to Claim 3.14. This claim is proved as Theorem 23.1. The idea of the proof is the following. Let D be a contravening decomposition star with graph $G(D)$. We assume that the graph $G(D)$ is not isomorphic to G_{fcc}, G_{hcp}, and G_{pent} and then prove that D is not contravening. This is a case-by-case argument, based on the explicit archival list of plane graphs.

To eliminate these remaining cases, more-or-less generic arguments can be used. A linear program is attached to each tame graph G. The linear program can be viewed as a linear relaxation of the nonlinear optimization problem of maximizing σ over all decomposition stars with a given tame graph G. Because it is obtained by relaxing the constraints on the nonlinear problem, the maximum of the linear problem is an upper bound on the maximum of the original nonlinear problem. Whenever the linear programming maximum is less than $8\,pt$, it can be concluded that there is no contravening decomposition star with the given tame graph G. This linear programming approach eliminates most tame graphs.

When a single linear program fails to give the desired bound, it is broken into a series of linear programming bounds, by branch and bound techniques. For every tame plane graph G other than G_{hcp}, G_{fcc}, and G_{pent}, we produce a series of linear programs that establish that there is no contravening decomposition star with graph G.

The proof is organized in the following way. Sections 4–7 introduce the basic definitions. Section 7 gives a proof of Claim 3.11. Section 8 proves Claim 3.16. Sections 9–14 present the fundamental estimates. Sections 18–19 give a proof of Claim 3.13. Sections 20–22 give a proof of Claim 3.12. Sections 23–25 give a proof of Claim 3.14. Claim 3.15 (Ferguson's thesis) appears in Paper V.

4. Construction of the Q-System

It is useful to separate the parts of space of relatively high packing density from the parts of space with relatively low packing density. The Q-system, which is developed in this section, is a crude way of marking off the parts of space where the density is potentially high. The Q-system is a collection of simplices whose vertices are points of the packing Λ. The Q-system is reminiscent of the Delaunay decomposition, in the sense of being a collection of simplices with vertices in Λ. In fact, the Q-system is the remnant of an earlier approach to the Kepler conjecture that was based entirely on the Delaunay decomposition (see [Ha2]). However, the Q-system differs from the Delaunay decomposition in crucial respects. The most fundamental difference is that the Q-system, while consisting of nonoverlapping simplices, does not partition all of space.

This section defines the set of simplices in the Q-system and proves that they do not overlap. In order to prove this, we develop a long series of lemmas that study the geometry of intersections of various edges and simplices. At the end of this section we give the proof that the simplices in the Q-system do not overlap.

4.1. Description of the Q-system

Fix a packing of balls of radius 1. We identify the packing with the set Λ of its centers. A packing is thus a subset Λ of $\mathbb{R}^3$ such that for all $v, w \in \Lambda$, $|v - w| < 2$ implies $v = w$. The centers of the balls are called *vertices*. The term "vertex" will be reserved for this technical usage. A packing is said to be *saturated* if for every $x \in \mathbb{R}^3$, there is some $v \in \Lambda$ such that $|x - v| < 2$. Any packing is a subset of a saturated packing. We assume that Λ is saturated. The set Λ is countably infinite.

Definition 4.1.   We define the *truncation parameter* to be the constant $t_0 = 1.255$. It is used throughout. Informal arguments that led to this choice of constant are described in Paper I.

Precise constructions that rely on the truncation parameter t_0 will appear below. We will regularly intersect Voronoi cells with balls of radius t_0 to obtain lower bounds on their volumes. We will regularly disregard vertices of the packing that lie at distance greater than $2t_0$ from a fixed $v \in \Lambda$ to obtain a finite subset of Λ (a finite cluster of balls in the packing) that is easier to analyze than the full packing Λ.

The truncation parameter is the first of many decimal constants that appear. Each decimal constant is an exact rational value, e.g., $2t_0 = 251/100$. They are not to be regarded as approximations of some other value.

Definition 4.2.   A *quasi-regular* triangle is a set $T \subset \Lambda$ of three vertices such that if $v, w \in T$ then $|w - v| \leq 2t_0$.

Definition 4.3.   A simplex is a set of four vertices. A *quasi-regular* tetrahedron is a simplex S such that if $v, w \in S$ then $|w - v| \leq 2t_0$. A *quarter* is a simplex whose edge lengths $y_1, \ldots, y_6$ can be ordered to satisfy $2t_0 \leq y_1 \leq \sqrt{8}, 2 \leq y_i \leq 2t_0, i = 2, \ldots, 6$. If a quarter satisfies the strict inequalities $2t_0 < y_1 < \sqrt{8}$, then we say that it is a strict quarter. We call the longest edge $\{v, w\}$ of a quarter its *diagonal*. When the quarter is strict, we also say that its diagonal is strict. When the quarter has a distinguished vertex, the quarter is *upright* if the distinguished vertex is an endpoint of the diagonal, and *flat* otherwise.

At times, we identify a simplex with its convex hull. We say, for example, that the circumcenter of a simplex is contained in the simplex to mean that the circumcenter is contained in the convex hull of the four vertices. Similar remarks apply to triangles, quasi-regular tetrahedra, quarters, and so forth. We write $|S|$ for the convex hull of S

when we wish to be explicit about the distinction between $|S|$ and its set of extreme points.

When we wish to give an order on an edge, triangle, simplex, etc., we present the object as an ordered tuple rather than a set. Thus, we refer to both $(v_1, \ldots, v_4)$ and $\{v_1, \ldots, v_4\}$ as simplices, depending on the needs of the given context.

Definition 4.4. Two manifolds with boundary *overlap* if their interiors intersect.

Definition 4.5. A set O of six vertices is called a *quartered octahedron* if there are four pairwise nonoverlapping strict quarters $S_1, \ldots, S_4$ all having the same diagonal, such that O is the union of the four sets S_i of four vertices. (It follows easily that the strict quarters S_i can be given a cyclic order with respect to which each strict quarter S_i has a face in common with the next, so that a quartered octahedron is literally a octahedron that has been partitioned into four quarters.)

Remark 4.6. A quartered octahedron may have more than one diagonal of length less than $\sqrt{8}$, so its decomposition into four strict quarters need not be unique. The choice of diagonal has no particular importance. Nevertheless, to make things canonical, we pick the diagonal of length less than $\sqrt{8}$ with an endpoint of smallest possible value with respect to the lexicographical ordering on coordinates; that is, with respect to the ordering $(y_1, y_2, y_3) < (y'_1, y'_2, y'_3)$, if $y_i = y'_i$, for $i = 1, \ldots, k$, and $y_{k+1} < y'_{k+1}$. This selection rule for diagonals is fully translation invariant in the sense that if one octahedron is a translate of another (whether or not they belong to the same saturated packing), then the selected diagonal of one is a translate of the selected diagonal of the other.

Definition 4.7. If $\{v_1, v_2\}$ is an edge of length between $2t_0$ and $\sqrt{8}$, we say that a vertex v ($\neq v_1, v_2$) is an *anchor* of $\{v_1, v_2\}$ if its distances to v_1 and v_2 are at most $2t_0$.

The two vertices of a quarter that are not on the diagonal are anchors of the diagonal, and the diagonal may have other anchors as well.

Definition 4.8. Let Q be the set of quasi-regular tetrahedra and strict quarters, enumerated as follows. This set is called the Q-system. It is canonically associated with a saturated packing Λ. (The Q stands for quarters and quasi-regular tetrahedra.)

1. All quasi-regular tetrahedra.
2. Every strict quarter such that none of the quarters along its diagonal overlaps any other quasi-regular tetrahedron or strict quarter.
3. Every strict quarter whose diagonal has four or more anchors, as long as there are not exactly four anchors arranged as a quartered octahedron.
4. The fixed choice of four strict quarters in each quartered octahedron.
5. Every strict quarter $\{v_1, v_2, v_3, v_4\}$ whose diagonal $\{v_1, v_3\}$ has exactly three anchors v_2, v_4, v_5 provided that the following hold (for some choice of indexing).
   (a) $\{v_2, v_5\}$ is a strict diagonal with exactly three anchors: v_1, v_3, v_4.
   (b) $d_{24} + d_{25} > \pi$, where d_{24} is the dihedral angle of the simplex $\{v_1, v_3, v_2, v_4\}$ along the edge $\{v_1, v_3\}$ and d_{25} is the dihedral angle of the simplex $\{v_1, v_3, v_2, v_5\}$ along the edge $\{v_1, v_3\}$.

No other quasi-regular tetrahedra or strict quarters are included in the Q-system $\mathcal{Q}$.

The following theorem is the main result of this section.

Theorem 4.9. *For every saturated packing, there exists a uniquely determined Q-system. Distinct simplices in the Q-system have disjoint interiors.*

While proving the theorem, we give a complete classification of the various ways in which one quasi-regular tetrahedron or strict quarter can overlap another.

Having completed our primary purpose of showing that the simplices in the Q-system do not overlap, we state the following small lemma. It is an immediate consequence of the definitions, but is nonetheless useful in the sections that follow.

Lemma 4.10. *If one quarter along a diagonal lies in the Q-system, then all quarters along the diagonal lie in the Q-system.*

Proof. This is true by construction. Each of the defining properties of a quarter in the Q-system is true for one quarter along a diagonal if and only if it is true of all quarters along the diagonal. $\square$

4.2. Geometric Considerations

Remark 4.11. The primary definitions and constructions of this paper are translation invariant. That is, if $\lambda \in \mathbb{R}^3$ and Λ is a saturated packing, then $\lambda + \Lambda$ is a saturated packing. If $A: \Lambda \to \mathbb{R}$ is a negligible fcc-compatible function for Λ, then $\lambda + v \mapsto A(v)$ is a negligible fcc-compatible function for $\lambda + \Lambda$. If $\mathcal{Q}$ is the Q-system of Λ, then $\lambda + \mathcal{Q}$ is the Q-system of $\lambda + \Lambda$. Because of general translational invariance, when we fix our attention on a particular $v \in \Lambda$, we often assume (without loss of generality) that the coordinate system is fixed in such a way that v lies at the origin.

Our simplices are generally assumed to come labeled with a distinguished vertex, fixed at the origin. (The origin will always be at a vertex of the packing.) We number the edges of each simplex $1, \ldots, 6$, so that edges 1, 2, and 3 meet at the origin, and the edges i and $i + 3$ are opposite, for $i = 1, 2, 3$. (See Fig. 4.1.) $S(y_1, y_2, \ldots, y_6)$ denotes a simplex whose edges have lengths y_i, indexed in this way. We refer to the endpoints away from the origin of the first, second, and third edges as the first, second, and third vertices.

Definition 4.12. In general, let dih(S) be the dihedral angle of a simplex S along its first edge. When we write a simplex in terms of its vertices (w_1, w_2, w_3, w_4), then $\{w_1, w_2\}$ is understood to be the first edge.

Definition 4.13. We define the *radial projection* of a set X to be the radial projection $x \mapsto x/|x|$ of $X \backslash 0$ to the unit sphere centered at the origin. We say the two sets *cross* if their radial projections to the unit sphere overlap.

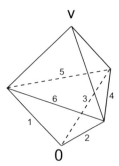

Fig. 4.1. $\mathcal{E}$ measures the distance between the vertices at 0 and v. The standard indexing of the edges of a simplex is marked on the lower simplex.

Definition 4.14. If S and S' are nonoverlapping simplices with a shared face F, we define $\mathcal{E}(S, S')$ as the distance between the two vertices (one on S and the other on S') that do not lie on F. We may express this as a function

$$\mathcal{E}(S, S') = \mathcal{E}(S(y_1, \ldots, y_6), y'_1, y'_2, y'_3)$$

of nine variables, where $S = S(y_1, \ldots, y_6)$ and $S' = S(y'_1, y'_2, y'_3, y_4, y_5, y_6)$, positioned so that S and S' are nonoverlapping simplices with a shared face F of edge-lengths (y_4, y_5, y_6). The function of nine variables is defined only for values (y_i, y'_i) for which the simplices S and S' exist (Fig. 4.1).

Several lemmas in this paper rely on calculations of lower bounds to the function $\mathcal{E}$ in the special case when the edge between the vertices 0 and v passes through the shared face F. If intervals containing $y_1, \ldots, y_6, y'_1, y'_2, y'_3$ are given, then lower bounds on $\mathcal{E}$ over that domain are generally easy to obtain. Detailed examples of calculations of the lower bound of this function can be found in Section 4 of [Ha6].

To work one example, we suppose we are asked to give a lower bound on $\mathcal{E}$ when the simplex $S = S(y_1, \ldots, y_6)$ satisfies $y_i \geq 2$ and $y_4, y_5, y_6 \leq 2t_0$ and $S' = S(y'_1, y'_2, y'_3, y_4, y_5, y_6)$ satisfies $y'_i \geq 2$, for $i = 1, \ldots, 3$. Assume that the edge $\{0, v\}$ passes through the face shared between S and S', and that $|v| < \sqrt{8}$, where v is the vertex of S' that is not on S. We claim that any pair S, S' can be deformed by moving one vertex at a time until S is deformed into $S(2, 2, 2, 2t_0, 2t_0, 2t_0)$ and S' is deformed into $S(2, 2, 2, 2t_0, 2t_0, 2t_0)$. Moreover, these deformations preserve the constraints (including that $\{0, v\}$ passes through the shared face), and are nonincreasing in $|v|$. From the existence of this deformation, it follows that the original $|v|$ satisfies

$$|v| \geq \mathcal{E}(S(2, 2, 2, 2t_0, 2t_0, 2t_0), 2, 2, 2).$$

We produce the deformation in this case as follows. We define the *pivot* of a vertex v with respect to two other vertices $\{v_1, v_2\}$ as the circular motion of v held at a fixed distance from v_1 and v_2, leaving all other vertices fixed. The *axis* of the pivot is the line through the two fixed vertices. Each pivot of a vertex can move in two directions. Let the vertices of S be $\{0, v_1, v_2, v_3\}$, labeled so that $|v_i| = y_i$. Let $S' = \{v, v_1, v_2, v_3\}$. We pivot v_1 around the axis through 0 and v_2. By choice of a suitable direction for the pivot, v_1 moves away from v and v_3. Its distance to 0 and v_2 remains fixed. We continue with this circular motion until $|v_1 - v_3|$ achieves its upper bound or the segment $\{v_1, v_3\}$ intersects the segment $\{0, v\}$ (which threatens the constraint that the segment $\{0, v\}$ must pass through the common face). (We leave it as an exercise[2] to check that the second possibility cannot occur because of the edge length upper bounds on both diagonals of $\sqrt{8}$. That is, there does not exist a convex planar quadrilateral with sides at least 2 and diagonals less than $\sqrt{8}$.) Thus, $|v_1 - v_3|$ attains its constrained upper bound $2t_0$. Similar pivots to v_2 and v_3 increase the lengths $|v_1 - v_2|$, $|v_2 - v_3|$, and $|v_3 - v_1|$ to $2t_0$. Similarly, v may be pivoted around the axis through v_1 and v_2 to decrease the distance to v_3 and 0 until the lower bound of 2 on $|v - v_3|$ is attained. Further pivots reduce all remaining edge lengths to 2. In this way we obtain a rigid figure realizing the absolute lower bound of $|v|$. A calculation with explicit coordinates gives $|v| > 2.75$.

Because lower bounds are generally easily determined from a series of pivots through arguments such as this one, we state them without proof. We state that these bounds were obtained by *geometric considerations*, to indicate that the bounds were obtained by the deformation arguments of this paragraph.

4.3. Incidence Relations

Lemma 4.15.   *Let v, v_1, v_2, v_3, and v_4 be distinct points in $\mathbb{R}^3$ with pairwise distances at least 2. Suppose that $|v_i - v_j| \leq 2t_0$ for $i \neq j$ and $\{i, j\} \neq \{1, 4\}$. Then v does not lie in the convex hull of $\{v_1, v_2, v_3, v_4\}$.*

Proof.   This lemma is proved in Lemma 3.5 of [Ha6].   □

Lemma 4.16.   *Let S be a simplex whose edges have length between 2 and $2\sqrt{2}$. Suppose that v has distance at least 2 from each of the vertices of S. Then v does not lie in the convex hull of S.*

Proof.   Assume for a contradiction that v lies in the convex hull of S. Place a unit sphere around v. The simplex S partitions the unit sphere into four spherical triangles, where each triangle is the intersection of the unit sphere with the cone over a face of S, centered at v. By the constraints on the lengths of edges, the arclength of each edge of the spherical triangle is at most $\pi/2$. ($\pi/2$ is attained when v has distance 2 to two of the vertices, and these two vertices have distance $2\sqrt{2}$ between them.) A spherical triangle with edges of arclength at most $\pi/2$ has area at most $\pi/2$. In fact, any such spherical triangle can be placed inside an octant of the unit sphere, and each octant has area $\pi/2$. This partitions the sphere of area 4π into four regions of area at most $\pi/2$. This is absurd.   □

Corollary 4.17.   *No vertex of the packing is contained in the convex hull of a quasi-regular tetrahedron or quarter (other than the vertices of the simplex).*

---

[2] Compare Lemma 4.21.

Proof. The corollary is immediate. □

Definition 4.18. Let $v_1, v_2, w_1, w_2, w_3 \in \Lambda$ be distinct. We say that an edge $\{v_1, v_2\}$ *passes through* the triangle $\{w_1, w_2, w_3\}$ if the convex hull of $\{v_1, v_2\}$ meets some point of the convex hull of $\{w_1, w_2, w_3\}$ and if that point of intersection is not any of the extreme points v_1, v_2, w_1, w_2, w_3.

Lemma 4.19. *An edge of length $2t_0$ or less cannot pass through a triangle whose edges have lengths $2t_0$, $2t_0$, and $\sqrt{8}$ or less.*

Proof. The distance between each pair of vertices is at least 2. Geometric considerations show that the edge has length at least

$$\mathcal{E}(S(2, 2, 2, 2t_0, 2t_0, \sqrt{8}), 2, 2, 2) > 2t_0.$$ □

Definition 4.20. Let $\eta(x, y, z)$ denote the circumradius of a triangle with edge-lengths x, y, and z.

Lemma 4.21. *Suppose that the circumradius of $\{v_1, v_2, v_3\}$ is less than $\sqrt{2}$. Then an edge $\{w_1, w_2\} \subset \Lambda$ of length at most $\sqrt{8}$ cannot pass through the face.*

Proof. Assume for a contradiction that $\{w_1, w_2\}$ passes through the triangle $\{v_1, v_2, v_3\}$. By geometric considerations, the minimal length for $\{w_1, w_2\}$ occurs when $|w_i - v_j| = 2$, for $i = 1, 2$, $j = 1, 2, 3$. This distance constraint places the circumscribing circle of $\{v_1, v_2, v_3\}$ on the sphere of radius 2 centered at w_1 (resp. w_2). If $r < \sqrt{2}$ is the circumradius of $\{v_1, v_2, v_3\}$, then for this extremal configuration we have the contradiction

$$\sqrt{8} \geq |w_1 - w_2| = 2\sqrt{4 - r^2} > \sqrt{8}.$$ □

Lemma 4.22. *If an edge of length at most $\sqrt{8}$ passes through a quasi-regular triangle, then each of the two endpoints of the edge is at most 2.2 away from each of the vertices of the triangle (see Fig. 4.2).*

Proof. Let the diagonal edge be $\{v_0, v_0'\}$ and let the vertices of the face be $\{v_1, v_2, v_3\}$. If $|v_i - v_0| > 2.2$ or $|v_i - v_0'| > 2.2$ for some $i > 0$, then geometric considerations give the contradiction

$$|v_0 - v_0'| \geq \mathcal{E}(S(2, 2, 2, 2t_0, 2t_0, 2t_0), 2, 2, 2.2) > \sqrt{8}.$$ □

Lemma 4.23. *Suppose S and S' are quasi-regular tetrahedra that share a face. Suppose that the edge e between the two vertices that are not shared has length at most $\sqrt{8}$. Then the convex hull of S and S' consists of three quarters with diagonal e. No other quarter overlaps S or S'.*

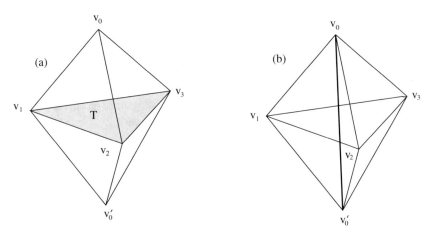

Fig. 4.2. (a) Two quasi-regular tetrahedra that share a face. (b) The same convex body viewed as three quarters that share a diagonal.

Proof. Suppose that S and S' are adjacent quasi-regular tetrahedra with a common face F. By Lemma 4.22, each of the six external faces of this pair of quasi-regular tetrahedra has circumradius at most $\eta(2.2, 2.2, 2t_0) < \sqrt{2}$. A diagonal of a quarter cannot pass through a face of this size by Lemma 4.21. This implies that no other quarter overlaps these quasi-regular tetrahedra.   □

Lemma 4.24. *Suppose an edge $\{w_1, w_2\}$ of length at most $\sqrt{8}$ passes through the face formed by a diagonal $\{v_1, v_2\}$ and one of its anchors. Then w_1 and w_2 are also anchors of $\{v_1, v_2\}$.*

Proof. This follows from the inequality

$$\mathcal{E}(S(2, 2, 2, \sqrt{8}, 2t_0, 2t_0), 2, 2, 2t_0) > \sqrt{8}$$

and geometric considerations.   □

Definition 4.25. Let Λ be a saturated packing. Assume that the coordinate system is fixed in such a way that the origin is a vertex of the packing. The *height* of a vertex is its distance from the origin.

Definition 4.26. We say that a vertex is *enclosed* over a figure if it lies in the interior of the cone at the origin generated by the figure.

Definition 4.27. An *adjacent pair of quarters* consists of two quarters sharing a face along a common diagonal. The common vertex that does not lie on the diagonal is called the *base point* of the adjacent pair. (When one of the quarters comes with a marked distinguished vertex, we do *not* assume that this marked vertex coincides with the base point of the pair.) The other four vertices are called the *corners* of the configuration.

Definition 4.28. If the two corners, v and w, that do not lie on the diagonal satisfy $|w - v| < \sqrt{8}$, then the base point and four corners can be considered as an adjacent pair in a second way, where $\{v, w\}$ functions as the diagonal. In this case we say that the original diagonal and the diagonal $\{v, w\}$ are *conflicting diagonals*.

Definition 4.29. A quarter is said to be *isolated* if it is not part of an adjacent pair. Two isolated quarters that overlap are said to form an *isolated pair*.

Lemma 4.30. *Suppose that there exist four nonzero vertices $v_1, \ldots, v_4$ of height at most $2t_0$ (that is, $|v_i| \leq 2t_0$) forming a skew quadrilateral. Suppose that the diagonals $\{v_1, v_3\}$ and $\{v_2, v_4\}$ have lengths between $2t_0$ and $\sqrt{8}$. Suppose the diagonals $\{v_1, v_3\}$ and $\{v_2, v_4\}$ cross. Then the four vertices are the corners of an adjacent pair of quarters with base point at the origin.*

Proof. Set $d_1 = |v_1 - v_3|$ and $d_2 = |v_2 - v_4|$. By hypothesis, d_1 and d_2 are at most $\sqrt{8}$. If $|v_1 - v_2| > 2t_0$, geometric considerations give the contradiction

$$\max(d_1, d_2) \geq \mathcal{E}(S(2t_0, 2, 2, 2t_0, \sqrt{8}, 2t_0), 2, 2, 2) > \sqrt{8} \geq \max(d_1, d_2).$$

Thus, $\{0, v_1, v_2\}$ is a quasi-regular triangle, as are $\{0, v_2, v_3\}$, $\{0, v_3, v_4\}$, and $\{0, v_4, v_1\}$ by symmetry. $\qquad\square$

Lemma 4.31. *If, under the same hypotheses as Lemma 4.30, there is a vertex w of height at most $\sqrt{8}$ enclosed over the adjacent pair of quarters, then $\{0, v_1, \ldots, v_4, w\}$ is a quartered octahedron.*

Proof. If the enclosed w lies over say $\{0, v_1, v_2, v_3\}$, then $|w - v_1|$, $|w - v_3| \leq 2t_0$ (Lemma 4.24), where $\{v_1, v_3\}$ is a diagonal. Similarly, the distance from w to the other two corners is at most $2t_0$. $\qquad\square$

Lemma 4.32. *Let v_1 and v_2 be anchors of $\{0, w\}$ with $2t_0 \leq |w| \leq \sqrt{8}$. If an edge $\{v_3, v_4\}$ passes through both faces, $\{0, w, v_1\}$ and $\{0, w, v_2\}$, then $|v_3 - v_4| > \sqrt{8}$.*

Proof. Suppose the figure exists with $|v_3 - v_4| \leq \sqrt{8}$. Label vertices so v_3 lies on the same side of the figure as v_1. Contract $\{v_3, v_4\}$ by moving v_3 and v_4 until $\{v_i, u\}$ has length 2, for $u = 0, w, v_{i-2}$, and $i = 3, 4$. Pivot w away from v_3 and v_4 around the axis $\{v_1, v_2\}$ until $|w| = \sqrt{8}$. Contract $\{v_3, v_4\}$ again. By stretching $\{v_1, v_2\}$, we obtain a square of edge two and vertices $\{0, v_3, w, v_4\}$. Short calculations based on explicit formulas for the dihedral angle and its partial derivatives give

$$\text{dih}(S(\sqrt{8}, 2, y_3, 2, y_5, 2)) > 1.075, \qquad y_3, y_5 \in [2, 2t_0], \tag{4.1}$$

$$\text{dih}(S(\sqrt{8}, y_2, y_3, 2, y_5, y_6)) > 1, \qquad y_2, y_3, y_5, y_6 \in [2, 2t_0]. \tag{4.2}$$

Then

$$\pi \geq \text{dih}(0, w, v_3, v_1) + \text{dih}(0, w, v_1, v_2) + \text{dih}(0, w, v_2, v_4) > 1.075 + 1 + 1.075 > \pi.$$

Therefore, the figure does not exist. $\qquad\square$

Lemma 4.33. *Two vertices w, w' of height at most $\sqrt{8}$ cannot be enclosed over a triangle $\{v_1, v_2, v_3\}$ satisfying $|v_1 - v_2| \le \sqrt{8}$, $|v_1 - v_3| \le 2t_0$, and $|v_2 - v_3| \le 2t_0$.*

Proof. For a contradiction, assume the figure exists. The long edge $\{v_1, v_2\}$ must have length at least $2t_0$ by Lemma 4.22. This diagonal has anchors $\{0, v_3, w, w'\}$. Assume that the cyclic order of vertices around the line $\{v_1, v_2\}$ is $0, v_3, w, w'$. We see that $\{v_1, w\}$ is too short to pass through $\{0, v_2, w'\}$, and w is not inside the simplex $\{0, v_1, v_2, w'\}$. Thus, the projections of the edges $\{v_2, w\}$ and $\{0, w'\}$ to the unit sphere at v_1 must intersect. It follows that $\{0, w'\}$ passes through $\{v_1, v_2, w\}$, or $\{v_2, w\}$ passes through $\{v_1, 0, w'\}$. However, $\{v_2, w\}$ is too short to pass through $\{v_1, 0, w'\}$. Thus, $\{0, w'\}$ passes through both $\{v_1, v_2, w\}$ and $\{v_1, v_2, v_3\}$. Lemma 4.32 gives the contradiction $|w'| > \sqrt{8}$. □

Lemma 4.34. *Let v_1, v_2, v_3 be anchors of $\{0, w\}$, where $2t_0 \le |w| \le \sqrt{8}$, $|v_1 - v_3| \le \sqrt{8}$, and the edge $\{v_1, v_3\}$ passes through the face $\{0, w, v_2\}$. Then $\min(|v_1 - v_2|, |v_2 - v_3|) \le 2t_0$. Furthermore, if the minimum is $2t_0$, then $|v_1 - v_2| = |v_2 - v_3| = 2t_0$.*

Proof. Assume $\min \ge 2t_0$. As in the proof of Lemma 4.32, we may assume that $(0, v_1, w, v_3)$ is a square. We may also assume, without loss of generality, that $|w - v_2| = |v_2| = 2t_0$. This forces $|v_2 - v_i| = 2t_0$, for $i = 1, 3$. This is rigid, and is the unique figure that satisfies the constraints. The lemma follows. □

4.4. Overlap of Simplices

This section gives a proof of Theorem 4.9 (simplices in the Q-system do not overlap). This is accomplished in a series of lemmas. The first of these treats quasi-regular tetrahedra.

Lemma 4.35. *A quasi-regular tetrahedron does not overlap any other simplex in the Q-system.*

Proof. Edges of quasi-regular tetrahedra are too short to pass through the face of another quasi-regular tetrahedron or quarter (Lemma 4.19). If a diagonal of a strict quarter passes through the face of a quasi-regular tetrahedron, then Lemma 4.23 shows that the strict quarter is one of three joined along a common diagonal. This is not one of the enumerated types of strict quarter in the Q-system. □

Lemma 4.36. *A quarter in the Q-system that is part of a quartered octahedron does not overlap any other simplex in the Q-system.*

Proof. By construction, the quarters that lie along a different diagonal of the octahedron do not belong to the Q-system. Edges of length at most $2t_0$ are too short to pass through an external face of the octahedron (Lemma 4.19). A diagonal of a strict quarter cannot pass through an external face either, because of Lemma 4.22. □

Lemma 4.37. *Let Q be a strict quarter that is part of an adjacent pair. Assume that Q is not part of a quartered octahedron. If Q belongs to the Q-system, then it does not overlap any other simplex in the Q-system.*

The proof of this lemma will give valuable details about how one strict quarter overlaps another.

Proof. Fix the origin at the base point of an adjacent pair of quarters. We investigate the local geometry when another quarter overlaps one of them. (This happens, for example, when there is a conflicting diagonal in the sense of Definition 4.27.)

Label the base point of the pair of quarters v_0, and the four corners v_1, v_2, v_3, v_4, with $\{v_1, v_3\}$ the common diagonal. Assume that $|v_1 - v_3| < \sqrt{8}$.

If two quarters overlap then a face on one of them overlaps a face on the other. By Lemmas 4.33 and 4.32, we actually have that some edge (in fact the diagonal) of each passes through a face of the other. This edge cannot exit through another face by Lemma 4.32 and it cannot end inside the simplex by Corollary 4.17. Thus, it must end at a vertex of the other simplex. We break the proof into cases according to which vertex of the simplex it terminates at. In Case 1 the edge has the base point as an endpoint. In Case 2, the edge has a corner as an endpoint.

Case 1. The edge $\{0, w\}$ passes through the triangle $\{v_1, v_2, v_3\}$, where $\{0, w\}$ is a diagonal of a strict quarter. Lemma 4.24 implies that v_1 and v_3 are anchors of $\{0, w\}$. The only other possible anchors of $\{0, w\}$ are v_2 or v_4, for otherwise an edge of length at most $2t_0$ passes through a face formed by $\{0, w\}$ and one of its anchors. If both v_2 and v_4 are anchors, then we have a quartered octahedron, which has been excluded by the hypotheses of the lemma. Otherwise, $\{0, w\}$ has at most three anchors: v_1, v_3, and either v_2 or v_4. In fact, it must have exactly three anchors, for otherwise there is no quarter along the edge $\{0, w\}$. So there are exactly two quarters along the edge $\{0, w\}$. There are at least four anchors along $\{v_1, v_3\}$: $0, w, v_2$, and v_4. The quarters along the diagonal $\{v_1, v_3\}$ lie in the Q-system. (None of these quarters is isolated.) The other two quarters, along the diagonal $\{0, w\}$, are not in the Q-system. They form an adjacent pair of quarters (with base point v_4 or v_2) that has conflicting diagonals, $\{0, w\}$ and $\{v_1, v_3\}$, of length at most $\sqrt{8}$.

Case 2. $\{v_2, v_4\}$ is a diagonal of length less than $\sqrt{8}$ (conflicting with $\{v_1, v_3\}$). (Note that if an edge of a quarter passes through the shared face of an adjacent pair of quarters, then that edge must be $\{v_2, v_4\}$, so that Cases 1 and 2 are exhaustive.) The two diagonals $\{v_1, v_3\}$ and $\{v_2, v_4\}$ do not overlap. By symmetry, we may assume that $\{v_2, v_4\}$ passes through the face $\{0, v_1, v_3\}$. Assume (for a contradiction) that both diagonals have an anchor other than 0 and the corners v_i. Let the anchor of $\{v_2, v_4\}$ be denoted v_{24} and that of $\{v_1, v_3\}$ be v_{13}. Assume the figure is not a quartered octahedron, so that $v_{13} \neq v_{24}$. By Lemma 4.19, it is impossible to draw the edges $\{v_1, v_{13}\}$ and $\{v_{13}, v_3\}$ between v_1 and v_3. In fact, if the edges pass outside the quadrilateral $\{0, v_2, v_{24}, v_4\}$, one of the edges of length at most $2t_0$ (that is, $\{0, v_2\}$, $\{v_2, v_{24}\}$, $\{v_{24}, v_4\}$, or $\{v_4, 0\}$) violates the lemma applied to the face $\{v_1, v_3, v_{13}\}$. If they pass inside the quadrilateral, one of the edges $\{v_1, v_{13}\}$, $\{v_{13}, v_3\}$ violates the lemma applied to the face $\{0, v_2, v_4\}$ or $\{v_{24}, v_2, v_4\}$. We conclude that at most one of the two diagonals has additional anchors.

If neither of the two diagonals has more than three anchors, we have nothing more than two overlapping adjacent pairs of quarters along conflicting diagonals. The two quarters along the lower edge $\{v_2, v_4\}$ lie in the Q-system. Another way of expressing this "lower-edge" condition is to require that the two adjacent quarters Q_1 and Q_2 satisfy $\operatorname{dih}(Q_1) + \operatorname{dih}(Q_2) > \pi$, when the dihedral angles are measured along the diagonal. The pair (Q'_1, Q'_2) along the upper edge will have $\operatorname{dih}(Q'_1) + \operatorname{dih}(Q'_2) < \pi$.

If there is a diagonal with more than three anchors, the quarters along the diagonal with more than three anchors lie in the Q-system. Any additional quarters along the diagonal $\{v_2, v_4\}$ belong to an adjacent pair. Any additional quarters along the diagonal $\{v_1, v_3\}$ cannot intersect the adjacent pair along $\{v_2, v_4\}$. Thus, every quarter intersecting an adjacent pair also belongs to an adjacent pair.

In both possibilities of Case 2, the two quarters left out of the Q-system correspond to a conflicting diagonal. □

Remark 4.38. We have seen in the proof of Lemma 4.37 that if a strict quarter Q overlaps a strict quarter that is part of an adjacent pair, then Q is also part of an adjacent pair. Thus, if an isolated strict quarter overlaps another strict quarter, then both strict quarters are necessarily isolated.

Lemma 4.39. *If an isolated strict quarter Q overlaps another strict quarter, then the diagonal of Q has exactly three anchors.*

The proof of the lemma will give detailed information about the geometrical configuration that is obtained when an isolated quarter overlaps another strict quarter.

Proof. Assume that there are two strict quarters Q_1 and Q_2 that overlap. Following Remark 4.38, assume that neither is adjacent to another quarter. Let $\{0, u\}$ and $\{v_1, v_2\}$ be the diagonals of Q_1 and Q_2. Suppose the diagonal $\{v_1, v_2\}$ passes through a face $\{0, u, w\}$ of Q_1. By Lemma 4.24, v_1 and v_2 are anchors of $\{0, u\}$. Again, either the length of $\{v_1, w\}$ is at most $2t_0$ or the length of $\{v_2, w\}$ is at most $2t_0$, say $\{w, v_2\}$ (by Lemma 4.34). It follows that $Q_1 = \{0, u, w, v_2\}$ and $|v_1 - w| \geq 2t_0$. (Q_1 is not adjacent to another quarter.) So w is not an anchor of $\{v_1, v_2\}$.

Let $\{v_1, v_2, w'\}$ be a face of Q_2 with $w' \neq 0, u$. If $\{v_1, w', v_2\}$ does not link $\{0, u, w\}$, then $\{v_1, w'\}$ or $\{v_2, w'\}$ passes through the face $\{0, u, w\}$, which is impossible by Lemma 4.19. So $\{v_1, v_2, w'\}$ links $\{0, u, w\}$ and an edge of $\{0, u, w\}$ passes through the face $\{v_1, v_2, w'\}$. It is not the edge $\{u, w\}$ or $\{0, w\}$, for they are too short by Lemma 4.19. So $\{0, u\}$ passes through $\{w', v_1, v_2\}$. The only anchors of $\{v_1, v_2\}$ (other than w') are u and 0 (by Lemma 4.32). Either $\{u, w'\}$ or $\{w', 0\}$ has length at most $2t_0$ by Lemma 4.34, but not both, because this would create a quarter adjacent to Q_2. By symmetry, $Q_2 = \{v_1, v_2, w', 0\}$ and the length of $\{u, w'\}$ is greater than $2t_0$. By symmetry, $\{0, u\}$ has no other anchors either. This determines the local geometry when there are two quarters that intersect without belonging to an adjacent pair of quarters (see Fig. 4.3). It follows that the two quarters form an isolated pair. □

Isolated quarters that overlap another strict quarter do not belong to the Q-system.

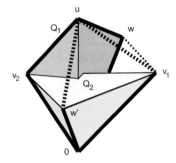

Fig. 4.3. An isolated pair. The isolated pair consists of two simplices $Q_1 = \{0, u, w, v_2\}$ and $Q_2 = \{0, w', v_1, v_2\}$. The six extremal vertices form an octahedron. This is not a quartered octahedron because the edges $\{u, w'\}$ and $\{w, v_1\}$ have length greater than $2t_0$.

We conclude with the proof of the main theorem of the section.

Proof of Theorem 4.9.   The rules defining the Q-system specify a uniquely determined set of simplices. The proof that they do not overlap is established by the preceding series of lemmas. Lemma 4.35 shows that quasi-regular tetrahedra do not overlap other simplices in the Q-system. Lemma 4.36 shows that the quarters in quartered octahedra are well-behaved. Lemma 4.37 shows that other quarters in adjacent pairs do not overlap other simplices in the Q-system. Finally, we treat isolated quarters in Lemma 4.39. These cases cover all possibilities since every simplex in the Q-system is a quasi-regular tetrahedron or strict quarter, and every strict quarter is either part of an adjacent pair or isolated.   □

5.   V-Cells

In the proof of the Kepler conjecture we make use of two quite different structures in space. The first structure is the Q-system, which was defined in the previous section. It is inspired by the Delaunay decomposition of space and consists of a nonoverlapping collection of simplices that have their vertices at the points of Λ. Historically, the construction of the nonoverlapping simplices of the Q-system grew out of a detailed investigation of the Delaunay decomposition.

The second structure is inspired by the Voronoi decomposition of space. In the Voronoi decomposition, the vertices of Λ are the centers of the cells. It is well known that the Voronoi decomposition and Delaunay decomposition are dual to one another. Our modification of Voronoi cells are called V-cells.

In general, it is not true that a Delaunay simplex is contained in the union of the Voronoi cells at its four vertices. This incompatibility of structures adds a few complications to Rogers's elegant proof of a sphere packing bound [Ro1]. In this section we show that V-cells are compatible with the Q-system in the sense that each simplex in the Q-system is contained in the union of the V-cells at its four vertices (Lemma 5.28). A second compatibility result between these two structures is proved in Lemma 5.29.

The purpose of this section is to define V-cells and to prove the compatibility results just mentioned. In the proof of the Kepler conjecture it is important to keep both structures (the Q-system and the V-cells) continually at hand. We frequently jump back and forth between these dual descriptions of space in the course of the proof. In Section 6 we define a geometric object (called the decomposition star) around a vertex that encodes both structures. The decomposition star will become our primary object of analysis.

5.1.   V-Cells

Definition 5.1.   The Voronoi cell $\Omega(v)$ around a vertex $v \in \Lambda$ is the set of points closer to v than to any other vertex.

Definition 5.2.   We construct a set of triangles $\mathcal{B}$ in the packing. The triangles in this set are called *barriers*. A triangle $\{v_1, v_2, v_3\}$ with vertices in the packing belongs to $\mathcal{B}$ if and only if one or more of the following properties hold:

1. The triangle is a quasi-regular.
2. The triangle is a face of a simplex in the Q-system.

Lemma 5.3.   *No two barriers overlap; that is, no two open triangular regions of $\mathcal{B}$ intersect.*

Proof.   If there is overlap, an edge $\{w_1, w_2\}$ of one triangle passes through the interior of another $\{v_1, v_2, v_3\}$. Since $|w_1 - w_2| < \sqrt{8}$, we have that the circumradius of $\{v_1, v_2, v_3\}$ is at least $\sqrt{2}$ by Lemma 4.21 and that the length $|w_1 - w_2|$ is greater than $2t_0$ by Lemma 4.19. If the edge $\{w_1, w_2\}$ belongs to a simplex in the Q-system, the simplex must be a strict quarter. If $\{v_1, v_2, v_3\}$ has edge lengths at most $2t_0$, then Lemma 4.22 implies that $|w_i - v_j| \leq 2.2$ for $i = 1, 2$ and $j = 1, 2, 3$. The simplices $\{v_1, v_2, v_3, w_1\}$ and $\{v_1, v_2, v_3, w_2\}$ form a pair of quasi-regular tetrahedra. We conclude that $\{v_1, v_2, v_3\}$ is a face of a quarter in the Q-system. Since, the simplices in the Q-system do not overlap, the edge $\{w_1, w_2\}$ does not belong to a simplex in the Q-system. The result follows. $\square$

Definition 5.4.   We say that a point y is *obstructed* at $x \in \mathbb{R}^3$ if the line segment from x to y passes through the interior of a triangular region in $\mathcal{B}$. Otherwise, y is unobstructed at x. The "obstruction" relation between x and y is clearly symmetric.

For each $w \in \Lambda$, let I_w be the cube of side 4, with edges parallel to the coordinate axes, centered at w. Thus,

$$I_0 = \{(y_1, y_2, y_3): |y_i| \leq 2, \ i = 1, 2, 3\}.$$

I_w has diameter $4\sqrt{3}$ and $I_w \subset B(w, 2\sqrt{3})$. Let $\mathbb{R}^{3\prime}$ be the subset of $x \in \mathbb{R}^3$ for which x is not equidistant from any two $v, w \in \Lambda(x, 2\sqrt{3}) = B(x, 2\sqrt{3}) \cap \Lambda$. The subset $\mathbb{R}^{3\prime}$ is dense in $\mathbb{R}^3$, and is obtained locally around a point x by removing finitely many planes (perpendicular bisectors of $\{v, w\}$, for $v, w \in B(x, 2\sqrt{3})$). For $x \in \mathbb{R}^{3\prime}$, the vertices of $\Lambda(x, 2\sqrt{3})$ can be strictly ordered by their distance to x.

Definition 5.5. Let Λ be a saturated packing. We define a map $\varphi \colon \mathbb{R}^{3\prime} \to \Lambda$ such that the image of x lies in $\Lambda(x, 2\sqrt{3})$. If $x \in \mathbb{R}^{3\prime}$, let

$$\Lambda_x = \{w \in \Lambda \colon x \in I_w \text{ and } w \text{ is unobstructed at } x\}.$$

If $\Lambda_x = \emptyset$, then let $\varphi(x)$ be the vertex of $\Lambda(x, 2\sqrt{3})$ closest to x. (Since Λ is saturated, $\Lambda(x, 2\sqrt{3})$ is nonempty.) If Λ_x is nonempty, then let $\varphi(x)$ be the vertex of Λ_x closest to x.

Definition 5.6. For $v \in \Lambda$, let $\mathrm{VC}(v)$ be defined as the closure of $\varphi^{-1}(v)$ in $\mathbb{R}^3$. We call it the *V-cell* at v.

Remark 5.7. In a saturated packing, the Voronoi cell at v will be contained in a ball centered at v of radius 2. Hence I_v contains the Voronoi cell at v. By construction, the V-cell at v is confined to the cube I_v. The cubes I_v were introduced into the definition of φ with the express purpose of forcing V-cells to be reasonably small. Had the cubes been omitted from the construction, we would have been drawn to frivolous questions such as whether the closest unobstructed vertex to some $x \in \mathbb{R}^3$ might be located in a remote region of the packing.

The set of V-cells is our promised decomposition of space.

Lemma 5.8. *V-cells cover space. The interiors of distinct V-cells are disjoint. Each V-cell is the closure of its interior.*

Proof. The sets $\varphi^{-1}(v)$, for $v \in \Lambda$, cover $\mathbb{R}^{3\prime}$. Their closures cover $\mathbb{R}^3$. The other statements in the lemma follow from the fact that a V-cell is a union of finitely many nonoverlapping, closed, convex polyhedra. This is proved in the following lemma. $\square$

Lemma 5.9. *Each V-cell is a finite union of nonoverlapping convex polyhedra.*

Proof. During this proof, we ignore sets of measure zero in $\mathbb{R}^3$ such as finite unions of planes. Thus, we present the proof as if each point belongs to exactly one Voronoi cell, although this fails on an inconsequential set of measure zero in $\mathbb{R}^3$.

It is enough to show that if $E \subset \mathbb{R}^3$ is an arbitrary unit cube, then the V-cell decomposition of space within E consists of finite unions of nonoverlapping convex polyhedra. Let X_E be the set of $w \in \Lambda$ such that I_w meets E. Included in X_E is the set of w whose Voronoi cells cover E. The rules for V-cells assign $x \in E$ to the V-cell centered at some $w \in X_E$.

Let d be an upper bound on the distance between a vertex in X_E and a point of E. By the pythagorean theorem, we may take $d = (1+2)\sqrt{3}$. Let B_E be the set of barriers with a vertex at most distance d from some point in E.

For each pair $\{u, v\}$ of distinct vertices of X_E, draw the perpendicular bisecting plane of $\{u, v\}$. Draw the plane through each barrier in B_E. Draw the plane through

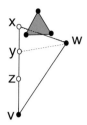

Fig. 5.1. A hypothetical arrangement that leads to a nonconvex V-cell at v.

each triple $\{u, v, w\}$, where $u \in X_E$ and $\{v, w\}$ are two of the vertices of a barrier in B_E. These finitely many planes partition E into finitely many convex polyhedra. The ranking of distances from x to the points of X_E is constant for all x in the interior of any fixed polyhedron. The set of $w \in X_E$ that are obstructed at x is constant on the interior of any fixed polyhedron. Thus, by the rules of construction of V-cells, for each of these convex polyhedra, there is a V-cell that contains it. The result follows.   □

Remark 5.10.   A number of readers of the first version of this manuscript presumed that V-cells were necessarily star-convex, in large part because of the inapt name "decomposition star" for a closely related object. The geometry of a V-cell is significantly more complex than that of a Voronoi cell. Nowhere do we make a general claim that all V-cells are convex, star-convex, or even connected. In Fig. 5.1 we depict a hypothetical case in which the V-cell at v is potentially disconnected. (This figure is merely hypothetical, because I have not checked whether it is possible to satisfy all the metric constraints needed for it to exist.) The shaded triangle represents a barrier. The point x is obstructed by the shaded barrier at w. If x and y lie closer to w than to v, if v is the closest unobstructed vertex to x, if w is the closest unobstructed vertex to y, if x, y, and z are all unobstructed at v, and if z lies closer to v than to w, then it follows that x and z lie in the V-cell at v, but that the intervening point y does not. Thus, if all of these conditions are satisfied, the V-cell at v is not star-shaped at v.

Remark 5.11.   Although we have not made a detailed investigation of the subtleties of the geometry of V-cells, we face a practical need to give explicit lower bounds on the volume of V-cells. Possible geometric pathologies are avoided in the proof by the use of truncation. (To obtain lower bounds on the volume of V-cells, parts of the V-cell can be discarded.) For example, Lemma 5.23 shows that inside $B(v, t_0)$, the V-cell and the Voronoi cell are equal.

In general, truncation will discard points x of V-cells where $\Lambda_x = \emptyset$. These estimates also discard points of the V-cell that are not part of a star-shaped subset of the V-cell (to be defined later).

Truncation will be justified later in Lemma 7.18, which shows that the term involving the volume of V-cells in the scoring function σ has a negative coefficient, so that by decreasing the volume through truncation, we obtain an upper bound on the function σ.

5.2. *Orientation*

We introduce the concept of the orientation of a simplex and study its basic properties. The orientation of a simplex will be used to establish various compatibilities between V-cells.

Definition 5.12. We say that the *orientation* of the face of a simplex is *negative* if the plane through that face separates the circumcenter of the simplex from the vertex of the simplex that does not lie on the face. The orientation is positive if the circumcenter and the vertex lie on the same side of the plane. The orientation is zero if the circumcenter lies in the plane.

Lemma 5.13. *At most one face of a quarter Q has negative orientation.*

Proof. The proof applies to any simplex with nonobtuse faces. (All faces of a quarter are acute.) Fix an edge and project Q orthogonally to a triangle in a plane perpendicular to that edge. The faces F_1 and F_2 of Q along the edge project to edges e_1 and e_2 of the triangular projection of Q. The line equidistant from the three vertices of F_i projects to a line perpendicular to e_i, for $i = 1, 2$. These two perpendiculars intersect at the projection of the circumcenter of Q. If the faces of Q are nonobtuse, the perpendiculars pass through the segments e_1 and e_2 respectively; and the two faces F_1 and F_2 cannot both be negatively oriented. $\qquad\square$

Definition 5.14. Define the polynomial χ by

$$\chi(x_1, \ldots, x_6) = x_1 x_4 x_5 + x_1 x_6 x_4 + x_2 x_6 x_5 + x_2 x_4 x_5 + x_5 x_3 x_6$$
$$+ x_3 x_4 x_6 - 2 x_5 x_6 x_4 - x_1 x_4^2 - x_2 x_5^2 - x_3 x_6^2.$$

In applications of χ, we have $x_i = y_i^2$, where $(y_1, \ldots, y_6)$ are the lengths of the edges of a simplex.

Lemma 5.15. *A simplex $S(y_1, \ldots, y_6)$ has negative orientation along the face indexed by $(4, 5, 6)$ if and only if $\chi(y_1^2, \ldots, y_6^2) < 0$.*

Proof. (This lemma is asserted without proof in [Ha6].) Let $x_i = y_i^2$. Represent the simplex as $S = \{0, v_1, v_2, v_3\}$, where $\{0, v_i\}$ is the ith edge. Write $n = (v_1 - v_3) \times (v_2 - v_3)$, a normal to the plane $\{v_1, v_2, v_3\}$. Let c be the circumcenter of S. We can solve for a unique $t \in \mathbb{R}$ such that $c + tn$ lies in the plane $\{v_1, v_2, v_3\}$. The sign of t gives the orientation of the face $\{v_1, v_2, v_3\}$. We find by direct calculation that

$$t = \frac{\chi(x_1, \ldots, x_6)}{\sqrt{\Delta(x_1, \ldots, x_6)} u(x_4, x_5, x_6)},$$

where the terms Δ and u in the denominator are positive whenever $x_i = y_i^2$, where $(y_1, \ldots, y_6)$ are the lengths of edges of a simplex (see Section 8.1 of [Ha6]). Thus, t and χ have the same sign. The result follows. $\qquad\square$

Lemma 5.16. *Let F be a set of three vertices. Assume that one edge between pairs of vertices has length between $2t_0$ and $\sqrt{8}$ and that the other two edges have length at most $2t_0$. Let v be any vertex not on Q. If the simplex (F, v) has negative orientation along F, then it is a quarter.*

Proof. The orientation of F is determined by the sign of the function χ (see Lemma 5.15). The face F is an acute or right triangle. Note that $\partial \chi / \partial x_1 = x_4(-x_4 + x_5 + x_6)$. By the law of cosines, $-x_4 + x_5 + x_6 \geq 0$ for an acute triangle. Thus, we have monotonicity in the variable x_1, and the same is true of x_2, and x_3. Also, χ is quadratic with a negative leading coefficient in each of the variables x_4, x_5, x_6. Thus, to check positivity, when any of the lengths is greater than $2t_0$, it is enough to evaluate

$$\chi(2^2, 2^2, 4t_0^2, x^2, y^2, z^2), \qquad \chi(2^2, 4t_0^2, 2^2, x^2, y^2, z^2), \qquad \chi(4t_0^2, 2^2, 2^2, x^2, y^2, z^2),$$

for $x \in [2, 2t_0]$, $y \in [2, t_0]$, and $z \in [2t_0, \sqrt{8}]$, and verify that these values are nonnegative. (The minimum, which must be attained at a corner of the domain, is 0.) $\quad\square$

Lemma 5.17. *Let $\{v_1, v_2, v_3\}$ be a quasi-regular triangle. Let v be any other vertex. If the simplex $S = \{v, v_1, v_2, v_3\}$ has negative orientation along $\{v_1, v_2, v_3\}$, then S is a quasi-regular tetrahedron and $|v - v_i| < 2t_0$.*

Proof. The proof is similar to the proof of Lemma 5.16. It comes down to checking that

$$\chi(2^2, 2^2, 4t_0^2, x^2, y^2, z^2) > 0,$$

for $x, y, z \in [2, 2t_0]$. $\quad\square$

Lemma 5.18. *If a face of a simplex has circumradius less than $\sqrt{2}$, then the orientation is positive along that face.*

Proof. If the face has circumradius less than $\sqrt{2}$, by monotonicity

$$\chi(y_1^2, \ldots, y_6^2) \geq \chi(4, 4, 4, y_4^2, y_5^2, y_6^2) = 2y_4^2 y_5^2 y_6^2 (2/\eta(y_4, y_5, y_6)^2 - 1) > 0.$$

(Here y_i are the edge-lengths of the simplex.) $\quad\square$

5.3. Interaction of V-cells with the Q-system

We study the structure of one V-cell, which we take to be the V-cell at the origin $v = 0$. Let Q be the set of simplices in the Q-system. For $v \in \Lambda$, let Q_v be the subset of those with a vertex at v.

Lemma 5.19. *If x lies in the (open) Voronoi cell at the origin, but not in the V-cell at the origin, then there exists a simplex $Q \in Q_0$, such that x lies in the cone (at 0) over Q. Moreover, x does not lie in the interior of Q.*

Proof. If x lies in the open Voronoi cell at the origin, then the segment $\{t\,x\colon 0 \leq t \leq 1\}$ lies in the Voronoi cell as well. By the definition of the V-cell, there is a barrier $\{v_1, v_2, v_3\}$ that the segment passes through. If the simplex $Q = \{0, v_1, v_2, v_3\}$ were to have positive orientation with respect to the face $\{v_1, v_2, v_3\}$, then the circumcenter of $\{0, v_1, v_2, v_3\}$ would lie on the same side of the plane $\{v_1, v_2, v_3\}$ as 0, forcing the intersection of the Voronoi cell with the cone over Q to lie in this same half-space. However, by assumption, x is a point of the Voronoi cell in the opposing half-space. Hence, the simplex Q has negative orientation along $\{v_1, v_2, v_3\}$.

By construction, the barriers are acute or right triangles. The function χ (which gives the sign of the orientations of faces) is monotonic in x_1, x_2, x_3 when these come from simplices (see the proof of Lemma 5.16.) We consider the implications of negative orientation for each kind of barrier. If the barrier is a quasi-regular triangle, then Lemma 5.17 gives that Q is a quasi-regular tetrahedron when $\chi < 0$. If the barrier is a face of a flat quarter in the Q-system, then Lemma 5.16 gives that Q is a flat-quarter in the Q-system as well. Hence $Q \in \mathcal{Q}_0$.

The rest is clear. $\square$

Lemma 5.20. *If x lies in the open ball of radius $\sqrt{2}$ at the origin, and if x is not in the closed cone over any simplex in $\mathcal{Q}_0$, then the origin is unobstructed at x.*

Proof. Assume for a contradiction that the origin is obstructed by the barrier $T = \{u, v, w\}$ at x, and $\{0, u, v, w\}$ is not in $\mathcal{Q}_0$. We show that every point in the convex hull of T has distance at least $\sqrt{2}$ from the origin. Since T is a barrier, each edge $\{u, v\}$ has length at most $\sqrt{8}$. Moreover, the heights $|u|$ and $|v|$ are at least 2, so that every point along each edge of T has distance at least $\sqrt{2}$ from the origin. Suppose that the closest point to the origin in the convex hull of T is an interior point p. Reflect the origin through the plane of T to get w'. The assumptions imply that the edge $\{0, w'\}$ passes through the barrier T and has length less than $\sqrt{8}$. If the barrier T is a quasi-regular triangle, then Lemma 4.22 implies that $\{0, u, v, w\}$ is a quasi-regular tetrahedron in $\mathcal{Q}_0$, which is contrary to the hypothesis. Hence T is the face of a quarter in $\mathcal{Q}_0$. By Lemma 4.34, one of the simplices $\{0, u, v, w\}$ or $\{w', u, v, w\}$ is a quarter. Since these are mirror images, both are quarters. Hence $\{0, u, v, w\}$ is a quarter and it is in the Q-system by Lemma 4.10. This contradicts the hypothesis of the lemma. $\square$

The following corollary is a V-cell analogue of a standard fact about Voronoi cells.

Corollary 5.21. *The V-cell at the origin contains the open unit ball at the origin.*

Proof. Let x lie in the open unit ball at the origin. If it is not in the cone over any simplex, then the origin is unobstructed by the lemma, and the origin is the closest point of Λ. Hence $x \in VC(0)$. A point in the cone over a simplex $\{0, v_1, v_2, v_3\} \in \mathcal{Q}_0$ lies in $VC(0)$ if and only if it lies in the set bounded by the perpendicular bisectors of v_i and the plane through $\{v_1, v_2, v_3\}$. The bisectors pose no problem. It is elementary to check that every point of the convex hull of $\{v_1, v_2, v_3\}$ has distance at least 1 from the origin. (Apply the reflection principle as in the proof of Lemma 5.20 and invoke Lemma 4.19.) $\square$

Lemma 5.22.   *If $x \in B(v, t_0)$, then x is unobstructed at v.*

Proof.   For a contradiction, supposed that the barrier T obstructs x from the v. As in the proof of Lemma 5.20, we find that every edge of T has distance at least $\sqrt{2}$ from the v. We may assume that the point of T that is closest to the origin is an interior point. Let w be the reflection of v through T. By Lemma 4.19, we have $|v - w| > 2t_0$. This implies that every point of T has distance at least t_0 from v. Thus T cannot obstruct $x \in B(0, t_0)$ from v. □

Lemma 5.23.   *Inside the ball of radius t_0 at the origin, the V-cell and Voronoi cell coincide:*

$$B(0, t_0) \cap VC(0) = B(0, t_0) \cap \Omega(0).$$

Proof.   Let $x \in B(0, t_0) \cap VC(0) \cap \Omega(v)$, where $v \neq 0$. By Lemma 5.22, the origin is unobstructed at x. Thus, $|x - v| \leq |x| \leq t_0$. By Lemma 5.22 again, v is unobstructed at x, so that $x \in VC(v)$, contrary to the assumption $x \in VC(0)$. Thus $B(0, t_0) \cap VC(0) \subset \Omega(0)$. Similarly, if $x \in B(0, t_0) \cap \Omega(0)$, then x is unobstructed at the origin, and $x \in VC(0)$. □

Definition 5.24.   For every pair of vertices v_1, v_2 such that $\{0, v_1, v_2\}$ is a quasi-regular triangle, draw a geodesic arc on the unit sphere with endpoints at the radial projections of v_1 and v_2. These arcs break the unit sphere into regions called *standard regions*, as follows. Take the complement of the union of arcs inside the unit sphere. The closure of a connected component of this complement is a standard region. We say that the standard region is triangular if it is bounded by three geodesic arcs, and say that it is nontriangular otherwise.

Lemma 5.25.   *Let v_1, v_2, v_3, and v_4 be distinct vertices such that $|v_i| \leq 2t_0$ for $i = 1, 2, 3, 4$ and $|v_1 - v_3|, |v_2 - v_4| \leq 2t_0$. Then the edges $\{v_1, v_3\}$ and $\{v_2, v_4\}$ do not cross. In particular, the arcs of Definition 5.24 do not meet except at endpoints.*

Proof.   Exchanging $(1, 3)$ with $(2, 4)$ if necessary, we may assume for a contradiction that the edge $\{v_1, v_3\}$ passes through the face $\{0, v_2, v_4\}$. Geometric considerations lead immediately to a contradiction:

$$2t_0 < \mathcal{E}(2, 2, 2, 2t_0, 2t_0, 2t_0, 2, 2, 2) \leq |v_1 - v_3| \leq 2t_0. \qquad □$$

Lemma 5.26.   *Each simplex in the Q-system with a vertex at the origin lies entirely in the closed cone over some standard region R.*

Proof.   Assume for a contradiction that $Q = \{0, v_1, v_2, v_3\}$ with v_1 in the open cone over R_1 and with v_2 in the open cone over R_2. Then $\{0, v_1, v_2\}$ and $\{0, w_1, w_2\}$ (a wall between R_1 and R_2) overlap; this is contrary to Lemma 5.3. □

Remark 5.27. The next two lemmas help to determine which V-cell a given point x belongs to. If x lies in the open cone over a simplex Q_0 in Q, then Lemma 5.28 describes the V-cell decomposition inside Q; beyond Q the origin is obstructed by a face of Q, so that such an x does not lie in the V-cell at 0. If x does not lie in the open cone over a simplex in Q, but lies in the open cone over a standard region R, then Lemma 5.29 describes the V-cell. It states, in particular, that, for unobstructed x, it can be determined whether x belongs to the V-cell at the origin by considering only the vertices w that lie in the closed cone over R (the standard region containing the radial projection of x). In this sense the intersection of a V-cell with the open cone over R is *local* to the cone over R.

Lemma 5.28. *If x lies in the interior of a simplex $Q \in Q$, and if it does not lie on the perpendicular bisector of any edge of Q, then it lies in the V-cell of the closest vertex of Q.*

Proof. The segment to any other vertex v crosses a face of the simplex. Such faces are barriers so that v is obstructed at x. Thus, the vertices of Q are the only vertices that are not obstructed at x.   □

Let B'_0 be the set of triangles T such that at least one of the following holds:

- T is a barrier at the origin, or
- $T = \{0, v, w\}$ consists of a diagonal of a quarter in the Q-system together with one of its anchors.

Lemma 5.29 (Decoupling Lemma). *Let $x \in I_0$, the cube of side 4 centered at the origin parallel to coordinate axes. Assume that the closed segment $\{x, w\}$ intersects the closed two-dimensional cone with center 0 over $F = \{0, v_1, v_2\}$, where $F \in B'_0$. Assume that the origin is not obstructed at x. Assume that x is closer to the origin than to both v_1 and v_2. Then $x \notin \mathrm{VC}(w)$.*

Remark 5.30. The Decoupling Lemma is a crucial result. It permits estimates of the scoring function in Section 7 to be made separately for each standard region. The estimates for separate standard regions are far easier to come by than estimates for the score of the full decomposition star. Eventually, the separate estimates for each standard will be reassembled with linear programming techniques in Section 23.

Proof. (This proof is a minor adaptation of Lemma 2.2 of [Ha7].) Assume for a contradiction that x lies in $\mathrm{VC}(w)$. In particular, we assume that w is not obstructed at x. Since the origin is not obstructed at x, w must be closer to x than x is to the origin: $x \cdot w \geq w \cdot w/2$. The line segment from x to w intersects the closed cone $C(F)$ of the triangle $F = \{0, v_1, v_2\}$.

Consider the set X containing x and bounded by the planes H_1 through $\{0, v_1, w\}$, H_2 through $\{0, v_2, w\}$, H_3 through $\{0, v_1, v_2\}$, $H_4 = \{x : x \cdot v_1 = v_1 \cdot v_1/2\}$, and $H_5 = \{x : x \cdot v_2 = v_2 \cdot v_2/2\}$. The planes H_4 and H_5 contain the faces of the Voronoi cell at

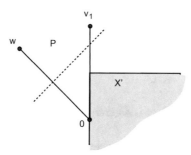

Fig. 5.2. The perpendicular bisector to $\{0, w\}$ (dashed line) cannot meet the quadrant X' (shaded).

0 defined by the vertices v_1 and v_2. The plane H_3 contains the triangle F. The planes H_1 and H_2 bound the set containing points, such as x, that can be connected to w by a segment that passes through $C(F)$.

Let $P = \{x: x \cdot w > w \cdot w/2\}$. The choice of w implies that $X \cap P$ is nonempty. We leave it as an exercise to check that $X \cap P$ is bounded. If the intersection of a bounded polyhedron with a half-space is nonempty, then some vertex of the polyhedron lies in the half-space. Thus, some vertex of X lies in P.

We claim that the vertex of X lying in P cannot lie on H_1. To see this, pick coordinates (x_1, x_2) on the plane H_1 with origin $v_0 = 0$ so that $v_1 = (0, z)$ (with $z > 0$) and $X \cap H_1 \subset X' := \{(x_1, x_2): x_1 \geq 0, \ x_2 \leq z/2\}$. See Fig. 5.2. If the quadrant X' meets P, then the point $v_1/2$ lies in P. This is impossible, because every point between 0 and v_1 lies in the Voronoi cell at 0 or v_1, and not in the Voronoi cell of w. (Recall that for every vertex v_1 on a barrier at the origin, $|v_1| < \sqrt{8}$.)

Similarly, the vertex of X in P cannot lie on H_2. Thus, the vertex must be the unique vertex of X that is not on H_1 or H_2, namely, the point of intersection of H_3, H_4, and H_5. This point is the circumcenter c of the face F. We conclude that the polyhedron $X_0 := X \cap P$ contains c. Since $c \in X_0$, the simplex $S = \{w, v_1, v_2, 0\}$ has nonpositive orientation along the face $\{0, v_1, v_2\}$. By Lemmas 5.16 and 5.17, the simplex S lies in $\mathcal{Q}_0$.

Let c be the circumcenter of the triangle $F = \{0, v_1, v_2\}$ and let c_2 be the circumcenter of the simplex $\{0, v_1, v_2, w\}$. Let C be the convex hull of $\{0, v_1/2, v_2/2, c, c_2\}$. The set C contains the set of points separated from w by the half-plane H_3, closer to w than to 0, and closer to 0 than both v_1 and v_2. The point x lies in this convex hull C. Since this convex hull is nonempty, the simplex S has negative orientation along the face $\{0, v_1, v_2\}$.

By assumption, w is not obstructed at x. Hence the segment from w to x does not pass through the face $\{0, v_1, v_2\}$. The set C' of points $y \in C$ such that the segment from w to y does not pass through the face $\{0, v_1, v_2\}$ is thus nonempty. The set C' must include the extreme point c_2 of C. This means that the plane $\{w, v_1, v_2\}$ separates c_2 from the origin, so that the simplex S has negative orientation also along the face $\{w, v_1, v_2\}$. This contradicts Lemma 5.13.   □

We draw out a simple consequence of the proof. Let $F = \{0, v_1, v_2\}$ with edges of length between 2 and $\sqrt{8}$. Let $S = \{0, w, v_1, v_2\}$, and assume that S has negative orientation along F. Let c be the circumcenter of the triangle $F = \{0, v_1, v_2\}$ and let c_2 be the circumcenter of the simplex $\{0, v_1, v_2, w\}$. Let C be the convex hull of $\{0, v_1/2, v_2/2, c, c_2\}$. The set C contains the set of points separated from w by the half-plane H_3, closer to w than to 0, and closer to 0 than both v_1 and v_2. Let x lie in this convex hull C.

Lemma 5.31. *In this context, w is obstructed at x.*

Proof. This is what the final paragraph of the previous proof proves by contradiction. □

6. Decomposition Stars

This section constructs a topological space DS such that each point of DS encodes the geometrical data surrounding a vertex in the packing. The points in this topological space are called decomposition stars. A decomposition star encodes all of the local geometrical information that will be needed in the local analysis of a sphere packing. These geometrical data are sufficiently detailed that it is possible to recover the V-cell at $v \in \Lambda$ from the corresponding point in the topological space. It is also possible to recover the simplices in the Q-system that have a vertex at $v \in \Lambda$. Thus, a decomposition star has a dual nature that encompasses both the Voronoi-like V-cell and the Delaunay-like simplices in the Q-system. By encoding both structures, the decomposition star becomes our primary geometric object of analysis.

It can be helpful at times to visualize the decomposition star as a polyhedral object formed by the union of the simplices at v in the Q-system with the V-cell at $v \in \Lambda$. Although such descriptions can be helpful to the intuition, the formal definition of a decomposition star is rather more combinatorial, expressed as a series of indexing sets that hold the data that are needed to reconstruct the geometry. The formal description of the decomposition star is preferred because it encodes more information than the polyhedral object.

The term "decomposition star" is derived from the earlier term "Delaunay star" that was used in [Ha2] as the name for the union of Delaunay simplices that shared a common vertex. Delaunay stars are star-convex. It is perhaps unfortunate that the term "star" has been retained, because (the geometric realization of) a decomposition star need not be star convex. In fact, Remark 5.10 suggests that V-cells can be rather poorly behaved in this respect.

6.1. Indexing Sets

We are ready for the formal description of decomposition stars.

Let $\omega = \{0, 1, 2, \ldots\}$. Pick a bijection $b: \omega \to \Lambda$ and use this bijection to index the vertices $b(i) = v_i \in \Lambda, i = 0, 1, 2 \ldots$. Define the following indexing sets:

- Let $I_1 = \omega$.
- Let I_2 be the set of unordered pairs of indices $\{i, j\}$ such that $|v_i - v_j| \leq 2t_0 = 2.51$.

- Let I_3 be the set of unordered tuples of indices $\{i, j, k, \ell\}$ such that the corresponding simplex is a strict quarter.
- Let I_4 be the set of unordered tuples $\{i, j, k, \ell\}$ of indices such that the simplex $\{v_i, v_j, v_k, v_\ell\}$ is in the Q-system.
- Let I_5 be the set of unordered triples $\{i, j, k\}$ of indices such that v_i is an anchor of a diagonal $\{v_j, v_k\}$ of a strict quarter in the Q-system.
- Let I_6 be the set of unordered pairs $\{i, j\}$ of indices such that the edge $\{v_i, v_j\}$ has length in the open interval $(2t_0, \sqrt{8})$. (This set includes all such pairs, whether or not they are attached to the diagonal of a strict quarter.)
- Let I_7 be the set of unordered triples $\{i, j, k\}$ of indices such that the triangle $\{v_i, v_j, v_k\}$ is a face of a simplex in the Q-system and such that the circumradius is less than $\sqrt{2}$.
- Let I_8 be the set of unordered quadruples $\{i, j, k, \ell\}$ of indices such that the corresponding simplex $\{v_i, v_j, v_k, v_\ell\}$ is a quasi-regular tetrahedron with circumradius less than 1.41.

The data are highly redundant, because some of the indexing sets can be derived from others. However, there is no need to strive for a minimal description of the data. Set $d_0 = 2\sqrt{2} + 4\sqrt{3}$. We recall that $\Lambda(v, d_0) = \{w \in \Lambda: |w - v| \le d_0\}$. Let

$$T' = \{i: v_i \in \Lambda(v, d_0)\}.$$

It is the indexing set for a neighborhood of v.

Fix a vertex $v = v_a \in \Lambda$. Let $I_0' = \{\{a\}\}$. Let

$$I_j' = \{x \in I_j: x \subset T'\}, \qquad \text{for} \quad 1 \le j \le 8.$$

Each I_j' is a finite set of finite subsets of ω. Hence $I_j' \in P(P(\omega))$, where $P(X)$ is the powerset of any set X.

Associate with each $v \in \Lambda$ the function $f: T' \to B(0, d_0)$ given by $f(i) = v_i - v$, and the tuple

$$t = (I_0', \dots, I_8') \in P(P(\omega))^9.$$

There is a natural action of the permutation group of ω on the set of pairs (f, t), where a permutation acts on the domain of f and on $P(P(\omega))$ through its action on ω. Let $[f, t]$ be the orbit of the pair (f, t) under this action. The orbit $[f, t]$ is independent of the bijection $b: \omega \to \Lambda$. Thus, it is canonically attached to (v, Λ).

Definition 6.1. Let DS° be the set of all pairs $[f, t]$ that come from some v in a saturated packing Λ.

Put a topology on all pairs (f, t) (as we range over all saturated packings Λ, all choices of indexing $b: \omega \to \Lambda$, and all $v \in \Lambda$) by declaring (f, t) to be close to (f', t') if and only if $t = t'$, domain$(f) = $ domain(f'), and for all $i \in$ domain(f), $f(i)$ is close to $f'(i)$. That is, we take the topology to be that inherited from the standard topology on $B(0, d_0)$ and the discrete topology on the finite indexing sets.

The topology on pairs (f, t) descends to the orbit space and gives a topology on DS°.

There is a natural compactification of DS° obtained by replacing open conditions by closed conditions. That is, for instance if $\{i, j\}$ is a pair in I_6, we allow $|f(i) - f(j)|$ to lie in the closed interval $[2t_0, \sqrt{8}]$. The conditions on each of the other indexing sets I_j are similarly relaxed so that they are closed conditions.

Compactness comes from the compactness of the closed ball $B(0, d_0)$, the closed conditions on indexing sets, and the finiteness of T'.

Definition 6.2. Let DS be the compactification given above of DS°. Call it the space of *decomposition stars*.

Definition 6.3. Let v be a vertex in a saturated packing Λ. We let $D(v, \Lambda)$ denote the decomposition star attached to (v, Λ).

Because of the discrete indexing sets, the space of decomposition stars breaks into a large number of connected components. On each connected component, the combinatorial data are constant. Motion within a fixed connected component corresponds to a motion of a finite set of sphere centers of the packing (in a direction that preserves all of the combinatorial structures).

In a decomposition star, it is no longer possible to distinguish some quasi-regular tetrahedra from quarters solely on the basis of metric relations. For instance, the simplex with edge lengths $2, 2, 2, 2, 2, 2t_0$ is a quasi-regular tetrahedron and is also in the closure of the set of strict quarters. The indexing set I_2', which is part of the data of a decomposition star, determines whether the simplex is treated as a quasi-regular tetrahedron or a quarter.

Roughly speaking, two decomposition stars $D(v, \Lambda)$ and $D(v', \Lambda')$ are close if the translations $\Lambda(v, d_0) - v$ and $\Lambda'(v', d_0) - v'$ have the same cardinality, and there is a bijection between them that respects all of the indexing sets I_j' and proximity of vertices.

6.2. Cells Attached to Decomposition Stars

To each decomposition star we can associate a V-cell centered at 0 by a direct adaptation of Definition 5.6.

Lemma 6.4. *The V-cell at v depends on Λ only through $\Lambda(v, d_0)$ and the indexing sets I_j'.*

Proof. We wish to decide whether a given x belongs to the V-cell at v or to another contender $w \in \Lambda$. We assume that $x \in I_v$, for otherwise x cannot belong to the V-cell at v. Similarly, we assume $x \in I_w$. We must determine whether v or w is obstructed at x. For this we must know whether barriers lie on the path between x and v (or w). Since $|x - w| \leq 2\sqrt{3}$ and $|x - v| \leq 2\sqrt{3}$, the point p of intersection of the barrier and the segment $\{x, v\}$ (or $\{x, w\}$) satisfies $|x - p| \leq 2\sqrt{3}$. All the vertices of the barrier then have distance at most $\sqrt{8}+2\sqrt{3}$ from x, and hence distance at most $d_0 = \sqrt{8}+4\sqrt{3}$ from v. The decomposition star $D(v, \Lambda)$ includes all vertices in $\Lambda(v, d_0)$ and the indexing sets of the decomposition star label all the barriers in $\Lambda(v, d_0)$. Thus, the decomposition star at v gives all the data that are needed to determine whether $x \in I_v$ belongs to the V-cell at v.   □

Corollary 6.5. *There is a V-cell $\mathrm{VC}(D)$ attached to each decomposition star D such that if $D = D(v, \Lambda)$, then $\mathrm{VC}(D) + v$ is the V-cell attached to (v, Λ) in Definition 5.6.*

Proof. By the lemma the map from (v, Λ) maps through the data determining the decomposition star $D(v, \Lambda)$. The definition of V-cell extends: the V-cell at 0 attached to $[f, t]$ is the set of points in $B(0, C_0)$ for which the origin is the unique closest unobstructed vertex of range(f). The barriers for the obstruction are to be reconstructed from the indexing data sets I_j of t.   □

Lemma 6.6. $\mathrm{VC}(D)$ *is a finite union of nonoverlapping convex polyhedra. Moreover, $D \mapsto \mathrm{vol}(\mathrm{VC}(D))$ is continuous.*

Proof. For the proof, we ignore sets of measure zero, such as finite unions of planes. We may restrict our attention to a single connected component of the space of decomposition stars. On each connected component, the indexing set for each barrier (near the origin) is fixed. The indexing set for the set of vertices near the origin is fixed. For each D the VC-cell breaks into a finite union of convex polyhedra by Lemma 5.9.

As the proof of that lemma shows, some faces of the polyhedra are perpendicular bisecting planes between two vertices near the origin. Such planes vary continuously on (a connected component) of DS. The other faces of polyhedra are formed by planes through three vertices of the packing near the origin. Such planes also vary continuously on DS. It follows that the volume of each convex polyhedron is a continuous function on DS. The sum of these volumes, giving the volume of $\mathrm{VC}(D)$, is also continuous.   □

Lemma 6.7. *Let Λ be a saturated packing. The Voronoi cell $\Omega(v)$ at v depends on Λ only through $\Lambda(v, d_0)$.*

Proof. Let x be an extreme point of the Voronoi cell $\Omega(v)$. The vertex v is one of the vertices closest to x. If the distance from x to v is at least 2, then there is room to place another ball centered at x into the packing without overlap. Then Λ is not saturated.

Thus, the distance from x to v is less than 2. The Voronoi cell lies in the ball $B(v, 2)$. The Voronoi cell is bounded by the perpendicular bisectors of segments $\{v, w\}$ for $w \in \Lambda$. If w has distance 4 or more from v, then the bisector cannot meet the ball $B(v, 2)$ and cannot bound the cell. Since $4 < d_0$, the proof is complete.   □

Corollary 6.8. *The vertex v and the decomposition star $D(v, \Lambda)$ determine the Voronoi cell at v. In fact, the Voronoi cell is determined by v and the first indexing set I_1' of $D(v, \Lambda)$.*

Definition 6.9. The Voronoi cell $\Omega(D)$ of $D \in \mathrm{DS}$ is the set containing the origin bounded by the perpendicular bisectors of $\{0, v_i\}$ for $i \in I_1'$.

Remark 6.10. It follows from Corollary 6.8 that

$$\Omega(D(v, \Lambda)) = v + \Omega(v).$$

In particular, they have the same volume.

Remark 6.11. From a decomposition star D, we can recover the set of vertices $U(D)$ of distance at most $2t_0$ from the origin, the set of barriers at the origin, the simplices of the Q-system having a vertex at the origin, the V-cell $\mathrm{VC}(D)$ at the origin, the Voronoi cell $\Omega(D)$ at the origin, and so forth. In fact, the indexing sets in the definition of the decomposition star were chosen specifically to encode these structures.

6.3. *Colored Spaces*

In Section 3 we introduced a function σ that will be formally defined in Definition 7.8. The details of the definition of σ are not needed for the discussion that follows. The function σ on the space DS of decomposition stars is continuous. This section gives an alternate description of the sense in which this function is continuous.

We begin with an example that illustrates the basic issues. Suppose that we have a discontinuous piecewise linear function on the unit interval $[-1, 1]$, as in Fig. 6.1. It is continuous, except at $x = 0$.

We break the interval in two at $x = 0$, forming two compact intervals $[-1, 0]$ and $[0, 1]$. We have continuous functions $f_-: [-1, 0] \to \mathbb{R}$ and $f_+: [0, 1]$, such that

$$f(x) = \begin{cases} f_-(x), & x \in [-1, 0], \\ f_+(x), & \text{otherwise.} \end{cases}$$

We have replaced the discontinuous function by a pair of continuous functions on smaller intervals, at the expense of duplicating the point of discontinuity $x = 0$. We view this pair of functions as a single function F on the compact topological space with two

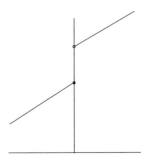

Fig. 6.1. A piecewise linear function.

components

$$[-1, 0] \times \{-\} \quad \text{and} \quad [0, 1] \times \{+\},$$

where $F(x, a) = f_a(x)$ and $a \in \{-, +\}$.

This is the approach that we follow in general with the Kepler conjecture. The function σ is defined by a series of case statements, and the function does not extend continuously across the boundary of the cases. However, in the degenerate cases that land precisely between two or more cases, we form multiple copies of the decomposition star for each case, and place each case into a separate compact domain on which the function σ is continuous.

This can be formalized as a *colored space*. A colored space is a topological space X together with an equivalence relation on X with the property that no point x is equivalent to any other point in the same connected component as x. We refer to the connected components as colors, and call the points of X *colored points*. We call the set of equivalence classes of X the underlying uncolored space of X. Two colored points are equal as uncolored points if they are equivalent under the equivalence relation.

In our example, there are two colors "$-$" and "$+$." The equivalence class of (x, a) is the set of pairs (x, b) with the same first coordinate. Thus, if $x \neq 0$, the equivalence class contains one element $(x, \text{sign}(x))$, and in the boundary case $x = 0$ there are two equivalent elements $(0, -)$ and $(0, +)$.

In our treatment of decomposition stars, there are various cases: whether an edge has length less than or greater than $2t_0$, less than or greater than $\sqrt{8}$, whether a face has circumradius less than or greater than $\sqrt{2}$, and so forth. By duplicating the degenerate cases (say an edge of exact length $2t_0$), creating a separate connected component for each case, and expressing the optimization problem on a colored space, we obtain a continuous function σ on a compact domain X.

The colorings have in general been suppressed in places from the notation. To obtain consistent results, a statement about $x \in [2, 2t_0]$ should be interpreted as having an implicit condition saying that x has the coloring induced from the coloring on the component containing $[2, 2t_0]$. A later statement about $y \in [2t_0, \sqrt{8}]$ deals with y of a different color, and no relation between x and y of different colors is assumed at the endpoint $2t_0$.

7. Scoring

This section is coauthored by Samuel P. Ferguson and Thomas C. Hales.

In earlier sections, we describe each packing of unit balls by its set $\Lambda \subset \mathbb{R}^3$ of centers of the packing. We showed that we may assume that our packings are saturated in the sense that there is no room for additional balls to be inserted into the packing without overlap. Lemma 3.3 shows that the Kepler conjecture follows if for each saturated packing Λ we can find a function $A \colon \Lambda \to \mathbb{R}$ with two properties: the function is fcc-compatible and it is saturated in the sense of Definition 3.2.

The purpose of the first part of this section is to define a function $A \colon \Lambda \to \mathbb{R}$ for every saturated packing Λ and to show that it is negligible. The formula defining A consists of a term that is a correction between the volume of the Voronoi cell $\Omega(v)$ and that of the V-cell $VC(v)$ and a further term coming from simplices of the Q-system that have a vertex at v.

A major theorem in this volume will be that this negligible function is fcc-compatible. The proof of fcc-compatibility can be expressed as a difficult nonlinear optimization problem over the compact topological space DS that was introduced in Section 6. In fact, we construct a continuous function A_0 on the space DS such that for each saturated packing Λ and each $v \in \Lambda$, the value of the function A at v is a value in the range of the function A_0 on DS. In this way, we are able to translate the fcc-compatibility of A into an extremal property of the function A_0 on the space DS.

The proof of fcc-compatibility is more conveniently couched as an optimization problem over a function that is related to the function A_0 by an affine rescaling. This new function is called the score and is denoted σ. (The exact relationship between A_0 and σ appears in Definition 7.12.) The function σ is a continuous function on the space DS. This function is defined in the final paragraphs of this section.

7.1. Definitions

For every saturated packing Λ, and $v \in \Lambda$, there is a canonically associated decomposition star $D(v, \Lambda)$. The negligible function $A: \Lambda \to \mathbb{R}$ that we define is a composite

$$A = A_0 \circ D(\cdot, \Lambda): \Lambda \to \text{DS} \to \mathbb{R}, \qquad v \mapsto D(v, \Lambda) \mapsto A_0(D(v, \Lambda)), \qquad (7.1)$$

where $A_0: \text{DS} \to \mathbb{R}$ is as defined by (7.2) and (7.6) below. Each simplex in the Q-system with a vertex at v defines by translation to the origin a simplex in the Q-system with a vertex at 0 attached to $D(v, \Lambda)$. Let $\mathcal{Q}_0(D)$ be this set of translated simplices at the origin. This set is determined by D.

Definition 7.1. Let Q be a quarter in $\mathcal{Q}_0(D)$. We say that the *context* of Q is (p, q) if there are p anchors and $p - q$ quarters along the diagonal of Q. Write $c(Q, D)$ for the context of $Q \in \mathcal{Q}_0(D)$.

Note that q is the number of "gaps" between anchors around the diagonal. For example, the context of a quarter in a quartered octahedron is $(4, 0)$. The context of a single quarter is $(2, 1)$.

The function A_0 will be defined to be a continuous function on DS of the form

$$A_0(D) = -\text{vol}(\Omega(D)) + \text{vol}(VC(D)) + \sum_{Q \in \mathcal{Q}_0(D)} A_1(Q, c(Q, D), 0). \qquad (7.2)$$

Thus, the function A_0 measures the difference in volume between the Voronoi cell and the V-cell, as well as certain contributions A_1 from the Q-system. The function $A_1(Q, c, v)$ depends on Q, its context c, and a vertex v of Q. The function $A_1(Q, c, v)$ will not depend on the second argument when Q is a quasi-regular tetrahedron. (The context is not defined for such simplices.)

Definition 7.2. An *orthosimplex* consists of the convex hull of $\{0, v_1, v_1 + v_2, v_1 + v_2 + v_3\}$, where v_2 is a vector orthogonal to v_1, and v_3 is orthogonal to both v_1 and v_2. We

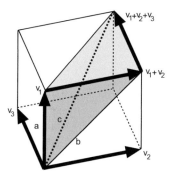

Fig. 7.1. The Rogers simplex is an orthosimplex.

can specify an orthosimplex up to congruence by the parameters $a = |v_1|$, $b = |v_1 + v_2|$, and $c = |v_1 + v_2 + v_3|$, where $a \le b \le c$. This parametrization of the orthosimplex departs from the usual parametrization by the lengths $|v_1|$, $|v_2|$, $|v_3|$. For $a \le b \le c$, the *Rogers simplex* $R(a, b, c)$ is an orthosimplex of the form

$$R(a, b, c) = S(a, b, c, \sqrt{c^2 - b^2}, \sqrt{c^2 - a^2}, \sqrt{b^2 - a^2}).$$

See Fig. 7.1.

Definition 7.3. Let R be a Rogers simplex. We define the *quoin* of R to be the wedge-like solid (a quoin) situated above R. It is defined as the solid bounded by the four planes through the faces of R and a sphere of radius c at the origin. (See Fig. 7.2.) We let quo(R) be the volume of the quoin over R. If $R = R(a, b, c)$ is a Rogers simplex, the volume quo(R) is given explicitly as follows:

$$6\,\mathrm{quo}(R) = (a + 2c)(c - a)^2 \arctan(e) + a(b^2 - a^2)e$$
$$- 4c^3 \arctan(e(b - a)/(b + c)), \tag{7.3}$$

where $e \ge 0$ is given by $e^2(b^2 - a^2) = (c^2 - b^2)$.

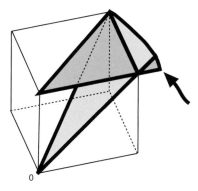

Fig. 7.2. The quoin above a Rogers simplex is the part of the shaded solid outside the illustrated box. It is bounded by the shaded planes, the plane through the front face of the box, and a sphere centered at the origin passing through the opposite corner of the box.

Let S be a simplex and let v be a vertex of that simplex. Let $VC(S, v)$ be the subset of $|S|$ consisting of points closer to v than to any other vertex of S. By Lemma 5.28, if $S \in \mathcal{Q}_0(D)$, then

$$VC(S, 0) = VC(D) \cap |S|.$$

Under the assumption that S contains its circumcenter and that every one of its faces contains its circumcenter, an explicit formula for the volume $\text{vol}(VC(S, v))$ has been calculated in Section 8.6.3 of [Ha6]. This volume formula is an algebraic function of the edge lengths of S, and may be analytically continued to give a function of S with chosen vertex v:

$$\text{vol } VC^{\text{an}}(S, v).$$

Lemma 7.4. *Let $B(0, t)$ be a ball of radius t centered at the origin. Let v_1 and v_2 be vertices. Assume that $|v_1| < 2t$ and $|v_2| < 2t$. Truncate the ball by cutting away the caps*

$$\text{cap}_i = \{x \in B(0, t) \colon |x - v_i| < |x|\}.$$

Assume that the circumradius of the triangle $\{0, v_1, v_2\}$ is less than t. Then the intersection of the caps, $\text{cap}_1 \cap \text{cap}_2$, is the union of four quoins.

Proof. This is true by inspection. See Fig. 7.3. Slice the intersection $\text{cap}_1 \cap \text{cap}_2$ into four pieces by two perpendicular planes: the plane through $\{0, v_1, v_2\}$, and the plane perpendicular to the first and passing through 0 and the circumcenter of $\{0, v_1, v_2\}$. Each of the four pieces is a quoin. □

Definition 7.5. Let $v \in \mathbb{R}^3$ and let X be a measurable subset of $\mathbb{R}^3$. Let $\text{sol}(X, v)$ be the area of the radial projection of $X \setminus \{0\}$ to the unit sphere centered at the origin. We call this area the *solid angle* of X (at v). When $v = 0$, we write the function as $\text{sol}(X)$.

Let $S = \{v_0, v_1, v_2, v_3\}$ be a simplex. Fix t in the range $t_0 \le t \le \sqrt{2}$. Assume that t is at most the circumradius of S. Assume that it is at least the circumradius of each of the

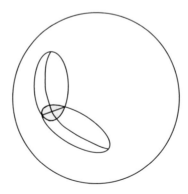

Fig. 7.3.   The intersection of two caps on the unit ball can be partitioned into four quoins (shaded).

faces of S. Let $VC_t(S, v_0)$ be the intersection of $VC(S, v_0)$ with the ball $B(v_0, t)$. Under the assumption that S contains its circumcenter and that every one of its faces contains it circumcenter, an explicit formula for the volume

$$\text{vol}(VC_t(S, v_0))$$

is calculated by means of Lemma 7.4 through a process of inclusion and exclusion. In detail, start with $|S| \cap B(v_0, t)$. Truncate this solid by caps: cap_1, cap_2, and cap_3 bounded by the sphere of radius t centered at v_0 and the perpendicular bisectors (respectively) of $\{v_0, v_1\}$, $\{v_0, v_1\}$, and $\{v_0, v_2\}$. If we subtract the volume of each cap, cap_i, then we must add back the volume of the doubly counted intersections of the caps. The intersections of caps are given as quoins (Lemma 7.4). This leads to the following formula. Let $h_i = |v_i|/2$ and $\eta_{ij} = \eta(0, v_i, v_j)$, and let S_3 be the group of permutations of $\{1, 2, 3\}$ in

$$\text{vol } VC_t(S, v_0) = \frac{\text{sol}(S)}{3} - \sum_{i=1}^{3} \frac{\text{dih}(S, v_i)}{2\pi} \text{vol cap}_i + \sum_{(i,j,k) \in S_3} \text{quo}(R(h_i, \eta_{ij}, t)). \quad (7.4)$$

We extend formula (7.4) by setting

$$\text{quo}(R(a, b, c)) = 0,$$

if the constraint $a < b < c$ fails to hold. Similarly, set $\text{vol cap}_i = 0$ if $|v_i| \geq 2t$. With these conventions, formula (7.4) extends to all simplices. We write the extension of $\text{vol } VC_t(S, v)$ as

$$\text{vol } VC_t^+(S, v).$$

Definition 7.6.   Let[3]

$$s\text{-vor}(S, v) = 4(-\delta_{\text{oct}} \text{vol } VC^{\text{an}}(S, v) + \text{sol}(S, v)/3),$$

$$s\text{-vor}(S, v, t) = 4(-\delta_{\text{oct}} \text{vol } VC_t^+(S, v) + \text{sol}(S, v)/3),$$

and

$$s\text{-vor}_0(S, v) = s\text{-vor}(S, v, t_0).$$

When it is clear from the context that the vertex v is fixed at the origin, we drop v from the notation of these functions. If $S = \{v_1, v_2, v_3, v_4\}$, we define $\Gamma(S)$ as the average

$$\Gamma(S) = \frac{1}{4} \sum_{i=1}^{4} s\text{-vor}(S, v_i). \quad (7.5)$$

The average $\Gamma(S)$ is called the *compression* of S.

---

[3] In [Ha1] the volumes in this definition were volumes of Voronoi cells, and hence the notation vor for the function was adopted. We retain vor in the notation, although this direct connection with Voronoi cells has been lost.

Definition 7.7. Let Q be a quarter. Let $\eta^+(Q)$ be the maximum of the circumradii of the two faces of Q along the diagonal of Q.

Let Q be a simplex in the Q-system. We define an involution $v \to \hat{v}$ on the vertices of Q as follows. If Q is a quarter and v is an endpoint of the diagonal, then let $\hat{v}$ be the opposite endpoint of the diagonal. In all other cases, set $\hat{v} = v$.

We are ready to complete the definition of the function $A: \Lambda \to \mathbb{R}$. The definition of A was reduced to that of A_0 in (7.1). The function A_0 was reduced in turn to that of A_1 in (7.2). To complete the definition, we define A_1.

Definition 7.8. Set

$$A_1(S, c, v) = -\text{vol VC}(S, v) + \frac{\text{sol}(S, v)}{3\delta_{\text{oct}}} - \frac{\sigma(S, c, v)}{4\delta_{\text{oct}}}, \tag{7.6}$$

where σ is given as follows:

1. When S is a quasi-regular tetrahedron:
   (a) If the circumradius of S is less than 1.41, set

$$\sigma(S, -, v) = \Gamma(S).$$

   (b) If the circumradius of S is at least 1.41, set

$$\sigma(S, -, v) = \text{s-vor}(S, v).$$

2. When S is a strict quarter:
   (a) If $\eta^+(S) < \sqrt{2}$:
      (i) If the context c is $(2, 1)$ or $(4, 0)$, set

$$\sigma(S, c, v) = \Gamma(S).$$

      (ii) If the context of S is anything else, set

$$\sigma(S, c, v) = \Gamma(S) + \frac{\text{s-vor}_0(S, v) - \text{s-vor}_0(S, \hat{v})}{2}.$$

   (b) If $\eta^+(S) \geq \sqrt{2}$:
      (i) If the context of S is $(2, 1)$, set

$$\sigma(S, c, v) = \text{s-vor}(S, v).$$

      (ii) If the context of S is $(4, 0)$, set

$$\sigma(S, c, v) = \frac{\text{s-vor}(S, v) + \text{s-vor}(S, \hat{v})}{2}.$$

      (iii) If the context of S is anything else, set

$$\sigma(S, c, v) = \frac{\text{s-vor}(S, v) + \text{s-vor}(S, \hat{v})}{2}$$
$$+ \frac{\text{s-vor}_0(S, v) - \text{s-vor}_0(S, \hat{v})}{2}.$$

When the context and vertex v are given, we often write $\sigma(S)$ or $\sigma(S, v)$ for $\sigma(S, c, v)$.

When $\eta^+ < \sqrt{2}$, we say that the quarter has compression type. Otherwise, we say it has Voronoi type. To say that a quarter has compression type means that $\Gamma(S)$ is one term of the function $\sigma(S, v)$. It does not mean that $\Gamma(S)$ is equal to $\sigma(S, v)$.

The definition of σ on quarters can be expressed a second way in terms of a function μ. If S is a quarter, set

$$\mu(S, v) = \begin{cases} \Gamma(S), & \text{if} \quad \eta^+(S) < \sqrt{2}, \\ \text{s-vor}(S, v), & \text{otherwise.} \end{cases} \tag{7.7}$$

If S is a flat quarter, we have $\sigma(S, c, v) = \mu(S, v)$, for all contexts c.

Suppose S is an upright quarter. Definition 7.8 can be expressed as follows:

- Context $(2, 1)$: set $\sigma(S, c, v) = \mu(S, v)$.
- Context $(4, 0)$: set $\sigma(S, c, v) = (\mu(S, v) + \mu(S, \hat{v}))/2$.
- Other contexts: set $\sigma(S, c, v) = (\mu(S, v) + \mu(S, \hat{v}) + \text{s-vor}_0(S, v) - \text{s-vor}_0(S, \hat{v}))/2$.

Lemma 7.9. $A_0\colon \text{DS} \to \mathbb{R}$ *is continuous.*

Proof. The continuity of $D \mapsto \text{vol VC}(D)$ is proved in Lemma 6.6. The continuity of $D \mapsto \text{vol }\Omega(D)$ is similarly proved. The terms vol $\text{VC}(S, v)$ and $\text{sol}(S, v)$ are continuous. To complete the proof we check that the function $\sigma(S, c, v)$ is continuous. It is not continuous when viewed as a function of the set of quarters, because of the various cases breaking at circumradius 1.41 and $\eta^+(S) = \sqrt{2}$. However, these cutoffs have been inserted into the data defining a decomposition star (in the indexing sets I_8 and I_9). Thus, the different cases in the definition of $\sigma(S, c, v)$ land in different connected components of the space DS and continuity is obtained. $\qquad\square$

We conclude this section with a result that will be of use in the next section.

Lemma 7.10. *Let $S = \{v_1, v_2, v_3, v_4\}$ be a simplex in the S-system, and let c be its* *context. Then*

$$\sum_{i=1}^{4} A_1(S, c, v_i) = 0.$$

Proof. By formula (7.6), this is equivalent to

$$\sum_{i=1}^{4} \sigma(S, c, v_i) = \sum_{i=1}^{4} \text{s-vor}(S, c, v_i). \tag{7.8}$$

Equation (7.8) is evident from Definition 7.8 for σ. In fact, the terms of the form s-vor$_0$ have opposing signs and cancel when we sum. The other terms are weighted averages of the terms s-vor(S, c, v_i). Equation (7.8) is thus established because a sum is unaffected by taking weighted averages of its terms. $\qquad\square$

7.2. Negligibility

Let $B(x, r)$ be the closed ball of radius $r \in \mathbb{R}$ centered at x. Let $\Lambda(x, r) = \Lambda \cap B(x, r)$.

Recall from Definition 3.2 that a function $A: \Lambda \to \mathbb{R}$ is said to be *negligible* if there is a constant C_1 such that for all $r \geq 1$,

$$\sum_{v \in \Lambda(x,r)} A(v) \leq C_1 r^2.$$

Recall the function $A: \Lambda \to \mathbb{R}$ given by (7.1). Explicitly, let

$$A(v) = A_0(D(v, \Lambda)),$$

where A_0 in turn depends on functions A_1 and σ, as determined by (7.2), (7.6), and Definition 7.8.

Theorem 7.11. *The function A of (7.1) is negligible.*

Proof. First we consider a simplification, where we replace A with A' defined by

$$A'(v, \Lambda) = -\text{vol}(\Omega(D(v, \Lambda))) + \text{vol}(\text{VC}(D(v, \Lambda))).$$

(That is, at first we ignore the function A_1.) The Voronoi cells partition $\mathbb{R}^3$, as do the V-cells. We have $\Omega(v, \Lambda) \subset B(v, 2)$ (by saturation) and $\text{VC}(v, \Lambda) \subset B(v, 2\sqrt{3})$ (by Definition 5.5). Hence the Voronoi cells with $v \in \Lambda(x, r)$ cover $B(x, r - 2)$. Moreover, the V-cells with $v \in \Lambda(x, r)$ are contained in $B(x, r + 2\sqrt{3})$. Hence

$$\sum_{v \in \Lambda(x,r)} A'(v) \leq -\text{vol } B(x, r - 2) + \text{vol } B(x, r + 2\sqrt{3}) \leq C_1' r^2$$

for some constant C_1'.

If we do not make the simplification, we must include the sum

$$\sum_{v \in \Lambda(x,r)} \sum_{Q \in \mathcal{Q}_v(D(v,\Lambda))} A_1(Q, c, v).$$

Each quarter $Q = \{v_1, v_2, v_3, v_4\}$ in the Q-system occurs in four sets $\mathcal{Q}_{v_i}(D(v_i, \Lambda))$. By Lemma 7.10 the sum cancels, except when some vertex of Q lies inside $\Lambda(x, r)$ and another lies outside. Each such simplex lies inside a shell of width $2\sqrt{8}$ around the boundary. The contribution of such boundary terms is again bounded by a constant times r^2. This completes the proof. $\square$

7.3. Fcc-Compatibility

We have constructed a negligible function A. The rest of this volume will prove that this function is fcc-compatible. This section translates fcc-compatibility into a property that will be easier to prove. To begin with, we introduce a rescaled version of the function A.

Definition 7.12. Let $\sigma: \mathrm{DS} \to \mathbb{R}$ be given by

$$\sigma(D) = -4\delta_{\mathrm{oct}}(\mathrm{vol}\,\Omega(D) + A_0(D)) + 16\pi/3.$$

It is called the *score* of the decomposition star.

Recall from Definition 3.6 the constant $pt \approx 0.05537$. This constant is called a point.

Lemma 7.13. *Let A_0, A, and σ be the functions defined by (7.1), (7.2), (7.6), and Definition 7.8. The following are equivalent:*

1. *The minimum of the function on DS given by*

$$D \mapsto \mathrm{vol}\,\Omega(D) + A_0(D)$$

   is $\sqrt{32}$.
2. *The maximum of σ on DS is $8\,pt$.*

Moreover, these statements imply

- *For every saturated packing Λ, the function A is fcc-compatible.*

(Eventually, we prove fcc-compatibility by proving $\sigma(D) \le 8\,pt$ for all $D \in \mathrm{DS}$.)

Proof. To see the equivalence of the first and second statements, use Definition 7.12, and the identity

$$8\,pt = -4\delta_{\mathrm{oct}}(\sqrt{32}) + 16\pi/3.$$

(Note that this identity is parallel in form to Definition 7.12 for σ.)

For a given saturated packing Λ, the function A has the form $A(v) = A_0(D(v, \Lambda))$. Also, $\Omega(D(v, \Lambda))$ is a translate of $\Omega(v)$, the Voronoi cell at v. In particular, they have the same volume. Thus, $\mathrm{vol}\,\Omega(v) + A(v)$ lies in the range of the function

$$\mathrm{vol}\,\Omega(D) + A_0(D)$$

on DS. The minimum of this function is $\sqrt{32}$ by the first of the equivalent statements. It now follows from the definition of fcc-compatibility, that $A: \Lambda \to \mathbb{R}$ is indeed fcc-compatible. □

Theorem 7.14. *If the maximum of the function σ on DS is $8\,pt$, then for every saturated packing Λ there exists a negligible fcc-compatible function A.*

Proof. This follows immediately from Theorem 7.11 and Lemma 7.13. □

7.4. *Scores of Standard Clusters*

The last section introduced a function σ called the score. We show that the function σ can be expressed as a sum over terms attached to each of the standard regions.

Definition 7.15. A *standard cluster* is a pair (R, D) where D is a decomposition star and R is one of its standard regions. A *quad cluster* is the standard cluster obtained when the standard region is a quadrilateral.

We break σ into a sum

$$\sigma(D) = \sum_R \sigma_R(D), \tag{7.9}$$

indexed by the standard clusters (R, D). Let

$$VC_R(D) = VC(D) \cap \text{cone}(R),$$

whenever R is a measurable subset of the unit sphere. Let

$$\mathcal{Q}_0(R, D) = \{Q \in \mathcal{Q}_0(D) : Q \subset \text{cone}(R)\}.$$

By Lemma 5.26, each Q is entirely contained in the cone over a single standard region.

Definition 7.16. Let R be a measurable subset of the unit sphere. Set

$$\text{vor}_R(D) = 4 \left(-\delta_{\text{oct}} \, \text{vol } VC_R(D) + \text{sol}(R)/3\right).$$

Let R be a standard region. Set

$$\sigma_R(D) = \text{vor}_R(D) - 4\delta_{\text{oct}} \sum_{Q \in \mathcal{Q}_0(R,D)} A_1(Q, c(Q, D), 0).$$

Lemma 7.17. $\sigma(D) = \sum_R \sigma_R(D)$, *where the sum runs over all standard regions R.*

Proof.

$$\sigma(D) = -4\delta_{\text{oct}}(\text{vol } \Omega(D) + A_0(D)) + 16\pi/3$$

$$= -4\delta_{\text{oct}}(\text{vol } VC(D) + \sum_{Q \in \mathcal{Q}_0(D)} A_1(Q, c(Q, D), 0)) + (4)(4\pi/3)$$

$$= \sum_R 4 \left(-\delta_{\text{oct}} \text{ vol } VC_R(D) - \delta_{\text{oct}} \sum_{Q \in \mathcal{Q}_0(R,D)} A_1(Q, c(Q, D), 0) + \text{sol}(R)/3\right). \quad \square$$

Also, we have

$$\text{vor}(D) = \sum_{R \in \mathcal{R}(D)} \text{vor}_R(D). \tag{7.10}$$

Lemma 7.18. *Let $R' \subset R$ be the part of a standard region that does not lie in any cone over any $Q \in Q_0(R, D)$. Then*

$$\sigma_R(D) = \text{vor}_{R'}(D) + \sum_{Q \in \mathcal{Q}_0(R,D)} \sigma(Q, c(Q, D), 0).$$

Proof. Substitute the definition of A_1 (7.6) into the definition of $\sigma_R(D)$, noting that $\mathrm{VC}(Q, 0) = \mathrm{VC}_{R''}(D)$, where R'' is the intersection of Q with the unit sphere. $\qquad\square$

Remark 7.19. Lemma 7.18 explains why we have chosen the same symbol σ for the functions $\sigma_R(D)$ and $\sigma(Q, c, v)$. We can view Lemma 7.18 as asserting a linear relation in the functions σ:

$$\sigma_R(D) = \sigma_{R'}(D) + \sum \sigma(Q, c, 0).$$

The sum runs over $Q \in \mathcal{Q}_0$ that lie in the cone over R.

7.5. Scores of Simplices and Cones

Many of the functions in this paper are defined by terms involving volumes of simple solids. To give estimates on the functions, it is often convenient to partition the solids into smaller pieces and then define corresponding functions on each of the pieces. For this reason, we define some variants of the functions vor and σ.

Remark 7.20. We now define a few more variants of the function vor. The function s-vor and its truncated version s-vor$(\cdot, t)$ have been defined already. The function vor$_R(D)$ will also be given a truncated version vor$_R(D, t)$, for a real truncation parameter $t \geq 0$. The special case vor$_R(D, t_0)$ will be abbreviated vor$_{0, R}(D)$. There will be another variant r-vor for Rogers simplices, and another c-vor for general sets. The general form of these functions is

$$\text{c-vor}(X) = 4(-\delta_{\text{oct}} \operatorname{vol}(X) + \operatorname{sol}(X)/3),$$

for any subset $X \subset \mathbb{R}^3$. The differences between the different versions of vor come from the different choices of the set X and the way they are parametrized.

Definition 7.21. Let $R = R(a, b, c)$ be a Rogers simplex. Assume that the vertex terminating the edges of lengths a, b, and c is situated at the origin. Let

$$\text{r-vor}(R) = 4(-\delta_{\text{oct}} \operatorname{vol}(R) + \operatorname{sol}(R)/3).$$

Definition 7.22. Let $C(h, t)$ denote the compact cone of height h and circular base. Set

$$\varphi(h, t) = 2(2 - \delta_{\text{oct}} t h(h + t))/3.$$

Then

$$\text{c-vor}(C(h, t)) = 2\pi(1 - h/t)\varphi(h, t). \tag{7.11}$$

Remark 7.23. Below, we introduce variants of the function σ. We have already encountered σ in Definitions 7.8, 7.12, and 7.16. Informally, we call σ (and various functions that are closely related to it) the *score*. Equation (7.11) represents the *score* of $C(h, t)$.

The solid angle of $C(h, t)$ is $2\pi(1 - h/t)$, so $\varphi(h, t)$ is the score per unit area. Also, $\varphi(t, t)$ is the score per unit area of a ball of radius t. That is, $\varphi(t, t) = 4(-\delta_{\text{oct}} \text{ vol}/ \text{sol} + \frac{1}{3})$.

We set

$$\text{s-vor}(S, t) = \text{sol}(S)\varphi(t, t) + \sum_{\substack{i = 1 \\ h_i \leq t}}^{3} d_i(1 - h_i/t)(\varphi(h_i, t) - \varphi(t, t))$$

$$- \sum_{(i, j, k) \in S_3} 4\delta_{\text{oct}} \text{ quo}(R(h_i, \eta(y_i, y_j, y_{k+3}), t)). \tag{7.12}$$

In the definition we adopt the convention that $\text{quo}(R) = 0$, if $R = R(a, b, c)$ does not exist (that is, if the condition $0 < a \leq b \leq c$ is violated). In the second sum, S_3 is the set of permutations on three letters. This definition is compatible with Definition 7.6.

Similarly, we define $\text{vor}_P(D, t)$ for arbitrary standard clusters (P, D). (We shift notation from R to P for a standard region to avoid conflict with Rogers simplices R in the following definition.) Extending the notation in an obvious way, we have

$$\text{vor}_P(D, t) = \text{sol}(P)\varphi(t, t) + \sum_{|v_i| \leq 2t} d_i(1 - |v_i|/(2t))(\varphi(|v_i|/2, t) - \varphi(t, t))$$

$$- \sum_R 4\delta_{\text{oct}} \text{ quo}(R). \tag{7.13}$$

The first sum runs over vertices in P of height at most $2t$. The second sum runs over Rogers simplices $R(|v_i|/2, \eta(F), t)$ in P, where $F = \{0, v_1, v_2\}$ is a face of circumradius $\eta(F)$ at most t, formed by vertices in P. The constant d_i is the total dihedral angle along $\{0, v_i\}$ of the standard cluster. The truncations $t = t_0 = 1.255$ and $t = \sqrt{2}$ will be of particular importance. Set $A(h) = (1 - h/t_0)(\varphi(h, t_0) - \varphi(t_0, t_0))$.

Remark 7.24. We have introduced both untruncated and truncated versions of functions vor and σ. The truncated versions are used to give upper bounds on the untruncated versions. For example, in the function $\sigma(D)$, the V-cell contributes through its volume, as in Remark 7.20. The volume appears with a negative coefficient $-4\delta_{\text{oct}}$. Thus, we obtain an upper bound on $\sigma(D)$ by discarding bits of volume from the V-cell. This suggests that we might try to give upper bounds on the score $\sigma(D)$ by truncating the V-cell in various ways. This is the reason for the truncated versions of these functions.

7.6. The Example of a Dodecahedron

Example 7.25. The following example illustrates why better bounds on the density of packings can be obtained with $\sigma(D)$ than with a naive approach based on the volume of Voronoi cells. By scoring quasi-regular tetrahedra with the compression function $\Gamma(S)$ rather than s-vor(S), we will find that the score is lowered below $8\,pt$ for configurations with many quasi-regular tetrahedra. To work one example, we assume that the decomposition star consists of twelve vertices located at distance 2 from the origin, at the vertices of a regular icosahedron. The score is approximately

$$20 \, \Gamma(S(2, 2, 2, 2.10292, 2.10292, 2.10292)) \approx 1.8 \, pt < 8 \, pt.$$

If s-vor(S) had been used, the score would violate Theorem 3.7:

$$20 \, \text{s-vor}(S) \approx 13.5493 \, pt > 8 \, pt.$$

(This is tied to the fact that the regular dodecahedron of inradius 1 has a smaller volume than the rhombic dodecahedron of inradius 1.)

References

[CS2]   J. H. Conway and N. J. A. Sloane, *Sphere Packings, Lattices and Groups*, third edition, Springer-Verlag, New York, 1998.

[Fej8] L. Fejes Tóth, *Lagerungen in der Ebene auf der Kugel und im Raum*, second edition, Springer-Verlag, Berlin, 1972.

[Ha1]   T. C. Hales, The sphere packing problem, *J. Comput. Appl. Math.*. **44** (1992), 41–76.

[Ha2]   T. C. Hales, Remarks on the density of sphere packings in three dimensions, *Combinatorica* **13** (1993), 181–187.

[Ha5]   T. C. Hales, A reformulation of the Kepler conjecture, Unpublished manuscript, Nov. 1996.

[Ha6]   T. C. Hales, Sphere packings, I, *Discrete Comput. Geom.* **17** (1997), 1–51.

[Ha7]   T. C. Hales, Sphere packings, II, *Discrete Comput. Geom.* **18** (1997), 135–149.

[Ha12] Thomas C. Hales, Cannonballs and honeycombs, *Notices Amer. Math. Soc.* **47**(4) (2000), 440–449.

[Ha13] T. C. Hales, Sphere Packings in 3 Dimensions, Arbeitstagung, 2001.

[Ha14] T. C. Hales, Some algorithms arising in the proof of the Kepler conjecture, in *Discrete and Computational Geometry*, Algorithms and Combinatorics, vol. 25, Springer-Verlag, Berlin, 2003, pp. 489–507.

[Ha16] T. C. Hales, Computer Resources for the Kepler Conjecture, http://annals.math.princeton.edu/keplerconjecture/. (The source code, inequalities, and other computer data relating to the solution are also found at http://xxx.lanl.gov/abs/math/9811078v1.)

[Hi]   D. Hilbert, Mathematische Probleme, *Archiv Math. Phys.* **1** (1901), 44–63, also in *Mathematical Developments Arising from Hilbert Problems*, Proceedings of Symposia in Pure Mathematics, vol. 28, 1976, pp. 1–34.

[Ro1]   C. A. Rogers, The packing of equal spheres, *Proc. London Math. Soc.* (3) **8** (1958), 609–620.

Received November 11, 1998, and in revised form September 12, 2003, and July 25, 2005.
Online publication February 27, 2006.

Sphere Packings III. Extremal Cases, by T. C. Hales

This is the third of six papers comprising the Kepler Conjecture proof. It contains Chapters 8–10 of the proof.

The asterisks added in the margin of some pages indicate that changes are required, as listed in the Hales list of errata given on pp. 372–373 of this volume.

Contents

Sphere Packings III. *Extremal Cases*, by T. C. Hales (Discrete Comput. Geom., **36** (2006), 71–110).

The original version of this chapter was revised. An erratum to this chapter can be found at http://dx.doi.org/10.1007/978-1-4614-1129-1_12

Discrete Comput Geom 36:71–110 (2006)
DOI: 10.1007/s00454-005-1212-0

Discrete & Computational
Geometry
© 2006 Springer Science+Business Media, Inc.

Sphere Packings, III. Extremal Cases

Thomas C. Hales

Department of Mathematics, University of Pittsburgh,
Pittsburgh, PA 15217, USA
hales@pitt.edu

Abstract. This paper is the third in a series of six papers devoted to the proof of the Kepler conjecture, which asserts that no packing of congruent balls in three dimensions has density greater than the face-centered cubic packing. In the previous paper in this series, a continuous function f on a compact space was defined, certain points in the domain were conjectured to give the global maxima, and the relation between this conjecture and the Kepler conjecture was established. This paper shows that those points are indeed local maxima. Various approximations to f are developed, that will be used in subsequent papers to bound the value of the function f. The function f can be expressed as a sum of terms, indexed by regions on a unit sphere. Detailed estimates of the terms corresponding to triangular and quadrilateral regions are developed.

Introduction

This paper has three objectives. The first is dealing with the two types of decomposition stars that attain the optimal Kepler conjecture bound. The second is obtaining general upper bounds on the score of decomposition star by truncation. The third is obtaining various upper bounds on the score associated to individual triangular and quadrilateral regions of a general decomposition star.

The first section contains a proof that the decomposition stars attached to the face-centered cubic (fcc) and hexagonal-close packing (hcp) give local maxima to the scoring function on the space of all decomposition stars. The proof describes precisely determined neighborhoods of these critical points. These special decomposition stars are shown to yield the global maximum of the scoring function on these restricted neighborhoods.

The second section gives an approximation to a decomposition star that provides an upper bound approximation to the scoring function σ. In the simplest cases, the approximation to the decomposition star is obtained by truncating the decomposition star at distance $t_0 = 1.255$ from the origin. More generally, we define a collection of simplices (that do not overlap any simplices in the Q-system), and define a somewhat different truncation for each type of simplex in the collection. For want of a more suggestive term, these simplices are said to form the $\mathcal{S}$-system.

When truncation at t_0 cuts too deeply, we reclaim a scrap of volume that lies outside the ball of radius t_0 but still inside the V-cell. This scrap is called a *crown*. These scraps are studied in that same section.

In a final section we develop a series of bounds on the score function in triangular and quadrilateral regions, for use in later papers.

8.   Local Optimality

The first several sections in the previous papers have established the fundamental definitions and constructions of this issue. This section establishes the local optimality of the function $\sigma \colon \mathrm{DS} \to \mathbb{R}$ in a neighborhood of the decomposition stars of the fcc and hcp.

8.1.   *Results*

Here is a sketch of the proof of local optimality. The fcc and hcp score precisely 8 *pt*. They also contain precisely eight tetrahedra around each vertex. In fact, the decomposition stars have eight quasi-regular tetrahedra and six other quad clusters. The proof shows that each of the eight quasi-regular tetrahedra scores at most 1 *pt*. Equality is obtained only when the tetrahedron is regular of side 2. Furthermore, the proof shows that each of six quad clusters have a nonpositive score. It will follows from these facts that any decomposition star with eight quasi-regular tetrahedra, six quad clusters, and no other standard clusters scores at most 8 *pt*. The case of equality is analyzed as well. The purpose of this section is to give a proof of the following theorem.

Theorem 8.1 (Local Optimality).   *Let D be a contravening decomposition star. Let $U(D)$ be the set of sphere packing vectors at distance at most $2t_0$ from the origin. Assume that*:

1. *The set $U(D)$ has twelve elements.*
2. *There is a bijection ψ between $U(D)$ and the kissing arrangement U_{fcc} of twelve tangent unit balls in the fcc configuration, or a bijection with U_{hcp} the twelve tangent unit balls in the hcp configuration; such that for all $v, w \in U(D), |w-v| \le 2t_0$ if and only if $|\psi(w) - \psi(v)| = 2$. That is, the proximity graph of $U(D)$ is the same as the contact graph of U_{fcc} or U_{hcp}.*

Then $\sigma(D) \le 8\,pt$. Equality holds if and only if U coincides with U_{fcc} or U_{hcp} up to a Euclidean motion. Decomposition stars D exist with $U(D) = U_{\mathrm{fcc}}$ and others exist with $U(D) = U_{\mathrm{hcp}}$.

Remark 8.2.   This theorem is one of the key claims of Section 3.3. This theorem is phrased slightly differently from Claim 3.15 in Section 3.3. The reason for this is that we have not formally introduced the plane graph $G(D)$ of a decomposition star. (This happens in Section 20.2.) Once $G(D)$ has been formally introduced, then Theorem 8.1 can be expressed more directly, as follows. We let G_{fcc} and G_{hcp} be the plane graphs attached to the decomposition stars of vertices in the fcc and hcp, respectively. (These graphs are independent of the vertices selected.)

Corollary 8.3 (Local Optimality—Second Version). *Contravening decomposition stars exist. If D is a contravening decomposition star, and if the plane graph of D is isomorphic to G_{fcc} or G_{hcp}, then $\sigma(D) = 8\,pt$. Moreover, up to Euclidean motion, $U(D)$ is the kissing arrangement of the twelve balls around a central ball in the fcc packing or the kissing arrangement of twelve balls in the hcp.*

The following theorem is also one of the main results of this section. It is a key part of the proof of local optimality.

Theorem 8.4. *A quad cluster scores at most 0, and that only for a quad cluster whose corners have height 2, forming a square of side 2. That is, $\sigma_R(D) \leq 0$. Other standard clusters have strictly negative scores: $\sigma_R(D) < 0$.*

The argument that the score of a quad cluster is nonpositive is general and can be used to prove that the score of any cluster attached to a non-triangular standard region (Definition 5.24) has nonpositive score.

8.2. Rogers Simplices

To prove Theorem 8.4, we chop the cluster (R, D) into small pieces and show that the "density" of each piece is at most δ_{oct}. To prepare for this proof, this section describes various small geometric solids that have a density at most δ_{oct}. The first of these is the Rogers simplex.

Lemma 8.5. *Let $R(a, b, c)$ be a Rogers simplex, with $1 \leq a < b < c$. It has a distinguished vertex (the terminal point of the edges of lengths a, b, and c), which we assume to be the origin. Let $A(a, b, c)$ be the volume of the intersection of $R(a, b, c)$ with a ball of radius 1 at the origin. Then the ratio*

$$A(a, b, c)/\text{vol}(R(a, b, c))$$

is monotonically decreasing in each variable.

Proof. This is Rogers's lemma, as formulated in Lemma 8.6 of [Ha6]. □

Lemma 8.6. *Consider the Rogers simplex $R(a, b, \sqrt{2})$ with vertex at the origin. Assume $1 \leq a \leq b$ and $\eta(2, 2, 2) \leq b \leq \sqrt{2}$. Let A be the volume of the intersection of the simplex with a closed ball of radius 1 at the origin. Then*

$$A \leq \delta_{\text{oct}}\, \text{vol}(R(a, b, \sqrt{2})).$$

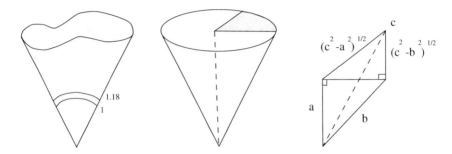

Fig. 8.1. Some sets of low density.

Equality is attained if and only if $a = 1$ and $b = \eta(2, 2, 2)$ or for a degenerate simplex of zero volume.

Proof. This is a special case of Lemma 8.5. See the third frame of Fig. 8.1. □

Lemma 8.7. *Consider the wedge of a cone*

$$W = W(\alpha, z_0) = \{t\,x \colon 0 \le t \le 1, x \in P(\alpha, z_0)\} \subset \mathbb{R}^3,$$

where $P(\alpha, z_0)$ has the form

$$P = \{(x_1, x_2, x_3) \colon x_3 = z_0, \ x_1^2 + x_2^2 + x_3^2 \le 2, \ 0 \le x_2 \le \alpha x_1\},$$

with $z_0 \ge 1$. Let A be the volume of the intersection of the wedge with $B(0, 1)$. Then

$$A \le \delta_{\mathrm{oct}} \operatorname{vol}(W).$$

Equality is attained if and only if W has zero volume.

Proof. This is calculated in Section 4 of [Ha7]. See the second frame of Fig. 8.1. □

Lemma 8.8. *Let C be the cone at the origin over a set P, where P is measurable and every point of P has distance at least 1.18 from the origin. Let A be the volume of the intersection of C with $B(0, 1)$. Then*

$$A \le \delta_{\mathrm{oct}} \operatorname{vol}(C).$$

Equality is attained if and only if C has zero volume.

Proof. The ratio $A/\operatorname{vol}(C)$ is at most $1/1.18^3 < \delta_{\mathrm{oct}}$. See the first frame of Fig. 8.1. □

8.3.   *Bounds on Simplices*

In this and future sections, we rely on some inequalities that are not proved in this paper. There is an archive of hundreds of inequalities that have been proved by computer. This full archive appears in [Ha16]. The justification of these inequalities appears in the same archive. (The proofs of these inequalities were executed by computer.) An explanation of how computers are able to prove inequalities can be found in [Ha14] and [Ha6]. Each inequality carries a nine-digit identifying number. To invoke an inequality, we state it precisely, and give its identifying number, e.g., CALC-123456789. The first of these appears in Lemma 8.10. Some results rely on a simple combination of inequalities, rather than a single inequality. To make it easier to reference a group of inequalities, the archive at [Ha16] gives a separate nine-digit identifier to certain groups of inequalities. This permits us to reference such a group by a single number.

Definition 8.9.   Recall that the constant pt, a *point*, is equal to $\sigma(S)$, where S is a regular tetrahedron with edges of length 2. We have $pt = 4\arctan(\sqrt{2}/5) - \pi/3 \approx 0.05537$.

Lemma 8.10.   *A quasi-regular tetrahedron S satisfies $\sigma(S) \leq 1\,pt$. Equality occurs if and only if the quasi-regular tetrahedron is regular of edge length 2.*

Proof.   This is CALC-586468779.   □

Remark 8.11.   The reader who wishes to dig more deeply into this particular proof may do so. An early published proof of this lemma was not fully automated (see Lemma 9.1.1 of [Ha6]). This early proof show by conventional means that $\sigma(S) \leq 1\,pt$ in an explicit neighborhood of $(2, 2, 2, 2, 2, 2)$.

Lemma 8.12.   *A quarter in the Q-system scores at most 0. That is, $\sigma(Q) \leq 0$. Equality is attained if and only if five edges have length 2 and the diagonal has length $\sqrt{8}$.*

Proof.   Throughout the proof of this lemma, we refer to quarters with five edges of length 2 and one of length $\sqrt{8}$ as *extremal* quarters. We make use of the definition of σ on quarters from Definition 7.8. The general context (that is, contexts other than $(2, 1)$ and $(4, 0)$) of upright quarters is established by the inequalities[4] that hold for all upright quarters Q with distinguished vertex v:

$$2\Gamma(Q) + \text{s-vor}_0(Q, v) - \text{s-vor}_0(Q, \hat{v}) \leq 0,$$

$$\text{s-vor}(Q, v) + \text{s-vor}(Q, \hat{v}) + \text{s-vor}_0(Q, v) - \text{s-vor}_0(Q, \hat{v}) \leq 0.$$

Equality is attained if and only if the quarter is extremal. For the remaining quarters (that is, contexts $(2, 1)$ and $(4, 0)$), it is enough to show that $\Gamma(Q) \leq 0$, if $\eta^+ \leq \sqrt{2}$ and s-vor$(Q, v) \leq 0$, if $\eta^+ \geq \sqrt{2}$.

---

[4] CALC-522528841 and CALC-892806084.

Consider the case $\eta^+ \leq \sqrt{2}$. If Q is a quarter such that every face has circumradius at most $\sqrt{2}$, then[5] $\Gamma(Q) \leq 0$. Equality is attained if and only if the quarter is extremal. Because of this, we may assume that the circumradius of Q is greater than $\sqrt{2}$. The inequality $\eta^+(Q) \leq \sqrt{2}$ implies that the faces of Q along the diagonal have nonnegative orientation. The other two faces have positive orientation, by Lemma 5.17. Since (Definition 7.6)

$$4\Gamma(Q) = \sum_{i=1}^{4} \text{s-vor}(Q, v_i),$$

it is enough to show that s-vor$(Q) < 0$. Since the orientation of every face is nonnegative and the circumradius is greater than $\sqrt{2}$, s-vor$(Q, \sqrt{2})$ is a strict truncation of the V-cell in Q, so that

$$\text{s-vor}(Q) < \text{s-vor}(Q, \sqrt{2}).$$

We show the right-hand side is nonpositive. Let v be the distinguished vertex of Q. Let A be $\frac{1}{3}$ the solid angle of Q at v. By the definition of s-vor$(Q, \sqrt{2})$, it is nonpositive if and only if

$$A \leq \delta_{\text{oct}} \text{vol}(\text{VC}(Q, v) \cap B(v, \sqrt{2})). \tag{8.1}$$

(VC$(Q, 0)$ is defined in Section 7.1.) The intersection VC$(Q, v) \cap B(v, \sqrt{2})$ consists of six Rogers simplices $R(a, b, \sqrt{2})$, three conic wedges (extending out to $\sqrt{2}$), and the intersection of $B(v, \sqrt{2})$ with a cone over v. By Lemmas 8.6–8.8, these three types of solids give inequalities like that of (8.1). Summing the inequalities from these lemmas, we get (8.1).

Consider the case $\eta^+ \geq \sqrt{2}$ and $\sigma = \text{s-vor}$. If the quarter is upright, then[6] s-vor$(Q) \leq 0$. The quarters achieving equality are extremal. Thus, we may assume the quarter is flat. If the orientation of a flat quarter is negative along the face containing the origin and the diagonal, then[7] s-vor$(Q) \leq 0$. The quarters achieving equality are extremal. In the remaining case, the only possible face along which the orientation is negative is the top face. This means that the analytic continuation defining s-vor(Q) is the same as

$$4(-\delta_{\text{oct}} \text{vol}(X) + \text{sol}(X)/3),$$

where X is the subset of the cone at v over Q consisting of points in that cone closer to v than to any other vertex of Q. The extreme point of X has distance at least $\sqrt{2}$ from v (since η^+ and hence the circumradius of Q are at least $\sqrt{2}$). Thus,

$$\text{s-vor}(Q) \leq \text{s-vor}(Q, \sqrt{2}).$$

We have s-vor$(Q, \sqrt{2}) \leq 0$ as in the previous paragraph, by Lemma 8.6–8.8. If equality is attained, the wedges and cones must have zero volume, and each Rogers simplex must

---

[5] CALC-346093004.
[6] CALC-40003553.
[7] CALC-5901405.

have the form $R(1, \eta(2, 2, 2), \sqrt{2})$ (or zero volume). This happens exactly when the flat quarter has five edges of length 2 and a diagonal of length $\sqrt{8}$. This completes the proof. □

Lemma 8.13. *Let S be a simplex all of whose faces have circumradius at most $\sqrt{2}$. Assume that S is not a quasi-regular tetrahedron or quarter. Then s-vor$(S) < 0$.*

Proof. The assumptions imply that the orientation is positive along each face. Let v be the distinguished vertex of S.

Assume first that there are at least two edges of length at least $2t_0$ at the origin or that there are two opposite edges of length at least $2t_0$. Then the circumradius b of each of the three faces at v is at least $\eta(2, 2t_0, 2) > 1.207$. By the monotonicity properties of the circumradius of S, the simplex S has circumradius at least that of the simplex $S(2, 2, 2, 2, 2, 2t_0)$, which a calculation shows is greater than 1.3045. By definition, s-vor$(S) < 0$ if and only if

$$\text{sol}(|S| \cap B(v, 1))/3 < \delta_{\text{oct}} \text{vol}(\text{VC}(S, 0)).$$

This inequality breaks into six separate inequalities corresponding to the six Rogers's simplices $R(a, b, c)$ constituting $\text{VC}(S, 0)$. Rogers's lemma, Lemma 8.5, shows each of the six Rogers's simplices has density at most that of $R(1, 1.207, 1.3045)$, which is less than δ_{oct}. The result follows in this case.

Now assume that there is at most one edge of length at least $2t_0$ at the origin, and that there is not a pair of opposite edges of length at most $2t_0$. There are four cases up to symmetry, depending on which edges have length at least $2t_0$, and which have shorter length. Let S be a simplex such that every face has circumradius at most $\sqrt{2}$. We have[8] s-vor$(S(y_1, y_2, \ldots, y_6)) < 0$ for $(y_1, \ldots, y_6)$ in any of the following four domains:

$$[2t_0, \sqrt{8}][2, 2t_0]^3[2t_0, \sqrt{8}][2, 2t_0], \qquad [2t_0, \sqrt{8}][2, 2t_0]^3[2t_0, \sqrt{8}]^2,$$
$$[2, 2t_0]^3[2t_0, \sqrt{8}]^2[2, 2t_0], \qquad [2, 2t_0]^3[2t_0, \sqrt{8}]^3. \qquad \square$$

8.4. *Breaking Clusters into Pieces*

As we stated above, the strategy in the proof of local optimality will be to break quad clusters into smaller pieces and then to show that each piece has density at most δ_{oct}. There are several preliminary lemmas that will be used to prove that this decomposition into smaller pieces is well-defined. These lemmas are presented in this section.

Lemma 8.14. *Let T be a triangle whose circumradius is less than $\sqrt{2}$. Assume that none of its edges passes through a barrier in $\mathcal{B}$. Then T does not overlap any barrier in $\mathcal{B}$.*

---

[8] CALC-629256313, CALC-917032944, CALC-738318844, and CALC-587618947.

Proof. By hypothesis no edge of T passes through an edge in the barrier. By Lemma 4.21, no edge of a barrier passes through T. Hence they do not overlap. □

Lemma 8.15. *Let $T = \{u, v, w\}$ be a set of three vertices whose circumradius is less than $\sqrt{2}$. Assume that one of its edges $\{v, w\}$ passes through a barrier $b = \{v_1, v_2, v_3\}$ in $\mathcal{B}$. Then:*

- *The edge $\{v, w\}$ has length between $2t_0$ and $\sqrt{8}$.*
- *The vertex u is a vertex of b.*
- *One of the endpoints $y \in \{v, w\}$ is such that $\{y, v_1, v_2, v_3\}$ is a simplex in $\mathcal{Q}$.*

Proof. The edge $\{v, w\}$ must have length at least $2t_0$ by Lemma 4.19. If the edge $\{u, v\}$ has length at least $2t_0$, it cannot pass through b because of Lemma 4.33. If it has length at most $2t_0$, it cannot pass through b because of Lemma 4.19. Hence $\{u, v\}$ and similarly $\{u, w\}$ do not pass through b. The edges of b do not pass through T. The only remaining possibility is for u to be a vertex of b.

If b is a quasi-regular triangle, Lemma 4.22 gives the result. If b is a face of a quarter in the Q-system, then Lemma 4.34 gives the result. □

Definition 8.16 (Law of Cosines). Consider a triangle with sides a, b, and c. The angle opposite the edge of length c is given as

$$\operatorname{arc}(a, b, c) = \arccos((a^2 + b^2 - c^2)/(2ab)) = \frac{\pi}{2} + \arctan \frac{c^2 - a^2 - b^2}{\sqrt{u(a^2, b^2, c^2)}}$$

$$\text{with} \quad u(x, y, z) = -x^2 - y^2 - z^2 + 2xy + 2yz + 2zx.$$

Lemma 8.17 (First Separation Lemma). *Let v be a vertex of height at most $\sqrt{8}$. Let v_2 and v_3 be such that*

- *$0, v, v_2$, and v_3 are distinct vertices, and*
- *$\eta(0, v_2, v_3) < \sqrt{2}$.*

Then the open cone at the origin over the set $B(0, \sqrt{2}) \cap B(v, \sqrt{2})$ does not meet the closed cone C at the origin over the convex hull of $\{v_2, v_3\}$.

Proof. Let D be the open disk spanning the circle of intersection of $B(0, \sqrt{2})$ and $B(v, \sqrt{2})$. It is enough to show that this disk does not meet C. This disk is contained in $B(v, \sqrt{2})$, and so we bound this ball away from the given cone.

Assume for a contradiction that these two sets meet. Let v' be the reflection of v through the plane $P = \{0, v_2, v_3\}$.

If the closest point p in P to v lies outside C, then the edge constraints $|v| \leq \sqrt{8}$ forces the closest point in C to lie along the edge $\{0, v_2\}$ or $\{0, v_3\}$. Since $|v_2|, |v_3| \leq \sqrt{8}$, this closest point has distance at least $\sqrt{2}$ from v. Thus, we may assume that the closest point in P to v lies in C.

Assume next that the closest point in P to v lies in the convex hull of $0, v_2$, and v_3. We obtain an edge $\{v, v'\}$ of length at most $\sqrt{8}$ that passes through a triangle of circumradius less than $\sqrt{2}$. This contradicts Lemma 4.21.

Assume finally that the closest point lies in the cone over $\{v_2, v_3\}$ but not in the convex hull of 0, v_2, v_3. By moving v toward C (preserving $|v|$), we may assume that $|v-v_2| = |v-v_3| = 2$. Stretching the edge $\{v_2, v_3\}$, we may assume that the circumradius of $\{0, v_2, v_3\}$ is precisely $\sqrt{2}$. Since the closest point in P is not in the convex hull of $\{0, v_2, v_3\}$, we may move v_2 and v_3 away from v while preserving the circumradius and increasing the lengths $|v - v_2|$ and $|v - v_3|$. By moving v again toward C, we may assume without loss of generality that $|v_2| = |v_3| = 2$ and $|v_2 - v_3| = \sqrt{8}$. We have reduced to a one-parameter family of arrangements, parametrized by $|v|$. We observe that the disk in the statement of the lemma is tangent to the segment $\{v_2, v_3\}$ at its midpoint, no matter what the value of $|v|$ is. Thus, in the extremal case, the open disk does not intersect the segment $\{v_2, v_3\}$ or the cone C that it generates. This completes the proof. □

Lemma 8.18 (Second Separation Lemma). *Let v_1 be a vertex of height at most $2t_0$. Let v_2 and v_3 be such that*

- *0, v_1, v_2, and v_3 are distinct vertices,*
- *$\{0, v_1, v_2, v_3\} \notin Q_0$, and*
- *$\{0, v_2, v_3\}$ is a barrier.*

Then the open cone at the origin over the set $B(0, \sqrt{2}) \cap B(v_1, \sqrt{2})$ does not meet the closed cone C at the origin over $\{v_2, v_3\}$.

Proof. Since v_1 has height at most $2t_0$, and $\{0, v_2, v_3\}$ is a barrier, it follows from Lemma 4.10 that $\{0, v_1, v_2, v_3\}$ is in the Q-system if $|v_1 - v_2| \leq 2t_0$ and $|v_1 - v_3| \leq 2t_0$. This is contrary to the hypothesis. Thus, we may assume without loss of generality that $|v_1 - v_2| > 2t_0$.

By arguing as in the proof of Lemma 8.17, we may assume that the orthogonal projection of v_1 to the plane P is a point in the cone C. Let v_1' be the reflection of v_1 through C. We have that either $\{v_2, v_3\}$ passes through $\{0, v_1, v_1'\}$ or $\{v_1, v_1'\}$ passes through $\{0, v_2, v_3\}$. We may assume for a contradiction that $|v_1 - v_1'| < \sqrt{8}$.

If $\{v_2, v_3\}$ passes through $\{0, v_1, v_1'\}$, then v_2 and v_3 are anchors of the diagonal $\{v_1, v_1'\}$ by Lemma 4.24. This gives the contradiction $|v_1 - v_2| \leq 2t_0$.

If $\{v_1, v_1'\}$ passes through $\{0, v_2, v_3\}$, then by Lemma 4.22 $\{0, v_2, v_3\}$ is a face of a quarter. Moreover, v_1 and v_1' are anchors of the diagonal of that quarter by Lemma 4.24. Since $|v_1 - v_2| > 2t_0$, the diagonal must not have v_2 as an endpoint, so that the diagonal is $\{0, v_3\}$. Lemma 4.34 forces one of $|v_1 - v_2|$ or $|v_1' - v_2|$ to be at most $2t_0$. However, these are both equal to $|v_1 - v_2| > 2t_0$, a contradiction. □

Definition 8.19. We define an enlarged set of simplices Q_0'. Let Q_0' be the set of simplices S with a vertex at the origin such that either $S \in Q_0$, or S is a simplex with a vertex at the origin and with circumradius less than $\sqrt{2}$ such that none of its edges passes through a barrier.

Lemma 8.20. *The simplices in Q_0' do not overlap one another.*

Proof. The simplices in $\mathcal{Q}_0$ are in the Q-system and do not overlap. No edge of length less than $\sqrt{8}$ passes through any edge of a simplex in $\mathcal{Q}'_0 \setminus \mathcal{Q}_0$, by Lemma 4.21. By construction, none of the edges of a simplex in $\mathcal{Q}'_0 \setminus \mathcal{Q}_0$ can pass through a barrier, and this includes all the faces of $\mathcal{Q}_0$. Thus, there is no overlap.   □

Definition 8.21. Let v be a vertex of height at most $2.36 = 2(1.18)$. Let $C(v)$ be the cone at the origin generated by the intersection $B(v, \sqrt{2}) \cap B(0, \sqrt{2})$. Define a subset $C'(v)$ of $C(v)$ by the conditions:

1. $x \in C(v)$.
2. x is closer to 0 than to v.
3. $x \in B(0, \sqrt{2})$.
4. x does not lie in the cone over any simplex in $\mathcal{Q}_0$.
5. For every vertex $u \neq 0, v$ such that the face $\{0, u, v\}$ is a barrier or has circumradius less than $\sqrt{2}$ and such that none of the edges of this face pass through a barrier, we have that x and v lie in the same half-space bounded by the plane perpendicular to $\{0, u, v\}$ and passing through 0 and the circumcenter of $\{0, u, v\}$.
6. For every simplex $\{0, v_1, v_2, v\} \in \mathcal{Q}_0$, the segment $\{x, v\}$ does not cross through the cone $C(\{0, v_1, v_2\})$.

Lemma 8.22. *For every vertex v of height at most 2.36, we have $C'(v) \subset \mathrm{VC}(0)$.*

Proof. Assume for a contradiction that $x \in C'(v) \cap \mathrm{VC}(u)$, with $u \neq 0$. Lemma 5.20 implies that x is unobstructed at 0. Thus $|x - u| < |x| \leq \sqrt{2}$.

Assume that the hypotheses of Condition 5 in Definition 8.21 are satisfied. This, together with $x \in C(v)$ implies that $\eta(\{0, u, v\}) < \sqrt{2}$. An element x that is closer to 0 than to v and in the same half-space as v (in the half-space bounded by the perpendicular plane to $\{0, u, v\}$ through 0 and the circumcenter of $\{0, u, v\}$) is closer to 0 than to u, which is contrary to $x \in \mathrm{VC}(u)$. This completes the proof, except in the case that an edge of the triangle $\{0, u, v\}$ passes through a barrier b. Assume that this is so.

The edge $\{0, v\}$ cannot pass through a barrier because it is too short (length less than $2t_0$).

Suppose that the edge $\{u, v\}$ passes through a barrier b. By Lemma 8.15 applied to $T = \{0, u, v\}$, the origin is a vertex of b. There are three possibilities:

1. x is obstructed from u by b.
2. x is obstructed from v by b.
3. x is not obstructed from either u or v by b.

The first possibility runs contrary to the hypothesis $x \in \mathrm{VC}(u)$. The second possibility, together with Lemma 8.18, implies that $\{v, b\}$ is a simplex in the Q-system. This is contrary to Condition 6 defining $C'(v)$.

The third possibility is eliminated as follows. Every point in the half-space containing v and bounded by the plane of b

- is obstructed at u by b, or
- has distance at least $\sqrt{2}$ from u (because each edge of b has this property).

Since x has neither of these properties, we find that x must lie in the same half-space bounded by the plane of b as u. Let S be the simplex formed by b and v. If $S \notin \mathcal{Q}_0$, then Lemma 8.18 shows that no part of the cone $C(v)$ lies in the same half-space as u. So $S \in \mathcal{Q}_0$. By Condition 6 on $C'(v)$, the line from x to v does not intersect the cone at the origin over b. However, then the arc-length of the geodesic on the unit sphere running from the projection of x to the projection of v is at least $\mathrm{arc}(|v|, \sqrt{8}, 2) \geq \mathrm{arc}(|v|, \sqrt{2}, \sqrt{2})$. This measurement shows that x lies outside the cone $C(v)$, which is contrary to assumption.

Suppose that the edge $\{0, u\}$ passes through the barrier b. By Lemma 8.15 applied to $T = \{0, u, v\}$, we get that v is a vertex of b. There are again three possibilities:

1. x is obstructed from u by b.
2. x is not obstructed from either u or 0 by b.
3. x is obstructed from 0 by b.

The first possibility runs contrary to the hypothesis $x \in \mathrm{VC}(u)$. The second places x outside the convex hull of 0, b, u and gives $|x - u| + |x| > \sqrt{8}$, which is contrary to $|x - u| \leq |x| \leq \sqrt{2}$. The third possibility cannot occur by the observation made at the beginning of the proof that x is unobstructed at 0. $\square$

It follows from the definition that $C'(v)$ is star convex at the origin. We make this more explicit in the following lemma.

Lemma 8.23. *Assume $|v| \leq 2.36$. Let $F(v)$ be the intersection of $\Omega(0) \cap \Omega(v)$; that is, the face of the Voronoi cell of $\Omega(0)$ associated with the vertex v. Let $F'(v)$ be the part of $F(v) \cap B(0, 1.18)$ that is not in the cone over any simplex in $\mathcal{Q}_0$. Let $H(v)$ be the closure of the union of segments from the origin to points of $F'(v)$. Let $C''(v)$ be the cone at the origin spanned by $B(0, 1.18) \cap B(v, 1.18)$. Then the closure of $C'(v) \cap C''(v)$ is equal to $H(v)$.*

Proof. We have $F'(v) \subset C''(v)$.

First we show that $F'(v)$ lies in the closure of $C'(v)$. For this, we check that points of $F'(v)$ satisfy the (closed counterparts of) Conditions 1–6 defining $C'(v)$ (see Definition 8.21). Conditions 1–4 are immediate from the definitions. If u is a vertex as in Condition 5, then the half-space it determines is that containing the origin and the edge of the Voronoi cell determined by u and v. Condition 5 now follows. Consider Condition 6. Suppose that $\{x, v\}$ crosses the cone $\{0, v_1, v_2\}$ and that $x \in F'(v)$. (The point of intersection has height at most $\sqrt{2}$ and hence lies in the convex hull of $\{0, v_1, v_2\}$.) This implies that x is obstructed at v. By Lemma 5.22, this implies that $|x - v| \geq t_0$. Since x is equidistant from v and the origin, we find that $|x| \geq t_0$, which is contrary to $x \in B(0, 1.18)$.

To finish the proof, we show that $C'(v) \cap C''(v) \subset H(v)$. For a contradiction, consider a point $x \in C'(v) \cap C''(v)$ that is not in $H(v)$. It must lie in the cone over some other face of the Voronoi cell; say that of u. The constraints force the circumradius of $T = \{0, v, u\}$ to be at most 1.18. The edges of T are too short to pass through a barrier. Thus, Condition 5 defining $C'(v)$ places a bounding plane that is perpendicular to T and that runs through the origin and the circumcenter of T. This prevents x from lying in the cone over the face of the Voronoi cell attached to u. $\square$

Remark 8.24. In the lemma, it is enough to consider simplices along $\{0, w\}$, because

$$\text{arc}(|v|, \sqrt{8}, 2) > \text{arc}(|v|, 1.18, 1.18).$$

Corollary 8.25. *If $x \in \text{VC}(0)$, with $0 < |x| \le 1.18$, if the point at distance 1.18 from 0 along the ray $(0, x)$ does not lie in $\text{VC}(0)$, and if x is not in the cone over any simplex of $\mathcal{Q}_0$, then there is some v such that $x \in C'(v)$, and $|v| \le 2.36$.*

Proof. If $x \in \text{VC}(0) \cap B(0, 1.18)$, then $x \in \Omega(0) \cap B(0, 1.18)$ by Lemma 5.23. Also, x lies in the cone over some face $F(v)$ of the Voronoi cell $\Omega(0)$. The hypotheses imply that x lies in the cone over $F'(v)$. Lemma 8.23 implies that $x \in C'(v)$.   □

Lemma 8.26. *Assume that $|u| \le 2.36$ and that $|v| \le 2.36$. The sets $C'(u), C'(v)$ do not overlap for $u \ne v$.*

Proof. If there is some x in the overlap, then the circumradius of $\{0, u, v\}$ is less than $\sqrt{2}$. If no edge of $\{0, u, v\}$ passes through a barrier, then the defining conditions of $C'(u)$ and $C'(v)$ separate them along the plane perpendicular to $\{0, u, v\}$ and passing through the origin and the circumcenter of $\{0, u, v\}$.

  If some edge of $\{0, u, v\}$ passes through a barrier, then an argument like that in the proof of Lemma 8.22 shows they do not overlap. In fact, the edges $\{0, u\}$ and $\{0, v\}$ are too short to pass through a barrier. Suppose the edge $\{u, v\}$ passes through a barrier b. By Lemma 8.15 applied to $T = \{0, u, v\}$, the origin is a vertex of b. If neither of the simplices $\{u, b\}$ and $\{v, b\}$ are in $\mathcal{Q}_0$, then the plane through b separates $C'(u)$ from $C'(v)$. Assume without loss of generality that $S = \{v, b\} \in \mathcal{Q}_0$. By Condition 6 of the definition of C' (Definition 8.21), the segment from x to v does not intersect the cone at the origin formed by b. As in the proof of Lemma 8.22, x lies outside the cone $C(v)$; unless x and v lie in the same half-space formed by the plane of b. The cone $C(u)$ intersects this half-space at x. By Lemma 8.18, we have $\{u, b\} \in \mathcal{Q}_0$. Condition 6 in the definition of C' now keeps x at distance at least $\sqrt{2}$ from u. This completes the proof. □

Lemma 8.27. *Let S be a simplex whose circumradius is less than $\sqrt{2}$. If five of the six edges of the simplex do not pass through a barrier, then the sixth edge e does not pass through a barrier either, unless both endpoints of the edge opposite e in S are vertices of the barrier.*

Proof. We leave this as an exercise. The point is that it is impossible to draw the barrier without having one of its edges pass through a face of S, which is ruled out by Lemma 4.21.   □

8.5. Proofs

We are finally prepared to give a proof of Theorem 8.4. We break the proof into two lemmas.

Lemma 8.28. *If R is a standard region that is not a triangle, then $\sigma_R(D) \leq 0$.*

Proof. This proof is an adaptation of the main result in [Ha7, Theorem 4.1]. We consider the V-cell at a vertex, which we take to be the origin. We will partition the V-cell into pieces. On each piece it will be shown that σ is nonpositive.

Throughout the proof we make use of the correspondence between $\sigma_R(D) \leq 0$ and the bound of δ_{oct} on densities, on standard regions R (away from simplices in the Q-system). This correspondence is evident from Lemma 7.18, which gives the formula

$$\sigma_R(D) = 4\left(-\delta_{\mathrm{oct}} \operatorname{vol} \operatorname{VC}_{R'}(D) + \operatorname{sol}(R')/3\right) + \sum_{Q \in \mathcal{Q}_0(R,D)} \sigma(Q, c(Q, D), 0).$$

If $\sigma(Q, c(Q, D), 0) \leq 0$, and $\operatorname{vol} \operatorname{VC}_{R'}(D) \neq 0$ then $\sigma_R(D) \leq 0$ follows from the inequality

$$(\operatorname{sol}(R')/3)/\operatorname{vol} \operatorname{VC}_{R'}(D) \leq \delta_{\mathrm{oct}}.$$

This is an assertion about the ratio of two volumes, that is, a bound δ_{oct} on the density of $\operatorname{VC}_{R'}(D)$.

The parts of $\operatorname{VC}(D)$ that lie in the cone over some simplex in $\mathcal{Q}_0$ are easily treated. If S is in $\mathcal{Q}_0$, then it is either a quasi-regular tetrahedron or a quarter. If it is a quasi-regular tetrahedron, it is excluded by the hypothesis of the lemma. If it is a quarter, $\sigma(S) \leq 0$ by Lemma 8.12. The parts of $\operatorname{VC}(D)$ that lie in the cone over some simplex in $\mathcal{Q}_0' \setminus \mathcal{Q}_0$ are also easily treated. The simplex $S = \{0, v, w, w'\}$ has circumradius less than $\sqrt{2}$. Use s-vor(S) on the simplex. Lemma 8.13 shows that s-vor$(S) < 0$ as desired.

Next we consider the parts of $\operatorname{VC}(D)$ that are not in any $C'(v)$ (with $|v| \leq 2.36$) and that are not in any cone over a simplex in $\mathcal{Q}_0$. (Note that by Lemmas 8.17 and 8.18, if a cone over some simplex in $\mathcal{Q}_0'$ meets $C'(v)$, then v must be a vertex of that simplex.) By Corollary 8.25, if x belongs to this set, then all the points out to radius 1.18 in the same direction belong to this set. By Lemma 8.8, the density of such parts is less than δ_{oct}.

Finally, we treat the parts of $\operatorname{VC}(D)$ that are in some $C'(v)$ but that lie outside all cones over simplices in $\mathcal{Q}_0'$.

Fix v of height at most 2.36. Let $w_1, w_2, \ldots, w_k$ be the vertices w near $\{0, v\}$ such that either $\{0, v, w\}$ is a barrier or it has circumradius less than $\sqrt{2}$, and such that none of its edges passes through a barrier. We view the triangles $\{0, v, w_i\}$ as a fan of triangles around the edge $\{0, v\}$. We assume that the vertices are indexed so that consecutive triangles in this fan have consecutive indices (modulo k). We will analyze the densities separately within each wedge, where a wedge is the intersection along the line $\{0, v\}$ of half-spaces bounded by the half-planes $\{0, v, w_i\}$ and $\{0, v, w_{i+1}\}$. Space is partitioned by these k different wedges. Fix i and write $w = w_i$, $w' = w_{i+1}$. Let $S = \{0, v, w, w'\}$.

Let F be the convex planar region in the perpendicular bisector of $\{0, v\}$ defined by the points inside the closure of $C'(v)$, inside the wedge between $\{0, v, w\}$ and $\{0, v, w'\}$, closer to v than to w, and closer to v than to w'. This planar region is illustrated in Fig. 8.2. The edge e lies in the line perpendicular to $\{0, v, w\}$ and through the circumcenter of $\{0, v, w\}$. It extends from the circumcenter out to distance $\sqrt{2}$ from the vertices 0, v, w. If the circumradius of $\{0, v, w\}$ is greater than $\sqrt{2}$, the edge e reduces to a point, and only the arc a at distance $\sqrt{2}$ from 0 and v appears. Similar comments apply to e'.

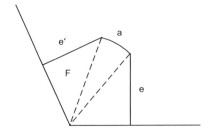

Fig. 8.2.   A planar region.

Case 1: *Circumradius of S Is Less than* $\sqrt{2}$. We show that this case does not occur. If none of the edges of this simplex pass through a barrier, then this simplex belongs to $\mathcal{Q}'_0$, a case already considered. By definition of the wedges, the edges $\{0, v\}$, $\{0, w\}$, $\{0, w'\}$, $\{v, w\}$, and $\{v, w'\}$ do not pass through a barrier. Since five of the six edges do not pass through a barrier, and since S is formed by consecutive triangles in the fan around $\{0, v\}$, the sixth does not pass through a barrier either, by Lemma 8.27.

Case 2: *Circumradius of S Is at Least* $\sqrt{2}$. Let $r \geq \sqrt{2}$ be the circumradius. We claim that the edge e cannot extend beyond the wedge through the half-plane through $\{0, v, w'\}$. In fact, the circumcenter of $\{0, v, w, w'\}$ lies on the extension (in one direction or the other) of the segment e to a point at distance r from the origin. If this circumcenter does not lie in the wedge, then the orientation is negative along one of the faces $\{0, v, w\}$ or $\{0, v, w'\}$. This face must have circumradius at least $\sqrt{2}$, by Lemma 5.18, and this forces the face to be a barrier. If the orientation is negative along a barrier, then the simplex $\{0, v, w, w'\}$ is a simplex in $\mathcal{Q}_0$ (Lemmas 5.16 and 5.17). This is contrary to our assumption above that $\{0, v, w, w'\}$ is not in $\mathcal{Q}_0$.

These comments show that Fig. 8.2 correctly represents the basic shape of F, with the understanding that the edges e and e' may degenerate to a point. By construction, every point x in the open convex hull $\{F, 0\}$ of F and 0 lies in $C'(v) \subset VC(0)$. The convex hull $\{F, 0\}$ is the union of three solids, two Rogers simplices along the triangles $\{0, v, w\}$ and $\{0, v, w'\}$ respectively, and the conic solid given by the convex hull of the arc a, $v/2$ and 0. By Lemmas 8.6 and 8.7, these solids have density at most δ_{oct}.

This completes the proof that $\sigma_R(D)$ is never positive on nontriangular standard regions R. Note that the decomposition into the parts of cones $C'(v)$ inside a wedge is compatible with the partition of the unit sphere into standard regions, so that the estimate holds over each standard region, and not just over the union of the standard regions. $\square$

Lemma 8.29.   *If R is a standard region that is not a triangle, and if $\sigma_R(D) = 0$, then (R, D) is a quad cluster. Moreover, the four corners of R in the quad cluster have height 2, forming a square of side 2.*

Proof.   To analyze the case of equality, first we note that any truncation at 1.18 produces a strict inequality (Lemma 8.8 is strict if the volume is nonzero), so that every point must lie over a simplex in $\mathcal{Q}'_0$ or over some $C'(v)$. We have s-vor$(S) < 0$ for simplices with circumradius less than $\sqrt{2}$. The only simplices in $\mathcal{Q}_0$ that produce equality are those with five edges of length 2 and a diagonal of length $\sqrt{8}$. Any nontrivial arc a produces strict inequality (see Lemma 8.7), so we must have that e and e' meet at exactly distance $\sqrt{2}$ from 0 and v. Moreover, if e does not degenerate to a point, the corresponding Rogers simplex gives strict inequality, unless $\{0, v, w\}$ is an equilateral triangle with side length 2. We conclude that the entire part of the V-cell over the standard region must be assembled from Rogers simplices $R(1, \eta(2, 2, 2), \sqrt{2})$, and quarters with lengths $(2, 2, 2, 2, 2, \sqrt{8})$. This forces each vertex v of height at most $2t_0$ to have height 2. It forces each pair of triangles $\{0, v_1, v_2\}$ $\{0, v_2, v_3\}$, that determine consecutive edges along the boundary of the standard region to meet at right angles:

$$\mathrm{dih}(0, v_2, v_1, v_3) = 0.$$

This forces the object to be a quad cluster of the indicated form. $\qquad\square$

We conclude the section with a proof of the main theorem. With all our preparations in place, the proof is short.

Proof of Theorem 8.1 (Local Optimality).   The hypothesis implies that there are six quad clusters and eight quasi-regular tetrahedra at the origin of the decomposition star. By Lemma 8.10, each quasi-regular tetrahedron scores at most $1\,pt$ with equality if and only if the tetrahedron is regular with edge-length 2. By Theorem 8.4, each quad cluster scores at most 0, with equality if and only if the corners of the quad cluster form a square with edge-length 2 at distance 2 from the origin. Thus, $\sigma(D)$ is at most $8\,pt$. In the case of equality, there are twelve vertices at distance 2 from the origin, forming eight equilateral triangles and six squares (all of edge-length 2). These conditions are satisfied precisely when the arrangement is U_{fcc} or U_{hcp} up to a Euclidean motion. $\qquad\square$

9.   The $\mathcal{S}$-System

9.1.   *Overview*

In this section we define a decomposition of a V-cell. Let VC be the V-cell at the origin. For any $t > 0$, let $V(t)$ be the intersection of VC with the ball $B(0, t)$ at the origin of radius t. We write VC as the disjoint union of $V(t_0)$ and its complement δ.

Assume that there is an upright quarter in the Q-system with diagonal $\{0, v\}$. As usual, we call $\{0, v\}$ an *upright diagonal*. We will define $\delta(v) \subset \delta$. It will be a subset of a set of the form $C(D_v) \cap \delta$ for some subset D_v of the unit sphere. The sets D_v will be defined so as not to overlap one another for distinct v. Then the sets $\delta(v)$ do not overlap one another either. We will give an explicit formula for the volume of $\delta(v)$.

We will define a set $\mathcal{S}$ of simplices, each having a vertex at the origin. (The letter "$\mathcal{S}$" is for simplex.) The vertices of the simplices will be vertices of the packing, and their edges will have length at most $2\sqrt{2}$. The sets $C(S)$, for distinct $S \in \mathcal{S}$, will not overlap. Over a simplex $S \in \mathcal{S}$, the V-cell will be truncated at a radius $t_S \geq t_0$. After defining the constants t_S, we will set

$$V_S(t_S) = C(S) \cap V(t_S) = C(S) \cap B(t_S) \cap \mathrm{VC}(0).$$

That is, $V_S(t_S)$ is the part of the V-cell at the origin, contained in the cone over S and in the ball of radius t_S. If $VC(0) \cap C(S) \subset B(t_S) \subset B(t'_S)$, then $V_S(t_S) = V_S(t'_S)$.

Since $t_S \geq t_0$, the sets $V_S(t_S)$ and δ may overlap. Nevertheless, we will show that $V_S(t_S)$ does not overlap any $\delta(v)$. Let $V^S(t_0)$ be the set of points in $V(t_0)$ that do not lie in $C(S)$, $S \in \mathcal{S}$. We will derive an explicit formula for the volume of $V^S(t_0)$.

In $VC(0)$, there are nonoverlapping sets

$$\delta(v), \quad V_S(t_S), \quad V^S(t_0).$$

Let δ' be the complement in $VC(0)$ of the union of these sets. These sets give a decomposition of $VC(0)$. Corresponding to this decomposition is a formula for $\sigma(D)$ of the form

$$\sigma(D) = \text{c-vor}(V^S(t_0)) + \sum_S \text{c-vor}(V_S(t_S)) - \sum_v 4\delta_{\text{oct}} \text{vol}(\delta(v)) - 4\delta_{\text{oct}}\text{vol}(\delta').$$

Since $\text{vol}(\delta') \geq 0$, we obtain an upper bound on $\sigma(D)$ by dropping the rightmost term.

9.2. The Set $\delta(v)$

Let $\{0, v\}$ be the diagonal of an upright quarter in $\mathcal{Q}_0$. We define $\delta(v) \subset C(D_v) \cap \delta$ for an appropriate subset D_v of the unit sphere.

Definition 9.1. Set $\eta_0(h) = \eta(2h, 2, 2t_0)$.

If $h \leq \sqrt{2}$, then $\eta_0(h) \leq \eta_0(\sqrt{2}) < 1.453$.

Let D_0 be the spherical cap on the unit sphere, centered along $\{0, v\}$ and having arcradius θ, where $\cos\theta = |v|/(2\eta_0(|v|/2))$.

The area of D_0 is $2\pi(1 - \cos\theta)$. Let $v_1, \ldots, v_k$ be the anchors around $\{0, v\}$ indexed cyclically. The radial projections of the edges $\{v, v_i\}$ (extended as necessary) slice the spherical cap into k wedges W_i, between $\{v, v_i\}$ and $\{v, v_j\}$, where $j \equiv i + 1 \mod k$, so that $D_0 = \bigcup W_i$.

Definition 9.2. Let $\mathcal{W}$ be the set of wedges $W = W_i$ such that either

1. W occupies more than half the spherical cap (so that its area is at least $\pi(1 - \cos\theta)$), or

2. $|v_i - v_j| \geq 2.77$, $\text{rad}(0, v, v_i, v_j) \geq \eta_0(|v|/2)$, and the circumradius of $\{0, v_i, v_j\}$ or $\{v, v_i, v_j\}$ is $\geq \sqrt{2}$.

Fix i, j, with $j \equiv i + 1 \mod k$. If $W = W_i$ is a wedge in $\mathcal{W}$, let $\{0, v_i, v\}^{\perp}$ be the plane through the origin and the circumcenter of $\{0, v_i, v\}$, perpendicular to $\{0, v_i, v\}$. Skip the following step if the circumradius of $\{0, v_i, v\}$ is greater than $\eta_0(|v|/2)$, but if the circumradius is at most this bound, let c_i be the intersection of $\{0, v_i, v\}^{\perp}$ with the circular boundary of W. Extend W by adding to W the spherical triangle with vertices the radial projections of v, v_i, and c_i. Similarly, extend W with the triangle from $\{v, v_j, c_j\}$,

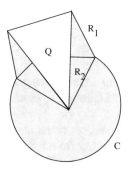

Fig. 9.1. An example of a set W^e (shaded region).

if the circumradius of $\{0, v_j, v\}$ permits. (An example of this is illustrated in Fig. 9.1.) Let W^e be the extension of the wedge obtained by adding these two spherical triangles.

We will define $\delta(v, W^e) \subset C(W^e) \cap \delta$. Then $\delta(v)$ is defined as the union of $\delta(v, W^e)$, for $W \in \mathcal{W}$. Let

$$E_w = \{x : 2x \cdot w \le w \cdot w\},$$

for $w = v, v_i, v_j$. These are half-spaces bounding the Voronoi cell. Set $E_\ell = E_{v_\ell}$.

If Condition 2 holds, we let c be the radial projection of the circumradius of $\{0, v_i, v_j, v\}$ to the unit sphere. The arclength from c to the radial projection of v is θ', where

$$\cos\theta' = |v|/(2\,\text{rad}) < |v|/(2\eta_0) = \cos\theta.$$

We conclude that $\theta' > \theta$ and c does not lie in D_0.

Definition 9.3. In both Conditions 1 and 2, set   ✳

$$\Delta(v, W^e) = (E_v \cap E_i \cap E_j \cap C(W^e)),$$
$$\delta(v, W^e) = \Delta(v, W^e) \backslash B(t_0).$$

Remark 9.4. There are some degenerate cases in this construction depending on the number of anchors. If there is no anchor, then $\Delta(v, W^e)$ is to be defined simply as $(E_v \cap C(W^e))$. If there is one anchor v_i, then

$$\Delta(v, W^e) = (E_v \cap E_i \cap C(W^e)).$$

Remark 9.5. The following remark applies when the points c_i and c_j have been constructed, and is irrelevant when that step was skipped in the construction described above. Observe that

$$E_v \cap E_i \cap E_j \cap C(W^e)$$

is the union of four Rogers simplices

$$R(|w|/2, \eta(0, v, v_\ell), \eta_0(|v|/2)), \qquad w = v, v_\ell, \quad \ell = i, j,$$

and a conic wedge over W between c_i and c_j. (The inequality $\theta' > \theta$ implies that the Rogers simplices do not overlap.)

In general, we break $\Delta(v, W^e)$ into an inner part $\Delta^-(v, W^e)$ (the part outside the Rogers simplices together with (as many as) two Rogers simplices along $(0, v)$), and the Rogers simplices R_w, for $w = v_i, v_j$. We take R_w to be the empty set, when there is no anchor w with $\eta(0, v, w) < \eta_0(|v|/2)$.

We present a series of lemmas that explore the geometry of the sets $\Delta(v, W^e)$. In the next few lemmas we make use of a function ε, which is defined as follows.

Definition 9.6. If Λ is a set of vertices containing v, let $\varepsilon_v(\Lambda, x) \in \Lambda$ be given as the vertex $w \in \Lambda \backslash \{v\}$ such that the ray from v through x intersects the perpendicular bisecting plane of $\{v, w\}$ before that of any other $w' \in \Lambda \backslash \{v\}$. If the ray from v through x does not intersect any of the planes, then we set ε to the default value v. In cases of ties, resolve the tie in any consistent manner. If $x \in \Omega(0)$ (the Voronoi cell at the origin), then x lies in the cone over the face attached to the vertex $\varepsilon_0(\Lambda, x) \in \Lambda$.

We define a function ε' in a similar fashion. Assume $\varepsilon_v(\Lambda, x) = w$, where the ray from v to x intersects the perpendicular bisector to $\{v, w\}$ at x'. Set

$$\varepsilon_v'(\Lambda, x) = \varepsilon_{w/2}(\Lambda \backslash \{w\}, x').$$

That is, move along the face of the Voronoi cell from x until another face is encountered. Let the corresponding vertex be the value of ε'. If $x \in \Omega(0)$ in the cone over the face attached to the vertex w, and if $w/2$ lies on that face, then x' lies in the sector of the face formed by the cone at $w/2$ generated by the edge of the Voronoi cell between the faces associated to w and $\varepsilon_0'(\Lambda, x)$.

Lemma 9.7. Let $S = \{0, v, w, u\}$ be a simplex. Assume that $\{0, v\}$ is an upright diagonal of a quarter in the Q-system, that w and v are anchors of $\{0, v\}$, and that $\mathrm{rad}(S) < \eta_0(|v|/2)$. Assume there is a wedge W of $\mathcal{W}$ along the face $\{0, v, w\}$ (on the same side of the face as u). Let R_w be the Rogers simplex $R(|w|/2, \eta(0, v, w), \eta_0(|v|/2))$ along the face $\{0, w, v\}$ along the edge of $\{0, w\}$ on the same side of the face as u. Then:

1. There exists an anchor w' between u and w with $|w - w'| \le 2t_0$, and $|w' - w| \ge 2.77$.
2. $\{0, v, u, w'\}$ is an upright quarter in the Q-system and its face $\{0, v, w'\}$ is a barrier.
3. The barrier $\{0, v, w'\}$ obstructs every point of R_w from u.

Proof. Since $\mathrm{rad}(S) < \eta_0(|v|/2)$ is contrary to the conditions defining wedges, the wedge must run from the face $\{0, v, w\}$ to a face $\{0, v, w'\}$, where w' is an anchor between w and u. By the hypotheses defining wedges $W \in \mathcal{W}$, we have that the length of $\{u, w'\}$ is at least 2.77. For the same reason, the circumradius of $\{0, v, w, w'\}$ is at least $\eta_0(|v|/2)$.

We claim that R_w lies in the convex hull of $S = \{0, v, w, w'\}$. Since $|w - w'| \ge 2.77$, we see that the orientation of each face of $\{0, v, w, w'\}$ is positive. Since $\mathrm{rad}(S) \ge \eta_0$, we have

$$R_w \subset R(|w|/2, \eta(0, v, w), \mathrm{rad}(S)) \subset S.$$

Thus it is enough to show that each point of the convex hull of S is obstructed.

For this, it is enough to show that the extreme point w is obstructed from u by the barrier $\{0, v, w'\}$. In other words, we show that the edge $\{w, v\}$ passes through $\{0, v, w'\}$. For this we show that no other geometrical configuration of points is possible.

w' is not in the convex hull of $\{0, v, w, u\}$ by Lemma 4.15. The vertex w' is not enclosed over $\{0, v, w, u\}$ because

$$\mathcal{E}(S(2, 2, 2t_0, 2t_0, 2t_0, 2(1.453)), 2, 2, 2) > 2t_0.$$

(The constant 1.453 is from Definition 9.1.) The edge $\{v, w'\}$ does not pass through $\{0, w, u\}$, for otherwise we would reach a contradiction

$$2\eta_0(|v|/2) > 2\,\mathrm{rad}(S) \geq |u - w| \geq \mathcal{E}(S(2, 2, 2, \sqrt{8}, 2, 2), 2, 2, 2.77)$$

$$\geq 2(1.453) \geq 2\eta_0(|v|/2).$$

We conclude that $\{u, w\}$ passes through $\{0, v, w'\}$.   □

Lemma 9.8. *Let $F = \{0, u_1, u_2\}$ be a quasi-regular triangle. Let $\{0, v\}$ be the diagonal of an upright quarter in the Q-system. The set $\Delta(v, W^e)$ does not overlap the cone at 0 over the triangle F.*

Proof. We prove the lemma for the subsets $\Delta^-(v, W^e)$ and R_w in two separate cases, beginning with $\Delta^-(v, W^e)$. Let S be the simplex $\{0, u_1, u_2, v\}$.

Assume that the orientation of S along F is negative. The simplex S is an upright quarter, so that u_1 and u_2 are anchors of v. This is contrary to the construction of the wedges W in $\mathcal{W}$. Thus, the orientation of F must be positive.

Assume that $\mathrm{rad}(S) < \eta_0(|v|/2)$. Then again, u_1 and u_2 are anchors. It follows that S is an upright quarter. As in the previous paragraph, this is contrary to construction. Thus, $\mathrm{rad}(S) \geq \eta_0(|v|/2)$.

We now have that the orientation of F is positive and that $\mathrm{rad}(S) \geq \eta_0(|v|/2)$. These two facts allow us to separate $\Delta^-(v, W^e)$ from $\{0, u_1, u_2, v\}$ as follows. Each interior point x of $\Delta^-(v, W^e)$ has

$$\varepsilon_0(\{0, u_1, u_2, v\}, x) = v.$$

Let F_0 be the set of points in the intersection of $\Omega(0)$ with the convex hull of F. Each point y in F_0 has

$$\varepsilon_0(\{0, u_1, u_2, v\}, y) \in \{u_1, u_2\}.$$

Since ε_0 takes distinct values on these two sets, they are disjoint.

Next consider the subset R_w, in the case that $w \in \{u_1, u_2\}$. To be definite, assume that $w = u_1$. The same argument as above establishes that the orientation of F is positive and that $\mathrm{rad}(S) \geq \eta_0(|v|/2)$. Each interior point x of R_w has

$$\varepsilon_0'(\{0, u_1, u_2, v\}, x) = v.$$

However, each point y in F_0 has

$$\varepsilon_0'(\{0, u_1, u_2, v\}, y) \in \{u_1, u_2\}.$$

This proves this case.

Until the end of the proof, we may assume that $w \notin \{u_1, u_2\}$. If the value of $\varepsilon_0(\{0, u_1, u_2, w\}, \cdot)$ is w on interior points of R_w and in $\{u_1, u_2\}$ on F_0, then we have separated the sets. Assume to the contrary, that ε takes value w at a point x of F_0. Let $S'' = \{0, u_1, u_2, w\}$. This assumption implies that the orientation of S'' is negative along F. This in turn implies that S'' is a quasi-regular tetrahedron. The vertex v is not enclosed over S'', because the simplices in the Q-system do not overlap. For similar reasons, the face $\{0, v, w\}$ does not overlap the simplex S''. By the previous case (when $w \in \{u_1, u_2\}$), the interior of R_w does not intersect the faces of the quasi-regular tetrahedron S'' along the edge $\{0, w\}$. These facts imply that the interior of R_w is disjoint from S''. In particular, it does not meet the cone at 0 over the triangle F.

Assume finally that ε takes value u_1 or u_2 at an interior point y of R_w. (Say u_1.) Let $S' = \{0, v, w, u_1\}$. Assume that R_w lies on the same side of $\{0, v, w\}$ as u_1. If $\mathrm{rad}(S') < \eta_0(|v|/2)$, then each point of R_w is obstructed from u_1 (by Lemma 9.7). However, no point of F_0 is obstructed from u_1. Thus, R_w and F_0 are disjoint in this case. Assume that $\mathrm{rad}(S') \geq \eta_0(|v|/2)$. If the orientation of $\{0, v, w\}$ in S' is negative, then S' is a quarter and the result follows. Assume that the orientation is positive. Now $\varepsilon_0(S', x) = w$ for $x \in R_w$, contrary to assumption.

We may now assume that R_w lies outside the simplex S' on the opposite side of the face $\{0, v, w\}$. This case is dismissed by Lemma 5.31, which guarantees that the interior of R_w with $\varepsilon = u_1$ are obstructed from u_1 by $\{0, v, w\}$. However, none of the points of F_0 with $\varepsilon = u_1$ are obstructed from u_1. $\qquad\square$

Corollary 9.9. *Each $\Delta(v, W^e)$ lies entirely in the cone over the standard region that contains $\{0, v\}$.*

Proof. The cone over a standard region is bounded by the cones over the quasi-regular triangles. $\qquad\square$

Lemma 9.10. *Let $F = \{0, u_1, u_2\}$ be a triangle. Assume that $|u_1| \leq 2t_0$, $|u_2| \leq 2t_0$, and $2t_0 \leq |u_1 - u_2| \leq \sqrt{8}$. Let $\{0, v\}$ be the diagonal of an upright quarter in the Q-system. Assume that if u_1 and u_2 are both anchors of v, then they are consecutive anchors around v. Under these conditions, the set $\Delta(v, W^e)$ does not overlap the cone at 0 over the triangle F.*

Proof. The proof is similar to that of Lemma 9.8. We prove the lemma for the subsets $\Delta^-(v, W^e)$ and R_w in two separate cases, beginning with $\Delta^-(v, W^e)$. Let S be the simplex $\{0, u_1, u_2, v\}$. The orientation of S along F is positive.

Assume that $\mathrm{rad}(S) < \eta_0(|v|/2)$. Then u_1 and u_2 are anchors. By the hypotheses of the lemma, they are consecutive anchors. By the rules defining $\Delta^-(v, W^e)$, there is no wedge W^e between u_1 and u_2. Thus, the result follows in this case.

We now have that the orientation of F is positive and that $\mathrm{rad}(S) \geq \eta_0(|v|/2)$. These two facts allow us to separate $\Delta^-(v, W^e)$ from the cone over $\{0, u_1, u_2, v\}$ as follows. Each interior point x of $\Delta^-(v, W^e)$ has

$$\varepsilon_0(\{0, u_1, u_2, v\}, x) = v.$$

Let F_0 be the intersection of $\Omega(0)$ with the convex hull of F. Each point y in F_0 has

$$\varepsilon_0(\{0, u_1, u_2, v\}, y) \in \{u_1, u_2\}.$$

Next consider the subset R_w, in the case that $w \in \{u_1, u_2\}$. To be definite, assume that $w = u_1$. The same argument as above establishes that the orientation of F is positive and that $\mathrm{rad}(S) \geq \eta_0(|v|/2)$. Each interior point x of R_w has

$$\varepsilon_0'(\{0, u_1, u_2, v\}, x) = v.$$

However, each point y in F_0 has

$$\varepsilon_0'(\{0, u_1, u_2, v\}, y) \in \{u_1, u_2\}.$$

This proves this case.

Consider the subset R_w, in the case that $w \notin \{u_1, u_2\}$. As in the previous proof, if the value of $\varepsilon_0(\{0, u_1, u_2, w\}, \cdot)$ is w on interior points of R_w and in $\{u_1, u_2\}$ on F_0, then we have separated the sets. Assume first to the contrary, that ε_0 takes value w at a point x of F_0. Let $S'' = \{0, u_1, u_2, w\}$. Our assumption on ε_0 implies that the orientation of S'' is negative along F, so that S'' is a flat quarter. The vertex v cannot be enclosed over S'', for otherwise w, u_1, and u_2 would all be anchors of v, which would mean that there is no region $W \in \mathcal{W}$. Similarly, the triangle $\{0, v, w\}$ cannot overlap the triangle $\{0, u_1, u_2\}$, for otherwise w, u_1, and u_2 would again be anchors, contrary to the hypothesis that u_1 and u_2 are consecutive anchors. Now we invoke Lemma 9.8, to establish that R_w does not intersect S'' and is therefore disjoint from F_0.

Assume finally, that ε_0 takes value u_1 or u_2 at an interior point y of R_w. (Say u_1.) This case is identical to the parallel case in the proof of Lemma 9.8. □

Lemma 9.11. *Let $F = \{0, u_1, u_2\}$ be a triangle. Assume that $2t_0 \leq |u_1| \leq \sqrt{8}$, $2 \leq |u_2| \leq 2t_0$, and $2 \leq |u_1 - u_2| \leq 2t_0$. Let $\{0, v\}$ be the diagonal of an upright quarter in the Q-system. Under these conditions, the set $\Delta(v, W^e)$ does not overlap the cone at 0 over the triangle F.*

Proof. Let $S = \{0, u_1, u_2, v\}$. The orientation of S along $\{0, u_1, u_2\}$ is positive. The circumradius $\mathrm{rad}(S)$ is at least $\eta_0(0, v, u_1) \geq \eta_0(|v|/2)$.

We now have that the orientation of F is positive and that $\mathrm{rad}(S) \geq \eta_0(|v|/2)$. We can then argue as in the proof of Lemmas 9.8 and 9.10, to get the result for $\Delta^-(v, W^e)$ and R_w (with $w = u_2$).

Consider the case R_w, with $w \neq u_2$. As in these earlier proofs, we may assume that ε_0 takes value w at a point x of F_0 (or that ε_0 takes value in $\{u_1, u_2\}$ at a point y of R_w).

In the case $\varepsilon_0 = w$ at $x \in F_0$, let $S'' = \{0, u_1, w, u_2\}$. We have that $\varepsilon_0 = w$ implies that the orientation of S'' along $\{0, u_1, u_2\}$ is negative. This in turn implies that S'' is an upright quarter. It is checked without difficulty that v is not enclosed over S'' and that the face $\{0, w, v\}$ does not cross the face $\{0, u_1, u_2\}$. It follows from Lemma 9.8 and the already treated cases of this lemma that R_w cannot intersect S''. Thus, it does not intersect the face $\{0, u_1, u_2\}$ of S''.

Finally, assume that ε_0 takes value in $\{u_1, u_2\}$ at a point y of R_w. The orientation of the face $\{0, v, w\}$ is positive in the simplex $\{0, v, w, u_1\}$ and the circumradius of

$\{0, v, w, u_1\}$ is greater than $\eta_0(|v|/2)$. This implies that ε_0 does not take the value u_1. Assume that $\varepsilon_0 = u_2$. This case is excluded in the same manner as the parallel case in the earlier Lemmas 9.8 and 9.10. □

Lemma 9.12. *Let $\{0, v\}$ be an upright diagonal of a quarter in the Q-system. If x lies in the interior of $\Delta(v, W^e)$, then x is unobstructed at 0.*

Proof. For a contradiction, assume that x is obstructed at 0 by barrier $T = \{u_1, u_2, u_3\}$.

The convex hull of T can be partitioned into three sets $T(i)$ depending on which vertex of T is closest to a given point in the convex hull. (Ties can be resolved in any consistent manner.) Let $y \in \Delta(v, W^e)$ be the point in the convex hull of T on the segment from 0 to x. Fix i so that $y \in T(i)$. If $v = u_i$, then each point y of $T(i)$ is closer to v than to 0. However, each point of $\Delta(v, W^e)$ is closer to 0 than to v. So x is not obstructed by T at 0.

We may now assume that $v \neq u_i$.

Partition $\mathbb{R}^3$ geometrically into three sets $V(u_i)$, $V(0)$, $V(v)$ according to which of $\{u_i, 0, v\}$ a point $z \in \mathbb{R}^3$ is closest to. (Again resolve ties in any consistent manner.)

∗ Assume further that $\max_j u_j \geq 2t_0$. This implies that $y \in T(i) \subset V(v) \cup V(u_i)$. On the other hand, we have by construction that $y \in \Delta(v, W^e) \subset V(0)$. (There are two cases involved in this conclusion, depending on whether u_i is an anchor of $\{0, v\}$.) However, the sets $V(\cdot)$ are disjoint; and we reach a contradiction. Thus, under these assumptions, x is unobstructed at 0.

∗ Next assume that $\max_j u_j < 2t_0$. Let $S = \{0, u_1, u_2, u_3\}$. Since T is a barrier, $S \in \mathcal{Q}_0$. By assumption, $\{0, v\}$ is a diagonal of an upright quarter in $\mathcal{Q}_0$. By the nonoverlap of quarters in $\mathcal{Q}_0$, we see that v is not enclosed over S. The wedge W^e on the unit sphere is spherically star convex with respect to the center $v/|v|$. Thus, if $\Delta(v, W^e)$ intersects the convex hull of T at y, then $\Delta(v, W^e)$ intersects the cone over a face $\{0, u_1, u_2\}$ of S at y'. (We can take $y'/|y'|$ to lie on the cone generated by the arc running from $v/|v|$ to $y/|y|$. This is impossible by Lemmas 9.8 and 9.10. □

Lemma 9.13. *Let $\{0, v\}$ be the upright diagonal of a quarter in the $\mathcal{Q}_0$-system. Then the interior of $\Delta(v, W^e)$ is a subset of $VC(0)$.*

Proof. We begin by showing that $\Delta^-(v, W^e) \subset VC(0)$. Suppose to the contrary, that a point x in the interior of Δ^- lies in $VC(w)$, with $w \neq 0$. Then x is closer to w than to 0. Thus, $\eta(0, v, w) < \eta_0(|v|/2)$, and w is an anchor of $\{0, v\}$. The face E_w in the construction $\Delta^-(v, W^e)$ prevents this from happening.

Now consider a point x of R_w, which we assume to lie in $VC(u)$, with $u \neq 0$. To avoid a trivial case, we may assume that $w \neq u$.

Assume that the orientation of $S = \{0, v, w, u\}$ is negative along the face $\{0, v, w\}$. Then S must be an upright quarter. By the construction of wedges $W \in \mathcal{W}$, we have that R_w must lie on the opposite side of the plane $\{0, v, w\}$ from u (for there is no wedge between the anchors of an upright quarter). The result now follows from Lemma 5.31.

If $\mathrm{rad}(S) < \eta_0(|v|/2)$, then u and w are anchors. In this case the result follows from Lemma 9.7.

Finally, if the orientation is positive and if $\operatorname{rad}(S) \geq \eta(|v|/2)$, then a point of R_w cannot be closer to u than to 0. $\qquad\qquad\square$

9.3. Overlap

Lemma 9.14. *The sets $\Delta(v, W^e)$ do not overlap one another.*

Proof. This is clear for two sets around the same vertex v. Consider the sets $\Delta(u, W^e)$ and $\Delta(v, W^e)$ at u and v.

To treat the points in $\Delta^-(u, W^e)$ and $\Delta^-(v, W^e)$, we may contract $\{u, v\}$ until $|u-v| = 2$. By the constraints on the edges of $\{0, u, v\}$, the circumcenter c of this triangle lies in the convex hull of the triangle. We have $\eta(0, u, v) \geq \eta_0(|v|/2)$ and $\eta(0, u, v) \geq \eta_0(|u|/2)$. So the plane through $\{0, c\}$ perpendicular to the plane $\{0, u, v\}$ separates $\Delta^-(u, W^e)$ from $\Delta^-(v, W^e)$.

Next we separate points in $\Delta^-(u, W^e)$ from points of $R_w^{(v)}$, where w is an anchor of v and $u \neq v$. Let $S = \{0, u, v, w\}$. The orientation of S along $\{0, v, w\}$ is positive. The circumradius of S satisfies

$$\operatorname{rad}(S) \geq \eta(0, u, v) > \eta_0(|v|/2).$$

Thus, $\varepsilon_0(S, \cdot)$ takes different values on $\Delta^-(u, W^e)$ and $R_w^{(v)}$, so that the sets are disjoint.

Next we separate points of $R_w^{(v)}$ from $R_w^{(u)}$. (Notice that we assume that the anchor is the same for the two Rogers simplices.) Let $S = \{0, u, v, w\}$. As above, we have

$$\operatorname{rad}(S) \geq \eta_0(|v|/2), \quad \eta_0(|w|/2).$$

The simplex S has positive orientation along the faces $\{0, u, w\}$ and $\{0, v, w\}$. Let c_u be the circumcenter of $\{0, u, w\}$, let c_v be the circumcenter of $\{0, v, w\}$, and let c be the circumcenter of S. Then $R_w^{(v)}$ lies in the convex hull of $\{0, w, c_v, c\}$, but $R_w^{(u)}$ lies in the convex hull of $\{0, w, c_u, c\}$. Thus, the sets are disjoint.

Finally, we separate points of $R_w^{(u)}$ from points of $R_{w'}^{(v)}$, where $w \neq w'$ and $u \neq v$. If the function $\varepsilon_0(\{0, w, w'\}, \cdot)$ separates the sets, we are done. Otherwise, we may assume say that $\varepsilon_0(\{0, w, w'\}, x) = w'$ from some $x \in R_w^{(u)}$. Let $S = \{0, u, w, w'\}$.

If w' is not an anchor of u, then $\operatorname{rad}(S) \geq \eta_0(|u|/2)$ and the orientation of S along $\{0, w, u\}$ is positive. In this case we have $\varepsilon_0 = w$ on $R_w^{(u)}$, which is contrary to the assumption. Thus, we may assume that w' is an anchor of u.

If the orientation of $\{0, u, w, w'\}$ is negative along $\{0, w, u\}$, then $\{0, u, w, w'\}$ is a quarter, contrary to the existence of $W \in \mathcal{W}$. So the orientation is positive. If $\operatorname{rad}(\{0, u, w, w'\}) < \eta_0(|u|/2)$, then Lemma 9.7 implies that each point of R_w is obstructed from w'. However, no point of $R_{w'}^{(v)}$ is obstructed from w. (In fact, a barrier that crosses $\Delta(v, W^e)$ is inconsistent with Lemmas 9.8, 9.10, and 9.11.) So $\operatorname{rad}(\{0, u, w, w'\}) \geq \eta_0(|u|/2)$. This is contrary to $\varepsilon_0(\{0, w, w'\}, x) = w'$ from some $x \in R_w^{(u)}$. $\qquad\qquad\square$

9.4. *The S-System Defined*

We consider three types of simplices: A, B, and C. Each type has its vertices at vertices of the packing. The edge lengths of these simplices are at most $2\sqrt{2}$.

Type A. This family consists of simplices $S(y_1, \ldots, y_6)$ whose edge lengths satisfy

$$y_1, y_2, y_3 \in [2, 2t_0], \qquad y_4, y_5 \in [2t_0, 2.77],$$

$$y_6 \in [2, 2t_0], \quad \text{and} \quad \eta(y_4, y_5, y_6) < \sqrt{2}.$$

(These conditions imply $y_4, y_5 < 2.697$, because $\eta(2.697, 2t_0, 2) > \sqrt{2}$.)

Type B. This family consists of certain flat quarters that are part of an isolated pair of flat quarters. It consists of those satisfying $y_2, y_3 \leq 2.23$, $y_4 \in [2t_0, 2\sqrt{2}]$.

Type C. This family consists of certain simplices $S(y_1, \ldots, y_6)$ with edge lengths satisfying $y_1, y_4 \in [2t_0, 2\sqrt{2}]$, $y_2, y_3, y_5, y_6 \in [2, 2t_0]$. We impose the condition that the first edge is the diagonal of some upright quarter in the Q-system, and that the upper endpoints of the second and third edges (that is, the second and third vertices of the simplex) are consecutive anchors of this diagonal. We also assume that $y_4 < 2.77$, or that both face circumradii of S along the fourth edge are less than $\sqrt{2}$.

Lemma 9.15. *If a vertex w is enclosed over a simplex S of type A, B, or C, then its height is greater than 2.77. Also, $\{0, w\}$ is not the diagonal of an upright quarter in the Q-system.*

Proof. In case A, $\eta(y_4, y_5, y_6) < \sqrt{2}$, so an enclosed vertex must have height greater than $2\sqrt{2}$. It is too long to be the diagonal of a quarter.

In case B we use the fact that the isolated quarter does not overlap any quarter in the Q-system. We recall that a function $\mathcal{E}$, defined in Section 4.2, measures the distance between opposing vertices in a pair of simplices sharing a face. An enclosed vertex has length at least

$$\mathcal{E}(S(2, 2, 2, 2\sqrt{2}, 2t_0, 2t_0), 2t_0, 2, 2) > 2.77.$$

By the symmetry of isolated quarters, this means that the diagonal of a flat quarter must also be at least 2.77.

In case C the same calculation gives that the enclosed vertex w has height at least 2.77. Let the simplex S be given by $\{0, v, v_1, v_2\}$, where $\{0, v\}$ is the upright diagonal. By Lemma 4.24, v_1 and v_2 are anchors of $\{0, w\}$. The edge between w and its anchor cannot cross $\{v, v_i\}$ by Lemma 4.19. (Recall that two sets are said to *cross* if their radial projections overlap.) The distance between w and v is at most $2t_0$ by Lemma 4.32. If $\{0, w\}$ is the diagonal of an upright quarter, the quarter takes the form $\{0, w, v_1, v_3\}$, or $\{0, w, v_2, v_3\}$ for some v_3, by Lemma 4.32. If both of these are quarters, then the diagonal $\{v_1, v_2\}$ has four anchors v, w, 0, and v_3. The selection rules for the Q-system place the quarters around this diagonal in the Q-system. So neither $\{0, w, v_1, v_3\}$ nor $\{0, w, v_2, v_3\}$ is in the Q-system. Suppose that $\{0, w, v_1, v_3\}$ is a quarter, but that $\{0, w, v_2, v_3\}$ is not. Then $\{0, w, v_1, v_3\}$ forms an isolated pair with $\{v_1, v_2, v, w\}$. In either case the quarters along $\{0, w\}$ are not in the Q-system. □

Remark 9.16. The proof of this lemma does not make use of all the hypotheses on C. The conclusion holds for any simplex $S(y_1, \ldots, y_6)$, with $y_1, y_4 \in [2t_0, 2\sqrt{2}]$, $y_2, y_3, y_5, y_6 \in [2, 2t_0]$.

9.5. *Disjointness*

Let $S = \{0, v_1, v_2, v_3\}$ be a simplex of type A, B, or C. An edge $\{v_4, v_5\}$ of length at most $2\sqrt{2}$ such that $|v_4|, |v_5| \le 2t_0$ cannot cross two of the edges $\{v_i, v_j\}$ of S. In fact, it cannot cross any edge $\{v_i, v_j\}$ with $|v_i|, |v_j| \le 2t_0$ by Lemma 4.30. The only possibility is that the edge $\{v_4, v_5\}$ crosses the two edges with endpoint v_1, with $|v_1| \ge 2t_0$ in case C. However, this too is impossible by Lemma 4.32.

Similar arguments show that the same conclusion holds for an edge $\{v_4, v_5\}$ of length at most $2t_0$ such that $|v_4| \le 2t_0$, $v_5 \le 2\sqrt{2}$. The only additional fact that is needed is that $\{v_4, v_5\}$ cannot cross the edge between the vertex v of an upright diagonal $\{0, v\}$ and an anchor (Lemma 4.19).

Lemma 9.17. *Consider two simplices S, S', each of type A, B, C, or a quarter in the Q-system. Assume that S and S' do not lie in the cone over a quadrilateral region. Then S and S' do not overlap.*

Proof. By hypothesis, the standard region is not a quadrilateral, and we thus exclude the case of conflicting diagonals in a quad cluster. We claim that no vertex w of S is enclosed over S'. Otherwise, w must have height at least $2t_0$, so that $\{0, w\}$ is the diagonal of an upright in the Q-system, and this is contrary to Lemma 9.15. Similarly, no vertex of S' is enclosed over S.

Let $\{v_1, v_2\}$ be an edge of S crossing an edge $\{v_3, v_4\}$ of S'. By the preceding remarks, neither of these edges can cross two edges of the other simplex. The endpoints of the edges are not enclosed over the other simplex. This means that one endpoint of each edge $\{v_1, v_2\}$ and $\{v_3, v_4\}$ is a vertex of the other simplex. This forces S and S' to have three vertices in common, say 0, v_2, and v_3. We have $S = \{0, v_1, v_3, v_2\}$ and $S' = \{0, v_3, v_2, v_4\}$. If $|v_2| \in [2t_0, 2\sqrt{2}]$, then we see that the anchors v_3, v_4 of $\{0, v_2\}$ are not consecutive. This is impossible for simplices of type C and upright quarters. Thus, v_2 and v_3 have height at most $2t_0$. We conclude, without loss of generality, that $|v_4| \in [2t_0, 2\sqrt{2}]$ and $|v_1 - v_2| \ge 2t_0$.

The heights of the vertices of S are at most $2t_0$, so it has type A or B, or it is a flat quarter in the Q-system. If S' is an upright quarter in the Q-system, then it does not overlap an isolated quarter or a flat quarter in the Q-system, so S has type A. This imposes the contradictory constraints on A:

$$2.77 \ge |v_1 - v_2| \ge \mathcal{E}(S(2t_0, 2, 2, 2\sqrt{2}, 2t_0, 2t_0), 2, 2, 2) > 2.77.$$

Thus S' has type C. This forces S to have type A. We reach the same contradiction $2.77 \ge \mathcal{E} > 2.77$. $\qquad\square$

9.6. *Separation of Simplices of Type A*

Let $V_S = VC(0) \cap C(S)$, for a simplex S of type A, B, or C. We truncate V_S to $V_S(t_S)$ by intersecting V_S with a ball of radius t_S. The parameters t_S depend on S.

If S has type A, we use $t_S = +\infty$ (no truncation). If v is enclosed over $S = \{0, v_1, v_2, v_3\}$, then since $\eta(v_1, v_2, v_3) < \sqrt{2}$, the face $\{v_1, v_2, v_3\}$ has positive orientation for S and $\{v, v_1, v_2, v_3\}$. This implies that the V-cells at v and 0 do not intersect, and there is no need to truncate. If a simplex adjacent to S has negative orientation along a face shared with A, then it must be a quarter $Q = \{0, v_4, v_1, v_2\}$ (Lemma 5.13) or a quasi-regular tetrahedron. It cannot be an isolated quarter because of the edge length constraint 2.77 on simplices of type A. If it is in the Q-system, the face between S and the adjacent simplex is a barrier, and it does not interfere with the V-cell over S. Assume that it is not in the Q-system. There must be a conflicting diagonal $\{0, w\}$, where w is enclosed over Q. (w cannot be enclosed over S by results of Lemma 9.17.) This shields the V-cell at v_4 from $C(S)$ by the two barriers $\{0, w, v_1\}$ and $\{0, w, v_2\}$ of quarters in the Q-system.

This shows that nothing external to a simplex of type A affects the shape of $V_S(t_S)$ (that is, $VC(0) \cap C(S)$ consists of points of S that are closer to 0 than to the other vertices of S). Thus, $V_S(t_S)$ can be computed from S alone. Similarly, $V_S(t_S)$ does not influence the external geometry, since all faces have positive orientation.

We also remark that $V_S(t_S)$ does not overlap any of the sets $\Delta(v, W^e)$. This is evident from Lemmas 9.8 and 9.10.

Our justification that $V_S(t_S)$ can be treated as an independently scored entity is now complete.

9.7. *Separation of Simplices of Type B*

If $S(y_1, \ldots, y_6)$ has type B, we label vertices so that the diagonal is the fourth edge, with length y_4. We set $t_S = 1.385$. The calculation of $\mathcal{E}$ in Lemma 9.15 shows that any enclosed vertex over S has height at least $2.77 = 2t_S$.

Vertices outside $C(S)$ cannot affect the shape of $V_S(t_S)$. In fact, such a vertex v' would have to form a quarter or quasi-regular tetrahedron with a face of S. The V-cell at v' cannot meet $C(S)$ unless it is a quarter that is not in the Q-system. However, by definition, an isolated quarter is not adjacent (along a face along the diagonal) to any other quarters.

To separate the scoring of $V_S(t_S)$ from the rest of the standard cluster, we also show that the terms of formula (7.12) for $V_S(t_S)$ are represented geometrically by solids that lie in the cone $C(S)$. This is more than a formality because S can have negative orientation along the face F formed by the origin and the diagonal (the fourth edge).

Definition 9.18. Let $\beta_\psi(\theta) \in [0, \pi/2]$ be defined by the equations

$$\cos^2 \beta_\psi = (\cos^2 \psi - \cos^2 \theta)/(1 - \cos^2 \theta), \qquad \text{for} \quad \psi \leq \theta.$$

Let p and q be points on the unit sphere separated by arclength θ. If we place a spherical cap of arcradius ψ on the unit sphere centered at p, then $\beta_\psi(\theta)$ is the angle at q between the arc (q, p) and the tangent to the cap which passes through q.

Let $S = \{0, v_1, v_2, v_3\}$, where v_i is the endpoint of the ith edge. We establish that the solids representing the conic and Rogers terms of formula (7.12) lie over $C(S)$ by showing[9] that $\beta_\psi(\text{arc}(y_1, y_3, y_5)) < \text{dih}_3(S(y_1, \ldots, y_6))$, where dih_3 is the dihedral angle along the third edge. We use $\cos \psi = y_1/2.77$ and assume $y_2, y_3 \in [2, 2.23]$.

The reasons given in Section 9.6 for the disjointness of $\delta(v)$ and $V_S(t_S)$ apply to simplices of type B as well. This completes the justification that $V_S(t_S)$ is an object that can be treated in separation from the rest of the local V-cell.

9.8. Separation of Simplices of Type C

If $S(y_1, \ldots, y_6)$ is of type C, we label vertices so that the upright diagonal is the first edge. We use $t_S = +\infty$ (no truncation). Each face of S has positive orientation by Lemma 5.13. So $V_S(t_S) \subset S$.

Vertices outside S cannot affect the shape of $V_S(t_S)$. Any vertex v' would have to form a quarter along a face of S. If the shared face lies along the first edge, it is a quarter Q in the Q-system, because one and hence all quarters along this edge are in the Q-system. The faces of this quarter are then barriers. If the shared face lies along the fourth edge, then its length is at most 2.77, so that the quarter cannot be part of an isolated pair. If it is not in the Q-system, there must be a conflicting diagonal. The two faces along this conflicting diagonal of the adjacent pair in the Q-system (that is, the pair taking precedence over Q in the Q-system) are barriers that shield the V-cell at v' from S.

The reasons given in Section 9.6 for the disjointness of $\delta(v)$ and $V_S(t_S)$ apply to simplices of type C as well. This completes the justification that $V_S(t_S)$ is an object that can be treated in separation from the rest of the local V-cell.

9.9. Simplices of Type C'

We introduce a small variation on simplices of type C, called type C'. We define a simplex $\{0, v, v_1, v_2\}$ of type C' to be one satisfying the following conditions:

1. The edge $\{0, v\}$ is an upright diagonal of an upright quarter in the Q-system.
2. $|v_2| \in [2.45, 2t_0]$.
3. v_1 and v_2 are anchors of v.
4. $|v - v_2| \in [2.45, 2t_0]$.
5. The edge $\{v_1, v_2\}$ is a diagonal of a flat quarter with face $\{0, v_1, v_2\}$.

It follows that v_1 and v_2 are consecutive anchors of $\{0, v\}$.

On simplices S of type C', we label vertices so that the upright diagonal is the first edge. We use $t_S = +\infty$ (no truncation). Each face of S has positive orientation by Lemma 5.13. So $V_S(t_S) \subset S$.

Simplices of type C' are separated from quarters in the Q-system and simplices of types A and B by procedures similar to those described for type C. The following lemma is helpful in this regard.

---

[9] CALC-193836552.

Lemma 9.19. *The flat quarter along the face $\{0, v_1, v_2\}$ is in the Q-system.*

Proof.

$$\mathcal{E}(S(2, 2, 2.45, 2\sqrt{2}, 2t_0, 2t_0), 2, 2, 2) > 2\sqrt{2},$$

so nothing is enclosed over the flat quarter.

$$\mathcal{E}(S(2, 2, 2, 2\sqrt{2}, 2t_0, 2t_0), 2t_0, 2.45, 2) > 2\sqrt{2},$$

so no edge between vertices of the packing can cross inside the anchored simplex. This implies that the flat quarter does not have a conflicting diagonal and is not part of an isolated pair.   □

Similar arguments show that there is not a simplex with negative orientation along the top face of S.

Unlike the other cases, there can in fact be overlap between $\Delta(v, W^e)$ and a simplex of type C', when the upright diagonal of the simplex is $\{0, v\}$. This is because the conditions defining a wedge $W \in \mathcal{W}$ are not incompatible with the conditions defining type C'. Nevertheless, except in the obvious case where the simplex of type C' and the wedge are both constructed between the same consecutive anchors of $\{0, v\}$, there can be no overlap of a $\Delta(v, W^e)$ with a simplex of type C'.

9.10. *Scoring*

The construction of the decomposition of the V-cell VC(0) is now complete. It consists of the pieces

- $\delta(v)$, for each diagonal $\{0, v\}$ of an upright quarter in the Q-system,
- truncations of Voronoi pieces $V_S(t_S)$ for simplices of type A, B, or C (and on rare occasion C'),
- $V^S(t_0)$, the truncation at t_0 of all parts of VC(0) that do not lie in any of the cones $C(S)$ over simplices of type A, B or C,
- δ', the part not lying in any of the preceding.

By the results of Sections 9.6–9.8, $\sigma(D)$ can be broken into a corresponding sum,

$$\sigma_R(D) = \sum_Q \sigma(Q) + \sigma(V_P), \qquad \text{for quarters } Q \text{ in the } Q\text{-system, where}$$

$$\sigma(V_P) = \text{c-vor}(V_P^S(t_0)) + \sum_{A,B,C} \text{c-vor}(V_S(t_S))$$

$$- \sum_v 4\delta_{\text{oct}} \text{vol}(\delta_P(v)) - 4\delta_{\text{oct}} \text{vol}(\delta_P').$$

By dropping the final term, $4\delta_{\text{oct}} \text{vol}(\delta_P')$, we obtain an upper bound on $\sigma(V_P)$. Because of the separation results of Sections 9.6–9.7, we may score $V_P^S(t_0)$ by formula (7.13). Bounds on the score of simplices of type B appear in CALC-193836552.

Lemma 9.20. *Let R be a standard region that is not a triangle in a decomposition start
D. $\tau_{0,R}(D) \geq 0$.*

Proof. Everything truncated at t_0 can be broken into three types of pieces: Rogers
simplices $R(a, b, t_0)$, wedges of t_0-cones, and spherical regions. (See Fig. 8.1.) The
wedges of t_0-cones and spherical regions can be considered as the degenerate cases $b = t_0$
and $a = b = t_0$ of Rogers simplices, so it is enough to show that $\tau(R(a, b, t_0)) \geq 0$. We
have $t_0 > \sqrt{3/2}$, so by Rogers's lemma [Ha6, Lemma 8.6.2],

$$\tau(R(a, b, t_0)) > \tau(R(1, \eta(2, 2, 2), \sqrt{3/2})).$$

The right-hand side is zero. (In fact, the vanishing of the right-hand side is essentially
Rogers's bound. When Rogers's bound is met, $\tau = 0$.)   □

10. Bounds on the Score in Triangular and Quadrilateral Regions

10.1. *The Function τ*

We consider the functions $\sigma_R(D) - \lambda \zeta \operatorname{sol}(R) pt$, for $\lambda = 0$, 1, or 3.2, where R is a
standard cluster. We write

$$\tau_R(D) = \operatorname{sol}(R)\zeta \, pt - \sigma_R(D).$$

We will see that $\tau_R(D)$ has a simple interpretation. If D is a decomposition star with
standard clusters $\{R\}$, set $\tau(D) = \sum_R \tau_R(D)$.

Lemma 10.1. $\tau_R(D) \geq 0$, *for all standard clusters R.*

Proof. If R is not a quasi-regular tetrahedron, then $\sigma_R(D) \leq 0$ by Theorem 8.4 and
$\operatorname{sol}(R) > 0$, so that the result is immediate. If R is a quasi-regular tetrahedron, the result
appears in the archive of inequalities CALC-53415898.   □

Lemma 10.2.

$$\sigma(D) = 4\pi \zeta \, pt - \tau(D).$$

Proof. Let $\{R\}$ be the standard clusters in D. Then

$$\sigma(D) = \sum_R \sigma_{R_i}(D) + \left(4\pi - \sum_R \operatorname{sol}(R_i)\right) \zeta \, pt = 4\pi \zeta \, pt - \sum_R \tau_{R_i}(D). \qquad □$$

Since $22.8 > 4\pi \zeta$ and $14.8 \, pt > 4\pi \zeta \, pt - 8 \, pt$, we find as an immediate corollary
that if there are standard clusters satisfying $\tau_{R_1}(D) + \cdots + \tau_{R_k}(D) \geq 14.8 \, pt$, then D
does not contravene.

The function $\tau_R(D)$ gives the amount *squandered* by a particular standard cluster
R. If nothing is squandered, then $\tau_{R_i}(D) = 0$ for every standard cluster, and the upper

bound is $4\pi\zeta\,pt \approx 22.8\,pt$. To say that a decomposition star does not contravene is to say that at least $(4\pi\zeta - 8)\,pt \approx 14.8\,pt$ are squandered.

Remark 10.3.   (This remark is not used elsewhere.) The bound $\sigma(D) \leq 4\pi\zeta\,pt$ implies Rogers's bound on density. It is the unattainable bound that would be obtained by packing regular tetrahedra around a common vertex with no distortion and no gaps. (More precisely, in the terminology of [Ha1], the score $s_0 = 4\pi\zeta\,pt$ corresponds to the *effective density* $16\pi\,\delta_{\mathrm{oct}}/(16\pi - 3s_0) = \sqrt{2}/\zeta \approx 0.7796$, which is Rogers's bound.) Every positive lower bound on some $\tau_R(D)$ translates into an improvement on Rogers's bound.

Lemma 10.4.   *A triangular standard region does not contain any enclosed vertices.*

Proof.   This fact is proved in [Ha6, Lemma 3.7].   □

10.2.   *Types*

Let v be a vertex of height at most $2t_0$. We say that v has *type* (p, q) if every standard region with a vertex at $\bar{v}$ (the radial projection of v) is a triangle or a quadrilateral, and if there are exactly p triangular faces and q quadrilateral faces that meet at $\bar{v}$. We write (p_v, q_v) for the type of v.

This section derives the bounds on the scores of the clusters around a given vertex as a function of the type of the vertex. Define constants $\tau_{\mathrm{LP}}(p, q)/pt$ by Table 10.1. The entries marked with an asterisk will not be needed.

Lemma 10.5.   *Let* $S_1, \ldots, S_p$ *and* $R_1, \ldots, R_q$ *be the tetrahedra and quad clusters around a vertex of type* (p, q). *Consider the constants of Table 10.1. Now,*

$$\sum^p \tau(S_i) + \sum^q \tau_{R_i}(D) \geq \tau_{\mathrm{LP}}(p, q).$$

Table 10.1. $\tau_{\mathrm{LP}}(p, q)/pt$

	q					
	0	1	2	3	4	5
$p = 0$	*	*	15.18	7.135	10.6497	22.27
1	*	*	6.95	7.135	17.62	32.3
2	*	8.5	4.756	12.9814	*	*
3	*	3.6426	8.334	20.9	*	*
4	4.1396	3.7812	16.11	*	*	*
5	0.55	11.22	*	*	*	*
6	6.339	*	*	*	*	*
7	14.76	*	*	*	*	*

Proof. Set

$$(d_i^0, t_i^0) = (\mathrm{dih}(S_i), \tau(S_i)), \qquad (d_i^1, t_i^1) = (\mathrm{dih}(R_i), \tau(R_i)).$$

The linear combination $\sum^p \tau(S_i) + \sum^q \tau_{R_i}(D)$ is at least the minimum of $\sum^p t_i^0 + \sum^q t_i^1$ subject to $\sum^p d_i^0 + \sum^q d_i^1 = 2\pi$ and to the system of linear inequalities CALC-830854305 and the system of linear inequalities CALC-940884472 (obtained by replacing τ and dihedral angles by t_i^j and d_i^j). The constant $\tau_{\mathrm{LP}}(p, q)$ was chosen to be slightly smaller than the actual minimum of this linear programming problem.

The entry $\tau_{\mathrm{LP}}(5, 0)$ is based on Lemma 10.6, $k = 1$. □

Lemma 10.6. *Let $v_1, \ldots, v_k$, for some $k \le 4$, be distinct vertices of a decomposition star of type $(5, 0)$. Let $S_1, \ldots, S_r$ be quasi-regular tetrahedra around the edges $\{0, v_i\}$, for $i \le k$. Then*

$$\sum_{i=1}^r \tau(S_i) > 0.55k\, pt$$

and

$$\sum_{i=1}^r \sigma(S_i) < r\, pt - 0.48k\, pt.$$

Proof. We have $\tau(S) \ge 0$, for any quasi-regular tetrahedron S. We refer to the edges y_4, y_5, y_6 of a simplex $S(y_1, \ldots, y_6)$ as its top edges. Set $\xi = 2.1773$.

The proof of the first inequalities relies on seven calculations.[10] Throughout the proof, we refer to these inequalities simply as inequality i, for $i = 1, \ldots, 7$.

We claim (Claim 1) that if $S_1, \ldots, S_5$ are quasi-regular tetrahedra around an edge $\{0, v\}$ and if $S_1 = S(y_1, \ldots, y_6)$, where $y_5 \ge \xi$ is the length of a top edge e on S_1 shared with S_2, then $\sum_1^5 \tau(S_i) > 3(0.55)\, pt$. This claim follows from inequalities 1 and 2 if some other top edge in this group of quasi-regular tetrahedra has length greater than ξ. Assuming all the top edges other than e have length at most ξ, the estimate follows from $\sum_1^5 \mathrm{dih}(S_i) = 2\pi$ and inequalities 3 and 4.

Now let $S_1, \ldots, S_8$ be the eight quasi-regular tetrahedra around two edges $\{0, v_1\}$, $\{0, v_2\}$ of type $(5, 0)$. Let S_1 and S_2 be the simplices along the face $\{0, v_1, v_2\}$. Suppose that the top edge $\{v_1, v_2\}$ has length at least ξ. We claim (Claim 2) that $\sum_1^8 \tau(S_i) > 4(0.55)\, pt$. If there is a top edge of length at least ξ that does not lie on S_1 or S_2, then this claim reduces to inequality 1 and Claim 1. If any of the top edges of S_1 or S_2 other than $\{v_1, v_2\}$ has length at least ξ, then the claim follows from inequalities 1 and 2. We assume all top edges other than $\{v_1, v_2\}$ have length at most ξ. The claim now follows from inequalities 3 and 5, since the dihedral angles around each vertex sum to 2π.

We prove the bounds for τ. The proof for σ is entirely similar, but uses the constant $\xi = 2.177303$ and seven new calculations[11] rather than the seven given above. Claims analogous to Claims 1 and 2 hold for the σ bound by this new group of seven inequalities.

---

[10] CALC-636208429.
[11] CALC-129662166.

Consider τ for $k = 1$. If a top edge has length at least ξ, this is inequality 1. If all top edges have length less than ξ, this is inequality 3, since dihedral angles sum to 2π.

We say that a top edge lies around a vertex v if it is an edge of a quasi-regular tetrahedron with vertex v. We do not require v to be the endpoint of the edge.

Take $k = 2$. If there is an edge of length at least ξ that lies around only one of v_1 and v_2, then inequality 1 reduces us to the case $k = 1$. Any other edge of length at least ξ is covered by Claim 1. So we may assume that all top edges have length less than ξ. Then the result follows easily from inequalities 3 and 6.

Take $k = 3$. If there is an edge of length at least ξ lying around only one of the v_i, then inequality 1 reduces us to the case $k = 2$. If an edge of length at least ξ lies around exactly two of the v_i, then it is an edge of two of the quasi-regular tetrahedra. These quasi-regular tetrahedra give $2(0.55)\,pt$, and the quasi-regular tetrahedra around the third vertex v_i give $0.55\,pt$ more. If a top edge of length at least ξ lies around all three vertices, then one of the endpoints of the edge lies in $\{v_1, v_2, v_3\}$, so the result follows from Claim 1. Finally, if all top edges have length at most ξ, we use inequalities 3, 6, and 7.

Take $k = 4$. Suppose there is a top edge e of length at least ξ. If e lies around only one of the v_i, we reduce to the case $k = 3$. If it lies around two of them, then the two quasi-regular tetrahedra along this edge give $2(0.55)\,pt$ and the quasi-regular tetrahedra around the other two vertices v_i give another $2(0.55)\,pt$. If both endpoints of e are among the vertices v_i, the result follows from Claim 2. This happens in particular if e lies around four vertices. If e lies around only three vertices, one of its endpoints is one of the vertices v_i, say v_1. Assume e is not around v_2. If v_2 is not adjacent to v_1, then Claim 1 gives the result. So taking v_1 adjacent to v_2, we adapt Claim 1, by using all seven inequalities, to show that the eight quasi-regular tetrahedra around v_1 and v_2 give $4(0.55)\,pt$. Finally, if all top edges have length at most ξ, we use inequalities 3, 6, and 7.   □

In a special case, the constant of Lemma 10.6 can be improved by a small amount. This small improvement will be used in Paper V.

Lemma 10.7. *Let v be a vertex of a decomposition star of type $(5, 0)$. Let $S_1, \ldots, S_5$ be quasi-regular tetrahedra around the edge $\{0, v\}$. Then*

$$\sum_{i=1}^{5} \sigma(S_i) < 4.52\,pt - 10^{-8}.$$

Proof. If any of the top edges has length greater than ξ, we use a slightly improved calculation[12] that yields this constant. Otherwise, the same calculation[13] that was used in the previous lemma gives the desired estimate

$$\sum \sigma < 5(0.31023815) - 2\pi(0.207045) < 4.52\,pt - 10^{-8}.$$ □

---

[12] CALC-241241504-1.
[13] CALC-82950290.

10.3.   Limitations on Types

Recall that a vertex of a planar map has type (p, q) if it is the vertex of exactly p triangles and q quadrilaterals. This section restricts the possible types that appear in a decomposition star.

Let t_4 denote the constant $0.1317 \approx 2.37838774 \, pt$.

Lemma 10.8.   *If R is a quad cluster, then*

$$\tau_R(D) \geq t_4.$$

Proof.   A calculation[14] asserts precisely this.   □

Lemma 10.9.   *The following eight types (p, q) are impossible:* (1) $p \geq 8$, (2) $p \geq 6$ *and* $q \geq 1$, (3) $p \geq 5$ *and* $q \geq 2$, (4) $p \geq 4$ *and* $q \geq 3$, (5) $p \geq 2$ *and* $q \geq 4$, (6) $p \geq 0$ *and* $q \geq 6$, (7) $p \leq 3$ *and* $q = 0$, (8) $p \leq 1$ *and* $q = 1$.

Proof.   Calculations[15] give a lower bound on the dihedral angle of p simplices and q quadrilaterals at $0.8638p + 1.153q$ and an upper bound of $1.874445p + 3.247q$. If the type exists, these constants must straddle 2π. One readily verifies in Cases 1–8 that these constants do not straddle 2π.   □

Lemma 10.10.   *If the type of any vertex of a decomposition star is one of* (4, 2), (3, 3), (1, 4), (1, 5), (0, 5), (0, 2), *then the decomposition star does not contravene.*

Proof.   According to Table 10.1, we have $\tau_{LP}(p, q) > (4\pi\zeta - 8) \, pt$, for $(p, q) = $ (4, 2), (3, 3), (1, 4), (1, 5), (0, 5), or (0, 2). By Lemma 10.2, the result follows in these cases.   □

Remark 10.11.   In summary of the preceding two lemmas, we find that we may restrict our attention to the following types of vertices:

$$
\begin{array}{lll}
(7, 0) & & \\
(6, 0) & & \\
(5, 0) & (5, 1) & \\
(4, 0) & (4, 1) & \\
& (3, 1) & (3, 2) \\
& (2, 1) & (2, 2) \quad (2, 3) \\
& & (1, 2) \quad (1, 3) \\
& & (0, 3) \quad (0, 4)
\end{array}
$$

It will be shown in Corollary 12.3, that the type (7, 0) does not occur in a contravening decomposition star.

---

[14] CALC-996268658.
[15] CALC-657406669, CALC-208809199, CALC-984463800, and CALC-277330628.

10.4. *Bounds on the Score in Quadrilateral Regions*

If the quad cluster has a diagonal of length at most $\sqrt{8}$ between two corners, there are three possible decompositions. (1) The two quarters formed by the diagonal lie in the Q-system so that the scoring rules for the Q-system are used. (2) There is a second diagonal of length at most $\sqrt{8}$, and we use the two quarters from the second diagonal for the scoring. (3) There is an enclosed vertex that makes the quad cluster into a quartered octahedron and the four upright quarters are in the Q-system.

Now suppose that neither diagonal is less than $\sqrt{8}$ and the quad cluster is not a quartered octahedron. If there is no enclosed vertex of length at most $\sqrt{8}$, the quad cluster contains no quarters. An upper bound on the score of the quad cluster (P, D) is $\mathrm{vor}_P(D, \sqrt{2})$. The remaining cases are called *mixed* quad clusters. Mixed quad clusters enclose a vertex of height at most $\sqrt{8}$ and do not contain flat quarters.

Lemma 10.12. *Assume a figure exists with vertices $v_1, \ldots, v_4, v$ subject to the constraints*

$$2 \leq |v_i| \leq 2t_0,$$

$$2 \leq |v_i - v_{i+1}| \leq 2t_0,$$

$$2 \leq |v_i - v_{i+2}|,$$

$$h_i \leq |v - v_i|,$$

$$2 \leq |v| \leq 2t_0, \qquad for \quad i = 1, \ldots, 4 \quad (mod\ 4),$$

where h_i are fixed constants that satisfy $h_i \in [2, 2\sqrt{2}]$. Let L be the quadrilateral on the unit sphere with vertices $v_i/|v_i|$ and edges running between consecutive vertices. Assume that v lies in the cone at the origin obtained by scaling L. Then another figure exists made of a (new) collection of vectors $v_1, \ldots, v_4$ and v subject to the constraints above together with the additional constraints

$$|v_i - v_{i+1}| = 2t_0,$$

$$|v_i| = 2, \qquad for \quad i = 1, \ldots, 4,$$

$$|v| = 2t_0.$$

Moreover, the quadrilateral L may be assumed to be convex.

Proof. This lemma in pure geometry is a special case of Lemma 4.3 of [Ha6]. □

Lemma 10.13. *A quadrilateral region does not enclose any vertices of height at most $2t_0$.*

Proof. Let $v_1, \ldots, v_4$ be the corners of the quad cluster, and let v be an enclosed vertex of height at most $2t_0$. We cannot have $|v_i - v| \leq 2t_0$ for two different vertices v_i, because two such inequalities would partition the region into two separate standard regions instead of a single quadrilateral region.

We apply Lemma 10.12 to assume

$$|v_i - v_{i+1}| = 2t_0, \qquad |v_i| = 2, \qquad |v| = 2t_0,$$

for $i = 1, \ldots, 4$. Reindexing and perturbing v as necessary, we may assume that $2 \leq |v_1 - v| \leq 2t_0$ and $|v_i - v| \geq 2t_0$, for $i = 2, 3, 4$. Moving v, we may assume it reaches the minimal distance to two adjacent corners (2 for v_1 or $2t_0$ for v_i, $i > 1$). Keeping v fixed at this minimal distance, perturb the quad cluster along its remaining degree of freedom until v attains its minimal distance to three of the corners. This is a rigid figure. There are four possibilities depending on which three corners are chosen. Pick coordinates to show that the distance from v to the remaining vertex violates its inequality. $\qquad \square$

Lemma 10.14. *The score of a mixed quad cluster is less than* $-1.04 \, pt$.

Proof. Any enclosed vertex in a quad cluster has length at least $2t_0$ by Lemma 10.13. ✱
In particular, the anchors of an enclosed vertex are corners of the quad cluster. There are no flat quarters.

We generally truncate the V-cell at $\sqrt{2}$ as in the proof of Theorem 8.4. By that lemma, it breaks the V-cell into pieces whose score is nonpositive. Thus, if we identify certain pieces that score less than $-1.04 \, pt$, the result follows. Nevertheless, a few simplices will be left untruncated in the following argument. We will leave a simplex untruncated only if we are certain that each of its faces has positive orientation and that the simplices sharing a face F with S either lie in the Q-system or have positive orientation along F. If these conditions hold, we may use[16] the function s-vor on S rather than truncation s-vor$_0$.

In this proof, by enclosed vertex, we mean one of height at most $2\sqrt{2}$. Let v be an enclosed vertex with the fewest anchors. If there are no anchors, the right circular cone $C(h, \eta_0(h))$ (aligned along $\{0, v\}$; see Definition 7.22) belongs to VC(0), where $\eta_0(h) = \eta(2h, 2, 2t_0)$ as in Definition 9.1 and $|v| = 2h$. In fact, if such a point lies in VC(u), with $u \neq v$, then u must be a corner of the quad cluster or an enclosed vertex of height at least $2t_0$. In either case the right circular cone belongs to VC(0). By formula (7.11), the score of this cone is $2\pi(1 - h/\eta_0(h))\varphi(h, \eta_0(h))$. An optimization in one variable gives an upper bound of $-4.52 \, pt$, for $t_0 \leq h \leq \sqrt{2}$. This gives the bound of $-1.04 \, pt$ in this case.

If there is one anchor, we cut the cone in half along the plane through $\{0, v\}$ perpendicular to the plane containing the anchor and $\{0, v\}$. The half of the cone on the far side of the anchor lies under the face at v of the V-cell. We get a bound of $-4.52 \, pt/2 < -1.04 \, pt$.

To treat the remaining cases, we define a function $K(S)$ on certain simplices S with circumradius at least $\sqrt{2}$. Let $S = S(y_1, y_2, \ldots, y_6)$. Let $R(a, b, c)$ denote a Rogers simplex. Set

$$K(S) = K_0(y_1, y_2, y_6) + K_0(y_1, y_3, y_5) + \mathrm{dih}(S)(1 - y_1/\sqrt{8})\varphi(y_1/2, \sqrt{2}), \quad (10.1)$$

---

[16] CALC-185703487, CALC-69785808, and CALC-104677697.

Fig. 10.1.   The set measured by the function $K(S)$.

where

$$K_0(y_1, y_2, y_6) = \text{r-vor}(R(y_1/2, \eta(y_1, y_2, y_6), \sqrt{2}))$$
$$+ \text{r-vor}(R(y_2/2, \eta(y_1, y_2, y_6), \sqrt{2}))$$
$$- \text{dih}(R(y_1/2, \eta(y_1, y_2, y_6), \sqrt{2}))(1 - y_1/\sqrt{8})\varphi(y_1/2, \sqrt{2}).$$

(If the given Rogers simplices do not exist because the condition $0 < a < b < c$ is violated, we set the corresponding terms in these expressions to 0.) The function $K(S)$ represents the part of the score coming from the four Rogers simplices along two of the faces of S, and the conic region extending out to $\sqrt{2}$ between the two Rogers simplices along the edge y_1 (Fig. 10.1). This region is closely related to the solids $\Delta(v, W^e)$ of Section 9.3, with the difference that the solids Δ lie in a ball of radius $\eta_0(|v|/2)$, but the solids here are truncated at $\sqrt{2}$.

In the remaining cases, each enclosed vertex has at least two anchors. Each anchor is a corner of the quad cluster. Fix an enclosed vertex v. Suppose that v_1, a corner, is an anchor of v. Assume that the face $\{0, v, v_1\}$ bounds at most one upright quarter. We sweep around the edge $\{0, v_1\}$, away from the upright quarter if there is one, until we come to another enclosed vertex v' such that $\{0, v_1, v'\}$ has circumradius less than $\sqrt{2}$ or such that v_1 is an anchor of $\{0, v'\}$. If such a vertex v' does not exist, we sweep all the way to v_2 a corner of the quad cluster adjacent to v_1.

If v' exists, then various calculations[17] give the bound $-1.04\,pt$, depending on the size of the circumradius of $\{0, v, v'\}$. This allows us to assume that we do not encounter such an enclosed vertex v' whenever we sweep away, as above, from the face formed by an anchor.

Now consider the simplex $S = \{0, v_1, v_2, v\}$, where v_1 is an anchor of $\{0, v\}$. We assume that it is not an upright quarter. There are three alternatives. The first is that S decreases the score of the quarter by at least $0.52\,pt$. Calculations[18] show that this occurs if the circumradius of the face $\{0, v, v_2\}$ is less than $\sqrt{2}$, or if the circumradius of the face is greater than $\sqrt{2}$, provided that the length of $\{v, v_1\}$ is at most 2.2. The second alternative[19] is that the face $\{0, v, v_1\}$ of S is shared with a quarter Q and that S and Q taken together bring the score down by $0.52\,pt$. In fact, if there are two such simplices S and S' along Q, then the three simplices Q, S, and S' pull the score[20] below

---

[17] CALC-104677697, CALC-69785808, CALC-586706757, and CALC-87690094.
[18] CALC-185703487 and CALC-441195992.
[19] CALC-848147403, CALC-969320489, and CALC-975496332.
[20] CALC-766771911.

$-1.04\,pt$. The third alternative is that there is a simplex $S' = \{0, v, v, v_3\}$ sharing the face $\{0, v, v_1\}$, which, like S, scores less than $-0.31\,pt$. In each case, S and the adjacent simplex through $\{0, v, v_1\}$ score less than $-0.52\,pt$. Since v has at least two anchors, the quad cluster scores less than $2(-0.52)\,pt = -1.04\,pt$. $\qquad\square$

10.5. A Volume Formula

In Definition 9.3 we found a solid $\delta(v, W^e)$ that lies outside the ball of radius t_0 at 0 but inside $VC(0)$. We now develop a formula for its volume.

Set $\varphi_0 = \varphi(t_0, t_0) \approx -0.5666$. We define

$$\text{crown}(h) = 2\pi(1 - h/\eta_0(h))(\varphi(h, \eta_0(h)) - \varphi_0). \tag{10.2}$$

It is equal to $-4\delta_{\text{oct}}$ times the volume of the region outside the sphere of radius t_0 and inside the finite cone $C(h, \eta_0(h))$. If v is an enclosed vertex of height $2h \in [2t_0, \sqrt{8}]$, such that every other vertex v' of the standard cluster satisfies

$$\eta(|v|, |v'|, |v - v'|) \geq \eta_0(h),$$

then the solid represented by $\text{crown}(|v|/2)$ lies outside the truncated V-cell, but inside the V-cell, so that if P is a quad cluster,

$$\text{c-vor}(V_P) < \text{c-vor}_0(V_P) + \text{crown}(|v|/2).$$

If a vertex v' satisfies $\eta(|v|, |v'|, |v - v'|) \leq \eta_0(h)$, then by the monotonicity of the circumradius of acute triangles, v' is an anchor of v. This anchor clips the crown just defined, and we add a correction term $\text{anc}(|v'|, |v|, |v-v'|)$ to account for this. Figure 10.2 illustrates the terms in the definition of anc().

Set

$$\begin{aligned}
\text{anc}(y_1, y_2, y_6) &= -\text{dih}(R_1)\,\text{crown}(y_1/2)/(2\pi) - \text{sol}(R_1)\varphi_0 + \text{r-vor}(R_1) \\
&\quad - \text{dih}(R_2)(1 - y_2/2t_0)(\varphi(y_2/2, t_0) - \varphi_0) \\
&\quad - \text{sol}(R_2)\varphi_0 + \text{r-vor}(R_2), \tag{10.3}
\end{aligned}$$

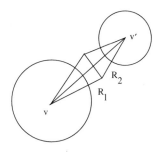

Fig. 10.2. An illustration of the terms anc.

where $R_i = R(y_i/2, \eta(y_1, y_2, y_6), \eta_0(y_1/2))$, for $i = 1, 2$. In general, there are Rogers simplices on both sides of the face $\{0, v, v'\}$, and this gives a factor of 2. For example, if v has a single anchor v', then

$$\text{c-vor}(V_P) < \text{c-vor}_0(V_P) + \text{crown}(|v|/2) + 2\,\text{anc}(|v|, |v'|, |v - v'|).$$

However, if the anchor gives a face of an upright quarter, only one side of the face lies in the V-cell, so that the factor of 2 is not required. For example, v' has context $(2, 1)$ with upright quarter Q, and if there are no other enclosed vertices, and if v', v'' are the anchors along the faces of the quarter, then

$$\text{c-vor}(V_P) \; < \; \text{c-vor}_0(V_P) + (1 - \text{dih}(Q)/(2\pi))\,\text{crown}(|v|/2)$$
$$+ \; \text{anc}(|v|, |v'|, |v - v'|) + \text{anc}(|v|, |v''|, |v - v''|).$$

In general, when there are multiple anchors around the same enclosed vertex v, we add a term $(2 - k)$ anc for each anchor, where $k \in \{0, 1, 2\}$ is the number of quarters bounded by the face formed by the anchor. We must be cautious (see Condition 2 in Definition 9.2 in the use of this formula. If the circumradius of $\{0, v, v', v''\}$ is less than $\eta_0(|v|/2)$, the Rogers simplices used to define the terms anc() at v' and v'' overlap. When this occurs, the geometric decomposition on which the correction terms anc() are based is no longer valid. In this case, other methods must be used.

If (P, D) is a mixed quad cluster, let (P, D') be the new quad cluster obtained by removing all the enclosed vertices. We define a V-cell $V(P, D')$ of (P, D') and the truncation of $V(P, D')$ at t_0. We take its score $\text{vor}_{0, P}(D')$ as we do for standard clusters. (P, D') does not contain any quarters.

Lemma 10.15. *If (P, D) is a mixed quad cluster, $\sigma_P(D') < \text{vor}_{0, P}(D)$.*

Remark 10.16. The special case of the proof where an upright quarter has context $c(Q) = (2, 1)$ will be applied in Section 11.2 in situations other than mixed quad clusters.

Proof. Suppose there exists an enclosed vertex that has context $(2, 1)$; that is, there is a single upright quarter $Q = S(y_1, y_2, \ldots, y_6)$ and no additional anchors. In this context $\sigma(Q) = \mu(Q)$. Let v be the enclosed vertex. To compare $\sigma_P(D)$ with $\text{vor}_{0, P}(D')$, consider the V-cell near Q. The quarter Q cuts a wedge of angle $\text{dih}(Q)$ from the crown at v. There is an anchor term for the two anchors of v along the faces of Q. Let V_P^v be the truncation at height t_0 of V_P near v and under the four Rogers simplices stemming from the two anchors. (Figure 10.3 shades the truncated parts of the quad cluster.) As a consequence

$$\text{c-vor}(V_P) \; < \; (1 - \text{dih}(Q)/(2\pi))\,\text{crown}(y_1/2) + \text{anc}(y_1, y_2, y_6)$$
$$+ \; \text{anc}(y_1, y_3, y_5) + \text{c-vor}(V_P^v). \qquad (10.4)$$

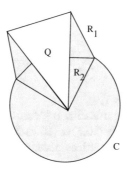

Fig. 10.3. The terms anc near an upright quarter.

Combining this inequality with calculations,[21] we find

$$\text{c-vor}(V_P) + \mu(Q) < \text{c-vor}(V_P^v) + \text{s-vor}_0(Q). \tag{10.5}$$

Now suppose there is an enclosed vertex v with context $(3, 1)$. Let the quad cluster have corners v_1, v_2, v_3, v_4, ordered consecutively. Suppose the two quarters along v are $Q_1 = \{0, v, v_1, v_2\}$ and $Q_2 = \{0, v, v_2, v_3\}$. We consider two cases.

Case 1: $\text{dih}(Q_1) + \text{dih}(Q_2) < \pi$ *or* $\text{rad}(0, v, v_1, v_3) \geq \eta(|v|, 2, 2t_0)$. In this case the use of correction terms to the crown are legitimate as in Definition 9.2. Proceeding as in context $(2, 1)$, we find that

$$\text{c-vor}(V_P) < (1 - (\text{dih}(Q_1) + \text{dih}(Q_2))/(2\pi)) \, \text{crown}(|v|/2)$$
$$+ \, \text{anc}(F_1) + \text{anc}(F_2) + \text{c-vor}(V_P^v). \tag{10.6}$$

Here V_P^v is defined by the truncation at height t_0 under the V-face determined by v and under the Rogers simplices stemming from the side of F_i that occur in the definition of anc. Also, $\text{anc}(F_i) = \text{anc}(y_i, y_j, y_k)$ for a face F_i with edges y_i along an upright quarter. By a calculation[22] applied to both Q_1 and Q_2, we have

$$\text{c-vor}(V_P) + \sum_{i=1}^{2} \sigma(Q_i) < \text{c-vor}(V_P^v) + \sum_{i=1}^{2} \text{s-vor}_0(Q_i). \tag{10.7}$$

That is, by truncating near v, and changing the scoring of the quarters to s-vor$_0$, we obtain an upper bound on the score.

Case 2: $\text{dih}(Q_1) + \text{dih}(Q_2) \geq \pi$ *and* $\text{rad}(0, v, v_1, v_3) \leq \eta_0(|v|/2)$. The anchor terms cannot be used here. In the mixed case, $\sqrt{8} < |v_1 - v_3|$, so

$$\sqrt{2} < \tfrac{1}{2}|v_1 - v_3| \leq \text{rad} \leq \eta_0(|v|/2),$$

---

[21] CALC-906566422, CALC-703457064, and CALC-175514843.
[22] CALC-554253147.

and this implies $|v| \geq 2.696$. We have[23]

$$\sum_{i=1}^{2} \sigma(Q_i) < \sum_{i=1}^{2} \text{s-vor}_0(Q_i) + \sum_{i=1}^{2} 0.01(\pi/2 - \text{dih}(Q_i)) < \sum_{i=1}^{2} \text{s-vor}_0(Q_i).$$

Inequality (10.7) holds, for $V_P^v = V_P$.

In the general case, we run over all enclosed vertices v and truncate around each vertex. For each vertex we obtain inequality (10.5) or (10.7). These inequalities can be coherently combined over multiple enclosed vertices because the V-faces were associated with different vertices v and none of the Rogers simplices used in the terms anc() overlap. More precisely, if Z is a set of enclosed vertices, set $V_P^Z = \bigcap_{v \in Z} V_P^v$, and $V_P^{v,Z} = V_P^Z \cap V_P^v$. Coherence means that we obtain valid inequalities by adding the superscript Z to V_P and V_P^v in inequalities (10.5) and (10.7), if $v \notin Z$. In sum, $\sigma_P(D) < \text{vor}_{0,P}(D)$.   □

References

[Ha1] T. C. Hales, The sphere packing problem, *J. Comput. Appl. Math.* **44** (1992), 41–76.

[Ha6] T. C. Hales, Sphere packings, I, *Discrete Comput. Geom.* **17** (1997), 1–51.

[Ha7] T. C. Hales, Sphere packings, II, *Discrete Comput. Geom.* **18** (1997), 135–149.

[Ha14] T. C. Hales, Some algorithms arising in the proof of the Kepler conjecture, *Discrete and Computational Geometry*, Algorithms and Combinatorics, vol. 25, Springer Verlag, Berlin, 2003, pp. 489–507.

[Ha16] T. C. Hales, Computer Resources for the Kepler Conjecture, http://annals.math.princeton.edu/keplerconjecture/. (The source code, inequalities, and other computer data relating to the solution are also found at http://xxx.lanl.gov/abs/math/ 9811078v1.)

Received November 11, 1998, and in revised form September 12, 2003, and July 25, 2005.
Online publication February 27, 2006.

---

[23] CALC-855677395.

6

Sphere Packings IV. Detailed Bounds, by T. C. Hales

This paper is the fourth of six papers giving the Kepler Conjecture proof. It contains Chapters 11–14 of the proof.

An addendum to the proof, which appeared in 2010, requires adding a new subsection to this paper, which would be the following:

12.14. Biconnected graphs

The material for this subsection appears as Section 8 of the paper T. C. Hales et al., "A Revision of the Proof of the Kepler Conjecture." This appears in Section III on pp. 361–371 of this volume.

The asterisks added in the margin of some pages indicate that changes are required, as listed in the Hales list of errata given on p. 373 of this volume. In addition, on p. 224 of this volume, $\delta_{\text{loop}}(4, 3) = 0.29426$, as given on p. 39 of the 1998 preprint T. C. Hales, *The Kepler Conjecture*, `arXiv:math/9811078v2`.

Contents

The original version of this chapter was revised. An erratum to this chapter can be found at
http://dx.doi.org/10.1007/978-1-4614-1129-1_12

Discrete Comput Geom 36:111–166 (2006)
DOI: 10.1007/s00454-005-1213-z

Discrete & Computational

Geometry

© 2006 Springer Science+Business Media, Inc.

Sphere Packings, IV. Detailed Bounds

Thomas C. Hales

Department of Mathematics, University of Pittsburgh,
Pittsburgh, PA 15217, USA
hales@pitt.edu

Abstract.   This paper is the fourth in a series of six papers devoted to the proof of the Kepler
conjecture, which asserts that no packing of congruent balls in three dimensions has density
greater than the face-centered cubic packing. In a previous paper in this series, a continuous
function f on a compact space was defined, certain points in the domain were conjectured to
give the global maxima, and the relation between this conjecture and the Kepler conjecture
was established. The function f can be expressed as a sum of terms, indexed by regions on
a unit sphere. In this paper detailed estimates of the terms corresponding to general regions
are developed. These results form the technical heart of the proof of the Kepler conjecture,
by giving detailed bounds on the function f. The results rely on long computer calculations.

Introduction

This paper contains the technical heart of the proof of the Kepler conjecture. Its primary
purpose is to obtain good bounds on the score $\sigma_R(D)$ when R is an arbitrary standard
region of a decomposition star D. This is particularly challenging, because we have no
a priori restrictions on the combinatorial type of the standard region R. It is not known
to be bounded by a simple polygon. It is not known to be simply connected. Moreover,
there are multitudes of possible geometrical configurations of upright and flat quarters,
each scored by a different rule. This paper deals with these complexities and bounds the
score $\sigma_R(D)$ in a way that depends on a simple numerical invariant $n(R)$ of R. When R
is bounded by a simple polygon, the numerical invariant is simply the number of sides
of the polygon. This bound on the score of a standard region represents the turning point
of the proof, in the sense that it caps the complexity of a contravening decomposition
star, and restrains the combinatorial possibilities. Later in the proof it is instrumental in
the complete enumeration of the plane graphs attached to contravening stars.

The first section proves a series of approximations for the score of upright quarters.
The strategy is to limit the number of geometrical configurations of upright quarters by

showing that a common upper bound (to the scoring function) can be found for quite disparate geometrical configurations of upright quarters. When a general upper bound can be found that is independent of the geometrical details of upright quarters, we say that the upright quarters can be *erased*. (A precise definition of what it means to erase an upright quarter appears below.) There are some upright quarters that cannot be treated in this manner; and this adds some complications to the proofs in this paper.

The second section states the main result of the paper (Theorem 12.1). An initial reduction reduces the proof to the case that the boundary of the given standard region is a polygon. A further argument is presented to reduce the proof to a convex polygon.

The third section completes the proof of the main theorem. This part of the proof relies on a new geometrical decomposition of the part of a V-cell over a standard region. The pieces in this decomposition are called *truncated corner cells*.

A final section in this paper collects miscellaneous further bounds that will be needed in later parts of the proof of the Kepler conjecture.

11. Upright Quarters

11.1. *Erasing Upright Quarters*

Definition 11.1.   A standard region is said to be *exceptional* if it is not a triangle or a quadrilateral. The pair (D, R) consisting of a decomposition star and an exceptional standard region is said to be an *exceptional cluster*. The vertices of the packing of height at most $2t_0$ that are contained in the closed cone over the standard region are called its *corners*.

Fix an exceptional cluster R. Throughout this paper, we assume that R lies on a star of score at least $8\,pt$. It is to be understood, when we say that a standard region does not exist, that we mean that there exists no such region on any star scoring more than $8\,pt$.

In this section we discuss how to eliminate many cases of upright diagonals. The results are summarized in Section 11.9.

If R is a standard region, we write $V_R(t)$ for the intersection of the local V-cell $V_R = VC(0) \cap C(R)$ with a ball $B(t)$, centered at the origin, of radius t. We usually take $t = t_0$. If $\{0, v\}$, of length between $2t_0$ and $2\sqrt{2}$, is not the diagonal of an upright quarter in the Q-system, then v does not affect the truncated cell $V_R(t_0)$ and may be disregarded. For this reason we confine our attention to upright diagonals that lie along an upright quarter in the Q-system.

We say that an upright diagonal $\{0, v\}$ can be *erased with penalty* $\pi_0 \geq 0$, if we have, in terms of the decomposition of Section 9,

$$\sum_Q \sigma(Q) + \sum_S \sigma(V_S(t_S)) - 4\delta_{\text{oct}}\,\text{vol}(\delta_P(v)) < \pi_0 + \sum_Q \text{s-vor}_0(Q) + \sum_S \text{s-vor}_0(S).$$

Here the sum over Q runs over the upright quarters around $\{0, v\}$. The scores $\sigma(Q)$ are context-dependent (see Section 7). The second sum runs over simplices S along $\{0, v\}$ of type C in the S-system. We define their score $\sigma(V_S(t_S))$ as in Section 9. Also, $\delta_P(v)$ is the piece of the decomposition defined in Section 9. The right-hand side is scored by the truncation function in Section 7 (formula (7.13)). When we erase without mention of a penalty, $\pi_0 = 0$ is assumed.

If the diagonal can be erased, an upper bound on the score is obtained by ignoring the upright diagonal and all of the structures around it coming from the decomposition of Section 9, and switching to the truncation at t_0. The current section shows that various vertices can be erased, and this will greatly reduce the number of combinatorial possibilities for an exceptional cluster.

11.2. Contexts

Each upright diagonal has a context (p, q), with p the number of anchors and $p - q$ the number of quarters around the diagonal (Definition 7.1). The dihedral angle of a quarter is less than[24] π, so the context $(2, 0)$ is impossible. There is at least one quarter, so $p \geq q + 1$, $p \geq 2$.

The context $(2, 1)$ is treated in Section 10.4. Lemma 10.15 shows that by removing the upright diagonal, and scoring the surrounding region by a truncated function vor_0, an upper bound on the score is obtained. In the remaining contexts, $p \geq 3$. We start with contexts satisfying $p = 3$. The context $(3, 0)$ is to be regarded as two quasi-regular tetrahedra sharing a face rather than as three quarters along a diagonal. In particular, by Definition 4.8, the upright quarters do not belong to the Q-system.

We recall that the score of an upright quarter is given by

$$\sigma(Q, v) = (\mu(Q, v) + \mu(Q, \hat{v}) + \text{s-vor}_0(Q, v) - \text{s-vor}_0(Q, \hat{v}))/2,$$

except in the contexts $(2, 1)$ and $(4, 0)$. Define $\nu(Q)$ to be the right-hand side of this equation. The context $(2, 1)$ has been treated, and the context $(4, 0)$ does not occur in exceptional clusters. Thus, for the remainder of this section, the scoring rule $\sigma(Q) = \nu(Q)$ is used.

We have several different variants on the score depending on the truncation, analytic continuation, and so forth. If f is any of the functions

$$\text{s-vor}_0, \quad \text{s-vor}, \quad \Gamma, \quad \nu,$$

we set τ_0, τ_V, τ_Γ, τ_ν, respectively, to

$$\tau_* = -f(S) + \text{sol}(S)\zeta \; pt.$$

We set $\tau(S, t) = -\text{s-vor}(S, t) + \text{sol}(S)\zeta \; pt$. The family of functions τ_* measure what is squandered by a simplex. We say that Q has *compression type* or *Voronoi type* according to the scoring of $\mu(Q)$. (See Section 7.1.)

Crowns and anchor correction terms are used in Section 10.4 to erase upright quarters. We imitate those methods here. The functions crown and anc are defined and discussed in Section 10.4. If $S = S(y_1, \ldots, y_6)$ is a simplex along $\{0, v\}$, set

$$\kappa(S(y_1, \ldots, y_6)) = \text{crown}(y_1/2)\,\text{dih}(S)/(2\pi) + \text{anc}(y_1, y_2, y_6) + \text{anc}(y_1, y_3, y_5).$$

---

[24] CALC-971555266.

$\kappa(S)$ is a bound on the difference in the score resulting from truncation around v. Assume that S is the simplex formed by $\{0, v\}$ and two consecutive anchors around $\{0, v\}$. Assume further that the circumradius of S is at least $\eta_0(y_1/2)$. Then we have

$$\kappa(S) = -4\delta_{oct} \, \text{vol}(\delta_P(W^e)),$$

where W^e is the extended wedge constructed in Section 9.2. To see this, it is a matter of interpreting the terms in κ. The function *crown* enters the volume through the region over the spherical cap D_0 of Section 9.2, lying outside $B(t_0)$. By multiplying by $\text{dih}(S)/(2\pi)$, we select the part of the spherical cap over the unextended wedge W between the anchors. The terms *anc* adjust for the four Rogers simplices lying above the extension W^e.

11.3. Three Anchors

Lemma 11.2. *The upright diagonal can be erased in the context* $(3, 2)$.

Proof. Let v_1 and v_2 be the two anchors of the upright diagonal $\{0, v\}$ along the quarter. Let the third anchor be v_3.

Assume first that $|v| \geq 2.696$. If Q is of compression type, then[25] the score is dominated by the truncated function s-vor_0. Assume Q is of Voronoi type. If $|v_1|, |v_2| \leq 2.45$, then a calculation[26] gives the result. Take $|v_2| \geq 2.45$. By symmetry, $|v - v_1|$ or $|v - v_2| \geq 2.45$. The case $|v - v_1| \geq 2.45$ is treated by another calculation.[27] We take $|v - v_2| \geq 2.45$. Let $S = \{0, v, v_2, v_3\}$. If S is of type C, the result follows.[28] S is of type C, if and only if $y_4 \leq 2.77$ (because $\eta_{456} \geq \eta(2.45, 2, 2.77) > \sqrt{2}$). If S is not of type C, we argue as follows. The function $h^2(\eta(2h, 2.45, 2.45)^{-2} - \eta_0(h)^{-2})$ is a quadratic polynomial in h^2 with negative values for $2h \in [2.696, 2\sqrt{2}]$. From this we find

$$\text{rad}(S) \geq \eta(2h, 2.45, 2.45) \geq \eta_0(h), \qquad \text{where} \quad 2h = |v|,$$

and this justifies the use of κ (see Section 9.2, Case (2)). That the truncated function dominates the score now follows from a calculation.[29]

Now assume that $|v| \leq 2.696$. If the simplices $\{0, v, v_1, v_3\}$ and $\{0, v, v_2, v_3\}$ are of type C, the bound follows from a calculation.[30,31] If say $S = \{0, v, v_2, v_3\}$ is not of type C, then

$$\text{rad}(S) \geq \sqrt{2} > \eta_0(2.696/2) \geq \eta_0(h),$$

justifying the use of κ. The bound follows from further calculations.[32-34] ($\Gamma + \kappa <$ octavor$_0$, etc.) □

[25] CALC-73974037.
[26] CALC-764978100.
[27] CALC-764978100.
[28] CALC-764978100.
[29] CALC-618205535.
[30] CALC-73974037.
[31] CALC-764978100.
[32] CALC-618205535.
[33] CALC-73974037.
[34] CALC-764978100.

Lemma 11.3.   *The upright diagonal can be erased in the context* (3, 1), *provided the three anchors do not form a flat quarter at the origin.*

Proof.   In the absence of a flat quarter, truncate, score, and remove the vertex v as in the context (3, 1) of Lemma 10.15. If there is a flat quarter, by the rules of Definition 4.8, v is enclosed over the flat quarter. We do nothing further with them for now. This unerased case appears in the summary, Section 11.9. See Lemma 11.27.   □

11.4.   *Six Anchors*

Lemma 11.4.   *An upright diagonal has at most five anchors.*

Proof.   The proof relies on constants and inequalities from two calculations.[35,36] If between two anchors there is a quarter, then the angle is greater than 0.956, but if there is not, the angle is greater than 1.23. So if there are k quarters and at least six anchors, they squander more than

$$k(1.01104) - [2\pi - (6 - k)1.23]0.78701 > (4\pi\zeta - 8)\,pt,$$

for $k \geq 0$.   □

11.5.   *Anchored Simplices*

Let $\{0, v\}$ be an upright diagonal, and let $v_1, v_2, \ldots, v_k = v_1$ be its anchors, ordered cyclically around $\{0, v\}$. This cyclic order gives dihedral angles between consecutive anchors around the upright diagonal. We define the dihedral angles so that their sum is 2π, even though this will lead us to depart from our usual conventions by assigning a dihedral angle greater than π when all the anchors are concentrated in some half-space bounded by a plane through $\{0, v\}$. When the dihedral angle of $S = \{0, v, v_i, v_{i+1}\}$ is at most π, we say that S is an *anchored simplex* if $|v_i - v_{i+1}| \leq 3.2$. (The constant 3.2 appears throughout this section.) All upright quarters are anchored simplices. If an upright diagonal is completely surrounded by anchored simplices, the upright diagonal is sometimes called a *loop*. If $|v_i - v_{i+1}| > 3.2$ and the angle is less than π, we say there is a *large gap* around $\{0, v\}$ between v_i and v_{i+1}.

   To understand how anchored simplices overlap we need a bound satisfied by vertices enclosed over an anchored simplex.

Lemma 11.5.   *A vertex w of height between 2 and $2\sqrt{2}$, enclosed in the cone over an anchored simplex $\{0, v, v_1, v_2\}$ with diagonal $\{0, v\}$ satisfies $|w - v| \leq 2t_0$. In particular, if $|w| \leq 2t_0$, then w is an anchor.*

---

[35] CALC-729988292.
[36] CALC-83777706.

Proof. As in Lemma 4.16, the vertex w cannot lie inside the anchored simplex. If $|v_1 - v_2| \leq 2\sqrt{2}$, the result follows from Lemma 5.16. In fact, if $|w| \leq 2\sqrt{2}$, the Voronoi cells at 0 and w meet, so that Lemma 5.16 forces $\{0, v_1, v_2, w\}$ to be a quarter. (This observation gives a second proof of Lemma 4.34.)

Assume that a figure exists with $|v_1 - v_2| > 2\sqrt{2}$. Suppose for a contradiction that $|v - w| > 2t_0$. Pivot v_1 around $\{0, v_2\}$ until $|v - v_1| = 2t_0$ and v_2 around $\{0, v_1\}$ until $|v - v_2| = 2t_0$. Rescale w so that $|w| = 2\sqrt{2}$. Set $x = |v_1 - v_2|$. If, through geometric considerations, w is not deformed into the plane of $\{0, v_2, v_1\}$, then we are left with the one-dimensional family $|w'| = |w' - w| = 2$, for $w' = v_2, v_1, |v - w| = |v| = |v_1 - v| = |v_2 - v| = 2t_0$, depending on x. This gives a contradiction

$$\pi \geq \text{dih}(v_2, v_1, 0, v) + \text{dih}(v_2, v_1, v, w)$$
$$= 2 \, \text{dih}(S(x, 2, 2t_0, 2t_0, 2t_0, 2)) > \pi,$$

for $x > 2\sqrt{2}$. (Equality is attained if $x = 2\sqrt{2}$.)

Thus, we may assume that w lies in the plane $P = \{0, v_1, v_2\}$. Take the circle in P at distance $2t_0$ from v. The vertices 0 and w lie on or outside the circle. The vertices v_1 and v_2 lie on the circle, so the diameter is at least $x > 2\sqrt{2}$. The distance from v to P is less than $x_0 = \sqrt{2t_0^2 - 2}$. The edge $\{0, w\}$ cannot pass through the center of the circle, because $|w|$ is less than the diameter. Reflect v through P to get v'. Then $|v - v'| < 2x_0$. Swapping v_1 and v_2 as necessary, we may assume that w is enclosed over $\{0, v, v', v_2\}$. The desired bound $|v - w| \leq 2t_0$ now follows from geometric considerations and the contradiction

$$2\sqrt{2} = |w| > \mathcal{E}(S(2, 2t_0, 2t_0, 2x_0, 2t_0, 2t_0), 2, 2t_0, 2t_0) = 2\sqrt{2}. \qquad \square$$

Corollary 11.6. *A vertex of height at most $2t_0$ is never enclosed over an anchored simplex.*

Proof. If so, it would be an anchor to the upright diagonal, contrary to the assumption that the anchored simplex is formed by consecutive anchors. $\qquad \square$

11.6. *Anchored Simplices Do Not Overlap*

Definition 11.7. Consider an upright diagonal that is not a loop. Let R be the standard region that contains the upright diagonal and its surrounding quarters. Assume we are in the context $(4, 1)$ or $(5, 1)$. In the context $(4, 1)$, suppose that there does not exist a plane through the upright diagonal such that all three quarters lie in the same half-space bounded by the plane. Then we say that the context is 3-*unconfined*. If such a plane exists, we say that the context is 3-*crowded*. We call the context $(5, 1)$ a 4-*crowded* upright diagonal. Sections 11.3 and 11.4 reduce everything to contexts with four or five anchors around each vertex. If there are five anchors, Lemma 11.14 and Remark 11.13 show that we can assume at most one large gap. This gives contexts $(5, 0)$ and $(5, 1)$. If there are four anchors, then Lemma 11.21 will dismiss all contexts except $(4, 0)$ and

(4, 1). Thus, every upright diagonal is exactly one of the following: a loop, 3-unconfined, 3-crowded, or 4-crowded.

Definition 11.8. The Cayley–Menger determinant expresses the volume of a simplex $S(y1, \ldots, y_6)$ in the form $\sqrt{\Delta(x_1, \ldots, x_6)}/12$, where $x_i = y_i^2$, and Δ is a polynomial with integer coefficients. The polynomial Δ will be used frequently.

This lemma is a consequence of the two others that follow. The context of the lemma is the set of anchored simplices that have not been erased by previous reductions.

Lemma 11.9. *Anchored simplices do not overlap.*

The remaining contexts have four or five anchors. Let w and the anchored simplex $S = \{0, v, v_1, v_2\}$ be as in Section 11.5. Our object is to describe the local geometry when an upright diagonal is enclosed over an anchored simplex. If $|v_1 - v_2| \leq 2\sqrt{2}$, we have seen in Lemma 4.32 that there can be no enclosed upright diagonal with four or more anchors over the anchored simplex S.

Assume $|v_1 - v_2| > 2\sqrt{2}$. Let $w_1, \ldots, w_k, k \geq 4$, be the anchors of $\{0, w\}$, indexed consecutively. The anchors of $\{0, w\}$ do not lie in $C(S)$, and the triangles $\{0, w, w_i\}$ and $\{0, v, v_j\}$ do not overlap. Thus, the plane $\{0, v_1, v_2\}$ separates w from $\{w_1, \ldots, w_k\}$. Set $S_i = \{0, w, w_i, w_{i+1}\}$. By a calculation[37]

$$\pi \geq \dih(S_1) + \cdots + \dih(S_{k-1}) \geq (k - 1)0.956.$$

Thus, $k = 4$. The common upright diagonal of the three simplices $\{S_i\}$ is *3-crowded*. We claim that $\{v_1, v_2\} = \{w_1, w_4\}$. Suppose to the contrary that, after reindexing as necessary, $S_0 = \{0, w, w_1, v_1\}$ is a simplex, with $v_1 \neq w_1$, that does not overlap $S_1, \ldots, S_3$. Then $\pi \geq \dih(S_0) + \cdots + \dih(S_3)$. So $0.28 \geq \pi - 3(0.956) \geq \dih(S_0)$. A calculation[38] now implies that $|w - v_1| \geq 2\sqrt{2}$.

Assume that $\{0, w, v_1, v_2\}$ are coplanar. Disregard the other vertices. We minimize $|v_1 - v_2|$ when

$$|w| = 2\sqrt{2}, \qquad |v_2| = |v_1| = |w - v_2| = 2, \qquad |w - v_1| = 2\sqrt{2}.$$

This implies $3.2 \geq |v_1 - v_2| \geq x$, where x is the largest positive root of the polynomial $\Delta(8, 4, 4, x^2, 4, 8)$. However, $x \approx 3.36$, a contradiction.

Since $\{0, w, v_1, v_2\}$ cannot be coplanar vertices, geometric considerations apply and

$$2\sqrt{2} \geq |w| \geq \mathcal{E}(S(2, 2, 2, 2, 2, 3.2), 2\sqrt{2}, 2, 2) > 2\sqrt{2}.$$

This contradiction establishes that $v_1 = w_1$.

Lemma 11.10. *Around a 3-crowded upright diagonal, all of the anchored simplices are quarters.*

---

[37] CALC-83777706.
[38] CALC-83777706.

Proof. The proof makes use of constants and inequalities from several different calculations.[39–41] The dihedral angles are at most $\pi - 2(0.956) < 1.23$. This forces $y_4 \leq 2t_0$, for each simplex S. So they are all quarters. □

Lemma 11.11. *If there is 3-crowded upright diagonal, then the three anchored simplices squander more than* 0.5606 *and score at most* -0.4339.

Proof. The proof makes use of constants and inequalities from several different calculations.[42–44] The three anchored simplices squander at least

$$3(1.01104) - \pi(0.78701) > 0.5606.$$

The bound on the score follows similarly from $v < -0.9871 + 0.80449 \, \text{dih}$. □

Lemma 11.12. *If a simplex at a 3-crowded upright diagonal overlaps an anchored simplex, the decomposition star does not contravene.*

Proof. Suppose that $\{0, v, v_1, v_2\}$ is an anchored simplex that another anchored simplex overlaps, with $\{0, v\}$ the upright diagonal. Let $\{0, w\}$ be a 3-crowded upright diagonal. We score the two simplices $S_i' = \{0, v, w, v_i\}$ by truncation at $\sqrt{2}$. Truncation at $\sqrt{2}$ is justified by face-orientation arguments or by geometric considerations:

$$\mathcal{E}(S(2, 2t_0, 2t_0, 2t_0, 2t_0, 2t_0), 2, 2, 2) > 2\sqrt{2}.$$

A calculation[45] gives

$$\tau_V(S_1', \sqrt{2}) + \tau_V(S_2', \sqrt{2}) \geq 2(0.13) + 0.2(\text{dih}(S_1') + \text{dih}(S_2') - \pi) > 0.26.$$

Together with the three simplices around the 3-crowded upright diagonal that squander at least 0.5606, we obtain the stated bound. □

11.7.   Five Anchors

When there are five anchors of an upright diagonal, each dihedral angle around the diagonal is at most $2\pi - 4(0.956) < \pi$.

Remark 11.13. There are at most two large gaps by the calculation[46]

$$3(1.65) + 2(0.956) > 2\pi.$$

---

[39] CALC-815492935.
[40] CALC-83777706.
[41] CALC-855294746.
[42] CALC-815492935.
[43] CALC-83777706.
[44] CALC-855294746.
[45] CALC-855294746.
[46] CALC-83777706.

Lemma 11.14. *If an upright diagonal has five anchors with two large gaps, then the three anchored simplices squander* $> (4\pi\zeta - 8)\, pt.$

Proof. By a calculation,[47] the anchored simplices are all quarters, $1.23 + 2(1.65) + 2(0.956) > 2\pi$. The dihedral angle is less than $2\pi - 2(1.65)$. The linear programming bound based on various inequalities[48] is greater than $0.859 > (4\pi\zeta - 8)\, pt.$ □

Definition 11.15. Define a *masked* flat quarter to be a flat quarter that is not in the Q-system because it overlaps an upright quarter in the Q-system. They can only occur in a very special setting.

Lemma 11.16. *Let* $\{0, v\}$ *be an upright diagonal with at least four anchors. If Q is a flat quarter that overlaps an anchored simplex along* $\{0, v\}$, *then the vertices of Q are the origin and three consecutive anchors of* $\{0, v\}$.

Proof. For there to be overlap, the diagonal $\{w_1, w_2\}$ of Q must pass through the face $\{0, v, v_1\}$ formed by some anchor v_1 (see Lemma 4.19). By Lemma 4.24, w_1 and w_2 are anchors of $\{0, v\}$. By Lemma 4.32, $w_2, v_1,$ and w_1 are consecutive anchors. If v_1 is a vertex of Q we are done. Otherwise, let $w_3 \neq 0, w_1, w_2$ be the remaining vertex of Q. The edges $\{v, v_1\}$ and $\{v_1, 0\}$ do not pass through the face $\{w_1, w_2, w_3\}$ by Lemma 4.19. Likewise, the edges $\{w_2, w_3\}$ and $\{w_3, w_1\}$ do not pass through the face $\{0, v, v_1\}$. Thus, v is enclosed over the quarter Q.

Let $w_3' \neq w_1, v_1, w_2$ be a fourth anchor of $\{0, v\}$. By Lemma 4.19, we have $w_3' = w_3$. □

Corollary 11.17 (of the proof). *If v is enclosed over a flat quarter, then* $\{0, v\}$ *has at most four anchors.*

When we are unable to erase the upright diagonal with five anchors and a large gap, we are able to obtain strong bounds on the score.

Lemma 11.18. *Suppose an upright diagonal in a decomposition star has five anchors and one large gap. The four anchored simplices score at most* -0.25. *The four anchored simplices squander at least* 0.4. *If any of the four anchored simplices is not an upright quarter then the decomposition star does not contravene.*

Proof. A list of inequalities[49] together with[50] dih > 1.65 give the bound -0.25. Further inequalities[51] give the bound 0.4. To get the final statement of the lemma, we again use a series of inequalities.[52,53] □

---

[47] CALC-83777706.
[48] CALC-729988292.
[49] CALC-815492935.
[50] CALC-83777706.
[51] CALC-729988292.
[52] CALC-628964355.
[53] CALC-187932932.

Corollary 11.19. *There is at most one 4-crowded upright diagonal in a contravening decomposition star.*

Proof. The crown along the large gap, with the bound of the lemma, gives[54] $0.4 - \kappa \geq 0.4 + 0.02274$ squandered by the upright quarters around a 4-crowded upright diagonal. The rest squanders a positive amount (see Lemma 9.20). If there are two 4-crowded upright diagonals, use $2(0.4 + 0.02274) > (4\pi\zeta - 8)\,pt$. □

Definition 11.20. We set $\xi_\Gamma = 0.01561, \xi_V = 0.003521, \xi'_\Gamma = 0.00935, \xi_\kappa = -0.029$, and $\xi_{\kappa,\Gamma} = \xi_\kappa + \xi_\Gamma = -0.01339$.

The first two constants appear in calculations[55,56] as penalties for erasing upright quarters of compression type, and Voronoi type, respectively. ξ'_Γ is an improved bound on the penalty for erasing when the upright diagonal is at least 2.57. Also, ξ_κ is an upper bound[57] on κ, when the upright diagonal is at most 2.57. If the upright diagonal is at least 2.57, then we still obtain the bound[58] $\xi_{\kappa,\Gamma} = -0.02274 + \xi'_\Gamma$ on the sum of κ with the penalty from erasing an upright quarter.

11.8. Four Anchors

Lemma 11.21. *If there are at least two large gaps around an upright diagonal with four anchors, then it can be erased.*

Proof. There are at least as many large gaps as upright quarters. Each large gap drops us by ξ_κ and each quarter lifts us by at most[59–61] ξ_Γ. We have $\xi_{\kappa,\Gamma} < 0$. □

Remark 11.22. Let $\{0, v\}$ be an enclosed vertex over a flat quarter. Then

$$|v| \geq \mathcal{E}(2, 2, 2, 2t_0, 2t_0, 2\sqrt{2}, 2, 2, 2) > 2.6.$$

If an edge of the flat quarter is sufficiently short, say $y_6 \leq 2.2$, then

$$|v| \geq \mathcal{E}(2, 2, 2, 2.2, 2t_0, 2\sqrt{2}, 2, 2, 2) > 2.7.$$

The two dihedral angles on the gaps are > 1.65. If the two quarters mask a flat quarter, we use the scoring of 2(c) in Section 11.9. We have $0.0114 < -2\xi_{\kappa,\Gamma}$.

---

[54] CALC-618205535.
[55] CALC-73974037.
[56] CALC-764978100.
[57] CALC-618205535.
[58] CALC-618205535.
[59] CALC-618205535.
[60] CALC-73974037.
[61] CALC-764978100.

When there is one large gap, we may erase with a penalty $\pi_0 = 0.008$.

Lemma 11.23. *Let v be an upright diagonal with four anchors. Assume that there is one large gap. The anchored simplices can be erased with penalty $\pi_0 = 0.008$. If any of the anchored simplices around v is not an upright quarter then we can erase with penalty $\pi_0 = 0.00222$.*

Moreover, if there is a flat quarter overlapping an upright quarter, then one of the following holds:

(1) *The truncated function s-vor$_0$ exceeds the score by at least 0.0063. The diagonal of the flat is at least 2.6, and the edge opposite the diagonal is at least 2.2.*
(2) *The truncated function exceeds the score by at least 0.0114. The diagonal of the flat is at least 2.7, and the edge opposite the diagonal is at most 2.2.*

Definition 11.24. Let a 3-unconfined upright diagonal be an upright diagonal that has four anchors and one large gap in a situation where there is no masked flat quarter.

Proof. The constants and inequalities used in this proof can be found in a series of calculations.[62–64]

First we establish the penalty 0.008. The truncated function s-vor$_0$ is an upper bound on the score of an anchored simplex that is not a quarter. By these inequalities, the result follows if the diagonal satisfies $y_1 \geq 2.57$.

Take $y_1 \leq 2.57$. If any of the upright quarters are of Voronoi type, the result follows from $(\xi_{\kappa,\Gamma} + \xi_\Gamma < 0.008)$. If the edges along the large gap are less than 2.25, the result follows from $(-0.03883 + 3\xi_\Gamma = 0.008)$. If all but one edge along the large gap are less than 2.25, the result follows from $(-0.0325 + 2\xi_\Gamma + 0.00928 = 0.008)$.

If there are at least two edges along the large gap of length at least 2.25, we consider two cases according to whether they lie on a common face of an upright quarter. The same group of inequalities gives the result. The bound 0.008 is now fully established.

Next we prove that we can erase with penalty 0.00222, when one of the anchored simplices is not a quarter. If $|v| \geq 2.57$, then we use

$$2\xi_\Gamma + \xi_V + \xi_\kappa \leq 0.00935 + 0.003521 - 0.2274 \leq 0. \qquad \text{}$$

If $|v| \leq 2.57$, we use

$$2(0.01561) - 0.029 \leq 0.00222.$$

Let $v_1 \ldots, v_4$ be the consecutive anchors of the upright diagonal $\{0, v\}$ with $\{v_1, v_4\}$ the large gap. Suppose $|v_1 - v_3| \leq 2\sqrt{2}$.

---

[62] CALC-618205535.
[63] CALC-73974037.
[64] CALC-764978100.

We claim the upright diagonal $\{0, v\}$ is not enclosed over $\{0, v_1, v_2, v_3\}$. Assume the contrary. The edge $\{v_1, v_3\}$ passes through the face $\{0, v, v_4\}$. Disregarding the vertex v_2, by geometric considerations, we arrive at the rigid figure

$$|v| = 2\sqrt{2}, \qquad |v_1| = |v_1 - v| = |v - v_3| = |v_3| = |v_3 - v_4| = 2,$$
$$|v - v_4| = |v_4| = 2t_0, \qquad |v_1 - v_4| = 3.2.$$

The dihedral angles of $\{0, v, v_1, v_4\}$ and $\{0, v, v_3, v_4\}$ are

$$\text{dih}(S(2\sqrt{2}, 2, 2t_0, 3.2, 2t_0, 2)) > 2.3, \qquad \text{dih}(S(2\sqrt{2}, 2, 2t_0, 2, 2t_0, 2)) > 1.16.$$

The sum is greater than π, contrary to the claim that the edge $\{v_1, v_3\}$ passes through the face $\{0, v, v_4\}$. (This particular conclusion leads to the corollary cited at the end of the proof.) Thus, $\{v_1, v_3\}$ passes through $\{0, v, v_2\}$ so that the simplices $\{0, v, v_1, v_2\}$ and $\{0, v, v_2, v_3\}$ are of Voronoi type.

To complete the proof of the lemma, we show that when there is a masked flat quarter, either (1) or (2) holds. Suppose we mask a flat quarter $Q' = \{0, v_1, v_2, v_3\}$. We have established that $\{v_1, v_3\}$ passes through the face $\{0, v, v_2\}$. To establish (1) assume that $|v_2| \geq 2.2$. The remark before the lemma gives

$$|v_1 - v_3| \geq \mathcal{E}(S(2, 2, 2, 2\sqrt{2}, 2t_0, 2t_0), 2, 2, 2) > 2.6.$$

The bound 0.0063 comes from

$$\xi_{\kappa,\Gamma} + 2\xi_V < -0.0063$$

To establish (2) assume that $|v_2| \leq 2.2$. The remark gives

$$|v_1 - v_3| \geq \mathcal{E}(S(2, 2, 2, 2\sqrt{2}, 2.2, 2t_0), 2, 2, 2) > 2.7.$$

If the simplex $\{0, v, v_3, v_4\}$ is of Voronoi type, then

$$\xi_\kappa + 3\xi_V < -0.0114.$$

Assume that $\{0, v, v_3, v_4\}$ is of compression type. We have

$$-0.004131 + \xi_{\kappa,\Gamma} + \xi_V \leq -0.0114. \qquad \square$$

Corollary 11.25 (of the proof).   *If there are four anchors and if the upright diagonal is enclosed over a flat quarter, then there are four anchored simplices and at least three quarters around the upright diagonal.*

11.9.   *Summary*

The following index summarizes the cases of upright quarters that have been treated in this section. If the number of anchors is the number of anchored simplices (no large gaps), the results appear in Section 13.12. Every other possibility has been treated.

- None, one, or two anchors.   (Section 11.2.)
- Three anchors: (Section 11.3.)
  - context $(3, 0)$,
  - context $(3, 1)$,
  - context $(3, 2)$,
  - context $(3, 3)$.
- Four anchors.   (Section 11.8.)
  - No gaps (Section 13.12),
  - One gap,
  - Two or more gaps.
- Five anchors: (Section 11.7),
  - No gaps (Section 13.12),
  - One gap (4-crowded),
  - Two or more gaps.
- Six or more anchors.   (Section 11.4.)

By truncation and various comparison lemmas, we have entirely eliminated upright diagonals except when there are between three and five anchors. We may assume that there is at most one large gap around the upright diagonal.

1. Consider an anchored simplex Q around a remaining upright diagonal. The score is $v(Q)$ if Q is a quarter, the analytic function s-vor(Q) if the simplex is of type C (Section 9.4), and the truncated function s-vor$_0(Q)$ otherwise.
2. Consider a flat quarter Q in an exceptional cluster. An upper bound on the score is obtained by taking the maximum of all of the following functions that satisfy the stated conditions on Q. Let y_4 denote the length of the diagonal and y_1 be the length of the opposite edge.
   (a) The function $\mu(Q)$.
   (b) s-vor$_0(Q) - 0.0063$, if $y_4 \geq 2.6$ and $y_1 \geq 2.2$. (Lemma 11.23.)
   (c) s-vor$_0(Q) - 0.0114$, if $y_4 \geq 2.7$ and $y_1 \leq 2.2$. (Lemma 11.23.)
   (d) $v(Q_1) + v(Q_2) + $s-vor$_x(S)$, if there is an enclosed vertex v over Q of height between $2t_0$ and $2\sqrt{2}$ that partitions the convex hull of (Q, v) into two upright quarters Q_1, Q_2 and a third simplex S. Here s-vor$_x = $ s-vor if S is of type C, and s-vor$_x = $ s-vor$_0$ otherwise. (Lemma 11.3.)
   (e) s-vor$(Q, 1.385)$ if the simplex is of type B (Section 9.4.)
   (f) s-vor$_0(Q)$ if the simplex is an isolated quarter with $\max(y_2, y_3) \geq 2.23$, ✱ $y_4 \geq 2.77$, and $\eta_{456} \geq \sqrt{2}$.
3. If S is a simplex of type A, its score is s-vor(S). (Section 9.4.)
4. Everything else is scored by the truncation vor$_0$. Formula (7.13) is used on these remaining pieces. On top of what is obtained for the standard cluster by summing all these terms, there is a penalty $\pi_0 = 0.008$ each time a 3-unconfined upright diagonal is erased.
5. The remaining upright diagonals that are not completely surrounded by anchored simplices are 3-unconfined, 3-crowded, or 4-crowded from Sections 11.6–11.8.

11.10. *Some Flat Quarters*

Recall that $\xi_V = 0.003521$, $\xi_\Gamma = 0.01561$, and $\xi_\Gamma' = 0.00935$. They are the penalties that result from erasing an upright quarter of Voronoi type, an upright quarter of compression type, and an upright quarter of compression type with diagonal ≥ 2.57. (See calculations.[65,66])

In the next lemma we score a flat quarter by any of the functions on the given domains

$$\hat\sigma = \begin{cases} \Gamma, & \eta_{234},\, \eta_{456} \leq \sqrt{2}, \\ \text{s-vor}, & \eta_{234} \geq \sqrt{2}, \\ \text{s-vor}_0, & y_4 \geq 2.6,\ y_1 \geq 2.2, \\ \text{s-vor}_0, & y_4 \geq 2.7, \\ \text{s-vor}_0, & \eta_{456} \geq \sqrt{2}. \end{cases}$$

Lemma 11.26. *$\hat\sigma$ is an upper bound on the functions in 2(a)–(f) of Section 11.9. That is, each function is dominated by some choice of $\hat\sigma$.*

Proof. The only case in doubt is the function of 3.10(d):

$$v(Q_1) + v(Q_2) + \text{s-vor}_x(S).$$

This is established by the following lemma.   □

We consider the context $(3, 1)$ that occurs when two upright quarters in the Q-system lie over a flat quarter. Let $\{0, v\}$ be the upright diagonal, and assume that $\{0, v_1, v_2, v_3\}$ is the flat quarter, with diagonal $\{v_2, v_3\}$. Let σ denote the score of the upright quarters and other anchored simplex lying over the flat quarter.

Lemma 11.27. $\sigma \leq \min(0, \text{s-vor}_0)$.

 Proof. The bound of 0 is established in Theorem 8.4.

By a calculation,[67] if $|v| \geq 2.69$, then the upright quarters satisfy

$$v < \text{s-vor}_0 + 0.01(\pi/2 - \text{dih}),$$

so the upright quarters can be erased. Thus we assume without loss of generality that $|v| \leq 2.69$.
We have

$$|v| \geq \mathcal{E}(S(2, 2, 2, 2t_0, 2t_0, 2\sqrt{2}), 2, 2, 2) > 2.6.$$

If $|v_1 - v_2| \leq 2.1$, or $|v_1 - v_3| \leq 2.1$, then

$$|v| \geq \mathcal{E}(S(2, 2, 2, 2.1, 2t_0, 2\sqrt{2}), 2, 2, 2) > 2.72,$$

---

[65] CALC-73974037.
[66] CALC-764978100.
[67] CALC-855677395.

contrary to assumption. So take $|v_1 - v_2| \geq 2.1$ and $|v_1 - v_3| \geq 2.1$. Under these conditions we have the interval calculation[68] $\nu(Q) <$ s-vor$_0(Q)$ where Q is the upright quarter. $\square$

Remark 11.28. If we have an upright diagonal enclosed over a masked flat quarter in the context $(4, 1)$, then there are three upright quarters. By the same argument as in the lemma, the two quarters over the masked flat quarter score $\leq$ s-vor$_0$. The third quarter can be erased with penalty ξ_V.

Define the *central vertex* v of a flat quarter to be the vertex for which $\{0, v\}$ is the edge opposite the diagonal.

Lemma 11.29. $\mu <$ s-vor$_0 +0.0268$ *for all flat quarters. If the central vertex has height ≤ 2.17, then* $\mu <$ s-vor$_0 +0.02$.

Proof. This is an interval calculation.[69] $\square$

We measure what is squandered by a flat quarter by $\hat{\tau} = $ sol $\zeta pt - \hat{\sigma}$.

Lemma 11.30. *Let v be a corner of an exceptional cluster at which the dihedral angle is at most 1.32. Then the vertex v is the central vertex of a flat quarter Q in the exceptional region. Moreover, $\hat{\tau}(Q) > 3.07\,pt$. If $\hat{\sigma} = $ s-vor$_0$ (and if $\eta_{456} \geq \sqrt{2}$), we may use the stronger constant $\tau_0(Q) > 3.07\,pt + \xi_V + 2\xi_\Gamma'$.*

Proof. Let $S = S(y_1, \ldots, y_6)$ be the simplex inside the exceptional cluster centered at v, with $y_1 = |v|$. The inequality dih ≤ 1.32 gives the interval calculation $y_4 \leq 2\sqrt{2}$, so S is a quarter. The result now follows by interval arithmetic.[70] $\square$

12. Bounds in Exceptional Regions

12.1. The Main Theorem

Let (R, D) be a standard cluster. Let U be the set of corners, that is, the set of vertices in the cone over R that have height at most $2t_0$. Consider the set E of edges of length at most $2t_0$ between vertices of U. We attach a multiplicity to each edge. We let the multiplicity be 2 when the edge projects radially to the interior of the standard region, and 0 when the edge projects radially to the complement of the standard region. The other edges, those bounding the standard region, are counted with multiplicity 1.

Let n_1 be the number of edges in E, counted with multiplicities. Let c be the number of classes of vertices under the equivalence relation $v \sim v'$ if there is a sequence of edges

---

[68] CALC-148776243.
[69] CALC-148776243.
[70] CALC-148776243.

in E from v to v'. Let $n(R) = n_1 + 2(c - 1)$. If the standard region under R is a polygon, then $n(R)$ is the number of sides.

Theorem 12.1. *Let D be a contravening decomposition star. $\tau_R(D) > t_n$, where $n = n(R)$ and*

$$t_4 = 0.1317, \qquad t_5 = 0.27113, \qquad t_6 = 0.41056,$$
$$t_7 = 0.54999, \qquad t_8 = 0.6045.$$

The decomposition star scores less than $8\,pt$, if $n(R) \geq 9$, for some standard cluster R. The scores satisfy $\sigma_R(D) < s_n$, for $5 \leq n \leq 8$, where

$$s_5 = -0.05704, \qquad s_6 = -0.11408, \qquad s_7 = -0.17112, \qquad s_8 = -0.22816.$$

Sometimes, it is convenient to calculate these bounds as a multiple of pt. We have

$$t_4 > 2.378\,pt, \qquad t_5 > 4.896\,pt, \qquad t_6 > 7.414\,pt,$$
$$t_7 > 9.932\,pt, \qquad t_8 > 10.916\,pt.$$

$$s_5 < -1.03\,pt, \qquad s_6 < -2.06\,pt, \qquad s_7 < -3.09\,pt, \qquad s_8 < -4.12\,pt.$$

Corollary 12.2. *Every standard region is a either a polygon or one shown in Fig. 12.1.*

In the cases that are not (simple) polygons, we call the *polygonal hull* the polygon obtained by removing the internal edges and vertices. We have $m(R) \leq n(R)$, where the constant $m(R)$ is the number of sides of the polygonal hull.

Proof. By the theorem, if the standard region is not a polygon, then $8 \geq n_1 \geq m \geq 5$. (Quad clusters and quasi-regular tetrahedra have no enclosed vertices. See Lemmas 10.4 and 5.13.) If $c > 1$, then $8 \geq n = n_1 + 2(c - 1) \geq 5 + 2(c - 1)$, so $c = 2$, and $n_1 = 5, 6$ (frames 2 and 5 of the figure).

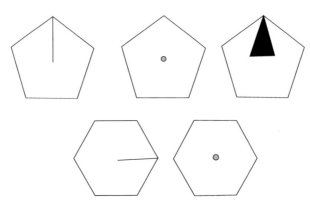

Fig. 12.1

Now take $c = 1$. Then $8 \geq n \geq 5 + (n - m)$, so $n - m \leq 3$. If $n - m = 3$, we get frame 3. If $n - m = 2$, we have $8 \geq m + 2 \geq 5 + 2$, so $m = 5, 6$ (frames 1 and 4). However, $n - m = 1$ cannot occur, because a single edge that does not bound the polygonal hull has even multiplicity. Finally, if $n - m = 0$, we have a polygon. □

Corollary 12.3. *If the type of a vertex of a decomposition star is $(7, 0)$, then it does not contravene.*

Proof. By Theorem 12.1, if there is a nontriangular region, we have

$$\tau(D) \geq \tau_{\mathrm{LP}}(7, 0) + t_4 > (4\pi\zeta - 8)\, pt.$$

Assume that all standard regions are triangular. If there is a vertex that does not lie on one of seven triangles, we have, by Lemma 10.5,

$$\tau(D) \geq \tau_{\mathrm{LP}}(7, 0) + 0.55\, pt > (4\pi\zeta - 8)\, pt.$$

Thus, all vertices lie on one of the seven triangles. The complement of these seven triangles is a region triangulation by five standard regions. There is some vertex of these five that does not lie on any of the other four standard regions in the complement. That vertex has type $(3, 0)$, which is contrary to Lemma 10.9. □

12.2. Nonagons

A few additional comments are needed to eliminate $n = 9$ and 10, even after the bounds t_9, t_{10} are established.

Lemma 12.4. *Let F be a set of one or more standard regions bounded by a simple polygon with at most nine edges. Assume that*

$$\sigma_F(D) \leq s_9 \quad \text{and} \quad \tau_F(D) \geq t_9,$$

where $s_9 = -0.1972$ and $t_9 = 0.6978$. Then D does not contravene.

Proof. Suppose that $n = 9$, and that R squanders at least t_9 and scores less than s_9. This bound is already sufficient to conclude that there are no other standard clusters except quasi-regular tetrahedra ($t_9 + t_4 > (4\pi\zeta - 8)\, pt$). There are no vertices of type $(4, 0)$ or $(6, 0)$: $t_9 + 4.14\, pt > (4\pi\zeta - 8)\, pt$ by Lemma 10.5. So all vertices not over the exceptional cluster are of type $(5, 0)$. Suppose that there are ℓ vertices of type $(5, 0)$. The polygonal hull of R has $m \leq 9$ edges. There are $m - 2 + 2\ell$ quasi-regular tetrahedra. If $\ell \leq 3$, then, by Lemma 10.6, the score is less than

$$s_9 + (m - 2 + 2\ell)\, pt - 0.48\ell\, pt < 8\, pt.$$

If, on the other hand, $\ell \geq 4$, the decomposition star squanders more than

$$t_9 + 4(0.55)\, pt > (4\pi\zeta - 8)\, pt.$$ □

The bound s_9 will be established as part of the proof of Theorem 12.1.

The case $n = 10$ is similar. If $\ell = 0$, the score is less than $(m - 2)\,pt \leq 8\,pt$, because the score of an exceptional cluster is strictly negative, Theorem 8.4. If $\ell > 0$, we squander at least $t_{10} + 0.55\,pt > (4\pi\zeta - 8)\,pt$ (Lemma 10.6).

12.3. *Distinguished Edge Conditions*

Take an exceptional cluster. We prepare the cluster by erasing upright diagonals, including those that are 3-unconfined, 3-crowded, or 4-crowded. The only upright diagonals that we leave unerased are loops. When the upright diagonal is erased, we score with the truncated function vor_0 away from flat quarters. Flat quarters are scored with the function $\hat{\sigma}$. The exceptional clusters in Sections 12 and 13 are assumed to be prepared in this way.

A simplex S is *special* if the fourth edge has length at least $2\sqrt{2}$ and at most 3.2, and the others have length at most $2t_0$. The fourth edge is called its diagonal.

We draw a system of edges between vertices. Each vertex will have height at most $2t_0$. The radial projections of the edges to the unit sphere will divide the standard region into subregions. We call an edge *nonexternal* if the radial projection of the edge lies entirely in the (closed) exceptional region.

1. Draw all nonexternal edges of length at most $2\sqrt{2}$ except those between non-consecutive anchors of a remaining upright diagonal. These edges do not cross (Lemma 4.30). These edges do not cross the edges of anchored simplices (Lemmas 4.22 and 4.24).
2. Draw all edges of (remaining) anchored upright simplices that are opposite the upright diagonal, except when the edge gives a special simplex. The anchored simplices do not overlap (Lemma 11.9), so these edges do not cross. These edges are nonexternal (Lemmas 11.5 and 4.19).
3. Draw as many additional nonexternal edges as possible of length at most 3.2 subject to not crossing another edge, not crossing any edge of an anchored simplex, and not being the diagonal of a special simplex.

We fix once and for all a maximal collection of edges subject to these constraints. Edges in this collection are called *distinguished* edges. The radial projection of the distinguished edges to the unit sphere gives the bounding edges of regions called the *subregions*. Each standard region is a union of subregions. The vertices of height at most $2t_0$ and the vertices of the remaining upright diagonals are said to form a *subcluster*.

By construction, the special simplices and anchored simplices around an upright quarter form a subcluster. Flat quarters in the Q-system, flat quarters of an isolated pair, and simplices of type A and B are subclusters. Other subclusters are scored by the function vor_0. For these subclusters, formula (7.13) extends without modification.

12.4. *Scoring Subclusters*

The terms of formula (7.13) defining $\mathrm{vor}_{0,P}(D) = \mathrm{vor}_P(D, t_0)$ have a clear geometric interpretation as quoins, wedges of t_0-cones, and solid angles (see Section 7). There is a quoin for each Rogers simplex. There is a somewhat delicate point that arises in connection with the geometry of subclusters. It is not true in general that the Rogers simplices entering into the truncation $vor_{0, P}(D)$ of (P, D) lie in the cone over P. Formula (7.13) should be viewed as an analytic continuation that has a nice geometric interpretation when things are nice, and which always gives the right answer when summed over all the subclusters in the cluster, but which may exhibit unusual behavior in general. The following lemma shows that the simple geometric interpretation of formula (7.13) is valid when the subregion is not triangular.

Lemma 12.5. *If a subregion is not a triangle and is not the subregion containing the anchored simplices around an upright diagonal, the cone of arcradius*

$$\psi = \arccos(|v|/(2t_0))$$

centered along $\{0, v\}$, where v is a corner of the subcluster, does not cross out of the subregion.

Proof. For a contradiction, let $\{v_1, v_2\}$ be a distinguished edge that the cone crosses. If both edges $\{v, v_1\}$ and $\{v, v_2\}$ have length less than $2t_0$, there can be no enclosed vertex w of height at most $2t_0$, unless its distance from v_1 and v_2 is less than $2t_0$:

$$\mathcal{E}(S(2, 2, 2, 2t_0, 2t_0, 3.2), 2t_0, 2, 2) > 2t_0.$$

In this case we can replace $\{v_1, v_2\}$ by an edge of the subregion closer to v, so without loss of generality we may assume that there are no enclosed vertices when both edges $\{v, v_1\}$ and $\{v, v_2\}$ have length less than $2t_0$.

The subregion is not a triangle, so $|v - v_1| \geq 2t_0$, or $|v - v_2| \geq 2t_0$, say $|v - v_1| \geq 2t_0$. Also $|v - v_2| \geq 2$. Pivot so that $|v_1 - v_2| = 3.2$, $|v - v_1| = 2t_0$, and $|v - v_2| = 2$. (The simplex $\{0, v_1, v_2, v\}$ cannot collapse $(\Delta \neq 0)$ as we pivot. For more details about why $\Delta \neq 0$, see inequality (12.2) in Section 12.7.) Then use[71] $\beta_\psi \leq dih_3$.   □

As a consequence, in nonspecial standard regions, the terms in the formula (7.13) for vor_0 retain their interpretations as quoins, Rogers simplices, t_0-cones, and solid angles, all lying in the cone over the standard region.

12.5. *Proof*

The proof of the theorem occupies the rest of the section. The inequalities for triangular and quadrilateral regions have already been proved. The bounds on t_3, t_4, s_3, and s_4 are found in Lemma 10.1, Section 11.1, Lemma 8.10, and Theorem 8.4, respectively. Thus, we may assume throughout the proof that the standard region is exceptional

We begin with a slightly simplified account of the method of the proof. Set $t_9 = 0.6978$, $t_{10} = 0.7891$, $t_n = (4\pi\zeta - 8)\,pt$, for $n \geq 11$. Set $D(n, k) = t_{n+k} - 0.06585\,k$, for $0 \leq k \leq n$, and $n + k \geq 4$. This function satisfies

$$D(n_1, k_1) + D(n_2, k_2) \geq D(n_1 + n_2 - 2, k_1 + k_2 - 2). \tag{12.1}$$

---

[71] CALC-193836552.

In fact, this inequality unwinds to $t_r + 0.13943 \geq t_{r+1}$, $D(3, 2) = 0.13943$, and $t_n = (0.06585)2 + (n - 4)D(3, 2)$, for $n = 4, 5, 6, 7$. These hold by inspection.

Call an edge between two vertices of height at most $2t_0$ *long* if it has length greater than $2t_0$. Add the distinguished edges to break the standard regions into subregions. We say that a subregion has *edge parameters* (n, k) if there are n bounding edges, where k of them are long. (We count edges with multiplicities as in Section 12.1, if the subregion is not a polygon.) Combining two subregions of edge parameters (n_1, k_1) and (n_2, k_2) along a long edge e gives a union with edge parameters $(n_1 + n_2 - 2, k_1 + k_2 - 2)$, where we agree not to count the internal edge e that no longer bounds. Inequality (12.1) localizes the main theorem to what is squandered by subclusters. Suppose we break the standard cluster into groups of subregions such that if the group has edge parameters (n, k), it squanders at least $D(n, k)$. Then by superadditivity (formula (12.1)), the full standard cluster R must squander $D(n, 0) = t_n$, $n = n(R)$, giving the result.

Similarly, define constants $s_4 = 0$, $s_9 = -0.1972$, $s_n = 0$, for $n \geq 10$. Set $Z(n, k) = s_{n+k} - k\varepsilon$, for $(n, k) \neq (3, 1)$, and $Z(3, 1) = \varepsilon$, where[72] $\varepsilon = 0.00005$. The function $Z(n, k)$ is subadditive:

$$Z(n_1, k_1) + Z(n_2, k_2) \leq Z(n_1 + n_2 - 2, k_1 + k_2 - 2).$$

In fact, this easily follows from $s_a + s_b \leq s_{a+b-4}$, for $a, b \geq 4$, and $\varepsilon > 0$. It will be enough in the proof of Theorem 12.1 to show that the score of a union of subregions with edge parameters (n, k) is at most $Z(n, k)$.

12.6. Preparation of the Standard Cluster

Fix a standard cluster. We return to the construction of subregions and distinguished edges, to describe the penalties. Take the penalty of 0.008 for each 3-unconfined upright diagonal. Take the penalty $0.03344 = 3\xi_\Gamma + \xi_{\kappa,\Gamma}$ for 4-crowded upright diagonals. Take the penalty $0.04683 = 3\xi_\Gamma$ for 3-crowded upright diagonals. Set $\pi_{max} = 0.06688$. The penalty in the next lemma refers to the combined penalty from erasing all 3-unconfined, 3-crowded, and 4-crowded upright diagonals in the decomposition star. The upright quarters that completely surround an upright diagonal (loops) are not erased.

Lemma 12.6. *The total penalty from a contravening decomposition star is at most* π_{max}.

Proof. Before any upright quarters are erased, each quarter squanders[73] > 0.033, so the star squanders $> (4\pi \zeta - 8)\, pt$ if there are twenty-five or more quarters. Assume there are at most twenty-four quarters. If the only penalties are 0.008, we have $8(0.008) < \pi_{max}$. If we have the penalty 0.04683, there are at most seven other quarters $(0.5606 + 8(0.033) > (4\pi\zeta - 8)\, pt)$ (Lemma 11.11), and no other penalties from this type or from 4-crowded upright diagonals, so the total penalty is at most $2(0.008) + 0.04683 < \pi_{max}$.

---

[72] Compare CALC-193836552.
[73] CALC-148776243.

Finally, if there is one 4-crowded upright diagonal, there are at most twelve other quarters (Section 11.7), and erasing gives the penalty $0.03344 + 4(0.008) < \pi_{\max}$. $\qquad$ □

The remaining upright diagonals are surrounded by anchored simplices. If the edge opposite the diagonal in an anchored simplex has length $\geq 2\sqrt{2}$, then there may be an adjacent special simplex whose diagonal is that edge. Section 13.12 will give bounds on the aggregate of these anchored simplices and special simplices. In all other contexts, the upright quarters have been erased with penalties.

Break the standard cluster into subclusters as in Section 12.3. If the subregion is a triangle, we refer to the bounds of Section 13.8. Sections 12.7–13.11 give bounds for subregions that are not triangles in which all the upright quarters have been erased. We follow the strategy outlined in Section 12.5, although the penalties will add certain complications.

We now assume that we have a subcluster without quarters and whose region is not triangular. The truncated function vor_0 is an upper bound on the score. Penalties are largely disregarded until Section 13.4.

We describe a series of deformations of the subcluster that increase $\mathrm{vor}_{0,P}(D)$ and decrease $\tau_{0,P}(D)$. These deformations disregard the broader geometric context of the subcluster. Consequently, we cannot claim that the deformed subcluster exists in any decomposition star D. As the deformation progresses, an edge $\{v_1, v_2\}$, not previously distinguished, can emerge with the properties of a distinguished edge. If so, we add it to the collection of distinguished edges, use it if possible to divide the subcluster into smaller subclusters, and continue to deform the smaller pieces. When triangular regions are obtained, they are set aside until Section 13.8.

12.7.   Reduction to Polygons

By deformation we can produce subregions whose boundary is a polygon. Let U be the set of vertices over the subregion of height $\leq 2t_0$. As in Section 12.1, the distinguished edges partition U into equivalence classes. Move the vertices in one equivalence class U_1 as a rigid body preserving heights until the class comes sufficiently close to form a distinguished edge with another subset. Continue until all the vertices are interconnected by paths of distinguished edges. vor_0 and τ_0 are unchanged by these deformations.

If some vertex v is connected to three or more vertices by distinguished edges, it follows from the connectedness of the open subregion that there is more than one connected component U_i (by paths of distinguished edges) of $U \setminus \{v\}$. Move $U_1 \cup \{v\}$ rigidly preserving heights and keeping v fixed until a distinguished edge forms with another component. Continue until the distinguished edges break the subregions into subregions with polygon boundaries. Again vor_0 and τ_0 are unchanged.

By the end of Section 12, we will deform all subregions into convex polygons.

Remark 12.7.   We will deform in such a way that the edges $\{v_1, v_2\}$ will maintain a length of at least 2. The proof that distances of at least 2 are maintained is given in Section 12.13.

We will deform in such a way that no vertex crosses a boundary of the subregion passing from the outside to the inside.

Edge length constraints prevent a vertex from crossing a boundary of the subregion from the inside to the outside. In fact, if v is to cross the edge $\{v_1, v_2\}$, the simplex $S = \{0, v_1, v, v_2\}$ attains volume 0. We may assume, by the argument of the proof of Lemma 12.4, that there are no vertices enclosed over S. Because we are assuming that the subregion is not a triangle, we may assume that $|v - v_1| > 2t_0$. We have $|v| \in [2, 2t_0]$. If v is to cross $\{v_1, v_2\}$, we may assume that the dihedral angles of S along $\{0, v_1\}$ and $\{0, v_2\}$ are acute. Under these constraints, by the explicit formulas of Section 8 of [Ha6], the vertex v cannot cross out of the subregion

$$\Delta(S) \geq \Delta(2t_0^2, 4, 4, 3.2^2, 4, 2t_0^2) > 0. \tag{12.2}$$

We say that a corner v_1 is *visible* from another v_2 if $\{v_1, v_2\}$ lies over the subregion. A deformation may make v_1 visible from v_2, making it a candidate for a new distinguished edge. If $|v_1 - v_2| \leq 3.2$, then as soon as the deformation brings them into visibility (obstructed until then by some v), inequality (12.2) shows that $|v_1 - v|, |v_2 - v| \leq 2t_0$. So v_1, v, v_2 are consecutive edges on the polygonal boundary, and $|v_1 - v_2| \geq 2\sqrt{4 - t_0^2} > \sqrt{8}$. By the distinguished edge conditions for special simplices, $\{v_1, v_2\}$ is too long to be distinguished. In other words, there can be no potentially distinguished edges hidden behind corners. They are always formed in full view.

12.8. *Some Deformations*

Definition 12.8. Consider three consecutive corners v_3, v_1, v_2 of a subcluster R such that the dihedral angle of R at v_1 is greater than π. We call such a corner *concave*. (If the angle is less than π, we call it *convex*.) Similarly, the angle of a subregion is said to be convex or concave depending on whether it is less than or greater than π.

Let $S = S(y_1, \ldots, y_6) = \{0, v_1, v_2, v_3\}$, $y_i = |v_i|$. Suppose that $y_6 > y_5$. Let $x_i = y_i^2$.

Lemma 12.9. *At a concave vertex, $\partial \operatorname{vor}_0 / \partial x_5 > 0$ and $\partial \tau_0 / \partial x_5 < 0$.*

Proof. As x_5 varies, $\operatorname{dih}_i(S) + \operatorname{dih}_i(R)$ is constant for $i = 1, 2, 3$. The part of formula (7.13) for vor_0 that depends on x_5 can be written

$$-B(y_1)\operatorname{dih}(S) - B(y_2)\operatorname{dih}_2(S) - B(y_3)\operatorname{dih}_3(S) - 4\delta_{\text{oct}}(\operatorname{quo}(R_{135}) + \operatorname{quo}(R_{315})),$$

where $B(y_i) = A(y_i/2) + \varphi_0$, $R_{135} = R(y_1/2, b, t_0)$, $R_{315} = R(y_3/2, b, t_0)$, $b = \eta(y_1, y_3, y_5)$, and $A(h) = (1 - h/t_0)(\varphi(h, t_0) - \varphi_0)$. Set $u_{135} = u(x_1, x_3, x_5)$ and $\Delta_i = \partial \Delta / \partial x_i$. (The notation comes from Section 8 of [Ha6] and Section 7 of this issue.) We have

$$\frac{\partial \operatorname{quo}(R(a, b, c))}{\partial b} = \frac{-a(c^2 - b^2)^{3/2}}{3b(b^2 - a^2)^{1/2}} \leq 0$$

and $\partial b/\partial x_5 \geq 0$. Also, $u \geq 0$, $\Delta \geq 0$ (see Section 8 of [Ha6]). So it is enough to show that

$$V_0(S) = u_{135}\Delta^{1/2}\frac{\partial}{\partial x_5}(B(y_1)\dih(S) + B(y_2)\dih_2(S) + B(y_3)\dih_3(S)) < 0.$$

By the explicit formulas of Section 8 of [Ha6], we have

$$V_0(S) = -B(y_1)y_1\Delta_6 + B(y_2)y_2u_{135} - B(y_3)y_3\Delta_4.$$

For τ_0, we replace B with $B - \zeta pt$. It is enough to show that

$$V_1(S) = -(B(y_1) - \zeta pt)y_1\Delta_6 + (B(y_2) - \zeta pt)y_2u_{135} - (B(y_3) - \zeta pt)y_3\Delta_4 < 0.$$

The lemma now follows from an interval calculation. We note that the polynomials V_i are linear in x_4, and x_6, and this may be used to reduce the dimension of the calculation. $\qquad\square$

We give a second form of the lemma when the dihedral angle of R is less than π, that is, at a convex corner.

Lemma 12.10. *At a convex corner, $\partial \mathrm{vor}_0 /\partial x_5 < 0$ and $\partial \tau_0/\partial x_5 > 0$, if y_1, y_2, $y_3 \in [2, 2t_0]$, $\Delta \geq 0$, and* (i) $y_4 \in [2\sqrt{2}, 3.2]$, y_5, $y_6 \in [2, 2t_0]$, *or* (ii) $y_4 \geq 3.2$, y_5, $y_6 \in [2, 3.2]$.

Proof. We adapt the proof of the previous lemma. Now $\dih_i(S) - \dih_i(R)$ is constant, for $i = 1, 2, 3$, so the signs change. vor_0 depends on x_5 through

$$\sum B(y_i)\dih_i(S) - 4\delta_{\mathrm{oct}}(\mathrm{quo}(R_{135}) + \mathrm{quo}(R_{315})).$$

So it is enough to show that

$$V_0 - 4\delta_{\mathrm{oct}}\Delta^{1/2}u_{135}\frac{\partial}{\partial x_5}(\mathrm{quo}(R_{135}) + \mathrm{quo}(R_{315})) < 0.$$

Similarly, for τ_0, it is enough to show that

$$V_1 - 4\delta_{\mathrm{oct}}\Delta^{1/2}u_{135}\frac{\partial}{\partial x_5}(\mathrm{quo}(R_{135}) + \mathrm{quo}(R_{315})) < 0.$$

By an interval calculation[74]

$$-4\delta_{\mathrm{oct}}u_{135}\frac{\partial}{\partial x_5}(\mathrm{quo}(R_{135}) + \mathrm{quo}(R_{315})) < 0.82, \qquad \text{on} \quad [2, 2t_0]^3,$$

$$< 0.5, \qquad \text{on} \quad [2, 2t_0]^3, \quad y_5 \geq 2.189.$$

The result now follows from the inequalities.[75] $\qquad\square$

---

[74] CALC-984628285.
[75] CALC-984628285.

Return to the situation of concave corner v_1. Let v_2, v_3 be the adjacent corners. By increasing x_5, the vertex v_1 moves away from every corner w for which $\{v_1, w\}$ lies outside the region. This deformation then satisfies the constraint of Remark 12.7. Stretch the shorter of $\{v_1, v_2\}$, $\{v_1, v_3\}$ until $|v_1 - v_2| = |v_1 - v_3| = 3.07$ (or until a new distinguished edge forms, etc.). Do this at all concave corners.

By stopping at 3.07, we prevent a corner crossing an edge from the outside to the inside. Let w be a corner that threatens to cross a distinguished edge $\{v_1, v_2\}$ as a result of the motion at a nonconvex vertex. To say that the crossing of the edge is from the outside to the inside implies more precisely that the vertex being moved is an endpoint, say v_1, of the distinguished edge. At the moment of crossing the simplex $\{0, v_1, v_2, w\}$ degenerates to a planar arrangement, with the radial projection of w lying over the geodesic arc connecting the radial projections of v_1 and v_2. To see that the crossing cannot occur, it is enough to note that the volume of a simplex with opposite edges of lengths at most $2t_0$ and 3.07 and other edges at least 2 cannot be planar. The extreme case is

$$\Delta(2^2, 2^2, (2t_0)^2, 2^2, 2^2, 3.07^2) > 0.$$

If $|v_1| \geq 2.2$, we can continue the deformations even further. We stretch the shorter of $\{v_1, v_2\}$ and $\{v_1, v_3\}$ until $|v_1 - v_2| = |v_1 - v_3| = 3.2$ (or until a new distinguished edge forms, etc.). Do this at all concave corners v_1 for which $|v_1| \geq 2.2$. To see that corners cannot cross an edge from the outside to the inside, we argue as in the previous paragraph, but replacing 3.07 with 3.2. The extreme case becomes

$$\Delta(2.2^2, 2^2, (2t_0)^2, 2^2, 2^2, 3.2^2) > 0.$$

12.9. Truncated Corner Cells

Because of the arguments in Section 12.8, we may assume without loss of generality that we are working with a subregion with the following properties. If v is a concave vertex and w is not adjacent to v, and yet is visible from v, then $|v - w| \geq 3.2$. If v is a concave corner, then $|v - w| \geq 3.07$ for both adjacent corners w. If v is a concave corner and $|v| \geq 2.2$, then $|v - w| \geq 3.2$ for both adjacent corners w. These hypotheses will remain in force through to the end of Section 12.

Recall from Definition 12.8 that we call a spherical region *convex* if its interior angles are all less than π. The case where the subregion is a convex triangle will be treated in Section 13.8. Hence, we may also assume in Sections 12.9–12.12 that the subregion is not a convex triangle.

We construct a *corner cell* at each corner. It depends on a parameter $\lambda \in [1.6, 1.945]$. In all applications, we take $\lambda = 1.945 = 3.2 - t_0$, $\lambda = 1.815 = 3.07 - t_0$, or $\lambda = 1.6 = 3.2/2$.

To construct the cell around the corner v, place a triangle along $\{0, v\}$ with sides $|v|$, t_0, λ (with λ opposite the origin). Generate the solid of rotation around the axis $\{0, v\}$. Extend to a cone over 0. Slice the solid by the perpendicular bisector of $\{0, v\}$, retaining the part near 0. Intersect the solid with a ball of radius t_0. The cones over the two boundary edges of the subregion at v make two cuts in the solid. Remove the slice that lies outside the cone over the subcluster. What remains is the corner cell at v with parameter λ.

Corner cells at corners separated by a distance less than 2λ may overlap. We define a truncation of the corner cell that has the property that the *truncated corner cells* at adjacent corners do not overlap. Let $\{0, v_i, v_j\}^\perp$ denote the plane perpendicular to the plane $\{0, v_i, v_j\}$ passing through the origin and the circumcenter of $\{0, v_i, v_j\}$.

Let v_1, v_2, v_3 be consecutive corners of a subcluster. Take the corner cell with parameter λ at the corner v_2. Slice it by the planes $\{0, v_1, v_2\}^\perp$ and $\{0, v_2, v_3\}^\perp$, and retain the part along the edge $\{0, v_2\}$. This is the truncated corner cell (tcc). By construction tcc's at adjacent corners are separated by a plane $(0, \cdot, \cdot)^\perp$. Tcc's at nonadjacent corners do not overlap if the corners are $\geq 2\lambda$ apart. Tcc's will only be used in subregions satisfying this condition. It will be shown in Section 12.11 that tcc's lie in the cone over the subregion (for suitable λ).

12.10. *Formulas for Truncated Corner Cells*

We will assign a score to tcc's, in such a way that the score of the subcluster can be estimated from the scores of the corner cells.

We write C_0 for a tcc. We write C_0^u for the corresponding untruncated corner cell. (Although we call this the untruncated corner cell to distinguish it from the corner cell, it is still truncated in the sense that it lies in the ball at the origin of radius t_0. It is untruncated in the sense that it is not cut by the planes $(\cdots)^\perp$.)

For any solid body X, we define the *geometric* truncated function by

$$\text{vor}_0^g(X) = 4(-\delta_{\text{oct}} \text{vol}(X) + \text{sol}(X)/3)$$

the counterpart for squander

$$\tau_0^g(X) = \zeta pt \, \text{sol}(X) - \text{vor}_0^g(X).$$

The solid angle is to be interpreted as the solid angle of the cone formed by all rays from the origin through nonzero points of X. We may apply these definitions to obtain formulas for $\text{vor}_0^g(C_0)$, and so forth.

The formula for the score of a tcc differs slightly according to the convexity of the corner. We start with a convex corner v, and let v_1, v, and v_2 be consecutive corners in the subregion.

Let $S = \{0, v, v_1, v_2\}$ be a simplex with $|v_1 - v_2| \geq 3.2$. The formula for the score of a tcc $C_0(S)$ simplifies if the face of C_0 cut by $\{0, v, v_1\}^\perp$ does not meet the face cut by $\{0, v, v_2\}^\perp$. We make that assumption in this subsection. Set $\chi_0(S) = \text{vor}_0^g(C_0(S))$. (The function χ_0 is unrelated to the function χ that was introduced in Definition 5.14 to measure the orientation of faces.)

$$\psi = \text{arc}(y_1, t_0, \lambda), \qquad h = y_1/2,$$
$$R'_{126} = R(y_1/2, \eta_{126}, y_1/(2\cos\psi)), \qquad R_{126} = R(y_1/2, \eta_{126}, t_0),$$
$$\text{sol}'(y_1, y_2, y_6) = +\text{dih}(R'_{126})(1 - \cos\psi) - \text{sol}(R'_{126}),$$

$$\chi_0(S) = \text{dih}(S)(1 - \cos\psi)\varphi_0$$
$$- \text{sol}'(y_1, y_2, y_6)\varphi_0 - \text{sol}'(y_1, y_3, y_5)\varphi_0$$
$$+ A(h)\,\text{dih}(S) - 4\delta_{\text{oct}}(\text{quo}(R_{126}) + \text{quo}(R_{135})).$$

In the three lines giving the formula for χ_0, the first line represents the score of the cone before it is cut by the planes $\{0, v, v_i\}^\perp$ and the perpendicular bisector of $\{0, v\}$. The second line is the correction resulting from cutting the tcc along the planes $\{0, v, v_i\}^\perp$. The face of the Rogers simplex R'_{126} lies along the plane $\{0, v, v_1\}^\perp$. The third line is the correction from slicing the tcc with the perpendicular bisector of $\{0, v\}$. This last term is the same as the term appearing for a similar reason in the formula for vor_0 in formula (7.13). In this formula R is the usual Rogers simplex and $\text{quo}(R_{ijk})$ is the quoin coming from a Rogers simplex along the face with edges (ijk).

The formula for the untruncated corner cell is obtained by setting "sol'" and "quo" to "0" in the expression for χ_0. Thus,

$$\text{vor}^{\text{g}}(C_0^{\text{u}}) = \text{dih}(S)[(1 - \cos\psi)\varphi_0 + A(h)].$$

The formula depends only on λ, the dihedral angle, and the height $|v|$. We write $C_0^{\text{u}} = C_0^{\text{u}}(|v|, \text{dih})$, and suppress λ from the notation. The dependence on $\text{dih}(S)$ is linear:

$$\tau_0^{\text{g}}(C_0^{\text{u}}(|v|, \text{dih})) = (\text{dih}/\pi)\tau_0^{\text{g}}(C_0^{\text{u}}(|v|, \pi)).$$

The dependence of χ_0 on the fourth edge $y_4 = |v_1 - v_2|$ comes through a term proportional to $\text{dih}(S)$. Since the dihedral angle is monotonic in y_4, so is χ_0. Thus, under the assumption that $|v_1 - v_2| \geq 3.2$, we obtain an upper bound on χ_0 at $y_4 = 3.2$. Our deformations will fix the lengths of the other five variables, and monotonicity gives us the sixth. Thus, the tcc's lead to an upper bound on vor_0^{g} (and a lower bound on τ_0^{g}) that does not require interval arithmetic.

At a concave vertex, the formula is similar. Replace "$\text{dih}(S)$" with "$(2\pi - \text{dih}(S))$" in the given expression for χ_0. We add a superscript minus to the name of the function at concave vertices, to denote this modification: $\chi_0^-(C_0)$.

12.11. Containment of Truncated Corner Cells

The assumptions made at the beginning of Section 12.9 remain in force.

Lemma 12.11. *Let v be a concave vertex with $|v| \geq 2.2$. The tcc at v with parameter $\lambda = 1.945$ lies in the truncated V-cell over R.*

Proof. Consider a corner cell at v and a distinguished edge $\{v_1, v_2\}$ forming the boundary of the subregion. The corner cell with parameter $\lambda = 1.945$ is contained in a cone of arcradius $\theta = \text{arc}(2, t_0, \lambda) < 1.21 < \pi/2$ (in terms of the function *arc* of Section 9.7). Take two corners w_1, w_2, visible from v, between which the given bounding edge appears. (We may have $w_i = v_i$.) The two visible edges, $\{v, w_i\}$, have length ≥ 3.2. (Recall that the distinguished edges at v have been deformed to length 3.2.) They have arc-length at least arc$(2t_0, 2t_0, 3.2) > 1.38$. The segment of the distinguished edge $\{v_1, v_2\}$ visible from v has arc-length at most arc$(2, 2, 3.2) < 1.86$.

We check that the corner cell cannot cross the visible portion of the edge $\{v_1, v_2\}$. Consider the spherical triangle formed by the edges $\{v, w_1\}, \{v, w_2\}$ (extended as needed) and the visible part of $\{v_1, v_2\}$. Let C be the radial projection of v and let AB be the radial projection of the visible part of $\{v_1, v_2\}$. Pivot A and B toward C until the edges AC and BC have arc-length 1.38. The perpendicular from C to AB has length at least

$$\arccos(\cos(1.38)/\cos(1.86/2)) > 1.21 > \theta.$$

This proves that the corner cell lies in the cone over the subregion.   □

Lemma 12.12. *Let v be a concave vertex. The truncated corner cell at v with parameter $\lambda = 1.815$ lies in the truncated V-cell over R.*

Proof. The proof proceeds along the same lines as the previous lemma, with slightly different constants. Replace 1.945 with 1.815, 1.38 with 1.316, and 1.21 with 1.1. Replace 3.2 with 3.07 in contexts giving a lower bound to the length of an edge at v, and keep it at 3.2 in contexts calling for an upper bound on the length of a distinguished edge. The constant 1.86 remains unchanged.   □

Lemma 12.13. *The truncated corner cells with parameter 1.6 in a subregion do not overlap.*

Proof. We may assume that the corners are not adjacent. If a nonadjacent corner w is visible from v, then $|w - v| \geq 3.2$, and an interior point intersection p is incompatible with the triangle inequality: $|p - v| \leq 1.6, |p - w| < 1.6$. If w is not visible, we have a chain $v = v_0, v_1, \ldots, v_r = w$ such that v_{i+1} is visible from v_i. Imagine a taut string inside the subregion extending from v to w. The radial projections of v_i are the corners of the string's path. The string bends in an angle greater than π at each v_i, so the angle at each intermediate v_i is greater than π. That is, they are concave. Thus, by our deformations $|v_i - v_{i+1}| \geq 3.07$. The string has arc-length at least r arc$(2t_0, 2t_0, 3.07) > r(1.316)$. However, the corner cells lie in cones of arcradius arc$(2, t_0, \lambda) < 1$. So $2(1.0) > r(1.316)$, or $r = 1$. Thus, w is visible from v.   □

Lemma 12.14. *The corner cell for $\lambda \leq 1.815$ does not overlap the t_0-cone wedge around another corner w.*

Proof. We take $\lambda = 1.815$. As in the previous proof, if there is overlap along a chain, then

$$\text{arc}(2, t_0, \lambda) + \text{arc}(2, t_0, t_0) > r \, \text{arc}(2t_0, 2t_0, 3.07),$$

and again $r = 1$. So each of the two vertices in question is visible from the other. However, overlap implies $|p - v| \leq 1.815$ and $|p - w| < t_0$, forcing the contradiction $|w - v| < 3.07$.   □

Lemma 12.15. *The corner cell for $\lambda \leq 1.945$ at a corner v satisfying $|v| \geq 2.2$ does not overlap the t_0-cone wedge around another corner w.*

Proof. We take $\lambda = 1.945$. As in the previous proof, if there is overlap along a chain, then

$$\mathrm{arc}(2, t_0, \lambda) + \mathrm{arc}(2, t_0, t_0) > r\,\mathrm{arc}(2t_0, 2t_0, 3.2),$$

and again $r = 1$. Then the result follows from

$$|w - v| \leq |p - v| + |p - w| < 1.945 + t_0 = 3.2. \qquad \square$$

Definition 12.16. By a *penalty-free* score, we mean the part of the scoring bound that does not include any of the penalty terms. We sometimes call the full score, including the penalty terms, the *penalty-inclusive* score.

Lemma 12.4 was stated in the context of a subregion before deformation, but a cursory inspection of the proof shows that the geometric conditions required for the proof remain valid by our deformations. (This assumes that the subregion is not a triangle, which we assumed at the beginning of Section 12.9.) In more detail, there is a solid $C P_0$ contained in the ball of radius t_0 at the origin, and lying over the cone of the subregion P such that a bound on the penalty-free subcluster score is $\mathrm{vor}_0^g(C P_0)$ and squander $\tau_0^g(C P_0)$.

Let $\{y_1, \ldots, y_r\}$ be a decomposition of the subregion into disjoint regions whose union is X. Then if we let $C P_0(y_i)$ denote the intersection of $C P_0(y_i)$ with the cone over y_i, we can write

$$\tau_0^g(C P_0) = \sum_i \tau_0^g(C P_0(y_i)).$$

These lemmas allow us to express bounds on the score (and squander) of a subcluster as a sum of terms associated with individual (truncated) corner cells. By Lemmas 12.11–12.15, these objects do not overlap under suitable conditions. Moreover, by the interpretation of terms provided by Section 12.4, the cones over these objects do not overlap, when the objects themselves do not. In other words, under the various conditions, we can take the (truncated) corner cells to be among the sets $C P_0(y_i)$.

To work a typical example, we place a tcc with parameter $\lambda = 1.6$ at each concave corner. We place a t_0-cone wedge X_0 at each convex corner. The cone over each object lies in the cone over the subregion. By Lemmas 12.5 and 9.20 (see the proof), the t_0-cone wedge X_0 squanders a positive amount. The part P' of the subregion outside all tcc's and outside the t_0-cone wedges squanders

$$\mathrm{sol}(P')(\zeta pt - \varphi_0) > 0,$$

where $\mathrm{sol}(P')$ is the part of the solid angle of the subregion lying outside the tccs. Dropping these positive terms, we obtain a lower bound on the penalty-free squander:

$$\tau_0^g(C P_0) \geq \sum_{C_0} \tau_0^g(C_0).$$

There is one summand for each concave corner of the subregion. Other cases proceed similarly.

12.12. Convexity

Lemma 12.17. *There are at most two concave corners.*

Proof. Use the parameter $\lambda = 1.6$ and place a tcc C_0 at each concave corner v. Let $C_0^u(|v|, \text{dih})$ denote the corresponding untruncated cell. The formula of Section 12.10 gives

$$\tau_0^g(C_0) = \tau_0^g(C_0^u(|v|, \text{dih})) - \text{sol}'(y_1, y_2, y_6)\varphi_0' - \text{sol}'(y_1, y_3, y_5)\varphi_0',$$

where $\varphi_0' = \zeta pt - \varphi_0 < 0.6671$. (The conditions $y_5 \geq 3.07$ and $y_6 \geq 3.07$ force the faces along the these edges to have circumradius greater than t_0, and this causes the "quo" terms in the formula to be zero.)

By monotonicity in dih, a lower bound on $\tau_0^g(C_0^u)$ is obtained at dih $= \pi$. $\tau_0(C_0^u(|v|, \pi))$ is an explicit monotone decreasing rational function of $|v| \in [2, 2t_0]$, which is minimized for $|v| = 2t_0$. We find

$$\tau_0(C_0^u(|v|, \text{dih})) \geq \tau_0(C_0^u(2t_0, \pi)) > 0.32.$$

The term $\text{sol}'(y_1, y_3, y_5)$ is maximized when $y_3 = 2t_0$, $y_5 = 3.07$, so that $\text{sol}' < 0.017$. (This was checked with interval arithmetic in Mathematica.) Thus,

$$\tau_0(C_0(v)) \geq 0.32 - 2(0.017)\varphi_0' > 0.297.$$

If there are three or more concave corners, then the penalty-free corner cells squander at least $3(0.297)$. The penalty is at most $\pi_{\max}$ (Section 12.6). So the penalty-inclusive squander is more than $3(0.297) - \pi_{\max} > (4\pi\zeta - 8)pt$. □

Lemma 12.18. *There are no concave corners of height at most 2.2.*

Proof. Suppose there is a corner of height at most 2.2. Place an untruncated corner cell $C_0^u(|v|, \text{dih})$ with parameter $\lambda = 1.815$ at that corner and a t_0-cone wedge at every other corner. The subcluster squanders at least $\tau_0(C_0(|v|, \pi)) - \pi_{\max}$. This is an explicit monotone decreasing rational function of one variable. The penalty-inclusive squander is at least

$$\tau_0(C_0^u(2t_0, \pi)) - \pi_{\max} > (4\pi\zeta - 8)pt. \qquad □ \qquad *$$

By the assumptions at the beginning of Section 12.9, the lemma implies that each concave corner has distance at least 3.2 from every other visible corner.

As in the previous lemma, when $\lambda = 1.945$, a lower bound on what is squandered by the corner cell is obtained for $|v| = 2t_0$, dih $= \pi$. The explicit formulas give penalty-free squander > 0.734. Two disjoint corner cells give penalty-inclusive squander $> (4\pi\zeta - 8)pt$. Suppose two at v_1, v_2 overlap. The lowest bound is obtained when $|v_1 - v_2| = 3.2$, the shortest distance possible.

We define a function $f(y_1, y_2)$ that measures what the union of the overlapping corner cells squander. Set $y_i = |v_i|$, $\ell = 3.2$, and

$$\alpha_1 = \text{dih}(y_1, t_0, y_2, \lambda, \ell, \lambda),$$

$$\alpha_2 = \text{dih}(y_2, t_0, y_1, \lambda, \ell, \lambda),$$

$$\text{sol} = \text{sol}(y_2, t_0, y_1, \lambda, \ell, \lambda),$$

$$\varphi_i = \varphi(y_i/2, t_0), \qquad i = 1, 2,$$

$$\lambda = 3.2 - t_0 = 1.945,$$

$$f(y_1, y_2) = 2(\zeta pt - \varphi_0)\,\text{sol} + 2\sum_1^2 \alpha_i (1 - y_i/(2t_0))(\varphi_0 - \varphi_i)$$

$$+ \sum_1^2 \tau_0(C(y_i, \lambda, \pi - 2\alpha_i)).$$

An interval calculation[76] gives $f(y_1, y_2) > (4\pi\zeta - 8)\,pt + \pi_{\max}$, for $y_1, y_2 \in [2, 2t_0]$.

We conclude that there is at most one concave corner. Let v be such a corner. If we push v toward the origin, the solid angle is unchanged and voro_0 is increased. Following this by the deformation of Section 12.8, we maintain the constraints $|v - w| - 3.2$, for adjacent corners w, while moving v toward the origin. Eventually $|v| = 2.2$. This is impossible by Lemma 12.18.

We verify that this deformation preserves the constraint $|v - w| \geq 2$, for all corners w such that $\{v, w\}$ lies entirely outside the subregion. If fact, every corner is visible from v, so that the subregion is star convex at v. We leave the details to the reader.

We conclude that all subregions can be deformed into convex polygons.

12.13. Proof that Distances Remain at Least 2

Remark 12.19. In Section 12.7, to allow for more flexible deformations, we drop all constraints on the lengths of (undistinguished) edges $\{v_1, v_2\}$ that cross the boundary of the subregion. We deform in such a way that the edges $\{v_1, v_2\}$ will maintain a length of at least 2.

Recall that we say that a vertex of a subregion is *convex* if its angle is less than π, and otherwise that is *concave* (Definition 12.8). In general, if P is a subregion and p_1 and p_2 are two vertices of P, there is a minimal curve joining p_1 and p_2 inside P. This curve is a finite sequence $e_1, \ldots, e_r$ of spherical geodesics. We refer to this sequence as the *sequence of arcs* from p_1 to p_2. The endpoint of each spherical arc is a vertex of P. All endpoints except possibly p_1 and p_2 are nonconvex. These endpoints are the radial

---

[76] CALC-984628285.

projections of corners of P: $v_0, v_1, \ldots, v_{r+1}$, with $p(v_0) = p_1$ and $p(v_{r+1}) = p_2$. The vertex p_1 is visible from p_2 if and only if $r = 1$.

Lemma 12.20. *This deformation of a subregion at a concave corner v maintains a distance of at least 2 to every other corner w.*

Proof. The proof is by contradiction. We may assume that $|v - w| < \sqrt{8}$. We may assume that v and w are the first corners to violate the condition of being at least 2 apart, so that distances between other pairs of corners are at least 2. A distinguished edge connects v and w, if w is visible from v. So assume that w is not visible. Let $e(v_1, v_2)$ be the first distinguished edge crossed by the geodesic arc g from $p(v)$ to $p(w)$. Let p_0 be the intersection of $e(v_1, v_2)$ and g. By construction, the deformation moves v into the subregion, and the subregion P is concave at the corner v, so that the arc from $p(v)$ to $p(w)$ begins in P, then crosses out at $e(v_1, v_2)$.

Geometric considerations show that $|v_1 - v_2| \geq 2.91$. In fact, geometric considerations show that the shortest possible distance for $|v_1 - v_2|$ under the condition that $|v - w| \leq 2$ is the length of the segment passing through the triangle of sides $2, 2t_0, 2t_0$ with both endpoints at distance exactly 2 from all three vertices of the triangle. This distance is greater than 2.91.

Let $e_1, \ldots, e_r$ be the sequence of arcs from $p(v)$ to $p(v_1)$, and let $f_1, \ldots, f_s$ be the sequence of arcs from $p(v)$ to $p(v_2)$. Since this sequence forms a minimal curve, the sum of the lengths of e_i is at most the sum of the lengths of $e(v, p_0)$ and $e(p_0, v_1)$, and the sum of the lengths of f_i is at most the sum of the lengths of $e(v, p_0)$ and $e(p_0, v_2)$.

Note that if $r + s \leq 4$, then one of the edge-lengths must be at least 3.2, for otherwise the sequence of arcs are all distinguished or diagonals of specials, and this would not permit the existence of a corner w. That is, we can fully enumerate the corners of the subregion, and each projects radially to an endpoint in the sequence of arcs, or is a vertex of a special simplex. None of these corners is separated from v by the plane $\{0, v_1, v_2\}$.

We have $r + s \leq 3$ by the following calculations; here $y \in [2, 2t_0]$:

$$5 \operatorname{arc}(2t_0, 2t_0, 2) > \operatorname{arc}(2, 2, 3.2) + 2 \operatorname{arc}(2, 2, 2),$$

$$3 \operatorname{arc}(2t_0, 2t_0, 2) + \operatorname{arc}(2t_0, y, 3.2) > \operatorname{arc}(y, 2, 3.2) + 2 \operatorname{arc}(2, 2, 2),$$

$$3 \operatorname{arc}(2t_0, 2t_0, 2) + \operatorname{arc}(2t_0, y, 3.2) > \operatorname{arc}(2, 2, 3.2) + 2 \operatorname{arc}(y, 2, 2).$$

First we prove the lemma in the special case that the distance from v to one of the endpoints, say v_1, of $\{v_1, v_2\}$ is at least 3.2. In this special case we claim that the constraints on the edge-lengths creates an impossible geometric configuration. The constraints are as follows. There are five points: $0, v_1, w, v, v_2$. The plane $\{0, v_1, v_2\}$ separates point w from v. The distance constraints are as follows:

$$2 \leq |u| \leq 2t_0,$$

for $u = v_1, w, v, v_2, |v - v_1| \geq 3.2, |v - w| \leq 2, |v - v_2| \geq 2, |w - v_1| \geq 2, |w - v_2| \geq 2$, and $2 \leq |v_1 - v_2| \leq 3.2$.

If the segment $\{v, w\}$ passes through the triangle $\{0, v_1, v_2\}$, then the desired impossibility proof follows by geometric considerations. Again, if the segment $\{v_1, v_2\}$ passes through the triangle $\{0, v, w\}$, then the desired impossibility proof follows by geometric considerations, provided that $\{0, v_1, v_2, w\}$ are not coplanar. Assume for a contradiction that $\{0, v_1, v_2, w\}$ lie in the plane P. We move back to the nonplanar case if $|v_2 - v|$ is not 2 (pivot v_2 around $\{0, w\}$ toward v), if $|v_1 - v|$ is not 3.2 (pivot v_1 around $\{0, w\}$ toward v), if $|w - v|$ is not 2 (pivot w around $\{v_1, v_2\}$ away from v), or v is not $2t_0$ (pivot v and w simultaneously preserving $|w - v|$ around $\{v_1, v_2\}$). Therefore, we may assume without loss of generality that $|v_2 - v| = 2$, $|v_1 - v| = 3.2$, $|w - v| = 2$, and $|v| = 2t_0$.

Let p be the orthogonal projection of v to the plane P. Let $h = |v - p|$. The distances from p to $u \in P$ is $f(|v-u|, h) = \sqrt{|v - u|^2 - h^2}$. We consider two cases depending on whether we can find a line in P through p dividing the plane into a half-plane containing v_1, 0, and v_2, or into a half-plane containing v_1, w, and v_2. In the first case we have

$$
\begin{aligned}
0 = {} & \mathrm{arc}(|p - v_1|, |p|, |v_1|) + \mathrm{arc}(|p - v_2|, |p|, |v_2|) \\
& - \mathrm{arc}(|p - v_1|, |p - v_2|, |v_1 - v_2|) \\
\geq {} & \mathrm{arc}(f(3.2, h), f(2t_0, h), 2) + \mathrm{arc}(f(2, h), f(2t_0, h), 2) \\
& - \mathrm{arc}(f(3.2, h), f(2, h), 3.2).
\end{aligned}
\tag{12.3}
$$

The function arc is monotonic in the arguments and from this it follows easily that this function of h is positive on its domain $0 \leq h \leq \sqrt{3}$. This is a contradiction. (The upper bound $\sqrt{3}$ is determined by the condition that the triangle $\{w, v_1, v\}$, which is equilateral in the extreme case, exists under the given edge constraints.) In the second case, we obtain the related contradiction

$$
\begin{aligned}
0 = {} & \mathrm{arc}(|p - v_1|, |p - w|, |v_1 - w|) + \mathrm{arc}(|p - v_2|, |p - w|, |v_2 - w|) \\
& - \mathrm{arc}(|p - v_1|, |p - v_2|, |v_1 - v_2|) \\
\geq {} & \mathrm{arc}(f(3.2, h), f(2, h), 2) + \mathrm{arc}(f(2, h), f(2, h), 2) \\
& - \mathrm{arc}(f(3.2, h), f(2, h), 3.2) \\
> {} & 0.
\end{aligned}
\tag{12.4}
$$

Now assume that the distances from v to the vertices v_1 and v_2 are at most 3.2.

If $r + s = 2$, then v_1 and v_2 are visible from v. Thus, they are distinguished or diagonals of special simplices. As $\{v_1, v_2\}$ is also distinguished, the corners of P are fully enumerated: v, v_1, v_2, and the vertices of special simplices. Since none of these are w, we conclude that w does not exist in this case.

If $r+s = 3$, then say $r = 1$ and $s = 2$. We have $\{v, v_1\}$ is distinguished or the diagonal of a special simplex. Let $p(v)$, $p(u)$ be the endpoints of f_1, for some corner u. We have $|u - v_1| \geq \sqrt{8}$ because $\{u, v_1\}$ is not distinguished, and $\max(|u - v|, |u - v_1|) \geq 3.2$, because otherwise we enumerate all vertices of P as in the case $r + s = 2$, and find that w is not among them. However, now geometric considerations lead to a contradiction: there does not exist a configuration of five points 0, u, v, v_1, v_2, with all distances at least 2 satisfying these constraints. (This can be readily solved by geometric considerations.) $\square$

13. Convex Polygons

13.1. Deformations

We divide the bounding edges over the polygon according to length $[2, 2t_0]$, $[2t_0, 2\sqrt{2}]$, $[2\sqrt{2}, 3.2]$. The deformations of Section 12.8 contract edges to the lower bound of the intervals ($2, 2t_0$, or $2\sqrt{2}$) unless a new distinguished edge is formed. By deforming the polygon, we assume that the bounding edges have length 2, $2t_0$, or $2\sqrt{2}$. (There are a few instances of triangles or quadrilaterals that do not satisfy the hypotheses needed for the deformations. These instances are treated in Sections 13.8 and 13.9.)

Lemma 13.1. *Let $S = S(y_1, \ldots, y_6)$ be a simplex, with $x_i = y_i^2$, as usual. Let $y_4 \geq 2$, $\Delta \geq 0$, $y_5, y_6 \in \{2, 2t_0, 2\sqrt{2}\}$. Fixing all the variables but x_1, let $f(x_1)$ be one of the functions s-vor$_0(S)$ or $-\tau_0(S)$. We have $f''(x_1) > 0$ whenever $f'(x_1) = 0$.*

Proof. This is an interval calculation.[77] $\qquad\square$

The lemma implies that f does not have an interior point local maximum for $x_1 \in [2^2, 2t_0^2]$. Fix three consecutive corners, v_0, v_1, v_2 of the convex polygon, and apply the lemma to the variable $x_1 = |v_1|^2$ of the simplex $S = \{0, v_0, v_1, v_2\}$. We deform the simplex, increasing f. If the deformation produces $\Delta(S) = 0$, then some dihedral angle is π, and the arguments for nonconvex regions bring us eventually back to the convex situation. Eventually y_1 is 2 or $2t_0$. Applying the lemma at each corner, we may assume that the height of every corner is 2 or $2t_0$. (There are a few cases where the hypotheses of the lemma are not met, and these are discussed in Sections 13.8 and 13.9.)

Lemma 13.2. *The convex polygon has at most seven sides.*

Proof. Since the polygon is convex, its perimeter on the unit sphere is at most a great circle 2π. If there are eight sides, the perimeter is at least $8 \operatorname{arc}(2t_0, 2t_0, 2) > 2\pi$. $\qquad\square$

13.2. Truncated Corner Cells

The following lemma justifies using tcc's at the corners as an upper bound on the score (and a lower bound on what is squandered). We fix the truncation parameter at $\lambda = 1.6$.

Lemma 13.3. *Take a convex subregion that is not a triangle. Assume edges between adjacent corners have lengths $\in \{2, 2t_0, 2\sqrt{2}, 3.2\}$. Assume nonadjacent corners are separated by distances ≥ 3.2. Then the truncated corner cell at each vertex lies in the cone over the subregion.*

---

[77] CALC-311189443.

Proof. Place a tcc at v_1. For a contradiction, let $\{v_2, v_3\}$ be an edge that the tcc overlaps. Assume first that $|v_1 - v_i| \geq 2t_0$, $i = 2, 3$. Pivot so that $|v_1 - v_2| = |v_1 - v_3| = 2t_0$. Write $S(y_1, \ldots, y_6) = \{0, v_1, v_2, v_3\}$. Set $\psi = \arc(y_1, t_0, 1.6)$. A calculation[78] gives $\beta_\psi(y_1, y_2, y_6) < \dih_2(S)$.

Now assume $|v_1 - v_2| < 2t_0$. By the hypotheses of the lemma, $|v_1 - v_2| = 2$. If $|v_1 - v_3| < 3.2$, then $\{0, v_1, v_2, v_3\}$ is triangular, contrary to hypothesis. So $|v_1 - v_3| \geq 3.2$. Pivot so that $|v_1 - v_3| = 3.2$. Then[79]

$$\beta_\psi(y_1, y_2, y_6) < \dih_2(S),$$

where $\psi = \arc(y_1, t_0, 1.6)$, provided $y_1 \in [2.2, 2t_0]$. Also, if $y_1 \in [2.2, 2t_0]$, then

$$\arc(y_1, t_0, 1.6) < \arc(y_1, y_2, y_6).$$

If $y_1 \leq 2.2$, then $\Delta_1 \geq 0$, so $\partial \dih_2 / \partial x_3 \leq 0$. Set $x_3 = 2t_0^2$. Also, $\Delta_6 \geq 0$, so $\partial \dih_2 / \partial x_4 \leq 0$. Set $x_4 = 3.2^2$.

Let c be a point of intersection of the plane $\{0, v_1, v_2\}^\perp$ with the circle at distance $\lambda = 1.6$ from v_1 on the sphere centered at the origin of radius t_0. The angle along $\{0, v_2\}$ between the planes $\{0, v_2, v_1\}$ and $\{0, v_2, c\}$ is

$$\dih(R(y_2/2, \eta_{126}, y_1/(2\cos\psi))).$$

This angle is less[80] than $\dih_2(S)$. Also, $\Delta_1 \geq 0$, $\partial \dih_3 / \partial x_2 \leq 0$, so set $x_2 = 2t_0^2$. Then $\Delta_5 < 0$, so $\dih_2 > \pi/2$. This means that $\{0, v_1, v_2\}^\perp$ separates the tcc from the edge $\{v_2, v_3\}$.   □

13.3.   *Analytic Continuation*

In this subsection we assume that $\lambda = 1.6$ and that the tcc under consideration lies at a convex vertex.

Assume that the face cut by $\{0, v, v_1\}^\perp$ meets the face cut by $\{0, v, v_2\}^\perp$. Let c_i be the point on the plane $\{0, v, v_i\}^\perp$ satisfying $|c_i - v| = 1.6$, $|c_i| = t_0$. (Pick the root within the wedge between v_1 and v_2.) The overlap of the two faces is represented in Fig. 13.1.

We let c_0 be the point of height t_0 on the intersection of the planes $\{0, v, v_1\}^\perp$ and $\{0, v, v_2\}^\perp$. We claim that c_0 lies over the truncated spherical region of the tcc, rather than the wedges of t_0-cones or the Rogers simplices along the faces $\{0, v, v_1\}$ and $\{0, v, v_2\}$. (This implies that c_0 cannot protrude beyond the corner cell as depicted in the second frame of the figure.) To see the claim, consider the tcc as a function of $y_4 = |v_1 - v_2|$. When y_4 is sufficiently large the claim is certainly true. Contract y_4 until $c_0 = c_0(y_4)$ meets the perpendicular bisector of $\{0, v\}$. Then c_0 is equidistant from 0, v, v_1 and v_2 so it is the circumcenter of $\{0, v, v_1, v_2\}$. It has distance t_0 from the origin, so the circumradius is t_0. This implies that $y_4 \leq 2t_0$.

---

[78] CALC-193836552.
[79] CALC-193836552.
[80] CALC-193836552.

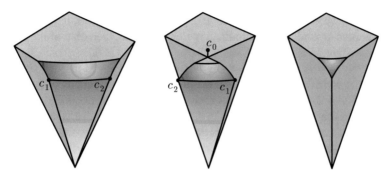

Fig. 13.1. Different forms of truncated corner cells are shown. The structure shown in the middle frame cannot occur.

The tcc is defined by the constraints represented in the third frame. The analytic continuation of the function $\chi_0(S) = \chi_0^{\mathrm{an}}(S)$, defined above, acquires a volume X, counted with a negative sign, lying under the spherical triangle (c_0, c_1, c_2). Extending our notation, we have an analytically defined function χ_0^{an} and a geometrically defined function χ_0^{g},

$$\chi_0^{\mathrm{an}}(S) = \chi_0^{\mathrm{g}}(S) - \text{c-vor}_0(X), \qquad \text{where}$$

$$\text{c-vor}_0(X) = 4(-\delta_{\mathrm{oct}}\,\mathrm{vol}(X) + \mathrm{sol}(X)/3) = \varphi_0\,\mathrm{sol}(X) < 0.$$

So $\chi_0^{\mathrm{an}} > \chi_0^{\mathrm{g}}$, and we may always use $\chi_0(S) = \chi_0^{\mathrm{an}}(S)$ as an upper bound on the score of a tcc.

For example, with $\lambda = 1.6$ and $S = S(2.3, 2.3, 2.3, 2.9, 2, 2)$, we have

$$\chi_0^{\mathrm{an}}(S) \approx -0.103981, \qquad \chi_0^{\mathrm{g}}(S) \approx -0.105102,$$

or, if $S = S(2, 2, 2t_0, 3.2, 2, 2t_0)$, then

$$\chi_0^{\mathrm{an}}(S) \approx -0.0718957, \qquad \chi_0^{\mathrm{g}}(S) \approx -0.0726143.$$

13.4.   Penalties

In Section 12.6 we determined the bound of $\pi_{\max} = 0.06688$ on penalties. In this section we give a more thorough treatment of penalties. Until now a penalty has been associated with a given standard region, but by taking the worst case on each subregion, we can move the penalties to the level of subregions. Roughly, each subregion should incur the penalties from the upright quarters that were erased along edges of that subregion. Each upright quarter of the original standard region is attached at an edge between adjacent corners of the standard cluster. The edges have lengths between 2 and $2t_0$. The deformations shrink the edges to length 2. We attach the penalty from the upright quarter to this edge of this subregion. In general, we divide the penalty evenly among the upright quarters along a common diagonal, without trying to determine a more detailed accounting. For example, the penalty 0.008 in Lemma 11.23 comes from three upright quarters. Thus, we give each of three edges a penalty of $0.008/3$, or, if there are only two upright quarters around the 3-unconfined upright diagonal, then each of the two upright quarters is assigned the penalty $0.00222/2$ (see Lemma 11.23).

The penalty $0.04683 = 3\xi_\Gamma$ in Section 12.6 comes from three upright quarters around a 3-crowded upright diagonal. Each of three edges is assigned a penalty of ξ_Γ. The penalty $0.03344 = 3\xi_\Gamma + \xi_{\kappa,\Gamma}$ comes from a 4-crowded upright diagonal of Section 11.7. It is divided among four edges. These are the only upright quarters that take a penalty when erased. (The case of two upright quarters over a flat quarter as in Lemma 11.3, are treated by a separate argument in Section 13.8. Loops are discussed in Section 13.12.)

The penalty can be reduced in various situations involving a masked flat quarter. For example, around a 3-crowded upright diagonal, if there is a masked flat quarter, two of the upright quarters are scored by the analytic function s-vor, so that the penalty plus adjustment is only[81,82] $0.034052 = 2\xi_V + \xi_\Gamma + 0.0114$. The adjustment 0.0114 reflects the scoring rules for masked flat quarters (Lemma 11.23). This we divide evenly among the three edges that carried the upright quarters. If e is an edge of the subregion R, let $\pi_0(R, e)$ denote the penalty and score adjustment along edge e of R.

In summary, we have the penalties,

$$\xi_\kappa, \; \xi_V, \; \xi_\Gamma, \; 0.008,$$

combined in various ways in the upright diagonals that are 3-unconfined, 3-crowded, or 4-crowded. There are score adjustments

$$0.0114 \quad \text{and} \quad 0.0063$$

from Section 11.9 for masked flat quarters. If the sum of these contributions is s, we set $\pi_0(R, e) = s/n$, for each edge e of R originating from an erased upright quarter of $S_n^{\pm}$.

13.5. Penalties and Bounds

Recall that the bounds for flat quarters we wish to establish from Section 12.5 are $Z(3, 1) = 0.00005$ and $D(3, 1) = 0.06585$. Flat quarters arise in two different ways. Some flat quarters are present before the deformations begin. They are scored by the rules of Section 11.9. Others are formed by the deformations. In this case they are scored by vor_0. Since the flat quarter is broken away from the subregion as soon as the diagonal reaches $2\sqrt{2}$, and then is not deformed further, the diagonal is fixed at $2\sqrt{2}$. Such flat quarters can violate our desired inequalities. For example,

$$Z(3, 1) < \text{s-vor}_0(S(2, 2, 2, 2\sqrt{2}, 2, 2)) \approx 0.00898,$$

$$\tau_0(S(2, 2, 2, 2\sqrt{2}, 2, 2)) \approx 0.0593.$$

On the other hand, as we will see, the adjacent subregion satisfies the inequality by a comfortable margin. Therefore, we define a transfer ε from flat quarters to the adjacent

---

[81] CALC-73974037.
[82] CALC-764978100.

subregion. (In an exceptional region, the subregion next to a flat quarter along the diagonal is not a flat quarter.)

For a flat quarter Q, set

$$\varepsilon_\tau(Q) = \begin{cases} 0.0066 & \text{(deformation)}, \\ 0 & \text{(otherwise)}, \end{cases}$$

$$\varepsilon_\sigma(Q) = \begin{cases} 0.009 & \text{(deformation)}, \\ 0 & \text{(otherwise)}. \end{cases}$$

The nonzero value occurs when the flat quarter Q is obtained by deformation from an initial configuration in which Q is not a quarter. The value is zero when the flat quarter Q appears already in the undeformed standard cluster. Set

$$\pi_\tau(R) = \sum_e \pi_0(R, e) + \sum_e \pi_0(Q, e) + \sum_Q \varepsilon_\tau(Q),$$

$$\pi_\sigma(R) = \sum_e \pi_0(R, e) + \sum_e \pi_0(Q, e) + \sum_Q \varepsilon_\sigma(Q).$$

The first sum runs over the edges of a subregion R. The second sum runs over the edges of the flat quarters Q that lie adjacent to R along the diagonal of Q.

The edges between corners of the polygon have lengths 2, $2t_0$, or $2\sqrt{2}$. Let k_0, k_1, and k_2 be the number of edges of these three lengths, respectively. By Lemma 13.2, we have $k_0 + k_1 + k_2 \leq 7$. Let $\tilde{\sigma}$ denote any of the functions of (a)–(f) of Section 11.9. Let $\tilde{\tau} = \text{sol } \zeta pt - \tilde{\sigma}$.

To prove Theorem 12.1, refining the strategy proposed in Section 12.5, we must show that for each flat quarter Q and each subregion R that is not a flat quarter, we have

$$\begin{aligned}
\tilde{\tau}(Q) &> D(3, 1) - \varepsilon_\tau(Q), \\
\tau_0(Q) &> D(3, 1) - \varepsilon_\tau(Q), \qquad \text{if} \quad y_4(Q) = 2\sqrt{2}, \\
\tau_V(R) &> D(3, 2) \qquad \text{(type A)}, \\
\tau_0(R) &> D(k_0 + k_1 + k_2, k_1 + k_2) + \pi_\tau(R),
\end{aligned} \tag{13.1}$$

where $D(n, k)$ is the function defined in Section 12.5. The first of these inequalities follows.[83–85] In general, we are given a subregion without explicit information about what the adjacent subregions are. Similarly, we have discarded all information about what upright quarters have been erased. Because of this, we assume the worst, and use the largest feasible values of π_τ.

Lemma 13.4. *We have $\pi_\tau(R) \leq 0.04683 + (k_0 + 2k_2 - 3)0.008/3 + 0.0066k_2$.*

---

[83] CALC-193836552.
[84] CALC-148776243.
[85] CALC-163548682.

Proof. The worst penalty $0.04683 = 3\xi_\Gamma$ per edge comes from a 3-crowded upright diagonal. The number of penalized edges not on a simplex around a 3-crowded upright diagonal is at most $k_0 + 2k_2 - 3$. For every three edges, we might have one 3-unconfined upright diagonal. The other cases such as 4-crowded upright diagonals or situations with a masked flat quarter are readily seen to give smaller penalties.   □

For bounds on the score, the situation is similar. The only penalties we need to consider are 0.008 from Lemma 11.23. If either of the other configurations of 3-crowded or 4-crowded upright diagonals occurs, then the score of the standard cluster is less than $s_8 = -0.228$, by Sections 11.6 and 11.7. This is the desired bound. So it is enough to consider subregions that do not have these upright configurations. Moreover, the penalty 0.008 does not occur in connection with masked flats. So we can take $\pi_\sigma(R)$ to be

$$(k_0 + 2k_2)0.008/3 + 0.009k_2.$$

If $k_0 + 2k_2 < 3$, we can strengthen this to $\pi_\sigma(R) = 0.009k_2$. Let $\tilde\sigma$ be any of the functions of (a)–(f) of Section 11.9. To prove Theorem 12.1, we will show

$$
\begin{aligned}
\tilde\sigma(Q) &< Z(3, 1) + \varepsilon_\sigma(Q), \\
\text{s-vor}_0(Q) &< Z(3, 1) + \varepsilon_\sigma(Q), \qquad \text{if} \quad y_4(Q) = 2\sqrt{2}, \\
\text{vor}_0(R) &< Z(3, 2) \qquad (\text{type } A), \\
\text{vor}_0(R) &< Z(k_0 + k_1 + k_2, k_1 + k_2) - \pi_\sigma(R).
\end{aligned}
\tag{13.2}
$$

The first of these inequalities follows.[86–88]

13.6. *Penalties*

Erasing an upright quarter of compression type gives a penalty of at most ξ_Γ and one of Voronoi type gives at most ξ_V. We take the worst possible penalty. It is at most $n\xi_\Gamma$ in an n-gon. If there is a masked flat quarter, the penalty is at most $2\xi_V$ from the two upright quarters along the flat quarter. We note in this connection that both edges of a polygon along a flat quarter lie on upright quarters, or neither does.

If an upright diagonal appears enclosed over a flat quarter, the flat quarter is part of a loop with context $(n, k) = (4, 1)$, for a penalty at most $2\xi'_\Gamma + \xi_V$. This is smaller than the bound on the penalty obtained from a loop with context $(n, k) = (4, 1)$, when the upright diagonal is not enclosed over the flat quarter:

$$\xi_\Gamma + 2\xi_V.$$

So we calculate the worst-case penalties under the assumption that the upright diagonals are not enclosed over flat quarters.

---

[86] CALC-193836552.
[87] CALC-148776243.
[88] CALC-163548682.

A loop of context $(n, k) = (4, 1)$ gives $\xi_\Gamma + 2\xi_V$ or $3\xi_\Gamma$. A loop of context $(n, k) = (4, 2)$ gives $2\xi_\Gamma$ or $2\xi_V$.

If we erase a 3-unconfined upright diagonal, there is a penalty of 0.008 (or 0 if it masks a flat quarter). This is dominated by the penalty $3\xi_\Gamma$ of context $(n, k) = (4, 1)$.

Suppose we have an octagonal standard region. We claim that a loop does not occur in context $(n, k) = (4, 2)$. If there are at most three vertices that are not corners of the octagon, then there are at most twelve quasi-regular tetrahedra, and the score is at most

$$s_8 + 12\, pt < 8\, pt.$$

Assume there are more than three vertices that are not corners over the octagon. We squander

$$t_8 + \delta_{\text{loop}}(4, 2) + 4\tau_{\text{LP}}(5, 0) > (4\pi\zeta - 8)\, pt. \qquad \text{\Large *}$$

As a consequence, context $(n, k) = (4, 2)$ does not occur.

So there are at most two upright diagonals and at most six quarters, and the penalty is at most $6\xi_\Gamma$. Let f be the number of flat quarters This leads to

$$\pi_F = \begin{cases} 6\xi_\Gamma, & f = 0, 1, \\ 4\xi_\Gamma + 2\xi_V, & f = 2, \\ 2\xi_\Gamma + 4\xi_V, & f = 3, \\ 0, & f = 4. \end{cases}$$

The 0 is justified by a parity argument. Each upright quarter occurs in a pair at each masked flat quarter. However, there is an odd number of quarters along the upright diagonal, so no penalty at all can occur.

Suppose we have a heptagonal standard region. Three loops are a geometric impossibility. Assume there are at most two upright diagonals. If there is no context $(n, k) = (4, 2)$, then we have the following bounds on the penalty:

$$\pi_F = \begin{cases} 6\xi_\Gamma, & f = 0, \\ 4\xi_\Gamma + 2\xi_V, & f = 1, \\ 3\xi_\Gamma, & f = 2, \\ \xi_\Gamma + 2\xi_V, & f = 3. \end{cases}$$

If an upright diagonal has context $(n, k) = (4, 2)$, then

$$\pi_F = \begin{cases} 5\xi_\Gamma, & f = 0, 1, \\ 3\xi_\Gamma + 2\xi_V, & f = 2, \\ \xi_\Gamma + 4\xi_V, & f = 3. \end{cases}$$

This gives the bounds used in the diagrams of cases.   \Large *

13.7. Constants

Theorem 12.1 now results from the calculation of a host of constants. Perhaps there are simpler ways to do it, but it was a routine matter to run through the long list of constants by computer. What must be checked is that the inequalities (13.1) and (13.2) of Section 13.5 hold for all possible convex subregions. Call these inequalities the D and Z inequalities. This section describes in detail the constants to check.

We begin with a subregion given as a convex n-gon, with at least four sides. The heights of the corners and the lengths of edges between adjacent edges have been reduced by deformation to a finite number of possibilities (lengths 2, $2t_0$, or lengths 2, $2t_0$, $2\sqrt{2}$, respectively). By Lemma 13.2, we may take $n = 4, 5, 6, 7$. Not all possible assignments of lengths correspond to a geometrically viable configuration. One constraint that eliminates many possibilities, especially heptagons, is that of Section 13.1: the perimeter of the convex polygon is at most a great circle. Eliminate all length-combinations that do not satisfy this condition. When there is a special simplex it can be broken from the subregion and scored[89] separately unless the two heights along the diagonal are 2. We assume in all that follows that all specials that can be broken off have been. There is a second condition related to special simplices. We have $\Delta(2t_0^2, 2^2, 2^2, x^2, 2^2, 2^2) < 0$, if $x > 3.114467$. This means that if the cluster edges along the polygon are $(y_1, y_2, y_3, y_5, y_6) = (2t_0, 2, 2, 2, 2)$, the simplex must be special ($y_4 \in [2\sqrt{2}, 3.2]$).

The easiest cases to check are those with no special simplices over the polygon. In other words, these are subregions for which the distances between nonadjacent corners are at least 3.2. In this case we approximate the score (and what is squandered) by tcc's at the corners. We use monotonicity to bring the fourth edge to length 3.2. We calculate the tcc constant bounding the score, checking that it is less than the constant $Z(k_0 + k_1 + k_2, k_1 + k_2) - \pi_\sigma$, from the Z inequalities. The D inequalities are verified in the same way.

When $n = 5, 6, 7$, and there is one special simplex, the situation is not much more difficult. By our deformations, we decrease the lengths of edges 2, 3, 5, 6 of the special to 2. We remove the special by cutting along its fourth edge e (the diagonal). We score the special with weak bounds.[90] Along the edge e, we then apply deformations to the $(n-1)$-gon that remains. If this deformation brings e to length $2\sqrt{2}$, then the $(n-1)$-gon may be scored with tcc's as in the previous paragraph. However there are other possibilities. Before e drops to $2\sqrt{2}$, a new distinguished edge of length 3.2 may form between two corners (one of the corners will be a chosen endpoint of e). The subregion breaks in two. By deformations, we eventually arrive at $e = 2\sqrt{2}$ and a subregion with diagonals of length at least 3.2. (There is one case that may fail to be deformable to $e = 2\sqrt{2}$, a pentagonal cases discussed further in Section 13.10.) The process terminates because the number of sides to the polygon drops at every step. A simple recursive computer procedure runs through all possible ways the subregion might break into pieces and checks that the tcc-bound gives the D and Z inequalities. The same argument works if there is a special simplex that overlaps each of the other special simplices in the subcluster.

When $n = 6, 7$ and there are two nonoverlapping special simplices, a similar argument can be applied. Remove both specials by cutting along the diagonals. Then deform both diagonals to length $2\sqrt{2}$, taking into account the possible ways that the subregion can break into pieces in the process. In every case the D and Z inequalities are satisfied.

---

[89] CALC-148776243.
[90] CALC-148776243.

There are a number of situations that arise that escape this generic argument and were analyzed individually. These include the cases involving more than two special simplices over a given subregion, two special simplices over a pentagon, or a special simplex over a quadrilateral. Also, the deformation lemmas are insufficient to bring all of the edges between adjacent corners to one of the three standard lengths $2, 2t_0, 2\sqrt{2}$ for certain triangular and quadrilateral regions. These are treated individually.

The next few sections describe the cases treated individually. The cases not mentioned in the sections that follow fall within the generic procedure just described.

13.8. *Triangles*

With triangular subregions there is no need to use any of the deformation arguments because the dimension is already sufficiently small to apply interval arithmetic directly to obtain our bounds. There is no need for the tcc-bound approximations.

Flat quarters and simplices of type A are treated by a computer calculation.[91] Other simplices are scored by the truncated function s-vor$_0$. We break the edges between corners into the cases $[2, 2t_0)$, $[2t_0, 2\sqrt{2})$, and $[2\sqrt{2}, 3.2]$. Let k_0, k_1, and k_2, with $k_0 + k_1 + k_2 = 3$, be the number of edges in the respective intervals.

If $k_2 = 0$, we can improve the penalties,

$$\pi_\tau = \pi_\sigma = 0.$$

To see this, first we observe that there can be no 3-crowded or 4-crowded upright diagonals. By placing three or more quarters around an upright diagonal, if the subregion is triangular, the upright diagonal becomes surrounded by anchored simplices, a case deferred until Section 13.12.

If $k_0 = k_1 = k_2 = 1$, we can take $\pi_\tau' = \xi_\Gamma + 2\xi_V + 0.0114 = 0.034052$. A few cases are needed to justify this constant. If there are no 3-crowded upright diagonals, π_τ' is at most

$$[\xi_\Gamma + 2\xi_V + \xi_{\kappa,\Gamma}]3/4 < 0.0254,$$

or

$$[\xi_\Gamma + 2\xi_V + \xi_{\kappa,\Gamma}]2/4 + 0.008/3 < 0.0254$$

If there are at most two edges in the subregion coming from a 3-crowded upright diagonal, then

$$(\xi_\Gamma + 2\xi_V + 0.0114)2/3 + 0.008/3 < 0.0254.$$

If three edges come from the simplices of a 3-crowded upright diagonal, we get 0.034052. To get somewhat sharper bounds, we consider how the edge k_2 was formed. If it is obtained by deformation from an edge in the standard region of length ≥ 3.2, then it becomes a distinguished edge when the length drops to 3.2. If the edge in the standard region already has length ≤ 3.2, then it is distinguished before the deformation process

---

[91] CALC-163548682.

begins, so that the subregion can be treated in isolation from the other subregions. We conclude that when $\pi'_\tau = 0.034052$ we can take $y_4 \geq 2.6$ or $y_5 = 3.2$ (Remark 11.22). The D and Z inequalities now follow.[92,93]

13.9. Quadrilaterals

We introduce some notation for the heights and edge lengths of a convex polygon. The heights will generally be 2 or $2t_0$, the edge lengths between consecutive corners will generally be 2, $2t_0$, or $2\sqrt{2}$. We represent the edge lengths by a vector

$$(a_1, b_1, a_2, b_2, \ldots, a_n, b_n),$$

if the corners of an n-gon, ordered cyclically have heights a_i and if the edge length between corner i and $i + 1$ is b_i. We say two vectors are equivalent if they are related by a different cyclic ordering on the corners of the polygon, that is, by the action of the dihedral group.

The vector of a polygon with a special simplex is equivalent to one of the form

$$(2, 2, a_2, 2, 2, \ldots).$$

If $a_2 = 2t_0$, then what we have is necessarily special (Section 13.7). However, if $a_2 = 2$, it is possible for the edge opposite a_2 to have length greater than 3.2.

Turning to quadrilateral regions, we use tcc scoring if both diagonals are greater than 3.2. Suppose that both diagonals are between $[2\sqrt{2}, 3.2]$, creating a pair of overlapping special simplices. The deformation lemma requires a diagonal longer than 3.2, so although we can bring the quadrilateral to the form

$$(a_1, 2, 2, 2, 2, 2, a_4, b_4),$$

the edges a_1, a_4, b_4 and the diagonal vary[94] continuously. We have bounds[95] on the score

$$\tau_0 > 0.235, \qquad \text{vor}_0 < -0.075, \qquad \text{if} \quad b_4 \in [2t_0, 2\sqrt{2}],$$
$$\tau_0 > 0.3109, \qquad \text{vor}_0 < -0.137, \qquad \text{if} \quad b_4 \in [2\sqrt{2}, 3.2].$$

We have $D(4, 1) = 0.2052$, $Z(4, 1) = -0.05705$. When $b_4 \in [2t_0, 2\sqrt{2}]$, we can take $\pi_\tau = \pi_\sigma = 0$. (We are excluding loops here.) When $b_4 \in [2\sqrt{2}, 3.2]$, we can take

$$\pi_\tau = \pi_{\max} + 0.0066,$$
$$\pi_\sigma = 0.008(5/3) + 0.009.$$

It follows that the D and Z inequalities are satisfied.

---

[92] CALC-852270725.
[93] CALC-819209129.
[94] CALC-148776243.
[95] CALC-128523606.

Suppose that one diagonal has length $[2\sqrt{2}, 3.2]$ and the other has length at least 3.2. The quadrilateral is represented by the vector

$$(2, 2, a_2, 2, 2, b_3, a_4, b_4).$$

The hypotheses of the deformation lemma hold, so that $a_i \in \{2, 2t_0\}$ and $b_j \in \{2, 2t_0, 2\sqrt{2}\}$. To avoid quad clusters, we assume $b_4 \geq \max(b_3, 2t_0)$. These are one-dimensional with a diagonal of length $[2\sqrt{2}, 3.2]$ as the parameter. The required verifications[96] have been made by interval arithmetic.

13.10. *Pentagons*

Some extra comments are needed when there is a special simplex. The general argument outlined above removes the special, leaving a quadrilateral. The quadrilateral is deformed, bringing the edge that was the diagonal of the special to $2\sqrt{2}$. This section discusses how this argument might break down.

Suppose first that there is a special and that both diagonals on the resulting quadrilateral are at least 3.2. We can deform using either diagonal, keeping both diagonals at least 3.2. The argument breaks down if both diagonals drop to 3.2 before the edge of the special reaches $2\sqrt{2}$ and both diagonals of the quadrilateral lie on specials. When this happens, the quadrilateral has the form

$$(2, 2, 2, 2, 2, 2, 2, b_4),$$

where b_4 is the edge originally on the special simplex. If both diagonals are 3.2, this is rigid, with $b_4 = 3.12$. We find its score to be

$$\text{s-vor}_0(S(2, 2, 2, b_4, 3.2, 2)) + \text{s-vor}_0(S(2, 2, 2, 3.2, 2, 2)) + 0.0461 \; < \; -0.205,$$

$$\tau_0(S(2, 2, 2, b_4, 3.2, 2)) + \tau_0(S(2, 2, 2, 3.2, 2, 2))2 \; > \; 0.4645.$$

So the D and Z inequalities hold easily.

If there is a special and there is a diagonal on the resulting quadrilateral ≤ 3.2, we have two nonoverlapping specials. It has the form

$$(2, 2, a_2, 2, 2, 2, a_4, 2, 2, b_5).$$

The edges a_2 and a_4 lie on the special. If $b_5 > 2$, cut away one of the special simplices. What is left can be reduced to a triangle, or a quadrilateral case and then treated[97] by computer. Assume $b_5 = 2$. We have a pentagonal standard region. We may assume that there is no 3-crowded or 4-crowded upright diagonal, for otherwise Theorem 12.1 follows trivially from the bounds in Section 9. A pentagon can then have at most a 3-unconfined upright diagonal for a penalty of 0.008.

---

[96] CALC-874876755.
[97] CALC-874876755.

If $a_2 = 2t_0$ or $a_4 = 2t_0$, we again remove a special simplex and produce triangles, quadrilaterals, or the special cases treated by computer.[98] We may impose the condition $a_2 = a_4 = b_5 = 2$. We score this full pentagonal arrangement by computer,[99] using the edge lengths of the two diagonals of the specials as variables. The inequalities follow.

13.11. *Hexagons and Heptagons*

We turn to hexagons. There may be three specials whose diagonals do not cross. Such a subcluster is represented by the vector

$$(2, 2, a_2, 2, 2, 2, a_4, 2, 2, 2, a_6, 2).$$

The heights a_{2i} are 2 or $2t_0$. Draw the diagonals between corners 1, 3, and 5. This is a three-dimensional configuration, determined by the lengths of the three diagonals, which is treated by computer.[100]

There is one case with a special simplex that did not satisfy the generic computer-checked inequalities for what is to be squandered. Its vector is

$$(a_1, 2, 2, 2, 2, 2, 2, b_4, 2, 2, 2, 2),$$

with $a_1 = b_4 = 2t_0$. A vertex of the special simplex has height $a_1 = 2t_0$ and all other corners have height 2. The subregion is a hexagon with one edge longer than 2. We have $D(6, 1) = 0.48414$. This is certainly obtained if the subregion contains a 3-crowded upright diagonal, squandering 0.5606. However, if this configuration does not appear, we can decrease π_r to $0.03344 + (2/3)0.008$, a constant coming from 4-crowded upright diagonals in Section 12.6. With this smaller penalty the inequality is satisfied.

Now turn to heptagons. The bound 2π on the perimeter of the polygon eliminates all but one equivalence class of vectors associated with a polygon that has two or more potentially specials simplices. The vector is

$$(2, 2, a_2, 2, 2, 2, a_4, 2, 2, 2, a_6, 2, a_7, 2),$$

$a_2 = a_4 = a_6 = a_7 = 2t_0$. In other words, the edges between adjacent corners are 2 and four heights are $2t_0$. There are two specials. This case is treated by the procedure outlined for subregions with two specials whose diagonals do not cross.

13.12. *Loops*

We now return to a collection of anchored simplices that surround the upright diagonal. This is the last case needed to complete the proof of Theorem 12.1. There are four or

---

[98] CALC-874876755.
[99] CALC-692155251.
[100] CALC-692155251.

five anchored simplices around the upright diagonal. There are linear inequalities[101–106] satisfied by the anchored simplices, broken up according to type: upright, type C, opposite edge > 3.2, etc. The anchored simplices are related by the constraint that the sum of the dihedral angles around the upright diagonal is 2π. We run a linear program in each case based on these linear inequalities, subject to this constraint to obtain bounds on the score and what is squandered by the anchored simplices.

When the edge opposite the diagonal of an anchored simplex has length $\in [2\sqrt{2}, 3.2]$ and the simplex adjacent to the anchored simplex across that edge is a special simplex, we use inequalities[107, 108] that run parallel to the similar system.[109, 110] It is not necessary to run separate linear programs for these. It is enough to observe that the constants for what is squandered improve on those from the similar system[111] and that the constants for the score in one system[112] differ with those of the other[113] by no more than 0.009.

When the dihedral angle of an anchored simplex is greater than 2.46, the simplex is dropped, and the remaining anchored simplices are subject to the constraint that their dihedral angles sum to at most $2\pi - 2.46$. There cannot be an anchored simplex with dihedral angle greater than 2.46 when there are five anchors: $2.46 + 4(0.956) > 2\pi$. There cannot[114] be two anchored simplices with dihedral angle greater than 2.46: $2(2.46 + 0.956) > 2\pi$.

The following table summarizes the linear programming results:

(n, k)	$D_{LP}(n, k)$	$D(n, k)$	$Z_{LP}(n, k)$	$Z(n, k)$
$(4, 0)$	0.1362	0.1317	0	0
$(4, 1)$	0.208	0.20528	−0.0536	−0.05709
$(4, 2)$	0.3992	0.27886	−0.2	−0.11418
$(4, 3)$	0.6467	0.35244	−0.424	−0.17127
$(5, 0)$	0.3665	0.27113	−0.157	−0.05704
$(5, 1)$	0.5941	0.34471	−0.376	−0.11413
$(5, \geq 2)$	0.9706	$(4\pi\zeta - 8)\,pt$	*	*

The bound for $D(4, 0)$ comes from Lemma 10.8. A few more comments are needed for $Z(4, 1)$. Let $S = S(y_1, \ldots, y_6)$ be the anchored simplex that is not a quarter. If $y_4 \geq 2\sqrt{2}$ or $\mathrm{dih}(S) \geq 2.2$, the linear programming bound is $< Z(4, 1)$. With this, if $y_1 \leq 2.75$, we have[115] $\sigma(S) < Z(4, 1)$. However, if $y_1 \geq 2.75$, the three upright quarters

[101] CALC-815492935.
[102] CALC-729988292.
[103] CALC-531888597.
[104] CALC-628964355.
[105] CALC-934150983.
[106] CALC-187932932.
[107] CALC-485049042.
[108] CALC-209361863.
[109] CALC-531888597.
[110] CALC-628964355.
[111] CALC-531888597.
[112] CALC-485049042.
[113] CALC-531888597.
[114] CALC-83777706.
[115] CALC-855294746.

along the upright diagonal satisfy

$$\nu < -0.3429 + 0.24573 \, \text{dih} \,.$$

With this stronger inequality, the linear programming bound becomes $< Z(4, 1)$. This completes the proof of Theorem 12.1.

Lemma 13.5. *Consider an upright diagonal that is a loop. Let R be the standard region that contains the upright diagonal and its surrounding simplices. Then the following contexts (m, k) are the only ones possible. Moreover, the constants that appear in the columns marked σ and τ are upper and lower bounds respectively for $\tau_R(D)$ when R contains one loop of that context.*

$n = n(R)$	(m, k)	σ	τ
4			
	(4, 0)	−0.0536	0.1362
5			
	(4, 1)	s_5	0.27385
	(5, 0)	−0.157	0.3665
6			
	(4, 1)	s_6	0.41328
	(4, 2)	−0.1999	0.5309
	(5, 1)	−0.37595	0.65995
7			
	(4, 1)	s_7	0.55271
	(4, 2)	−0.25694	0.67033
8			
	(4, 1)	s_8	0.60722
	(4, 2)	−0.31398	0.72484

Proof. In context (m, k), and if $n = n(R)$, we have

$$\sigma_R(D) < s_n + Z_{\text{LP}}(m, k) - Z(m, k), \qquad \tau_R(D) > t_n + D_{\text{LP}}(m, k) - D(m, k).$$

The result follows.   □

In the context $(n, k) = (4, 3)$, the standard region R must have at least seven sides, $n(R) \geq 7$. Then

$$\tau(D) \geq t_7 + \delta_{\text{loop}}(4, 3)$$
$$> (4\pi\zeta - 8)\,pt.$$

Thus, we may assume that this context does not occur.
If the context $(5, 1)$ appears in an octagon, we have

$$\tau(D) > \delta_{\text{loop}}(5, 1) + t_8 > (4\pi\zeta - 8)\,pt.$$

If this appears in a heptagon, we have

$$\tau(D) > \delta_{\text{loop}}(5, 1) + t_7 + 0.55\,pt > (4\pi\zeta - 8)\,pt,$$

because there must be a vertex that is not a corner of the heptagon. It cannot appear on a pentagon.

14. Further Bounds in Exceptional Regions

14.1. *Small Dihedral Angles*

Recall that Section 12.1 defines an integer $n(R)$ that is equal to the number of sides if the region is a polygon. Recall that if the dihedral angle along an edge of a standard cluster is at most 1.32, then there is a flat quarter along that edge (Lemma 11.30).

Lemma 14.1. *Let R be an exceptional cluster with a dihedral angle ≤ 1.32 at a vertex v. Then R squanders $> t_n + 1.47\,pt$, where $n = n(R)$.*

Proof. In most cases we establish the stronger bound $t_n + 1.5\,pt$. In the proof of Theorem 12.1 we erase all upright diagonals, except those completely surrounded by anchored simplices. The contribution to t_n from the flat quarter Q at v in that proof is $D(3, 1)$ (Section 12.5 and inequalities (13.1)). Note that $\varepsilon_\tau(Q) = 0$ here because there are no deformations. If we replace $D(3, 1)$ with $3.07\,pt$ from Lemma 11.30, then we obtain the bound. Now suppose the upright diagonal is completely surrounded by anchored simplices. Analyzing the constants of Section 13.12, we see that $D_{\mathrm{LP}}(n, k) - D(n, k) > 1.5\,pt$ except when $(n, k) = (4, 1)$.

Here we have four anchored simplices around an upright diagonal. Three of them are quarters. We erase and take a penalty. Two possibilities arise. If the upright diagonal is enclosed over the flat quarter, its height is ≥ 2.6 by geometric considerations and the top face of the flat quarter has circumradius at least $\sqrt{2}$. The penalty is $2\xi_\Gamma' + \xi_V$, so the bound holds by the last statement of Lemma 11.30.

If, on the other hand, the upright diagonal is not enclosed over the flat diagonal, the penalty is $\xi_\Gamma + 2\xi_V$. In this case we obtain the weaker bound $1.47\,pt + t_n$:

$$3.07\,pt > D(3, 1) + 1.47\,pt + \xi_\Gamma + 2\xi_V. \qquad \square$$

Remark 14.2. If there are r nonadjacent vertices with dihedral angles ≤ 1.32, we find that R squanders $t_n + r(1.47)\,pt$.

In fact, in the proof of the lemma, each $D(3, 1)$ is replaced with $3.07\,pt$ from Lemma 11.30. The only questionable case occurs when two or more of the vertices are anchors of the same upright diagonal (a loop). Referring to Section 13.12, we have the following observations about various contexts:

- $(4, 1)$ can mask only one flat quarter and it is treated in the lemma.
- $(4, 2)$ can mask only one flat quarter and $D_{\mathrm{LP}}(4, 2) - D(4, 2) > 1.47\,pt$.
- $(5, 0)$ can mask two flat quarters. Erase the five upright quarters, and take a penalty $4\xi_V + \xi_\Gamma$. We get

$$D(3, 2) + 2(3.07)\,pt > t_5 + 4\xi_V + \xi_\Gamma + 2(1.47)\,pt.$$

- $(5, 1)$ can mask two flat quarters, and $D_{\mathrm{LP}}(5, 1) - D(5, 1) > 2(1.47)\,pt$.

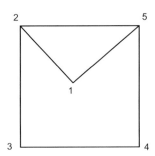

Fig. 14.1. A 4-circuit.

14.2. *A Particular 4-Circuit*

This subsection bounds the score of a particular 4-circuit on a contravening plane graph. The interior of the circuit consists of two faces: a triangle and a pentagon. The circuit and its enclosed vertex are show in Fig. 14.1 with vertices marked $p_1, \ldots, p_5$. The vertex p_1 is the enclosed vertex, the triangle is (p_1, p_2, p_5) and the pentagon is $(p_1, \ldots, p_5)$.

Suppose that D is a decomposition star whose associate graph contains such triangular and pentagonal standard regions. Recall that D determines a set $U(D)$ of vertices in Euclidean 3-space of distance at most $2t_0$ from the origin, and that each vertex p_i can be realized geometrically as a point on the unit sphere at the origin, obtained as the radial projection of some $v_i \in U(D)$.

Lemma 14.3. *One of the edges $\{v_1, v_3\}$, $\{v_1, v_4\}$ has length less than $2\sqrt{2}$. Both of them have lengths less than 3.02. Also, $|v_1| \geq 2.3$.*

Proof. This is a standard exercise in geometric considerations as introduced in Section 4.2. (The reader should review that section for the framework of the following argument.) We deform the figure using pivots to a configuration $v_2, \ldots, v_5$ at height 2, and $|v_i - v_j| = 2t_0$, $(i, j) = (2, 3), (3, 4), (4, 5), (5, 2)$. We scale v_1 until $|v_1| = 2t_0$. We can also take the distance from v_1 to v_5 and to v_2 to be 2. If we have $|v_1 - v_3| \geq 2\sqrt{2}$, then we stretch the edge $|v_1 - v_4|$ until $|v_1 - v_3| = 2\sqrt{2}$. The resulting configuration is rigid. Pick coordinates to find that $|v_1 - v_4| < 2\sqrt{2}$. If we have $|v_1 - v_3| \geq 2t_0$, follow a similar procedure to reduce to the rigid configuration $|v_1 - v_3| = 2t_0$, to find that $|v_1 - v_4| < 3.02$. The estimate $|v_1| \geq 2.3$ is similar. □

There are restrictive bounds on the dihedral angles of the simplices $\{0, v_1, v_i, v_j\}$ along the edge $\{0, v_1\}$. The quasi-regular tetrahedron has a dihedral angle of at most[116] 1.875. The dihedral angles of the simplices $\{0, v_1, v_2, v_3\}$, $\{0, v_1, v_5, v_4\}$ adjacent to it are

---

[116] CALC-984463800.

at most[117] 1.63. The dihedral angle of the remaining simplex $\{0, v_1, v_3, v_4\}$ is at most[118] 1.51. This leads to lower bounds as well. The quasi-regular tetrahedron has a dihedral angle that is at least $2\pi - 2(1.63) - 1.51 > 1.51$. The dihedral angles adjacent to the quasi-regular tetrahedron is at least $2\pi - 1.63 - 1.51 - 1.875 > 1.26$. The remaining dihedral angle is at least $2\pi - 1.875 - 2(1.63) > 1.14$.

A decomposition star D determines a set of vertices $U(D)$ that are of distance at most $2t_0$ from the center of D. Three consecutive vertices p_1, p_2, and p_3 of a standard region are determined as the projections to the unit sphere of three corners v_1, v_2, and v_3, respectively in $U(D)$. By Lemma 11.30, if the interior angle of the standard region is less than 1.32, then $|v_1 - v_3| \leq \sqrt{8}$.

Lemma 14.4. *These two standard regions $F = \{R_1, R_2\}$ give $\tau_F(D) \geq 11.16\,pt$.*

Proof. Let dih denote the dihedral angle of a simplex along a given edge. Let S_{ij} be the simplex $\{0, v_1, v_i, v_j\}$, for $(i, j) = (2, 3), (3, 4), (4, 5), (2, 5)$. We have $\sum_{(4)} \mathrm{dih}(S_{ij}) = 2\pi$. Suppose one of the edges $\{v_1, v_3\}$ or $\{v_1, v_4\}$ has length $\geq 2\sqrt{2}$. Say $\{v_1, v_3\}$.
We have[119]

$$\tau(S_{25}) - 0.2529\,\mathrm{dih}(S_{25}) > -0.3442,$$
$$\tau_0(S_{23}) - 0.2529\,\mathrm{dih}(S_{23}) > -0.1787,$$
$$\hat{\tau}(S_{45}) - 0.2529\,\mathrm{dih}(S_{45}) > -0.2137,$$
$$\tau_0(S_{34}) - 0.2529\,\mathrm{dih}(S_{34}) > -0.1371.$$

We have a penalty ξ_Γ for erasing, so that

$$\tau(D) \geq \sum_{(4)} \tau_x(S_{ij}) - 5\xi_\Gamma$$
$$> 2\pi(0.2529) - 0.3442 - 0.1787 - 0.2137 - 0.1371 - 5\xi_\Gamma$$
$$> 11.16\,pt,$$

where $\tau_x = \tau, \hat{\tau}, \tau_0$ as appropriate.

Now suppose $\{v_1, v_3\}$ and $\{v_1, v_4\}$ have length $\leq 2\sqrt{2}$. If there is an upright diagonal that is not enclosed over either flat quarter, the penalty is at most $3\xi_\Gamma + 2\xi_V$. Otherwise, the penalty is smaller: $4\xi_\Gamma' + \xi_V$. We have

$$\tau(D) \geq \sum_{(4)} \tau(S_{ij}) - (3\xi_\Gamma + 2\xi_V)$$
$$> 2\pi(0.2529) - 0.3442 - 2(0.2137) - 0.1371 - (3\xi_\Gamma + 2\xi_V)$$
$$> 11.16\,pt. \qquad \square$$

---

[117] CALC-821707685.
[118] CALC-115383627.
[119] CALC-572068135, CALC-723700608, CALC-560470084, and CALC-535502975.

14.3.   A Particular 5-Circuit

Lemma 14.5.   *Assume that R is a pentagonal standard region with an enclosed vertex v of height at most $2t_0$. Assume further that:*

- $|v_i| \leq 2.168$ *for each of the five corners.*
- *Each interior angle of the pentagon is at most 2.89.*
- *If v_1, v_2, v_3 are consecutive corners over the pentagonal region, then*

$$|v_1 - v_2| + |v_2 - v_3| < 4.804.$$

- $\sum_5 |v_i - v_{i+1}| \leq 11.407.$

Then $\sigma_R(D) < -0.2345$ or $\tau_R(D) > 0.6079$.

Proof.   Since -0.4339 is less than this the lower bound, a 3-crowded upright diagonal does not occur. Similarly, since -0.25 is less than the lower bound, a 4-crowded upright diagonal does not occur (Lemma 11.18 and Definition 11.7).

Suppose that there is a loop in context $(n, k) = (4, 2)$. Again by Lemma 13.5 (with $n(R) = 7$),

$$\sigma_R(D) < -0.2345.$$

We conclude that all loops have context $(n, k) = (4, 1)$.

Case 1. The vertex $v = v_{12}$ has distance at least $2t_0$ from the five corners of $U(D)$ over the pentagon. The penalty to switch the pentagon to a pure vor_0 score is at most $5\xi_\Gamma$ (see Section 12.6). There cannot be two flat quarters because then

$$|v_{12}| > \mathcal{E}(S(2, 2, 2, 2t_0, 2\sqrt{2}, 2\sqrt{2}), 2t_0, 2t_0, 2t_0) > 2t_0.$$

Case 1a. Suppose there is one flat quarter, $|v_1 - v_4| \leq 2\sqrt{2}$. There is a lower bound of 1.2 on the dihedral angles of the simplices $\{0, v_{12}, v_i, v_{i+1}\}$. This is obtained as follows. The proof relies on the convexity of the quadrilateral region. We leave it to the reader to verify that the following pivots can be made to preserve convexity. Disregard all vertices except v_1, v_2, v_3, v_4, v_{12}. We give the argument that dih$(0, v_{12}, v_1, v_4) > 1.2$. The others are similar. Disregard the length $|v_1 - v_4|$. We show that

$$sd := \text{dih}(0, v_{12}, v_1, v_2) + \text{dih}(0, v_{12}, v_2, v_3)$$
$$+ \text{dih}(0, v_{12}, v_3, v_4) < 2\pi - 1.2.$$

Lift v_{12} so $|v_{12}| = 2t_0$. Maximize sd by taking $|v_1 - v_2| = |v_2 - v_3| = |v_3 - v_4| = 2t_0$. Fixing v_3 and v_4, pivot v_1 around $\{0, v_{12}\}$ toward v_4, dragging v_2 toward v_{12} until $|v_2 - v_{12}| = 2t_0$. Similarly, we obtain $|v_3 - v_{12}| = 2t_0$. We now have $sd \leq 3(1.63) < 2\pi - 1.2$, by a calculation.[120]

---

[120] CALC-821707685.

Return to the original figure and move v_{12} without increasing $|v_{12}|$ until each simplex $\{0, v_{12}, v_i, v_{i+1}\}$ has an edge (v_{12}, v_j) of length $2t_0$. Interval calculations[121] show that the four simplices around v_{12} squander

$$2\pi(0.2529) - 3(0.1376) - 0.12 > (4\pi\zeta - 8)\,pt + 5\xi_\Gamma.$$

Case 1b. Assume there are no flat quarters. By hypothesis, the perimeter satisfies

$$\sum |v_i - v_{i+1}| \leq 11.407.$$

We have $\mathrm{arc}(2, 2, x)'' = 2x/(16-x^2)^{3/2} > 0$. The arclength of the perimeter is therefore at most

$$2\,\mathrm{arc}(2, 2, 2t_0) + 2\,\mathrm{arc}(2, 2, 2) + \mathrm{arc}(2, 2, 2.387) < 2\pi.$$

There is a well-defined interior of the spherical pentagon, a component of area $< 2\pi$. If we deform by decreasing the perimeter, the component of area $< 2\pi$ does not get swapped with the other component.

Disregard all vertices but $v_1, \ldots, v_5, v_{12}$. If a vertex v_i satisfies $|v_i - v_{12}| > 2t_0$, deform v_i as in Section 12.8 until $|v_{i-1} - v_i| = |v_i - v_{i+1}| = 2$, or $|v_i - v_{12}| = 2t_0$. If at any time, four of the edges realize the bound $|v_i - v_{i+1}| = 2$, we have reached an impossible situation, because it leads to the contradiction[122]

$$2\pi = \overset{(5)}{\sum} \mathrm{dih} < 1.51 + 4(1.16) < 2\pi.$$

(This inequality relies on the observation, which we leave to the reader, that in any such assembly, pivots can by applied to bring $|v_{12} - v_i| = 2t_0$ for at least one edge of each of the five simplices.)

The vertex v_{12} may be moved without increasing $|v_{12}|$ so that eventually by these deformations (and reindexing if necessary) we have $|v_{12} - v_i| = 2t_0$, $i = 1, 3, 4$. (If we have $i = 1, 2, 3$, the two dihedral angles along $\{0, v_2\}$ satisfy[123] $< 2(1.51) < \pi$, so the deformations can continue.)

There are two cases. In both cases $|v_i - v_{12}| = 2t_0$, for $i = 1, 3, 4$.

(i) $|v_{12} - v_2| = |v_{12} - v_5| = 2t_0$.
(ii) $|v_{12} - v_2| = 2t_0$, $|v_4 - v_5| = |v_5 - v_1| = 2$.

Case (i) follows from interval calculations[124]

$$\sum \tau_0 \geq 2\pi(0.2529) - 5(0.1453) > 0.644 + 7\xi_\Gamma.$$

In case (ii) we have again

$$2\pi(0.2529) - 5(0.1453).$$

---

[121] CALC-467530297 and CALC-135427691.
[122] CALC-115383627 and CALC-603145528.
[123] CALC-115383627.
[124] CALC-312132053.

In this interval calculation we have assumed that $|v_{12} - v_5| < 3.488$. Otherwise, setting $S = (v_{12}, v_4, v_5, v_1)$, we have

$$\Delta(S) < \Delta(3.488^2, 4, 4, 8, (2t_0)^2, (2t_0)^2) < 0,$$

and the simplex does not exist. ($|v_4 - v_1| \geq 2\sqrt{2}$ because there are no flat quarters.) This completes Case 1.

Case 2: The vertex v_{12} has distance at most $2t_0$ from the vertex v_1 and distance at least $2t_0$ from the others. Let $\{0, v_{13}\}$ be the upright diagonal of a loop $(4, 1)$. The vertices of the loop are not $\{v_2, v_3, v_4, v_5\}$ with v_{12} enclosed over $\{0, v_2, v_5, v_{13}\}$ by Lemma 11.5. The vertices of the loop are not $\{v_2, v_3, v_4, v_5\}$ with v_{12} enclosed over $\{0, v_1, v_2, v_5\}$ because this would lead to a contradiction:

$$y_{12} \geq \mathcal{E}(S(2, 2, 2, 2t_0, 2t_0, 3.2), 2t_0, 2t_0, 2) > 2t_0$$

or

$$y_{12} \geq \mathcal{E}(S(2, 2, 2, 2t_0, 2t_0, 3.2), 2, 2t_0, 2) > 2t_0.$$

We get a contradiction for the same reasons unless $\{v_1, v_{12}\}$ is an edge of some upright quarter of every loop of type $(4, 1)$.

We consider two cases. (Case 2a) There is a flat quarter along an edge other than $\{v_1, v_{12}\}$. That is, the central vertex is $v_2, v_3, v_4,$ or v_5. (Recall that the *central vertex* of a flat quarter is the vertex other than the origin that is not an endpoint of the diagonal.) (Case 2b) Every flat quarter has central vertex v_1.

Case 2a. We erase all upright quarters including those in loops, taking penalties as required. There cannot be two flat quarters by geometric considerations

$$\mathcal{E}(S(2, 2, 2, 2\sqrt{2}, 2\sqrt{2}, 2t_0), 2t_0, 2t_0, 2) > 2t_0,$$
$$\mathcal{E}(S(2, 2, 2, 2\sqrt{2}, 2\sqrt{2}, 2t_0), 2, 2t_0, 2t_0) > 2t_0.$$

The penalty is at most $7\xi_\Gamma$. We show that the region (with upright quarters erased) squanders $> 7\xi_\Gamma + 0.644$. We assume that the central vertex is v_2 (Case 2a(i)) or v_3 (Case 2a(ii)). In Case 2a(i), we have three types of simplices around v_{12}, characterized by the bounds on their edge lengths. Let $\{0, v_{12}, v_1, v_5\}$ have type A, $\{0, v_{12}, v_5, v_4\}$ and $\{0, v_{12}, v_4, v_3\}$ have type B, and let $\{0, v_{12}, v_3, v_1\}$ have type C. In Case 2a(ii) there are also three types. Let $\{0, v_{12}, v_1, v_2\}$ and $\{0, v_{12}, v_1, v_5\}$ have type A, $\{0, v_{12}, v_5, v_4\}$ type B, and $\{0, v_{12}, v_2, v_4\}$ type D. (There is no relation here between these types and the types of simplices A, B, and C defined in Section 9.) Upper bounds on the dihedral angles along the edge $\{0, v_{12}\}$ are given as calculations.[125] These upper bounds come as a result of a pivot argument similar to that establishing the bound 1.2 in Case 1a.

These upper bounds imply the following lower bounds. In Case 2a(i),

$$\text{dih} > 1.33 \quad (A),$$
$$\text{dih} > 1.21 \quad (B),$$
$$\text{dih} > 1.63 \quad (C),$$

---

[125] CALC-821707685, CALC-115383627, CALC-576221766, and CALC-122081309.

and in Case 2a(ii),

$$\text{dih} > 1.37 \quad (\text{A}),$$
$$\text{dih} > 1.25 \quad (\text{B}),$$
$$\text{dih} > 1.51 \quad (\text{D}),$$

In every case the dihedral angle is at least 1.21. In Case 2a(i), the inequalities give a lower bound on what is squandered by the four simplices around $\{0, v_{12}\}$. Again, we move v_{12} without decreasing the score until each simplex $\{0, v_{12}, v_i, v_{i+1}\}$ has an edge satisfying $|v_{12} - v_j| \leq 2t_0$. Interval calculations[126] give

$$\sum_{(4)} \tau_0 > 2\pi(0.2529) - 0.2391 - 2(0.1376) - 0.266$$

$$> 0.808.$$

In Case 2a(ii), we have[127]

$$\sum_{(4)} \tau_0 > 2\pi(0.2529) - 2(0.2391) - 0.1376 - 0.12$$

$$> 0.853.$$

So we squander more than $7\xi_\Gamma + 0.644$, as claimed.

Case 2b. We now assume that there are no flat quarters with central vertex $v_2, \ldots, v_5$. We claim that v_{12} is not enclosed over $\{0, v_1, v_2, v_3\}$ or $\{0, v_1, v_5, v_4\}$. In fact, if v_{12} is enclosed over $\{0, v_1, v_2, v_3\}$, then we reach the contradiction[128]

$$\pi < \text{dih}(0, v_{12}, v_1, v_2) + \text{dih}(0, v_{12}, v_2, v_3)$$
$$< 1.63 + 1.51 < \pi.$$

We claim that v_{12} is not enclosed over $\{0, v_5, v_1, v_2\}$. Let $S_1 = \{0, v_{12}, v_1, v_2\}$ and $S_2 = \{0, v_{12}, v_1, v_5\}$. We have by hypothesis,

$$y_4(S_1) + y_4(S_2) = |v_1 - v_2| + |v_1 - v_5| < 4.804.$$

An interval calculation[129] gives

$$\sum_{(2)} \text{dih}(S_i) \leq \sum_{(2)} (\text{dih}(S_i) + 0.5(0.4804/2 - y_4(S_i)))$$

$$< \pi.$$

So v_{12} is not enclosed over $\{0, v_1, v_2, v_5\}$.

---

[126] CALC-644534985, CALC-467530297, and CALC-603910880.
[127] CALC-135427691.
[128] CALC-821707685 and CALC-115383627.
[129] CALC-69064028.

Erase all upright quarters, taking penalties as required. Replace all flat quarters with s-voro-scoring taking penalties as required. (Any flat quarter has v_1 as its central vertex.) We move v_{12} keeping $|v_{12}|$ fixed and not decreasing $|v_{12} - v_1|$. The only effect this has on the score comes through the quoins along $\{0, v_1, v_{12}\}$. Stretching $|v_{12} - v_1|$ shrinks the quoins and increases the score. (The sign of the derivative of the quoin with respect to the top edge is computed in the proof of Lemma 12.9.)

If we stretch $|v_{12} - v_1|$ to length $2t_0$, we are done by Case 1 and Case 2a. (If deformations produce a flat quarter, use Case 2a, otherwise use Case 1.) By the claims, we can eventually arrange (reindexing if necessary) so that

(i) $|v_{12} - v_3| = |v_{12} - v_4| = 2t_0$, or
(ii) $|v_{12} - v_3| = |v_{12} - v_5| = 2t_0$.

We combine this with the deformations of Section 12.8 so that in case (i) we may also assume that if $|v_5 - v_{12}| > 2t_0$, then $|v_4 - v_5| = |v_5 - v_1| = 2$ and that if $|v_2 - v_{12}| > 2t_0$, then $|v_1 - v_2| = |v_2 - v_3| = 2$. In case (ii) we may also assume that if $|v_4 - v_{12}| > 2t_0$, then $|v_3 - v_4| = |v_4 - v_5| = 2$ and that if $|v_2 - v_{12}| > 2t_0$, then $|v_1 - v_2| = |v_2 - v_3| = 2$.

Break the pentagon into subregions by cutting along the edges (v_{12}, v_i) that satisfy $|v_{12} - v_i| \leq 2t_0$. So for example in case (i), we cut along (v_{12}, v_3), (v_{12}, v_4), (v_{12}, v_1), and possibly along (v_{12}, v_2) and (v_{12}, v_5). This breaks the pentagon into triangular and quadrilateral regions.

In case (ii) if $|v_4 - v_{12}| > 2t_0$, then the argument used in Case 1 to show that $|v_4 - v_{12}| < 3.488$ applies here as well. In case (i) or (ii) if $|v_{12} - v_2| > 2t_0$, then for similar reasons, we may assume

$$\Delta(|v_{12} - v_2|^2, 4, 4, 8, (2t_0)^2, |v_{12} - v_1|^2) \geq 0.$$

This justifies the hypotheses for the calculations[130] that we use. We conclude that

$$\sum \tau_0 \geq 2\pi(0.2529) - 3(0.1453) - 2(0.2391) > 0.6749.$$

If the penalty is less than $0.067 = 0.6749 - 0.6079$, we are done.

We have ruled out the existence of all loops except $(4, 1)$. Note that a flat quarter with central vertex v_1 gives penalty at most 0.02 by Lemma 11.29. If there is at most one such flat quarter and at most one loop, we are done:

$$3\xi_\Gamma + 0.02 < 0.067.$$

Assume there are two loops of context $(n, k) = (4, 1)$. They both lie along the edge $\{v_1, v_{12}\}$, which precludes any unmasked flat quarters. If one of the upright diagonals has height ≥ 2.696, then the penalty is at most $3\xi_\Gamma + 3\xi_V < 0.067$. Assume both heights are at most 2.696. The total interior angle of the exceptional face at v_1 is at least four times the dihedral angle of one of the flat quarters along $\{0, v_1\}$, or $4(0.74)$ by an interval calculation.[131] This is contrary to the hypothesis of an interior angle < 2.89. This completes Case 2. This shows that heptagons with pentagonal hulls do not occur. □

---

[130] CALC-312132053 and CALC-644534985.
[131] CALC-751442360.

Lemma 14.6. *Let R be an exceptional standard region. Let V be a set of vertices of R. If $v \in V$, let p_v be the number of triangular regions at v and let q_v be the number of quadrilateral regions at v. Assume that V has the following properties:*

1. *No two vertices in V are adjacent.*
2. *No two vertices in V lie on a common quadrilateral.*
3. *If $v \in V$, then there are five standard regions at v.*
4. *If $v \in V$, then the corner over v is a central vertex of a flat quarter in the cone over R.*
5. *If $v \in V$, then $p_v \geq 3$. That is, at least three of the five standard regions at v are triangular.*
6. *If $R' \neq R$ is an exceptional region at v, and if R has interior angle at least 1.32 at v, then R' also has interior angle at least 1.32 at v.*
7. *If $(p_v, q_v) = (3, 1)$, then the internal angle at v of the exceptional region is at most 1.32.*

Define $a \colon \mathbb{N} \to \mathbb{R}$ by

$$
a(n) = \begin{cases} 14.8, & n = 0, 1, 2, \\ 1.4, & n = 3, \\ 1.5, & n = 4, \\ 0, & \text{otherwise.} \end{cases}
$$

Let $\{F\}$ be the union of $\{R\}$ with the set of triangular and quadrilateral regions that have a vertex at some $v \in V$. Then

$$
\sum_F \tau_F(D) > \sum_{v \in V} (p_v d(3) + q_v d(4) + a(p_v)) \, pt.
$$

Proof. We erase all upright diagonals in the Q-system, except for those that carry a penalty: loops, 3-unconfined, 3-crowded, and 4-crowded diagonals.

We assume that if $(p_v, q_v) = (3, 1)$, then the internal angle is at most 1.32. Because of this, if we score the flat quarter by vor_0, then the flat quarter Q satisfies (Lemma 11.30)

$$
\mathrm{vor}_0(Q) > 3.07 \, pt > 1.4 \, pt + D(3, 1) + 2\xi_V + \xi_\Gamma. \tag{14.1}
$$

Every flat quarter that is masked by a remaining upright quarter in the Q-system has $y_4 \geq 2.6$. Moreover, $y_1 \geq 2.2$ or $y_4 \geq 2.7$. Let $\pi_v = 2\xi_V + \xi_\Gamma$ if the flat quarter is masked, and $\pi_v = 0$ otherwise.

We claim that the flat quarter (scored by vor_0) together with the triangles and quadrilaterals at a given vertex v squander at least

$$
(p_v d(3) + q_v d(4) + a(p_v)) \, pt + D(3, 1) + \pi_v. \tag{14.2}
$$

If $p_v = 4$, this is CALC-314974315. If $p_v = 3$, we may assume by the preceding remarks that there are two exceptional regions at v. If the internal angle of R at v is at most 1.32, then we use inequality (14.1). If the angle is at least 1.32, then by hypothesis, the angle R' at v is at least 1.32. We then appeal to the calculations CALC-675785884 and CALC-193592217.

To complete the proof of the lemma, it is enough to show that we can erase the upright quarters masking a flat quarter at v without incurring a penalty greater than π_v. For then, by summing the inequality (14.2) over v, we obtain the result.

If the upright diagonal is enclosed over the masked flat quarter, then the upright quarters can be erased with penalty at most ξ_V (by Remark 11.28). Assume the upright diagonal is not enclosed over the masked flat quarter.

If there are at most three upright quarters, the penalty is at most $2\xi_V + \xi_\Gamma$. Assume four or more upright quarters. If the upright diagonal is not a loop, then it must be 4-crowded. This can be erased with penalty

$$2\xi_V + 2\xi_\Gamma - \kappa < 2\xi_V + \xi_\Gamma.$$

Finally, assume that the upright quarter is a loop with four or more upright quarters. Lemma 13.5 limits the possibilities to parameters $(5, 0)$ or $(5, 1)$. In the case of a loop $(5, 1)$, there is no need to erase because $|V| \leq 3$ and by Lemma 13.5 the hexagonal standard region squanders at least

$$t_6 + 3a(p_v)\,pt$$

as required by the lemma. In the case of a loop $(5, 0)$ in a pentagonal region, if $|V| = 1$ then there is no need to erase (again we appeal to Lemma 13.5). If $|V| = 2$, then the two vertices share a penalty of $4\xi_V + \xi_\Gamma$, with each receiving

$$2\xi_V + \xi_\Gamma/2 < 2\xi_V + \xi_\Gamma. \qquad \square$$

References

[Ha6] T. C. Hales, Sphere packings, I, *Discrete Comput. Geom.* **17** (1997), 1–51.

Received November 11, 1998, and in revised form September 12, 2003, and July 25, 2005. Online publication February 27, 2006.

7

Sphere Packings V. Pentahedral Prisms, by S. P. Ferguson

This paper is the fifth of six papers giving the Kepler Conjecture proof. It contains Chapters 15–17 of the proof.

The asterisks added in the margin of some pages indicate that changes are required, as listed in the Hales list of errata given on pp. 373–374 of this volume.

Contents

The original version of this chapter was revised. An erratum to this chapter can be found at
http://dx.doi.org/10.1007/978-1-4614-1129-1_12

Discrete Comput Geom 36:167–204 (2006)
DOI: 10.1007/s00454-005-1214-y

Discrete & Computational

© 2006 Springer Science+Business Media, Inc.

Sphere Packings, V. Pentahedral Prisms

Samuel P. Ferguson

5960 Millrace Court B-303,
Columbia, MD 21045, USA
samf2@comcast.net

Abstract. This paper is the fifth in a series of papers devoted to the proof of the Kepler conjecture, which asserts that no packing of congruent balls in three dimensions has density greater than the face-centered cubic packing.

In this paper we prove that decomposition stars associated with the plane graph of arrangements we term pentahedral prisms do not contravene. Recall that a contravening decomposition star is a potential counterexample to the Kepler conjecture. We use interval arithmetic methods to prove particular linear relations on components of any such contravening decomposition star. These relations are then combined to prove that no such contravening stars exist.

Introduction

Pentahedral prisms come remarkably close to achieving the optimal score of 8 *pt*, that achieved by the decomposition stars of the face-centered cubic lattice packing. In this sense we consider pentahedral prisms to be "worst-case" decomposition stars.

Pentahedral prisms constituted a counterexample to an early version of Hales's approach to a proof of the Kepler conjecture, and have always been a somewhat thorny obstacle to the proof of the conjecture. Relations required to treat pentahedral prisms are delicate in contrast to the more general bounds which suffice to treat other decomposition stars.

15. Pentahedral Prisms

Recall that a *contravening decomposition star* is a potential counterexample to the Kepler conjecture. The subject of this paper is a particular class of potentially contravening decomposition stars.

We use the term *pentahedral prisms* to refer to this class of potentially contravening decomposition stars, and refer to a decomposition star in this class as a *pentahedral prism*. This class is defined by the plane graph in Fig. 15.1.

An example of an arrangement with such a graph is depicted in Fig. 15.2.

A pentahedral prism is characterized by the arrangement and combinatorics of its standard regions. It is composed of ten triangular standard regions, and five quadrilateral standard regions.

The ten triangles are arranged in two *pentahedral caps*, five triangles arranged around a common vertex. The five quadrilaterals lie in a band between the two caps. See Fig. 15.3.

Recall that the standard cluster attached to a triangular standard region is a quasi-regular tetrahedron. Likewise, the standard cluster attached to a quadrilateral is a quad cluster. We use the term pentahedral cap to refer to both the standard regions and the quasi-regular tetrahedrons which comprise it.

15.1. The Main Theorem

We begin by recalling various definitions from the second paper, Formulation [HF]. The constant *pt* is introduced in Definition 3.6. Similarly, *score* is defined in Theorem 3.5, as well as Definition 7.8 and Remark 7.20. We denote the score of a region R by $\sigma(R)$.

Theorem 15.1. *Each pentahedral prism P satisfies*

$$\sigma(P) \leq (8 - \varepsilon_0)\,pt$$

for $\varepsilon_0 = 10^{-8}$. Hence there are no contravening pentahedral prisms.

The next section introduces a series of propositions which prove the main theorem. The first proposition restricts our attention to a set of potentially contravening pentahedral prisms. Subsequent propositions provide a collection of relations which we use to prove the main theorem.

15.2. Propositions

The function sol(·) is introduced in Definition 7.5. The function dih(·) is introduced in Definition 4.12.

We present computations using auxiliary bounds which imply the main result of the paper, that the score of any pentahedral prism is strictly less than 8 *pt*.

Recall from Section 7.4 that the score decomposition for a decomposition star S takes the form

$$\sigma(S) = \sum_{R} \sigma(R),$$

where R runs over the standard clusters in S.

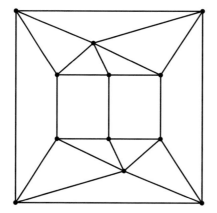

Fig. 15.1. The plane graph of a pentahedral prism.

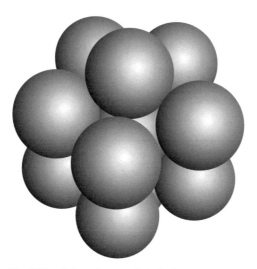

Fig. 15.2. Spheres in a pentahedral prism arrangement.

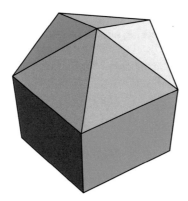

Fig. 15.3. The faces of a pentahedral prism.

In the case of a pentahedral prism P, the score $\sigma(P)$ decomposes as

$$\sigma(P) = \sum_{i=1}^{10} \sigma(T_i) + \sum_{j=1}^{5} \sigma(Q_j)$$

with the triangular regions T_i numbered so that the two pentahedral caps C_i consist of $\{T_i: 1 \le i \le 5\}$, $\{T_i: 6 \le i \le 10\}$ and Q_j denotes the quad clusters. Thus

$$\sigma(C_1) = \sum_{i=1}^{5} \sigma(T_i), \qquad \sigma(C_2) = \sum_{i=6}^{10} \sigma(T_i).$$

The following proposition gives basic inequalities which we use to form a restricted set of pentahedral prisms.

Proposition 15.2. *A pentahedral prism P satisfies the bound*

$$\sigma(P) \le (8 - \varepsilon_0)\, pt$$

for $\varepsilon_0 = 10^{-8}$ provided that any one of the following conditions holds:

1. *P contains a tetrahedron T such that*

$$\sigma(T) \le -0.52\, pt.$$

2. *P contains a quad cluster Q such that*

$$\sigma(Q) \le -1.04\, pt.$$

3. *P contains a pentahedral cap C such that*

$$\sigma(C) \le 3.48\, pt.$$

Proof. We use the following scoring bounds proved earlier for any admissible decomposition star.

First, Lemma 8.10 states that any quasi-regular tetrahedron T satisfies

$$\sigma(T) \le 1\, pt.$$

Theorem 8.4 states that any quad cluster Q satisfies

$$\sigma(Q) \le 0.$$

Next, a pentahedral cap C consists of five quasi-regular tetrahedra T_i sharing a common distinguished edge. At one end of the distinguished edge is the distinguished vertex $v = 0$ which is the center of the decomposition star P. Each T_i has context $c_i = (5, 0)$.

Lemmas 10.6 (with $k = 1$ and $r = 5$) and 10.7 state that any pentahedral cap C_i satisfies

$$\sigma(C_i) = \sum_{i=1}^{5} \sigma(T_i, c_i, v) \leq (4.52 - \varepsilon_0)\,pt$$

with $\varepsilon_0 = 10^{-8}$.

1. Suppose that some $\sigma(T) \leq -0.52\,pt$, with T contained in a pentahedral cap C_1. Then the inequalities above give

$$\sigma \leq -0.52\,pt + 4(1\,pt) + \sigma(C_2) + \sum_{j=1}^{5} \sigma(Q_j)$$

$$\leq 3.48\,pt + (4.52 - \varepsilon_0)\,pt + 5(0) = (8 - \varepsilon_0)\,pt.$$

2. Suppose that some quad cluster $\sigma(Q_j) \leq -1.04\,pt$. Then

$$\sigma(P) \leq \sigma(C_1) + \sigma(C_2) + (-1.04\,pt) + 4(0)$$

$$\leq 2((4.52 - \varepsilon_0)\,pt) - (1.04\,pt) = (8 - 2\varepsilon_0)\,pt.$$

3. Suppose that some pentahedral cap C_1 has $\sigma(C_1) \leq 3.48\,pt$. Then the inequalities above give

$$\sigma(P) \leq (3.48\,pt) + (4.52 - \varepsilon_0)\,pt + 5(0) = (8 - \varepsilon_0)\,pt.$$

This completes the proof. $\square$

Definition 15.3.   A *PC pentahedral prism* is a pentahedral prism such that:

1. All tetrahedra T have $\sigma(T) \geq -0.52\,pt$.
2. All quad clusters have $\sigma(Q) \geq 1.04\,pt$.
3. All pentahedral caps have $\sigma(C) \geq 3.48\,pt$.

The configuration arises as a pointwise limit of configurations in which (1)–(3) hold with strict inequality. A *strict PC pentahedral prism* is one in which (1)–(3) each hold with strict inequality.

All remaining propositions apply to PC pentahedral prisms. This restriction improves the quality of the bounds which we are able to prove on components of a pentahedral prism.

The following two propositions contain linear relations which imply the main theorem. We defer their proofs to the next section.

Proposition 15.4.   *For a quasi-regular tetrahedron T in a PC pentahedral prism, the following linear inequality holds between $\sigma(T)$, the spherical angle $\mathrm{sol}(T)$ (at the central vertex common to the five tetrahedra in the pentahedral cap), and the dihedral angle* dih(T) *associated with the first edge of the tetrahedron (that is, the edge common to the five tetrahedra in a pentahedral cap):*

$$\sigma(T) + m\,\text{sol}(T) + a\left(\text{dih}(T) - \frac{2\pi}{5}\right) - b_c \le 0.$$

Here a $= 0.0739626$, $b_c = 0.253095$, *and m* $= 0.3621$.

Proposition 15.5. *For a quad cluster Q in a PC pentahedral prism, the following linear inequality holds between* $\sigma(Q)$ *and the spherical angle* sol(Q):

$$\sigma(Q) + m\,\text{sol}(Q) - b_q \le 0.$$

Here $b_q = 0.49246$ *and again m* $= 0.3621$.

From Propositions 15.4 and 15.5 we can deduce the following theorem.

Theorem 15.6. *Each PC pentahedral prism P satisfies the score bound*

$$\sigma(P) \le 7.9997\,pt.$$

Proof. Propositions 15.4 and 15.5 provide linear relations on all of the standard clusters in a PC pentahedral prism P. We combine these relations to prove the required score bound.

Invoking Proposition 15.4 for the five quasi-regular tetrahedrons $\{T_i : i = 1 \ldots 5\}$ from a pentahedral cap, we find

$$\sum_{i=1}^{5}\sigma(T_i) + m\sum_{i=1}^{5}\text{sol}(T_i) + a\sum_{i=1}^{5}\left(\text{dih}(T_i) - \frac{2\pi}{5}\right) - 5b_c \le 0.$$

Summing over both pentahedral caps and using the relation that the sum of the five dihedral angles in a pentahedral cap is 2π,

$$\sum_{i=1}^{5}\text{dih}(T_i) = 2\pi,$$

we find

$$\sum_{i=1}^{10}\sigma(T_i) + m\sum_{i=1}^{10}\text{sol}(T_i) - 10b_c \le 0.$$

We represent the tetrahedra from the second pentahedral cap by the indices $i = 6 \ldots 10$.

Invoking Proposition 15.5 for the five quad clusters $\{Q_i : i = 11 \ldots 15\}$, and using the fact that the sum of the solid angles is 4π,

$$\sum_{i=1}^{10}\text{sol}(Q_i) + \sum_{j=11}^{15}\text{sol}(Q_j) = 4\pi,$$

we find

$$\sum_{i=1}^{10} \sigma(T_i) + \sum_{j=11}^{15} \sigma(Q_j) + 4\pi m - 5b - 10b_c \le 0. \qquad \ast$$

Therefore,

$$\sigma(P) \le 5b + 10b_c - 4\pi m. \qquad \ast$$

Substituting the values of $b, b_c, m,$ and pt, we find that the score of a PC pentahedral $\ast$
prism is less than $7.9997\, pt$. $\qquad \square$

Assuming Proposition 15.2 and Theorem 15.6 we can prove Theorem 15.1.

Proof of Theorem 15.1.   Given a pentahedral prism P, it is either PC or it is not. In the
former case its score is bounded by $7.9997\, pt$. In the latter case its score is bounded by
$(8 - 10^{-8})\, pt$. In both cases its score is bounded by $(8 - 10^{-8})\, pt$. $\qquad \square$

Remark 15.7.   The score bound in Theorem 15.1 is weaker than what is possible to
prove. In the interest of simplifying the exposition as well as the required computations,
we establish this weaker bound which suffices for this part of the proof of the Kepler
conjecture.

16.   The Main Propositions

In the first section we recall the definition of score, and introduce some local notation.
In the next section we recall the notion of dimension reduction, and prove its validity for
some relevant cases. In the following section we prove Proposition 15.4. In the remaining
sections we prove Proposition 15.5.

16.1.   *Scoring*

The development of a scoring function is central to the proof of the Kepler conjecture.
Its definition is therefore somewhat complicated. Fortunately, in our treatment of the
pentahedral prism we are able to restrict our attention to only a few cases in the scoring
system.

   Recall that *score* is defined in Theorem 3.5, as well as Definitions 7.6 and 7.8 and
Remark 7.20. See Remark 7.23 for a simplified version of the scoring function for
quarters.

   In our context, the score $\sigma(\cdot)$ breaks into four different scoring types: $gma(\cdot)$, $vor(\cdot)$,
$octavor(\cdot)$, and Voronoi.

   $gma(\cdot)$ applies to quasi-regular tetrahedrons and quarters, and is introduced as $\Gamma(\cdot)$
in Definition 7.6. We frequently use the term *compression* as an alias for $gma(\cdot)$. This
alias was introduced in Section 7.6.

vor($\cdot$) is the score determined by the analytic continuation of the Voronoi volume associated with the distinguished vertex of a tetrahedron, and corresponds to s-vor($\cdot$) in Definition 7.6.

We let octavor($\cdot$) denote the score of an upright quarter in context $(4, 0)$ which is not scored by compression. In this case, octavor($\cdot$) is the average of two vor($\cdot$) scores.

Voronoi scoring, which we also refer to as pure Voronoi scoring, is $\text{vor}_R(D)$ from Remark 7.20.

16.2. *Dimension Reduction*

The relations on tetrahedra required for the scoring bound on decomposition stars are typically six-dimensional, as they are formulated in terms of the edge lengths of a tetrahedron. For a quad cluster, they can be even higher-dimensional. For high-dimensional relations, the method of subdivision becomes very expensive, computationally speaking.

We define a simplification which reduces the dimension of the required computations. This simplification therefore reduces the computational expense of the verification of a relation.

We refer to this simplification as dimension-reduction. We apply this simplification for three different scoring types: compression, vor analytic, and Voronoi. These scoring functions are introduced in Definitions 7.6 and 7.8. See Remark 7.23 for a simplified version of the scoring function for quarters.

Theorem 16.1 (Dimension-Reduction). *Given a tetrahedron T with a fixed scoring type (one of compression, vor analytic, or Voronoi), the deformation consisting of moving a vertex v_i along the edge $(0, v_i)$ towards the origin increases the score of the tetrahedron.*

Note that this deformation holds the solid angle at the origin fixed. See Fig. 16.1. Since the reduction may be performed until either a scoring system or an edge-length constraint is met, this argument reduces the number of free parameters for the verification, thus reducing the dimension and complexity of the verification of a relation.

Proof. There are three cases to consider: compression scoring, vor analytic scoring and Voronoi scoring. This technique was introduced in Proposition 8.7.1 of [Ha6] for compression-scoring, and is proved there in the compression case.

Next we consider vor analytic-scored tetrahedra. The validity of the same reduction for vor analytic-scored tetrahedra is obvious if the tip of the Voronoi cell does not protrude. If the tip does protrude, we must use the analytic continuation for the Voronoi volume. In this case the validity of the reduction is not obvious.

The geometric constraint of moving a vertex along an edge can easily be stated analytically in terms of the original edge lengths, $(y_1, y_2, y_3, y_4, y_5, y_6)$. This action depends on a single parameter, the distance of the vertex v_1 from the origin, which we

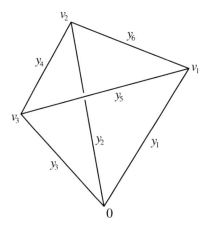

Fig. 16.1. Tetrahedron with distinguished vertex and labeled edges.

call t. The new edge lengths are given by

$$\left(t, y_2, y_3, y_4, \sqrt{t^2 + y_3^2 - \frac{t(y_1^2 + y_3^2 - y_5^2)}{y_1}}, \sqrt{t^2 + y_2^2 - \frac{t(y_1^2 + y_2^2 - y_6^2)}{y_1}} \right).$$

Recall from Section 8.6.3 of [Ha6] that the formula for the analytic Voronoi volume is a rational function of χ, u, $\sqrt{\Delta}$, and x_i, where $x_i = y_i^2$. Further recall that χ, u, and Δ are all polynomial functions in x_i that are defined in Sections 8.1 and 8.2 of [Ha6].

Substituting the computed edge lengths in the formula for the analytic Voronoi volume, taking the partial derivative with respect to t, replacing t with y_1, multiplying by the positive term

$$8\sqrt{\Delta} u(x_1, x_3, x_5) u(x_1, x_2, x_6)/y_1,$$

and then simplifying, we end up with a large homogeneous polynomial in x_i of degree 6, which is too ugly to exhibit here (having ninety-one terms).

Evaluating this polynomial over all possible quasi-regular tetrahedrons and quarters, we find that it is positive.

Therefore the volume is increasing in t, so to increase the score we should push the vertex in along the edge. The verification of the sign of the polynomial is found in Calculation 17.4.5.1. This completes the case of vor analytic-scored tetrahedra.

The final case is Voronoi scoring. The deformation does not change the solid angle of the tetrahedron. The only term of the Voronoi score that changes is a negative constant times a volume. The contraction of the tetrahedron decreases this volume, and increases the score. The validity of a similar reduction argument for Voronoi scoring of a tetrahedron is now obvious. $\qquad\square$

16.3. *Proof of Proposition* 15.4

It suffices to prove Proposition 15.4 for strict PC pentahedral prisms. Each nonstrict PC pentahedral prism is a pointwise limit of strict ones of the same combinatorial type, so the inequality in the conclusion of the proposition will hold for nonstrict PC pentahedral prisms by continuity.

We use three separate computations to construct and prove Proposition 15.4. First, we prove a relation between dihedral angle and score. We then show that if the dihedral angle of a tetrahedron in a pentahedral cap exceeds a certain bound, then the associated pentahedral prism is not a strict PC pentahedral prism. We call such a bound a *dihedral cutoff*. This cutoff allows us to prove the final bound.

In the following discussion, $\mathrm{dih}(T)$ refers to the dihedral angle associated with the first edge of a quasi-regular tetrahedron T, $\sigma(T)$ refers to the compression score of the tetrahedron, and $\mathrm{sol}(T)$ refers to the solid angle at the distinguished vertex. We restrict our attention to quasi-regular tetrahedrons whose score exceeds $-0.52\,pt$, as otherwise the associated pentahedral prism cannot contravene.

The first relation is

$$\sigma(T) \leq a_1 \,\mathrm{dih}(T) - a_2, \tag{16.1}$$

where $a_1 = 0.3860658808124052$ and $a_2 = 0.4198577862$. Calculation 17.4.1.1 provides the verification of this relation.

Lemma 16.2. *If a pentahedral prism has a pentahedral cap that contains a quasi-regular tetrahedron T with dihedral angle $\mathrm{dih}(T) \geq d_0$, where $d_0 = 1.4674$, then it is not a strict PC pentahedral prism.*

Proof. Applying relation (16.1) to four quasi-regular tetrahedrons T_i forming part of a strict PC pentahedral prism, we find

$$\sum_{i=1}^{4} \sigma(T_i) \leq a_1 \sum_{i=1}^{4} \mathrm{dih}(T_i) - 4a_2. \tag{16.2}$$

Applying the relation

$$\mathrm{dih}(T_5) = 2\pi - \sum_{i=1}^{4} \mathrm{dih}(T_i) \tag{16.3}$$

and adding $\sigma(T_5)$ to both sides of relation (16.2), we find

$$\sum_{i=1}^{5} \sigma(T_i) \leq \sigma(T_5) + a_1(2\pi - \mathrm{dih}(T_5)) - 4a_2. \tag{16.4}$$

The left-hand side represents the score of the pentahedral cap. If the right-hand side does not exceed $3.48\,pt$, then the pentahedral prism is not a strict PC pentahedral prism.

We assert that if $\mathrm{dih}(T) \geq d_0$, the right-hand side

$$\sigma(T_5) + a_1(2\pi - \mathrm{dih}(T_5)) - 4a_2$$

does not exceed $3.48\,pt$. Equivalently, we prove that $\mathrm{dih}(T) \geq d_0$ implies

$$\sigma(T) - a_1\,\mathrm{dih}(T) \leq 3.48\,pt - 2\pi a_1 + 4a_2,$$

which is verified in Calculation 17.4.1.2.

We conclude that if $\mathrm{dih}(T) \geq d_0$ then the pentahedral prism cannot be a strict PC pentahedral prism.   □

Hence we may restrict our attention to quasi-regular tetrahedrons whose dihedral angle does not exceed the dihedral cutoff d_0.

Using the dihedral cutoff, we establish the final relation,

$$\sigma(T) + m\,\mathrm{sol}(T) + a(\mathrm{dih}(T) - 2\pi/5) - b_c \leq 0$$

via Calculation 17.4.1.3. This completes the proof of Proposition 15.4.

16.4. *Proof of Proposition* 15.5: *Top Level*

It suffices to prove Proposition 15.5 for strict PC pentahedral prisms, by a similar argument to that used for Proposition 15.4.

Recall from Definition 7.15 that a quad cluster is a standard region that is a quadrilateral. Quad clusters can be classified as follows:

1. Flat quad clusters.
2. Octahedra.
3. Pure Voronoi quad clusters.
4. Mixed quad clusters.

We subdivide the third class into acute and obtuse types. See Section 10.4 for a discussion on the classification of quad clusters. By Lemma 10.14, the score of a mixed quad cluster is less than $-1.04\,pt$. A PC pentahedral prism therefore cannot contain a mixed quad cluster, so the bound of Proposition 15.5 holds trivially for this class.

We treat the remaining classes in the following sections.

16.5. *Proof of Proposition* 15.5: *Flat Quad Clusters*

Recall that a *flat* quarter is a quarter whose long edge is opposite its distinguished vertex.

Lemma 16.3.   *Given a flat quarter Q with $\sigma(Q) \geq -1.04\,pt$, the relation*

$$\sigma(Q) \leq -m\,\mathrm{sol}(Q) + b/2 \qquad\qquad (16.5)$$

holds.

Proof.   Label the diagonal of a flat quarter y_6.

Flat quarters may be scored using either compression or vor scoring. We treat each case separately.

First, suppose that we wish to prove the bound for compression scored quarters. This means that the circumradii of the two faces adjacent to the long diagonal do not exceed $\sqrt{2}$. We subdivide the verification into Calculation 17.4.2.1, a computation where we apply dimension-reduction and partial derivative information, and Calculation 17.4.2.2, a boundary verification, where we restrict our attention to cells which lie on the boundary between compression and vor scoring.

Second, we treat the vor-scoring case. In this case we prove the bound for vor-scored quarters. This means that at least one of the circumradii of the two faces adjacent to the long diagonal is at least $\sqrt{2}$. This verification is somewhat more complex than the compression case. We subdivide the verification into

1. Verification that the first three partials are negative on a small cell containing the corner (Calculation 17.4.2.3).
2. Verification of the bound on that small cell containing the corner, using the property that the first three partials are negative (Calculation 17.4.2.4).
3. A computation where we apply dimension-reduction and partial derivative reduction, omitting the corner cell (Calculation 17.4.2.5).
4. A boundary verification, where we restrict our attention to cells which lie on the boundary between compression and vor scoring, again omitting the corner cell (Calculation 17.4.2.6).

These calculations complete the proof of Lemma 16.3.   □

We are now prepared to prove Proposition 15.5 for flat quad clusters.

Flat quad clusters are composed of two flat quarters, whose common face includes the long edge.

By Proposition 15.2, we restrict our attention to flat quarters whose score exceeds $-1.04\,pt$, recalling the fact that the score of flat quarters is nonpositive.

Invoking Lemma 16.3 for each flat quarter and adding the relations, we arrive at the desired bound for flat quad clusters. This completes the proof.

16.6. *Proof of Proposition 15.5: Octahedra*

Recall that quartered octahedra, a type of quad cluster, are composed of four upright quarters arrayed around their common long edge (called the *diagonal*) so that each face containing the common edge is shared by two quarters.

We are required to prove a relation of the form

$$\sigma(H) + m\,\mathrm{sol}(H) - b \le 0,$$

where $\sigma(H)$ denotes the score of an octahedron H, $\mathrm{sol}(H)$ denotes the solid angle associated with the distinguished vertex, and m and b are positive constants. By Proposition 15.2, we restrict our attention to octahedra whose score exceeds $-1.04\,pt$.

Our treatment of octahedra, as usual, is comprised of a number of auxiliary computations. We prove bounds on upright quarters which are part of an octahedron, and then combine these bounds to deduce the required bound on octahedra in general.

The scoring function $\sigma(\cdot)$ for upright quarters is either compression (denoted by $gma(\cdot)$) or an average of two $vor(\cdot)$ scores, which we continue to refer to as vor-scoring. See Remark 7.23 for a simplified version of the scoring rules.

Due to the complex nature of octahedra, we consider a number of subcases. These cases are partitioned according to the length of the diagonal and the scoring system applied to the upright quarters.

Using a dihedral summation argument, we eliminate octahedra whose diagonal lies in the range [2.51, 2.716].

Next, we treat the case where the diagonal lies in the range $[2.716, 2\sqrt{2}]$. Using a dihedral correction term, we prove the bound for octahedra which are completely compression-scored, and octahedra which are completely vor-scored.

The remaining cases consist of octahedra which contain either two or three vor-scored quarters. (Since a quarter is vor-scored if one of the faces containing the diagonal has circumradius $\sqrt{2}$ or greater, it is not possible for an octahedron to contain only one vor-scored quarter.) We treat these cases using an additional correction term.

In all computations involving octahedra, we label the diagonal y_1.

In order to simplify the computations, we first prove an auxiliary cutoff bound. This first bound reduces the size of the cell over which we must conduct our search, as per Proposition 15.2.

Lemma 16.4. *If an upright quarter contains an edge numbered 2, 3, 5, or 6 whose length is not less than 2.2, then its score is less than or equal to* $-0.52\,pt$.

Proof. This is Calculation 17.4.3.1.   □

Since such an edge is shared by another upright quarter in the same octahedron, the score of the associated octahedron must fall below $-1.04\,pt$.

We restrict our search accordingly.

Lemma 16.5. *The score of an octahedron H with upright diagonal in the range* [2.51, 2.716] *is less than or equal to* $-1.04\,pt$.

Proof. In Calculation 17.4.3.2 we prove a bound of the form

$$\sigma(S) + c\,\mathrm{dih}(S) \le d$$

on upright quarters S, where $c = 0.1533667634670977$ and $d = 0.2265$. Adding the bound for four quarters S_i forming an octahedron, we find

$$\sum_{i=1}^{4} \sigma(S_i) + c \sum_{i=1}^{4} \mathrm{dih}(S_i) \le 4d.$$

Using the fact that the sum of the dihedral angles is 2π, we find that

$$\sigma(H) \le -2\pi c + 4d$$

for such an octahedron H.

A computation involving the constants c and d shows that the score is less than $-1.04\,pt$.   □

Again invoking Proposition 15.2, we need only consider octahedra whose diagonal lies in the range $[2.716, 2\sqrt{2}]$.

Using this assumption, we prove bounds of the form

$$\sigma(S) + m\,\text{sol}(S) + \alpha\,\text{dih}(S) \leq \frac{b}{4} + \alpha\frac{\pi}{2} \tag{16.6}$$

and

$$\sigma(S) + m\,\text{sol}(S) + \alpha\,\text{dih}(S) + \beta x_1 \leq \frac{b}{4} + \alpha\frac{\pi}{2} + 8\beta, \tag{16.7}$$

where $\text{dih}(S)$ refers to the dihedral angle associated with the diagonal, $\sigma(S)$ refers to the scoring scheme appropriate for a particular upright quarter S, and x_1 refers to the square of the length of the diagonal. We choose α and β according to the scoring scheme.

Appropriate values for the correction terms involving α and β were determined by experimentation.

Choosing $\alpha = 0.14$, we prove (16.6) for compression-scored quarters with diagonal in the interval $[2.716, 2\sqrt{2}]$ (Calculation 17.4.3.3). Using the same α, we prove (16.6) for vor-scored quarters with diagonal in the range $[2.716, 2.81]$ (Calculation 17.4.3.4).

Choosing $\alpha = 0.054$, $\beta = 0.00455$, we prove (16.7) for compression-scored quarters with diagonal in $[2.81, 2\sqrt{2}]$ (Calculation 17.4.3.5). Choosing the same α, but $\beta = -0.00455$, we prove (16.7) for vor-scored quarters with diagonal in $[2.81, 2\sqrt{2}]$ (Calculation 17.4.3.6).

Note that for vor-scored quarters, the first inequality is a relaxation of the second, since β is negative.

The verification of each of these inequalities involves a computation where we apply dimension-reduction and partial derivative information, and a boundary verification, where we restrict our attention to cells which lie on the boundary between compression and vor analytic scoring. Note that the dimension-reduction step for relation (16.7) is complicated by the presence of the βx_1 term.

Lemma 16.6. *Proposition 15.5 holds for octahedra with upright diagonals in the range* $[2.716, 2\sqrt{2}]$.

Proof. Summing inequality (16.6) over the four quarters S_i of an octahedron, we find

$$\sum_{i=1}^{4} \sigma(S_i) + m \sum_{i=1}^{4} \text{sol}(S_i) + \alpha \sum_{i=1}^{4} \text{dih}(S_i) \leq b + 2\alpha\pi.$$

Using the fact that the dihedral angles sum to 2π, we find

$$\sigma(H) + m\,\text{sol}(H) \leq b,$$

so octahedra H with diagonals in the range $[2.716, 2.81]$ satisfy the requisite bound.

Summing inequality (16.6) over a consistently scored octahedron (either all compression or all vor) with diagonal in the range $[2.81, 2\sqrt{2}]$, we again arrive at the desired bound.

The remaining cases involve octahedra which contain both compression and vorscored quarters, and whose diagonals lie in the range $[2.81, 2\sqrt{2}]$. For this case we use inequality (16.7).

The summation involving inequality (16.7) is identical to inequality (16.6) save for the presence of the β terms. If there are two vor-scored quarters and two compression-scored quarters, the beta terms cancel, giving the relation as before.

If there are three vor-scored quarters and one compression-scored quarter, we note that the same relation for vor-scored quarters holds if we replace β by $\beta/3$ (since we have now relaxed the bound). Summing the inequalities, the term involving β vanishes again, leaving the desired inequality.   □

Lemmas 16.5 and 16.6 prove Proposition 15.5 for octahedra.

16.7. *Proof of Proposition* 15.5: *Pure Voronoi Quad Clusters*

The next class of quad clusters which we treat are the pure Voronoi quad clusters. We will define a truncation operation on these quad clusters. Truncation will simplify the geometry of the quad clusters, and will provide a convenient scoring bound. We divide our treatment of pure Voronoi quad clusters into two cases in order to simplify the analysis and numerical verifications as much as possible.

Recall from the classification of quad clusters that a pure Voronoi quad cluster consists of the intersection of a V-cell at the origin with the cone at the origin over a quadrilateral standard region. We refer to the restriction of the V-cell to the cone over the quadrilateral as either the V-cell or the Voronoi cell of the quad cluster. Figure 16.2 describes the geometry of a simple V-cell.

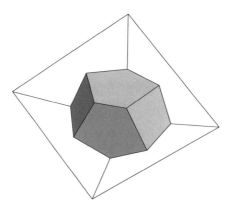

Fig. 16.2. A pure Voronoi quad cluster.

In addition, recall that a vertex lying in the cone over a pure Voronoi quad cluster must have height greater than $2\sqrt{2}$. Such vertices can significantly complicate the geometry of the V-cell, affecting its shape and volume.

We remove the effect of vertices lying above a pure Voronoi quad cluster by removing all points from the V-cell which have height greater than $\sqrt{2}$. We call this operation *truncation at* $\sqrt{2}$. Truncation decreases the volume of the quad cluster. This decrease in volume increases the score of the quad cluster, bringing it closer to the proposed bound.

We refer to truncated pure Voronoi quad clusters as *truncated* quad clusters.

We define a scoring operation on pure Voronoi quad clusters which we call *truncated Voronoi scoring*. This operation consists of truncation at $\sqrt{2}$, followed by the usual Voronoi scoring.

Each diagonal across the face of a cluster must have length greater than $2\sqrt{2}$, otherwise we could form two flat quarters, contradicting the decomposition. We choose the shorter of the two possible diagonals, and consider that diagonal in the analysis which follows.

We decompose the cluster into two tetrahedrons along the chosen diagonal. The face dividing the tetrahedrons is either acute or it is obtuse. We treat each case separately.

We must prove

$$\sigma(Q) + m\,\mathrm{sol}(Q) - b \le 0,$$

where $\sigma(Q)$ denotes the score of a pure Voronoi quad cluster Q, $\mathrm{sol}(Q)$ denotes the solid angle associated with the distinguished vertex, and m and b are positive constants. We call this relation a *bound* on the solid angle and score of a quad cluster. Invoking Proposition 15.2, we restrict our attention to quad clusters whose score exceeds $-1.04\,pt$.

16.8. Pure Voronoi Quad Clusters: Acute Case

Lemma 16.7. *If an acute quad cluster is divided along an acute separating face, then the score of each half is nonpositive.*

 Proof. This is a consequence of the arguments of Theorem 8.4. $\square$

We therefore restrict our attention to halves whose score exceeds $-1.04\,pt$, by Proposition 15.2.

If the separating face is acute, we prove

$$\sigma(S_i) + m\,\mathrm{sol}(S_i) - b/2 \le 0 \tag{16.8}$$

for each half S_i independently, and deduce the desired bound by adding the bounds for each half.

Lemma 16.8. *Let T_0 denote the tetrahedron with edge lengths $(2, 2, 2, 2, 2, 2\sqrt{2})$. Let $\mathrm{sol}(T_0)$ denote the solid angle of the tetrahedron T_0. Given a tetrahedron T, if $\mathrm{sol}(T) < \mathrm{sol}(T_0)$, then relation (16.8) holds.*

Proof. If $\mathrm{sol}(T) < \mathrm{sol}(T_0)$, then $m\,\mathrm{sol}(T) < m\,\mathrm{sol}(T_0)$, hence

$$m\,\mathrm{sol}(T) - b/2 < m\,\mathrm{sol}(T_0) - b/2 \le 0$$

and

$$\sigma(T) + m \operatorname{sol}(T) - b/2 < \sigma(T) \le 0,$$

by Lemma 16.7. □

We therefore may restrict our attention to halves whose solid angle is at least $\operatorname{sol}(T_0)$. In addition, we restrict our attention to halves for which the dividing face is acute.

Lemma 16.9. *The relation*

$$\sigma(T) + m \operatorname{sol}(T) - b/2 \le 0$$

holds for a tetrahedron T forming half of an acute quad cluster with score exceeding $-1.04\, pt$.

Proof. The required verifications for each half of an acute quad cluster are somewhat difficult to achieve directly, so we subdivide into a number of different cases in an attempt to reduce the complexity of the calculations. First, we show that the bound holds for all halves whose diagonal is at least 2.84 (Calculation 17.4.4.1). Using this information, we then prove the bound everywhere but in a small corner cell (Calculation 17.4.4.2). We then restrict our attention to the small corner cell (Calculation 17.4.4.3). These computations involve the use of partial derivative information, and include the required boundary computations. □

Invoking Lemma 16.9 for each half and adding them proves Proposition 15.5 for the acute case.

16.9. *Pure Voronoi Quad Clusters: Obtuse Case*

If the separating face is obtuse, the analysis becomes significantly harder. It is no longer possible to prove the desired bound on each half independently. The dimension of the full bound, even using the usual dimension-reduction techniques, is too high to make the verification tractable numerically. Therefore we adopt a different approach.

Using the dimension-reduction technique, we push each vertex along its edge until the distance from each vertex to the origin is 2. We call the resulting quad cluster a *squashed* cluster. Observe that the solid angle of the cluster is unchanged, while the volume of the Voronoi cell has decreased, thereby increasing the score of the cluster.

Since the central face is still obtuse, the length of the diagonal after this perturbation must still exceed $2\sqrt{2}$. Note, however, that the other edge lengths in the quad cluster can be as small as 4/2.51.

The geometry of the V-cell of a squashed cluster, assuming that there is no truncation from vertices of the packing lying above the quad cluster, is that of Fig. 16.2. When the V-cell is truncated at $\sqrt{2}$ from the origin, two potential arrangements arise. In the first arrangement, the truncated region is connected, as in Fig. 16.3. In second potential arrangement, the truncated region is formed of two disjoint pieces, as in Fig. 16.4.

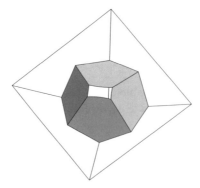

Fig. 16.3.   A typical truncated quad cluster.

Lemma 16.10.   *The disjoint case cannot arise for squashed quad clusters.*

Proof.   Suppose that it could. Pick an untruncated point along the central ridge of the V-cell (see Fig. 16.4). The distance of this point from the origin is then less than $\sqrt{2}$, but due to its location on the central ridge, it is equidistant from the two nearest vertices and the origin. This implies that the circumradius of the resulting triangle must be less than $\sqrt{2}$, which contradicts the fact that the diagonals have length at least $2\sqrt{2}$.   □

16.9.1. *A Geometric Argument.*   We introduce a simplification which reduces the complexity of the obtuse case. This simplification consists of a perturbation of the upper edge lengths of a squashed quad cluster. This perturbation increases the score while holding the solid angle of the quad cluster fixed.

   This simplification is based on a geometric decomposition of the truncated Voronoi cell. We describe the decomposition, and then describe a construction which will ultimately simplify the analysis.

   While our arguments will extend to treat a general squashed and truncated Voronoi cell associated with a general standard cluster, we restrict our attention to truncated Voronoi cells associated with quad clusters.

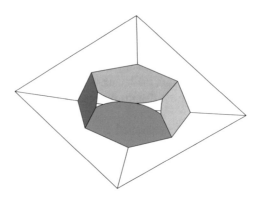

Fig. 16.4.   An impossible arrangement.

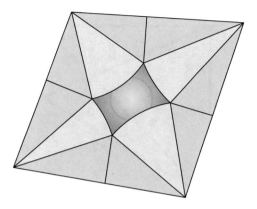

Fig. 16.5. A representation of a truncated quad cluster.

To begin, we consider the decomposition of a truncated Voronoi cell into its fundamental components. A truncated Voronoi cell is formed of three elements: a central spherical section (formed by the truncation), *wedges* of a right circular cone, and tetrahedrons called *Rogers simplices*.

We choose a representation of a truncated quad cluster composed of the radial projection of each element to a plane passing close to the four corners of the quad cluster. This decomposition is represented in Fig. 16.5.

16.9.2. *Rogers Simplices.* We now consider the geometry of the Rogers simplices.

Consider a face with edge lengths $(2, 2, t)$ associated with a side of a truncated quad cluster. Let b represent the circumradius of the face, and let r represent the orthogonal extension of a Rogers simplex from the face, as in Fig. 16.6.

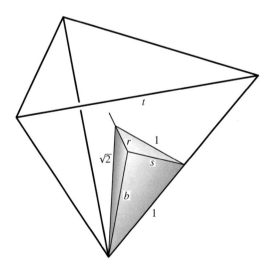

Fig. 16.6. Detail of truncated Voronoi decomposition.

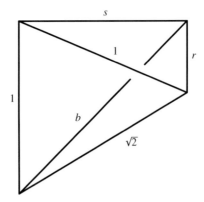

Fig. 16.7. Detail of Rogers simplex.

Then

$$b = \frac{4}{\sqrt{16 - t^2}},$$

$$r = \sqrt{2 - b^2} = \sqrt{\frac{16 - 2t^2}{16 - t^2}},$$

and

$$s = \sqrt{b^2 - 1} = \frac{t}{\sqrt{16 - t^2}}.$$

See Fig. 16.7.

16.9.3. *The Geometric Construction.* We now present the geometric construction which implies the simplification.

We represent the geometry of the truncated Voronoi cell associated with one-half of a quad cluster in Fig. 16.8.

We can simplify the representation by extending the wedges to enclose the Rogers simplices. See Fig. 16.9. This process adds an extra volume term.

The overlap between the wedges is slightly complicated. We simplify the overlap as follows. Take the cone over the overlap. Intersect it with a ball of radius $\sqrt{2}$ at the origin. We call the spherical sections produced by this construction *flutes*. This construction is represented in Fig. 16.10. Figure 16.11 is a planar representation of this construction.

To form each flute, we have added two extra pieces of volume (per flute) to our construction. We call these pieces *quoins*. We attach each quoin to a Rogers simplex. See Fig. 16.12.

16.9.4. *A Solid Angle Invariant.* We now require some notation for the volumes which enter into this construction. Let c denote the volume of the central spherical angle. Let r denote the volume of the Rogers simplices. Let w denote the volume of the wedges. Let w' denote the volume of the extended wedges. Let q denote the volume of the quoins.

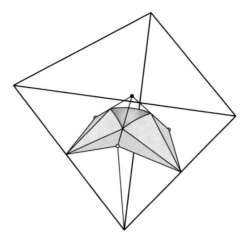

Fig. 16.8. Decomposition of a truncated Voronoi cell.

Let f denote the volume of the flutes. Finally, let v denote the volume of the truncated Voronoi cell.

Lemma 16.11. *If we hold the solid angle fixed, the volume of a truncated squashed Voronoi cell depends only on q, the volume of the quoins.*

Proof. By the original decomposition,

$$v = c + r + w.$$

By our construction,

$$v = c + w' + q - f.$$

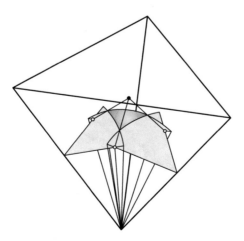

Fig. 16.9. Wedges extended to include the Rogers simplices.

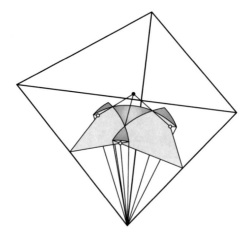

Fig. 16.10. Decomposition with flutes.

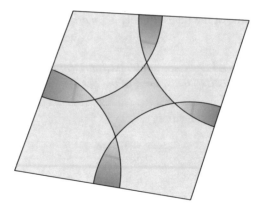

Fig. 16.11. Planar representation with flutes.

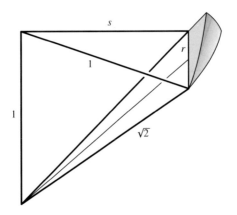

Fig. 16.12. Detail of Rogers simplex with quoin.

Recall that the solid angle s of the quad cluster is the sum of the dihedral angles minus 2π. The dihedral angles to which we refer are those associated with the edges between each corner of the quad cluster and the origin.

Our perturbation will hold the solid angle s of the quad cluster fixed. Therefore, the sum of the dihedral angles must also be fixed. This fixes w'.

Take the cone over each extended wedge and intersect it with a ball of radius $\sqrt{2}$ centered at the origin. Let t denote the sum of these volumes. Since the sum of the dihedral angles is fixed, t is also fixed.

Further, note that

$$\frac{2\sqrt{2}}{3}s = c + t - f.$$

This relation implies that $c - f$ is fixed. Combining this with the previous relations, we find that if we hold the solid angle fixed, the volume of the truncated Voronoi cell depends only on q, the volume of the quoins. □

16.9.5. *Variation of the Volume of a Quoin.* Consider a face $(2, 2, t)$ of a truncated quad cluster. Two Rogers simplices are associated with this face, as suggested in Fig. 16.6. Observe that the volume of the quoin associated with one of these Rogers simplices is increasing in $r = \sqrt{(16 - 2t^2)/(16 - t^2)}$. Next, observe that r is in turn decreasing in t. Therefore increasing t decreases the volume of the squashed quad cluster, if we hold the solid angle fixed (by varying the length of another edge of the squashed quad cluster).

Each half of a squashed quad cluster has two variable edge lengths (not counting the shared diagonal). We label the variable edge lengths of one-half of the squashed quad cluster y_1 and y_2. We label the length of the diagonal d. Holding the solid angle fixed, we may perturb one half by shrinking the larger and increasing the shorter length. We wish to establish the following lemma, that increasing the short length reduces the volume of the truncated Voronoi cell more than decreasing the longer length increases the volume.

Lemma 16.12. *Holding the solid angle fixed for one-half of a squashed quad cluster, shrinking the longer upper edge (while increasing the shorter edge appropriately) reduces the volume of the squashed quad cluster (increasing the score).*

To prove Lemma 16.12, we establish a variational formula for the volume of a quoin. We then verify that the volume of the quoin associated with the shorter edge is decreasing faster under this perturbation than the volume of the quoin associated with the longer edge is increasing.

In other words, we wish to show that $y_1 < y_2$ implies that $V(y_1) + V(y_2(y_1))$ is decreasing in y_1, or equivalently,

$$V_t(y_1) + V_t(y_2(y_1))\frac{dy_2}{dy_1} < 0,$$

where $V(t)$ is the volume of the quoin, $V_t(t)$ is the derivative of the volume, and y_2 is an implicit function of y_1.

We construct the volume of a quoin by integrating the area of a slice. We place the quoin in a convenient coordinate system. See Figs. 16.13 and 16.14. The truncating

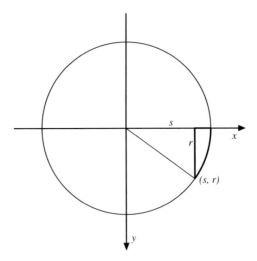

Fig. 16.13. Top view of quoin.

sphere has equation $x^2 + y^2 + z^2 = 2$. At the base of the quoin, $z = 1$, so $x = \sqrt{1 - y^2}$ gives the location of the right-boundary of the quoin. The plane forming the left face of the quoin is given by the equation $x = sz$, so the ridge of the quoin is given by the curve (su, y, u), where $u = \sqrt{(2 - y^2)/(1 + s^2)}$.

Hence the area of a slice parallel to the x-z plane is given by the formula

$$A(t, y) = \tfrac{1}{2}(su - s)(u - 1) + \int_{su}^{\sqrt{1-y^2}} (\sqrt{2 - x^2 - y^2} - 1)\, dx.$$

The volume of a quoin is therefore given by the formula

$$V(t) = \int_0^r A(t, y)\, dy.$$

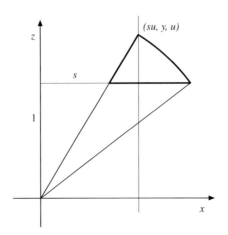

Fig. 16.14. Side view of quoin.

We actually only need to compute $V_t(t)$, which is fortunate, since the explicit formula for $V(t)$ is somewhat complicated. We have

$$V_t(t) = \int_0^r A_t(t, y)\, dy + A(t, r) r_t,$$

but $A(t, r) = 0$, so

$$V_t(t) = \int_0^r A_t(t, y)\, dy.$$

So in addition, we only need $A_t(t, y)$,

$$A_t(t, y) = \left(\frac{s}{2}(u^2 + 1) - \sqrt{1 - y^2} + \int_{t/4\sqrt{2 - y^2}}^{\sqrt{1 - y^2}} \sqrt{2 - x^2 - y^2}\, dx \right),$$

so

$$A_t(t, y) = \left(\frac{s}{2}(u^2 + 1) \right)_t - \sqrt{2 - \frac{t^2}{16}(2 - y^2) - y^2} \frac{1}{4}\sqrt{2 - y^2},$$

which simplifies to

$$A_t(t, y) = \frac{8}{(16 - t^2)^{3/2}} - \frac{2 - y^2}{2\sqrt{16 - t^2}}.$$

Hence

$$V_t(t) = \frac{8r}{(16 - t^2)^{3/2}} - \frac{r}{\sqrt{16 - t^2}} + \frac{r^3}{6\sqrt{16 - t^2}},$$

which simplifies to

$$V_t(t) = \frac{-2\sqrt{2}(8 - t^2)^{3/2}}{3(16 - t^2)^2}.$$

We are now prepared to prove Lemma 16.12.

Proof.   Holding the solid angle fixed, y_2 is an implicit function of y_1. We wish to prove that $y_1 < y_2$ implies

$$V_t(y_1) + V_t(y_2)\frac{dy_2}{dy_1} < 0. \tag{16.9}$$

We derive a formula for dy_2/dy_1, using the solid angle constraint

$$\mathrm{sol}(2, 2, 2, y_1, y_2, d) = c, \tag{16.10}$$

where c is a constant. Using formulas from [Ha6], (16.10) becomes

$$2 \arctan\left(\frac{\sqrt{\Delta}}{2a} \right) = c.$$

Let $x_1 = y_1^2$, $x_2 = y_2^2$, and $b = d^2$. Then

$$\Delta = -4b^2 - 4(x_1 - x_2)^2 + b(x_1(8 - x_2) + 8x_2)$$

and

$$a = 32 - d - x_1 - x_2.$$

So

$$\frac{-4b^2 - 4(x_1 - x_2)^2 + b(x_1(8 - x_2) + 8x_2)}{(32 - d - x_1 - x_2)^2} = c_1.$$

Therefore

$$\frac{dx_2}{dx_1} = -\frac{(16 - x_2)(x_2 + b - x_1)}{(16 - x_1)(x_1 + b - x_2)}$$

and

$$\frac{dy_2}{dy_1} = \frac{y_1}{y_2} \frac{dx_2}{dx_1},$$

hence

$$\frac{dy_2}{dy_1} = -\frac{y_1(16 - x_2)(x_2 + b - x_1)}{y_2(16 - x_1)(x_1 + b - x_2)}. \qquad (16.11)$$

We substitute the formula for dy_2/dy_1 into (16.9). Letting $x_i = y_i^2$, and noting that all the denominators are positive, we obtain on clearing denominators that the desired relation (16.9) is equivalent to

$$-(8 - x_1)^{3/2}(16 - x_2)y_2(x_1 + b - x_2) + (8 - x_2)^{3/2}(16 - x_1)y_1(x_2 + b - x_1) < 0,$$

or

$$(16 - x_1)^2 x_1(8 - x_2)^3(b - x_1 + x_2)^2 < (16 - x_2)^2 x_2(8 - x_1)^3(b + x_1 - x_2)^2.$$

If we define

$$g(x_1, x_2) = (16 - x_1)^2 x_1(8 - x_2)^3(b - x_1 + x_2)^2,$$

then the desired inequality is equivalent to $g(x_1, x_2) < g(x_2, x_1)$ for $x_1 < x_2$. There are several ways to prove this monotonicity relation. One is to prove that the polynomial

$$\frac{g(x_1, x_2) - g(x_2, x_1)}{8(x_1 - x_2)}$$

is positive for all allowable values for x_1, x_2, and b. Unfortunately, the resulting polynomial has degree 6, so the verification is somewhat unwieldy, although easy enough using interval methods.

A simpler method involves a factorization of g into g_1 and g_2. We show that g_1 and g_2 each satisfy the monotonicity relation, and the relation then follows for g.

Define

$$g_1(x_1, x_2) = (16 - x_1)x_1(8 - x_2)(b - x_1 + x_2)$$

and

$$g_2(x_1, x_2) = (16 - x_1)(8 - x_2)^2(b - x_1 + x_2).$$

Clearly $g = g_1 g_2$. We then construct the polynomials

$$p_1 = \frac{g_1(x_1, x_2) - g_1(x_2, x_1)}{x_1 - x_2}$$

and

$$p_2 = \frac{g_2(x_1, x_2) - g_2(x_2, x_1)}{x_1 - x_2}.$$

Simplifying p_1 and p_2, we find that

$$p_1 = 128b - 128x_1 - 8bx_1 + 8x_1^2 - 128x_2$$
$$+ 32x_1x_2 + bx_1x_2 - x_1^2x_2 + 8x_2^2 - x_1x_2^2$$

and

$$p_2 = -2048 + 192b + 320x_1 - 16bx_1 - 16x_1^2 + 320x_2$$
$$- 16bx_2 - 32x_1x_2 + bx_1x_2 + x_1^2x_2 - 16x_2^2 + x_1x_2^2.$$

These polynomials are quadratic in x_1 and x_2, and linear in b. The coefficient of b in p_1 is

$$128 - 8x_1 - 8x_2 + x_1x_2.$$

The coefficient of b in p_2 is

$$192 - 16x_1 - 16x_2 + x_1x_2.$$

Both coefficients are positive for x_1 and x_2 in $[16/2.51^2, 2.51^2]$. Therefore, the minimum values of p_1 and p_2 occur when b is at a minimum, $b = 8$.

The minimum value of each polynomial for values of x_1 and x_2 in the range $[16/2.51^2, 2.51^2]$ is now easily computed. Making the appropriate computations, we find that each polynomial is indeed positive. Hence the desired relation follows. □

16.9.6. *Final Simplification.*

Lemma 16.13. *Obtuse quad clusters satisfy the bound of Proposition 15.5.*

Proof. We begin with a squashed quad cluster with consecutive upper edge lengths (y_1, y_2, y_3, y_4) and diagonal d adjacent to the first two upper edges.

Recall that we chose the diagonal of the quad cluster to be the shorter of the two possible diagonals. We refer to the other possible diagonal as the *cross-diagonal*. Recall that the reduction fixes the length of the diagonal.

If the length of the cross-diagonal does not drop to $2\sqrt{2}$ under the perturbation of Lemma 16.12, we arrive at the configuration with edge lengths (y_1', y_1', y_2', y_2') with diagonal d.

If the length of the cross-diagonal does drop to $2\sqrt{2}$, then stop the perturbation. This gives a quadrilateral (y_1', y_2', y_3', y_4') with diagonal $2\sqrt{2}$. Applying the perturbation to each half independently, we find that the score of each half is maximized by the configuration $(y_1'', y_1'', y_2'', y_2'')$ with diagonal $2\sqrt{2}$. We verify the relation for this arrangement in Calculation 17.4.4.5.

If the length of the cross-diagonal did not drop to $2\sqrt{2}$, switch to the cross-diagonal and repeat the process. If the (new) cross-diagonal does not drop to $2\sqrt{2}$, we have arrived at the configuration (y, y, y, y) with diagonal d'. Choose a new diagonal d'' to be the shorter of the two possible diagonals. We verify the desired relation for this arrangement in Calculation 17.4.4.4.

Finally, we make a few comments about extra constraints in the verifications.

Since the score of a quad cluster is nonpositive, and $m(2\,\mathrm{sol}(T_0)) - b \le 0$ where $\mathrm{sol}(T_0) = \mathrm{sol}(2, 2, 2, 2, 2, 2\sqrt{2})$, we need only consider quad clusters for which the solid angle exceeds $2\,\mathrm{sol}(T_0)$.

The maximum length of the diagonal is $2.51\sqrt{2}$, since otherwise the triangles in the quadrilateral would be obtuse, forcing the cross-diagonal to be shorter than the diagonal. This would contradict our original choice of the shortest diagonal.

In Calculation 17.4.4.4 we assume that d is the shortest diagonal. Adding this constraint directly is tedious, since the formula for the cross-diagonal of the quad cluster is somewhat complicated. We apply a simpler but weaker constraint, that the diagonal d of a planar quadrilateral with edge lengths (y, y, y, y) is shorter than d', the other planar diagonal. The constraint $d \le d'$ gives the constraint $d^2 \le 2y^2$. Since the cross-diagonal of the quad cluster is shorter than the cross-diagonal of the planar quadrilateral, this constraint is weaker.   □

Lemmas 16.9 and 16.13 prove Proposition 15.5 for pure Voronoi quad clusters.

17. Calculations

The verifications of the relations required in this paper appear intractable using traditional methods. Therefore, we use a relatively new proof technique, interval arithmetic via floating-point computer calculations.

17.1. *Interval Arithmetic*

We review the basic notions of interval arithmetic.

Suppose that the value of a function $f(x)$ lies in the interval $[a, b]$. Further, suppose that $g(x)$ lies in the interval $[c, d]$. Then $f(x) + g(x)$ must lie in $[a + c, b + d]$. While it may be the case that we could produce better bounds than this for the function $f + g$, these interval bounds give crude control over the behavior of the function. Interval arithmetic provides a mechanism for formalizing arithmetic on these bounds.

We represent an interval t as $[\underline{t}, \bar{t}]$. Then for intervals a and b,

$$a + b = [\underline{a} + \underline{b}, \bar{a} + \bar{b}].$$

Likewise,

$$a - b = \left[\underline{a} - \overline{b}, \overline{a} - \underline{b}\right].$$

Multiplication is somewhat more complicated. Define

$$C = \{\underline{ab}, \underline{a}\overline{b}, \overline{a}\underline{b}, \overline{a}\overline{b}\}.$$

Then

$$a * b = [\min(C), \max(C)].$$

Division is similar, as long as the dividing interval does not contain zero.

Similarly, we can define the operation of a monotonic function on an interval. For example,

$$\arctan(a) = [\arctan(\underline{a}), \arctan(\overline{a})].$$

Using interval arithmetic, we can produce rigorous bounds for polynomials evaluated on intervals. Likewise, we can produce rigorous bounds for rational functions evaluated on intervals. Finally, we add the composition of monotonic functions. This allows us to produce interval bounds for functions such as sol($\cdot$) and vor($\cdot$) over quasi-regular tetrahedrons, quarters, or quad clusters.

17.2. The Method of Subdivision

The relations on tetrahedra and quad clusters required for our approach typically have the form $g(y) \leq 0$ for $y \in I$, where I is a product of closed intervals. As g is usually continuous, the existence of a maximum is trivial. However, bounds on the behavior of g over all of I computed directly via interval arithmetic are generally poor.

We define a *cell* to be a product of closed intervals. By subdividing I into sufficiently small cells, the quality of the computed bounds on each cell usually improves enough to prove the relation for each cell, and hence for the original domain I.

If in fact $g(y) \leq c < 0$, this approach works very well. However, if the bound is tight at a point y_0, i.e., $g(y_0) = 0$, then pure subdivision will usually fail, since the computed upper bound on g over any cell containing y_0 will typically be positive.

If y_0 is not an interior maximum, we turn to the partial derivatives of g. If we can show that the partials of g on a small cell containing y_0 have fixed sign (bounded away from zero), then the maximum value of g on that cell is easily computed. It is typically the case that a cell must be very small before we can determine the sign of the partials via interval-arithmetic bounds.

17.3. Numerical Considerations

Most real numbers are not representable in computer floating-point format. However, floating-point intervals may be found which contain any real number. Although the magnitude of real numbers representable in fixed-length floating-point format is finite, the format also provides for $\pm\infty$, which allows for interval containment of all reals.

These intervals may be added, multiplied, etc., and the resulting intervals will contain the result of the operation applied to the real numbers which they represent.

Since floating-point arithmetic is not exact, interval arithmetic conducted using floating-point arithmetic is not optimal, in the sense that the interval resulting from an operation will usually be larger than the true resultant interval, due to roundoff. However, barring hardware or software errors (implementation errors, not roundoff errors), floating-point interval arithmetic, unlike floating-point arithmetic, is correct, in the sense that it provides correct interval bounds on the value of a computation, while floating-point arithmetic alone only provides an approximation to the correct value of a computation. We may therefore use interval arithmetic to prove mathematical results. Floating-point arithmetic alone, in the absence of rigorous error analysis, cannot constitute a proof.

We implement floating-point interval arithmetic routines via the IEEE 754 Standard for floating-point arithmetic [IEEE].

Implementation of interval arithmetic is straightforward using directed rounding. In addition to arithmetic functions, we require interval implementations of the square root and arctangent functions. Fortunately, the IEEE standard provides the square root function. However, the arctangent function is somewhat problematic, since the standard mathematical libraries do not provide explicit error bounds for their implementations of the arctangent function. In theory, they should provide an accuracy for the arctangent routine of 0.7 ulps, meaning that the error is less than one unit in the last place. I add interval padding of the form $[v - \varepsilon, v + \varepsilon]$, where v is the computed value, and $\varepsilon = 2^{-49}$. This should be sufficient to guarantee proper interval containment, assuming that the library routines are correctly implemented.

Armed with standard interval arithmetic and interval arithmetic implementations of sqrt and arctan, we can implement interval arithmetic versions of all the special functions required for proving the sphere packing relations.

Evaluating these functions on cells, we get bounds. Unfortunately, these bounds are not very good. The bounds which we get from interval versions of the partial derivative functions are even worse. This means that cells have to be very small before we can draw conclusions about the signs of the partials. These bad bounds are due to the inherent nature of interval arithmetic—it produces worst-case results by design.

These bad bounds increase the complexity of the verifications tremendously. Some verifications, using these bounds, require the consideration of billions or trillions of cells, or worse. Therefore, we needed a method for producing better bounds than those which direct interval methods could provide.

The method which we eventually discovered is to use Taylor series. We compute explicit second (mixed) partial bounds for the major special functions, and use these bounds to produce very good interval bounds. These bounds are computed in Calculations 17.4.6.1–17.4.6.8. Essentially, the Taylor method postpones the error bound until the end of the computation, eliminating the error bound explosion which occurs with a straightforward interval method implementation.

17.4. *Calculations*

The following inequalities have been proved by computer using interval methods. Let $S = S(y) = S(y_1, \ldots, y_6)$ denote a tetrahedron parametrized by the edge lengths

$(y_1, \ldots, y_6)$. In addition, we often parametrize by the squares of the edge lengths $(x_1, \ldots, x_6)$.

Recall from Section 15.1 that $m = 0.3621$, $b = 0.49246$, $a = 0.0739626$, and $b_c = 0.253095$.

Recall that for our purposes, the scoring function $\sigma(\cdot)$ is given by one of four functions: $gma(\cdot)$, $vor(\cdot)$, $octavor(\cdot)$, or truncated Voronoi. See Remark 7.23 for a simplified version of the scoring function.

The scoring rules depend on $\eta(\cdot)$, the circumradius of a face, introduced in Definition 4.20.

17.4.1. Quasi-Regular Tetrahedra. Define $C = [2, 2.51]^6$, and recall

$$a_1 = 0.3860658808124052, \qquad a_2 = 0.4198577862, \qquad d_0 = 1.4674.$$

Calculation 17.4.1.1. *Either*

$$gma(S) \le a_1 \operatorname{dih}(S) - a_2$$

or

$$gma(S) \le -0.52\,pt$$

for $y \in C$, using dimension-reduction.

Calculation 17.4.1.2. *Either*

$$gma(S) - a_1 \operatorname{dih}(S) \le 3.48\,pt - 2\pi a_1 + 4a_2$$

or

$$\operatorname{dih}(S) < d_0$$

or

$$gma(S) \le -0.52\,pt$$

for $y \in C$, using dimension-reduction.

Calculation 17.4.1.3. *Either*

$$gma(S) + m \operatorname{sol}(S) + a(\operatorname{dih}(S) - 2\pi/5) - b_c \le 0$$

or

$$\operatorname{dih}(S) > d_0$$

or

$$gma(S) \le -0.52\,pt$$

for $y \in C$, using dimension-reduction.

17.4.2. *Flat Quad Clusters.* Define $I = [2, 2.51]^5[2.51, 2\sqrt{2}]$, and define the corner cell

$$C = [2, 2 + 0.51/16]^5[2\sqrt{2} - (2\sqrt{2} - 2.51)/16, 2\sqrt{2}].$$

Calculation 17.4.2.1. *Either*

$$gma(S) + m\,sol(S) \le b/2$$

or

$$\eta(y_1, y_2, y_6)^2 > 2$$

or

$$\eta(y_4, y_5, y_6)^2 > 2$$

or

$$gma(S) \le -1.04\,pt$$

for $y \in I$, using dimension reduction.

Calculation 17.4.2.2. *Either*

$$gma(S) + m\,sol(S) \le b/2$$

or

$$\eta(y_1, y_2, y_6)^2 = 2 \quad with \quad \eta(y_4, y_5, y_6)^2 \le 2$$

or

$$\eta(y_4, y_5, y_6)^2 = 2 \quad with \quad \eta(y_1, y_2, y_6)^2 \le 2$$

or

$$gma(S) \le -1.04\,pt$$

for $y \in I$, not using dimension-reduction.

Calculation 17.4.2.3. $d\,vor(S)/dy_i < 0$ *for $i = 1, 2, 3$ and $y \in C$.*

Calculation 17.4.2.4. *This computation is somewhat tricky, since the scoring constraint depends on both faces. The partial derivative information gives $y_3 = 2$. The rest of the analysis depends on which face is assumed to be large.*

If the (y_1, y_2, y_6) face is large, the partial derivative information implies that the face constraint is tight, so $\eta(y_1, y_2, y_6)^2 = 2$. Therefore solve for y_1 in terms of y_2 and y_6. Apply partial derivative information for y_4 and y_5. In this case,

$$vor(S) + m\,sol(S) \le b/2$$

for $y_3 = 2$, $y \in C$.

If the (y_4, y_5, y_6) face is large, assume that $y_1 = y_2 = 2$. Then either

$$\text{vor}(S) + m\,\text{sol}(S) \leq b/2$$

or

$$\eta(y_4, y_5, y_6)^2 < 2$$

for $y_1 = y_2 = y_3 = 2$, $y \in C$.

Calculation 17.4.2.5. *Either*

$$\text{vor}(S) + m\,\text{sol}(S) \leq b/2$$

or

$$\eta(y_1, y_2, y_6)^2 < 2 \quad \text{and} \quad \eta(y_4, y_5, y_6)^2 < 2$$

or

$$\text{vor}(S) \leq -1.04\,pt$$

for $y \in I$, $y \notin C$, using dimension-reduction and partial derivative information.

Calculation 17.4.2.6. *Either*

$$\text{vor}(S) + m\,\text{sol}(S) \leq b/2,$$

with

$$\eta(y_1, y_2, y_6)^2 = 2 \quad \text{or} \quad \eta(y_4, y_5, y_6)^2 = 2,$$

or

$$\text{vor}(S) \leq -1.04\,pt$$

for $y \in I$, $y \notin C$, not using dimension-reduction.

17.4.3. *Octahedra*

Calculation 17.4.3.1. $\sigma(S) \leq -0.52\,pt$, *for each (appropriately scored) upright quarter with edge lengths in the cell $[2.51, 2\sqrt{2}][2.2, 2.51][2, 2.51]^4$.*

Calculation 17.4.3.2. *Recall $c = 0.1533667634670977$ and $d = 0.2265$. Either*

$$gma(S) + c\,\text{dih}(S) \leq d$$

or

$$gma(S) \leq -1.04\,pt$$

for $y \in [2.51, 2.716][2, 2.2]^5$, using dimension-reduction. Note that for both faces adjacent to the diagonal,

$$\max \eta^2 = \eta(2.2, 2.2, 2.716)^2 < 2,$$

so all quarters in this cell are compression-scored.

Calculation 17.4.3.3. *Either*

$$gma(S) + m\,sol(S) + \alpha\,dih(S) \leq \frac{b}{4} + \alpha\frac{\pi}{2}$$

or

$$gma(S) \leq -1.04\,pt$$

for all compression-scored quarters $S(y)$, *where* $\alpha = 0.14$,

$$y \in [2.716, 2\sqrt{2}][2, 2.2]^2[2, 2.51][2, 2.2]^2,$$

using dimension-reduction.

Calculation 17.4.3.4. *Either*

$$octavor(S) + m\,sol(S) + \alpha\,dih(S) \leq \frac{b}{4} + \alpha\frac{\pi}{2}$$

or

$$octavor(S) \leq -1.04\,pt$$

for all vor analytic-scored quarters $S(y)$, *where* $\alpha = 0.14$,

$$y \in [2.716, 2.81][2, 2.2]^2[2, 2.51][2, 2.2]^2.$$

Calculation 17.4.3.5. *Either*

$$gma(S) + m\,sol(S) + \alpha\,dih(S) + \beta x_1 \leq \frac{b}{4} + \alpha\frac{\pi}{2} + 8\beta$$

or

$$gma(S) \leq -1.04\,pt$$

for all compression-scored quarters $S(y)$, *where* $\alpha = 0.054$, $\beta = 0.00455$, $x_1 = y_1^2$, *and*

$$y \in [2.81, 2\sqrt{2}][2, 2.2]^2[2, 2.51][2, 2.2]^2,$$

using some dimension-reduction.

Calculation 17.4.3.6. *Either*

$$octavor(S) + m\,sol(S) + \alpha\,dih(S) + \beta x_1 \leq \frac{b}{4} + \alpha\frac{\pi}{2} + 8\beta$$

or

$$octavor(S) \leq -1.04\,pt$$

for all vor analytic-scored quarters $S(y)$, *where* $\alpha = 0.054$, $\beta = -0.00455$, $x_1 = y_1^2$, *and*

$$y \in [2.81, 2\sqrt{2}][2, 2.2]^2[2, 2.51][2, 2.2]^2.$$

17.4.4. *Pure Voronoi Quad Clusters.* Recall $\text{sol}(T_0)$ denotes the solid angle of the tetrahedron $(2, 2, 2, 2, 2, 2\sqrt{2})$.

Define the corner cell $C = [2, 2+0.51/8]^5[2\sqrt{2}, 2.84]$. We denote truncated Voronoi scoring by σ. The constraint that the dividing face be acute translates into $x_1+x_2-x_6 \geq 0$. In each computation we apply dimension-reduction.

We begin with the acute case.

Calculation 17.4.4.1. *Either*

$$\sigma(S) + m\,\text{sol}(S) - b/2 \leq 0$$

or

$$\text{sol}(S) < \text{sol}(T_0)$$

or

$$x_1 + x_2 - x_6 < 0$$

or

$$\sigma(S) \leq -1.04\,pt$$

for $y \in [2, 2.51]^5[2.84, 4]$.

Calculation 17.4.4.2. *Either*

$$\sigma(S) + m\,\text{sol}(S) - b/2 \leq 0$$

or

$$\text{sol}(S) < \text{sol}(T_0)$$

or

$$x_1 + x_2 - x_6 < 0$$

or

$$\sigma(S) \leq -1.04\,pt$$

for $y \in [2, 2.51]^5[2\sqrt{2}, 2.84]$ with $y \notin C$.

Calculation 17.4.4.3. *Either*

$$\sigma(S) + m\,\text{sol}(S) - b/2 \leq 0$$

or

$$\text{sol}(S) < \text{sol}(T_0)$$

or

$$x_1 + x_2 - x_6 < 0,$$

$y \in C$.

Finally, we consider the obtuse case.

Calculation 17.4.4.4. *Either*

$$\sigma(S) + m \operatorname{sol}(S) - b/2 \le 0$$

or

$$\operatorname{sol}(S) < \operatorname{sol}(T_0)$$

or

$$\sigma(S) \le -0.52\,pt$$

or

$$2y^2 < d^2$$

for a symmetric pure Voronoi quad cluster composed of two copies of S, where

$$S = (2, 2, 2, y, y, d),$$

$y \in [4/2.51, 2.51]$ *and* $d \in [2\sqrt{2}, 2.51\sqrt{2}]$.

Calculation 17.4.4.5. *Either*

$$\sigma(S_1) + \sigma(S_2) + m(\operatorname{sol}(S_1) + \operatorname{sol}(S_2)) - b \le 0$$

or

$$\sigma(S_1) + \sigma(S_2) \le -1.04\,pt$$

or

$$\operatorname{sol}(S_1) + \operatorname{sol}(S_2) < 2\operatorname{sol}(T_0)$$

for a pure Voronoi quad cluster composed of two tetrahedrons S_1 and S_2, where

$$S_i = (2, 2, 2, y_i, y_i, 2\sqrt{2}),$$

$y_i \in [4/2.51, 2.51]$.

17.4.5. *Dimension Reduction*

Calculation 17.4.5.1. *The polynomial derived for the dimension-reduction argument is positive for $x \in [4, 2.51^2]^6$ and $x \in [4, 2.51^2]^5[4, 8]$.*

17.4.6. *Second Partial Bounds.* We compute all second partials $d^2/dx_i dx_j$ in terms of x_i, the squares of the edge lengths. We do each computation twice, once for quasi-regular tetrahedrons and once for quarters. We compute the second partials of dih($\cdot$), sol($\cdot$), compression volume, and Voronoi volume (the vor analytic volume). Since the scoring functions are linear combinations of sol($\cdot$) and the volume terms, we may derive second partial bounds for $gma(\cdot)$ and vor($\cdot$) from these.

With the application of additional computer power, these bounds could be improved. These bounds were computed using sixteen subdivisions. While using thirty-two subdivisions would improve the bounds by a factor of 2, perhaps, the time required for the computations increases by a factor of 64.

Calculation 17.4.6.1. *For quasi-regular tetrahedrons T, the second partials of* dih(T) *lie in*

$$[-0.0926959464, 0.0730008897].$$

Calculation 17.4.6.2. *For quarters Q, the second partials of* dih(Q) *lie in*

$$[-0.2384125007, 0.169150875].$$

Calculation 17.4.6.3. *For quasi-regular tetrahedrons T, the second partials of* sol(T) *lie in*

$$[-0.0729140255, 0.088401996].$$

Calculation 17.4.6.4. *For quarters Q, the second partials of* sol(Q) *lie in*

$$[-0.1040074557, 0.1384785805].$$

Calculation 17.4.6.5. *For quasi-regular tetrahedrons T, the second partials of gma(T) volume lie in*

$$[-0.0968945273, 0.0512553817].$$

Calculation 17.4.6.6. *For quarters Q, the second partials of gma(Q) volume lie in*

$$[-0.1362100221, 0.1016538923].$$

Calculation 17.4.6.7. *For quasi-regular tetrahedrons T, the second partials of* vor(T) *volume lie in*

$$[-0.1856683356, 0.1350478467].$$

Calculation 17.4.6.8. *For quarters Q, the second partials of* vor(Q) *volume lie in*

$$[-0.2373892383, 0.1994181009].$$

The computed $gma(\cdot)$ second partials then lie in

$$[-0.2119591984, 0.2828323141],$$

for quasi-regular tetrahedrons and quarters.

Likewise, the computed $vor(\cdot)$ second partials then lie in

$$[-0.7137209962, 0.8691765157],$$

for quasi-regular tetrahedrons and quarters.

Acknowledgments

This paper is a revised version of the author's Ph.D. thesis [Fer] at the University of Michigan. The author thanks Tom Hales, Jeff Lagarias, and the referees for their many contributions to this revision.

References

[Fer] S. P. Ferguson, Sphere Packings, V, Thesis, University of Michigan, 1997.
[Ha6] T. C. Hales, Sphere packings, I, *Discrete Comput. Geom.* **17** (1997), 1–51.
[HF] T. C. Hales and S. P. Ferguson, A formulation of the Kepler conjecture, *Discrete Comput. Geom.*, this issue, pp. 21–70.
[IEEE] *IEEE Standard for Binary Floating-Point Arithmetic, ANSI/IEEE Std.* 754–1985, IEEE, New York.

Received November 11, 1998, *and in revised form September* 12, 2003, *and July* 25, 2005.
Online publication February 27, 2006.

8

Sphere Packings VI. Tame Graphs and Linear Programs, by T. C. Hales

This paper is the sixth of six papers giving the Kepler Conjecture proof. It contains Chapters 18–25 of the proof.

The asterisks added in the margin of some pages indicate that changes are required, as listed in the Hales list of errata given on p. 374 of this volume.

Contents

Sphere Packings VI. *Tame Graphs and Linear Programs*, by T. C. Hales (Discrete Comput. Geom., **36** (2006), 205–266).

The original version of this chapter was revised. An erratum to this chapter can be found at http://dx.doi.org/10.1007/978-1-4614-1129-1_12

Discrete Comput Geom 36:205–265 (2006)
DOI: 10.1007/s00454-005-1215-x

Sphere Packings, VI. Tame Graphs and Linear Programs

Thomas C. Hales

Department of Mathematics, University of Pittsburgh,
Pittsburgh, PA 15217, USA
hales@pitt.edu

Abstract. This paper is the sixth and final part in a series of papers devoted to the proof of the Kepler conjecture, which asserts that no packing of congruent balls in three dimensions has density greater than the face-centered cubic packing. In a previous paper in this series, a continuous function f on a compact space is defined, certain points in the domain are conjectured to give the global maxima, and the relation between this conjecture and the Kepler conjecture is established. In this paper we consider the set of all points in the domain for which the value of f is at least the conjectured maximum. To each such point, we attach a planar graph. It is proved that each such graph must be isomorphic to a *tame* graph, of which there are only finitely many up to isomorphism. Linear programming methods are then used to eliminate all possibilities, except for three special cases treated in earlier papers: pentahedral prisms, the face-centered cubic packing, and the hexagonal-close packing. The results of this paper rely on long computer calculations.

Introduction

This paper is the last in the series of paper devoted to the proof of the Kepler conjecture. The first several sections prove a result that asserts that "all contravening graphs are tame." A contravening graph is one that is attached to a potential counterexample to the Kepler conjecture. Contravening graphs by nature are elusive and are studied by indirect methods. In contrast, the defining properties of tame graphs lend themselves to direct examination. (By definition, tame graphs are planar graphs such that the degree of every vertex is at least two and at most six, the length of every face is at least 3 and at most 8, and such that other similar explicit properties hold true.)

It is no coincidence that contravening graphs all turn out to be tame. The definition of a tame graph has been tailored to suit the situation at hand. We set out to prove explicit properties of contravening graphs, and when we are satisfied with what we have proved, we brand a graph with these properties a tame graph.

277

The first section of this paper gives the definition of a tame graph. The second section gives the classification of all tame graphs. There are several thousand such graphs. The classification was carried out by computer. This classification is one of the main uses of a computer in the proof of the Kepler conjecture. A detailed description of the algorithm that is used to find all tame graphs is presented in this section.

The third section of this paper gives a review of results from earlier parts of the paper that are relevant to the study of tame plane graphs. In the abridged version of the proof [Ha15], the results cited in this section are treated as axioms. This section thus serves as a guide to the results that are proved in this volume, but not in the abridged version of the proof.

This section also contains a careful definition of what it means to be a contravening plane graph. The first approximation to the definition is that it is the combinatorial plane graph associated with the net of edges on the unit sphere bounding the standard regions of a contravening decomposition star. The precise definition is somewhat more subtle because we wish ensure that every face of a contravening plane graph is a simple polygon. To guarantee that this property holds, we simplify the net of edges on the unit sphere whenever necessary.

The fourth and fifth sections of this paper contain the proof that all contravening plane graphs are tame. These sections complete the first half of this paper.

The second half of this paper is about linear programming. Linear programs are used to prove that with the exception of three tame graphs (those attached to the face-centered cubic packing, the hexagonal-close packing, and the pentahedral prism), a tame graph cannot be a contravening graph. This result reduces the proof of the Kepler conjecture to a close examination of three graphs. Pentahedral prism graphs are treated in the fifth paper. The face-centered cubic and hexagonal-close packing graphs are treated in Section 8 of the third paper. The linear programming results together with these earlier results complete the proof of the Kepler conjecture.

The sixth section of this paper describes how to attach a linear program to a tame plane graph. The output from this linear program is an upper bound on the score of all decomposition stars associated with the given tame plane graph. The seventh section of this paper shows how to use linear programs to eliminate what are called the *aggregate* tame plane graphs. The *aggregates* are those cases where the net of edges formed by the edges of standard regions was simplified to ensure that every face of a contravening plane graph is a polygon. By the end of this section, we have a proof that every standard region in a contravening decomposition star is bounded by a simple polygon.

The final section of this paper gives a long list of special strategies that are used when the output from the linear program in the sixth section does not give conclusive results. The general strategy is to partition the original linear program into a collection of refined linear programs with the property that the score is no greater than the maximum of the outputs from the linear programs in the collection. These branch and bound strategies are described in this final section. Linear programming shows that every decomposition star with a tame plane graph (other than the three mentioned above) has a score less than that of the decomposition stars attached to the face-centered cubic packing. This and earlier results imply the Kepler conjecture.

18. Tame Graphs

This section defines a class of plane graphs. Graphs in this class are said to be *tame*. In the next section we give a complete classification of all tame graphs. This classification of tame graphs was carried out by computer and is a major step of the proof of the Kepler conjecture.

18.1. Basic Definitions

Definition 18.1. An *n-cycle* is a finite set C of cardinality n, together with a cyclic permutation s of C. We write s in the form $v \mapsto s(v, C)$, for $v \in C$. The element $s(v, C)$ is called the *successor* of v (in C). A *cycle* is an *n*-cycle for some natural number n. By abuse of language, we often identify C with the cycle. The natural number n is the *length* of the cycle.

Definition 18.2. Let G be a nonempty finite set of cycles (called faces) of length at least 3. The elements of faces are called the *vertices* of G. An unordered pair of vertices $\{v, w\}$ such that one element is the successor of the other in some face is called an *edge*. The vertices v and w are then said to be *adjacent*. The set G is a *plane graph* if four conditions hold:

1. If an element v has successor w in some face F, then there is a unique face (call it $s'(F, v)$) in G for which v is the successor of w. (Thus, $v = s(w, s'(F, v))$, and each edge occurs twice with opposite orientation.)
2. For each vertex v, the function $F \mapsto s'(F, v)$ is a cyclic permutation of the set of faces containing v.
3. Euler's formula holds relating the number of vertices V, the number of edges E, and the number of faces F:

$$V - E + F = 2.$$

4. The set of vertices is connected. That is, the only nonempty set of vertices that is closed under $v \mapsto s(v, C)$ for all C is the full set of vertices.

Remark 18.3. The set of vertices and edges of a plane graph form a planar graph in the usual graph-theoretic sense of admitting an embedding into the plane. Every planar graph carries an orientation on its faces that is inherited from an orientation of the plane. (Use the right-hand rule on the face, to orient it with the given outward normal of the oriented plane.) For us, the orientation is built into the definition, so that properly speaking, we should call these objects *oriented plane graphs*. We follow the convention of distinguishing between planar graphs (which admit an embedding into the plane) and plane graphs (for which a choice of embedding has been made). Our definition is more restrictive than the standard definition of plane graph in the literature, because we require all faces to be simple polygons with at least three vertices. Thus, a graph with a single edge does not comply with our narrow definition of plane graph. Other graphs

Fig. 18.1. Some examples of graphs that are excluded from the narrow definition of plane graph, as defined in this section.

that are excluded by this definition are shown in Fig. 18.1. Standard results about plane graphs can be found in any of a number of graph theory textbooks. However, this paper is written in such a way that it should not be necessary to consult outside graph theory references.

Definition 18.4. Let *len* be the length function on faces. Faces of length 3 are called *triangles*, those of length 4 are called *quadrilaterals*, and so forth. Let tri(v) be the number of triangles containing a vertex v. A face of length at least 5 is called an *exceptional* face.

Two plane graphs are *properly* isomorphic if there is a bijection of vertices inducing a bijection of faces. For each plane graph, there is an opposite plane graph G^{op} obtained by reversing the cyclic order of vertices in each face. A plane graph G is isomorphic to another if G or G^{op} is properly isomorphic to the other.

Definition 18.5. The *degree* of a vertex is the number of faces it belongs to. An *n*-*circuit* in G is a cycle C in the vertex-set of G, such that for every $v \in C$, it forms an edge in G with its successor: that is, $(v, s(v, C))$ is an edge of G.

In a plane graph G we have a combinatorial form of the Jordan curve theorem: each *n*-circuit determines a partition of G into two sets of faces.

Definition 18.6. The *type* of a vertex is defined to be a triple of nonnegative integers (p, q, r), where p is the number of triangles containing the vertex, q is the number of quadrilaterals containing it, and r is the number of exceptional faces. When $r = 0$, we abbreviate the type to the ordered pair (p, q).

18.2. *Weight Assignments*

We call the constant tgt $= 14.8$, which arises repeatedly in this section, the *target*. (This constant arises as an approximation to $4\pi\zeta - 8 \approx 14.7947$, where $\zeta = 1/(2\arctan(\sqrt{2}/5))$.)

Define $a: \mathbb{N} \to \mathbb{R}$ by

$$
a(n) = \begin{cases}
14.8, & n = 0, 1, 2, \\
1.4, & n = 3, \\
1.5, & n = 4, \\
0, & \text{otherwise.}
\end{cases}
$$

Define $b: \mathbb{N} \times \mathbb{N} \to \mathbb{R}$ by $b(p, q) = 14.8$, except for the values in the following table (with tgt $= 14.8$):

p	q 0	1	2	3	4
0	tgt	tgt	tgt	7.135	10.649
1	tgt	tgt	6.95	7.135	tgt
2	tgt	8.5	4.756	12.981	tgt
3	tgt	3.642	8.334	tgt	tgt
4	4.139	3.781	tgt	tgt	tgt
5	0.55	11.22	tgt	tgt	tgt
6	6.339	tgt	tgt	tgt	tgt

Define $c: \mathbb{N} \to \mathbb{R}$ by

$$
c(n) = \begin{cases}
1, & n = 3, \\
0, & n = 4, \\
-1.03, & n = 5, \\
-2.06, & n = 6, \\
-3.03, & \text{otherwise.}
\end{cases}
$$

Define $d: \mathbb{N} \to \mathbb{R}$ by

$$
d(n) = \begin{cases}
0, & n = 3, \\
2.378, & n = 4, \\
4.896, & n = 5, \\
7.414, & n = 6, \\
9.932, & n = 7, \\
10.916, & n = 8, \\
\text{tgt} = 14.8, & \text{otherwise.}
\end{cases}
$$

A set V of vertices is called a *separated* set of vertices if the following four conditions hold:

1. For every vertex in V there is an exceptional face containing it.
2. No two vertices in V are adjacent.
3. No two vertices in V lie on a common quadrilateral.
4. Each vertex in V has degree 5.

A *weight assignment* of a plane graph G is a function $w: G \to \mathbb{R}$ taking values in the set of nonnegative real numbers. A weight assignment is *admissible* if the following properties hold:

1. If the face F has length n, then $w(F) \geq d(n)$.

2. If v has type (p, q), then

$$\sum_{F:\ v \in F} w(F) \geq b(p, q).$$

3. Let V be any set of vertices of type $(5, 0)$. If the cardinality of V is $k \leq 4$, then

$$\sum_{F:\ V \cap F \neq \emptyset} w(F) \geq 0.55k.$$

4. Let V be any separated set of vertices. Then

$$\sum_{F:\ V \cap F \neq \emptyset} (w(F) - d(\text{len}(F))) \geq \sum_{v \in V} a(\text{tri}(v)).$$

The sum $\sum_F w(F)$ is called the *total weight* of w.

18.3. Plane Graph Properties

We say that a plane graph is *tame* if it satisfies the following conditions:

1. The length of each face is at least 3 and at most 8.
2. Every 3-circuit is a face or the opposite of a face.
3. Every 4-circuit surrounds one of the cases illustrated in Fig. 18.2.
4. The degree of every vertex is at least 2 and at most 6.
5. If a vertex is contained in an exceptional face, then the degree of the vertex is at most 5.
6. $$\sum_F c(\text{len}(F)) \geq 8.$$
7. There exists an admissible weight assignment of total weight less than the target, $\text{tgt} = 14.8$.
8. There are never two vertices of type $(4, 0)$ that are adjacent to each other.

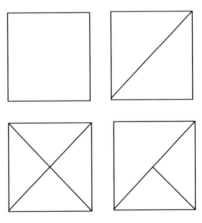

Fig. 18.2.   Tame 4-circuits.

It follows from the definitions that the abstract vertex-edge graph of G has no loops or multiple joins. Also, by construction, every vertex lies in at least two faces. Property 6 implies that the graph has at least eight triangles.

Remark 18.7.   We pause to review the strategy of the proof of the Kepler conjecture as described in Section 3.2. The decomposition stars that violate the main inequality $\sigma(D) \geq 8\,pt$ are said to contravene. A plane graph is associated with each contravening decomposition star. These are the contravening plane graphs. The main object of this paper is to prove that the only two contravening graphs are G_{fcc} and G_{hcp}, the graphs associated with the face-centered cubic and hexagonal-close packings.

   We have defined a set of plane graphs, called *tame graphs*. The next section will give a classification of tame plane graphs. (There are several thousand.) Section 20 gives a proof that all contravening plane graphs are tame. By the classification result, this reduces the possible contravening graphs to an explicit finite list. Case-by-case linear programming arguments will show that none of these tame plane graphs is a contravening graph (except G_{fcc} and G_{hcp}). Having eliminated all possible graphs, we arrive at the resolution of the Kepler conjecture.

19.   Classification of Tame Plane Graphs

19.1.   *Statement of the Theorem*

A list of several thousand plane graphs appears in [Ha16]. The following theorem is listed as one of the central claims in the proof in Section 3.3.

Theorem 19.1.   *Every tame plane graph is isomorphic to a plane graph in this list.*

   The results of this section are not needed except in the proof of Theorem 19.1.
   Computers are used to generate a list of all tame plane graphs and to check them against the archive of tame plane graphs. We describe a finite state machine that produces all tame plane graphs. This machine is not particularly efficient, and so we also include a description of pruning strategies that prevent a combinatorial explosion of possibilities.

19.2.   *Basic Definitions*

In order to describe how all tame plane graphs are generated, we need to introduce *partial plane graphs* that encode an incompletely generated tame graph. A partial plane graph is itself a graph, but marked in such a way to indicate that it is in a transitional state that will be used to generate further plane graphs.

Definition 19.2.   A *partial plane graph* is a plane graph with additional data: every face is marked as "complete" or "incomplete." We call a face *complete* or *incomplete* according to the markings. We require the following condition:

- *No two incomplete faces share an edge.*

Each unmarked plane graph is identified with the marked plane graph in which every face is complete. We represent a partial plane graph graphically by deleting one face (the face at infinity) and drawing the others and shading those that are complete.

A *patch* is a partial plane graph P with two distinguished faces F_1 and F_2, such that the following hold:

- Every vertex of P lies in F_1 or F_2.
- The face F_2 is the only complete face.
- F_1 and F_2 share an edge.
- Every vertex of F_2 that is not in F_1 has degree 2.

F_1 and F_2 will be referred to as the distinguished incomplete and the distinguished complete faces, respectively.

Patches can be used to modify a partial plane graph as follows. Let F be an incomplete face of length n in a partial plane graph G. Let P be a patch whose incomplete distinguished face F_1 has length n. Replace P with a properly isomorphic patch P' in which the image of F_1 is equal to F^{op} and in which no other vertex of P' is a vertex of G. Then

$$G' = \{F' \in G \cup P': F' \neq F^{op}, F' \neq F\}$$

is a partial plane graph. Intuitively, we cut away the faces F and F_1 from their plane graphs, and glue the holes together along the boundary (Fig. 19.1). (It is immediate that the condition in the definition of partial plane graphs (Definition 19.2) is maintained by this process.) There are n distinct proper ways of identifying F_1 with F^{op} in this construction, and we let φ be this identification. The isomorphism class of G' is uniquely determined by the isomorphism class of G, the isomorphism class of P, and φ (ranging over proper bijections $\varphi: F_1 \mapsto F^{op}$).

19.3. A Finite State Machine

For a fixed N we define a finite state machine as follows. The states of the finite state machine are isomorphism classes of partial plane graphs G with at most N vertices. The

Fig. 19.1. Patching a plane graph.

transitions from one state G to another are isomorphism classes of pairs (P, φ) where P is a patch, and φ pairs an incomplete face of G with the distinguished incomplete face of P. However, we exclude a transition (P, φ) at a state if the resulting partial plane graphs contains more than N vertices. Figure 19.1 shows two states and a transition between them.

The initial states I_n of the finite state machine are defined to be the isomorphism classes of partial plane graphs with two faces:

$$\{(1, 2, \ldots, n), (n, n-1, \ldots, 1)\},$$

where $n \le N$, one face is complete, and the other is incomplete. In other words, they are patches with exactly two faces.

A terminal state of this finite state machine is one in which every face is complete. By construction, these are (isomorphism classes of) plane graphs with at most N vertices.

Lemma 19.3. *Let G be a plane graph with at most N vertices. Then its state in the machine is reachable from an initial state through a series of transitions.*

Proof. Pick a face in G of length n and identify it with the complete face in the initial state I_n. At any stage at state G', we have an identification of all of the vertices of the plane graph G' with some of the vertices of G, and an identification of all of the complete faces of G' with some of the faces of G (all faces of G are complete). Pick an incomplete face F of G' and an oriented edge along that face. We let F' be the complete face of G with that edge, with the same orientation on that edge as F. Create a patch with distinguished faces $F_1 = F^{\mathrm{op}}$ and $F_2 = F'$. (F_1 and F_2 determine the patch up to isomorphism.) It is immediate that the conditions defining a patch are fulfilled. Continue in this way until a graph isomorphic to G is reached. $\square$

Remark 19.4. It is an elementary matter to generate all patches P such that the distinguished faces have given lengths n and m. Patching is also entirely algorithmic, and thus by following all paths through the finite state machine, we obtain all plane graphs with at most N vertices.

19.4. *Pruning Strategies*

Although we reach all graphs in this manner, it is not computationally efficient. We introduce pruning strategies to increase the efficiency of the search. We can terminate our search along a path through the finite state machine, if we can determine one of the following:

1. Every terminal graph along that path violates one of the defining properties of tameness.
2. An isomorphic terminal graph will be reached by some other path that will not be terminated early.

Here are some pruning strategies of type 1 above. They are immediate consequences of the conditions of the defining properties of tameness.

- If the current state contains an incomplete face of length 3, then eliminate all transitions, except for the transition that carries the partial plane graph to a partial plane graph that is the same in all respects, except that the face has become complete.
- If the current state contains an incomplete face of length 4, then eliminate all transitions except those that lead to the possibilities of Section 18.3, Property 3, where in Property 3 each depicted face is interpreted as being complete.
- Remove all transitions with patches whose complete face has length greater than 8.
- It is frequently possible to conclude from the examination of a partial plane graph that no matter what the terminal position, any admissible weight assignment will give total weight greater than the target (tgt $= 14.8$). In such cases, all transitions out of the partial plane graph can be pruned.

To take a simple example of the last item, we observe that weights are always non-negative, and that the weight of a complete face of length n is at least $d(n)$. Thus, if there are complete faces $F_1, \ldots, F_k$ of lengths $n_1, \ldots, n_k$, then any admissible weight assignment has total weight at least $\sum_{i=1}^{k} d(n_i)$. If this number is at least the target, then no transitions out of that state need be considered.

More generally, we can apply all of the inequalities in the definition of admissible weight assignment to the complete portion of the partial plane graph to obtain lower bounds. However, we must be careful, in applying Property 4 of admissible weight assignments, because vertices that are not adjacent at an intermediate state may become adjacent in the complete graph. Also, vertices that do not lie together in a quadrilateral at an intermediate state may do so in the complete graph.

Here are some pruning strategies of type 2:

- At a given state it is enough to fix one incomplete face and one edge of that face and then to follow only the transitions that patch along that face and add a complete face along that edge. (This is seen from the proof of Lemma 19.3.)
- In leading out from the initial state I_n, it is enough to follow paths in which every added complete face has length at most n. (A graph with a face of length m, for $m > n$, will be also be found downstream from I_m.)
- Make a list of all type (p, q) with $b(p, q) < \text{tgt} = 14.8$. Remove the initial states I_3 and I_4, and create new initial states $I_{p,q}$ ($I'_{p,q}$, $I''_{p,q}$, etc.) in the finite state machine. Define the state $I_{p,q}$ to be one consisting of $p + q + 1$ faces, with p complete triangles and q complete quadrilaterals all meeting at a vertex (and one other incomplete face away from v). (If there is more than one way to arrange p triangles and q quadrilaterals, create states $I_{p,q}$, $I'_{p,q}$, $I''_{p,q}$, for each possibility. See Fig. 19.2.) Put a linear order on states $I_{p,q}$. In state transitions downstream from $I_{p,q}$ disallow any transition that creates a vertex of type (p', q'), for any (p', q') preceding (p, q) in the imposed linear order.

This last pruning strategy is justified by the following lemma, which classifies vertices of type (p, q).

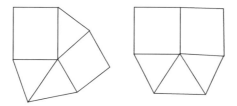

Fig. 19.2. States $I_{3,2}$ and $I'_{3,2}$.

Lemma 19.5. *Let A and B be triangular or quadrilateral faces that have at least two vertices in common in a tame graph. Then the faces have exactly two vertices in common, and an edge is shared by the two faces.*

Proof. Exercise. Some of the configurations that must be ruled out are shown in Fig. 19.3. Some properties that are particularly useful for the exercise are Properties 2 and 3 of tameness, and Property 2 of admissibility.   □

Once a terminal position is reached it is checked to see whether it satisfies all the properties of tameness.

Duplication is removed among isomorphic terminal plane graphs. It is not an entirely trivial procedure for the computer to determine whether an isomorphism between two plane graphs exists. This is accomplished by computing a numerical invariant of a vertex that depends only on the local structure of the vertex. If two plane graphs are properly isomorphic then the numerical invariant is the same at vertices that correspond under the proper isomorphism. If two graphs have the same number of vertices with the same numerical invariants, they become candidates for an isomorphism. All possible numerical-invariant preserving bijections are attempted until a proper isomorphism is found, or until it is found that none exist. If there is no proper isomorphism, the same procedure is applied to the opposite plane graph to find any possible orientation-reversing isomorphism.

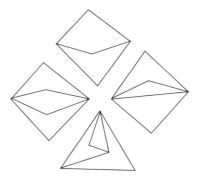

Fig. 19.3. Some impossibilities.

This same isomorphism-producing algorithm is used to match each terminal graph with a graph in the archive. It is found that each terminal graph matches with one in the archive. (The archive was originally obtained by running the finite state machine and making a list of all the terminal states up to isomorphism that satisfy the given conditions.)

In this way Theorem 19.1 is proved.

20. Contravening Graphs

We have seen that a system of points and arcs on the unit sphere can be associated with a decomposition star D. The points are the radial projections of the vertices of $U(D)$ (those at distance at most $2t_0 = 2.51$ from the origin). The arcs are the radial projections of edges between $v, w \in U(D)$, where $|v - w| \leq 2t_0$. If we consider this collection of arcs combinatorially as a graph, then it is not always true that these arcs form a plane graph in the restrictive sense of Section 18.

The purpose of this section is to show that if the original decomposition star contravenes, then minor modifications can be made to the system of arcs of the graph so that the resulting combinatorial graph has the structure of a plane graph in the sense of Section 18. These plane graphs are called contravening plane graphs, or simply contravening graphs.

20.1. A Review of Earlier Results

Let $\zeta = 1/(2 \arctan(\sqrt{2}/5))$. Let $\mathrm{sol}(R)$ denote the solid angle of a standard region R. We write τ_R for the following modification of σ_R:

$$\tau_R(D) = \mathrm{sol}(R)\zeta \, pt - \sigma_R(D) \tag{20.1}$$

and

$$\tau(D) = \sum \tau_R(D) = 4\pi \zeta \, pt - \sigma(D). \tag{20.2}$$

Since $4\pi \zeta \, pt$ is a constant, τ and σ contain the same information, but τ is often more convenient to work with. A contravening decomposition star satisfies

$$\tau(D) \leq 4\pi \zeta \, pt - 8 \, pt = (4\pi \zeta - 8) \, pt. \tag{20.3}$$

The constant $(4\pi \zeta - 8) \, pt$ (and its upper bound tgt pt where tgt $= 14.8$) occurs repeatedly in the discussion that follows.

Recall that a standard cluster is a pair (R, D) consisting of a decomposition star D and one of its standard regions R. If F is a finite set (or finite union) of standard regions, let

$$\sigma_F(D) = \sum_R \sigma_R(D), \qquad \tau_F(D) = \sum_R \tau_R(D), \tag{20.4}$$

where the sum runs over all the standard regions in F. When the sum runs over all standard regions,

$$\sigma(D) = \sum \sigma_R(D), \qquad \tau(D) = \sum \tau_R(D). \tag{20.5}$$

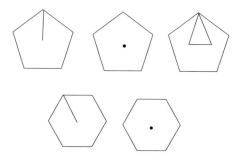

Fig. 20.1. Nonpolygonal standard regions ($n(R) = 7, 7, 8, 8, 8$).

A natural number $n(R)$ is associated with each standard region. If the boundary of that region is a simple polygon, then $n(R)$ is the number of sides. If the boundary consists of k disjoint simple polygons, with $n_1, \ldots, n_k$ sides then

$$n(R) = n_1 + \cdots + n_k + 2(k - 1).$$

Lemma 20.1. *Let R be a standard region in a contravening decomposition star D. The boundary of R is a simple polygon with at most eight edges, or one of the configurations of Fig. 20.1.*

Proof. This is Theorem 12.1 and Corollary 12.2.   □

Lemma 20.2. *Let R be a standard region. We have $\tau_R(D) \geq t_n$, where $n = n(R)$, and*

$$t_3 = 0, \qquad t_4 = 0.1317, \qquad t_5 = 0.27113,$$
$$t_6 = 0.41056, \qquad t_7 = 0.54999, \qquad t_8 = 0.6045.$$

Furthermore, $\sigma_R(D) \leq s_n$, for $5 \leq n \leq 8$, where

$$s_3 = 1\,pt, \qquad s_4 = 0, \qquad s_5 = -0.05704,$$
$$s_6 = -0.11408, \qquad s_7 = -0.17112, \qquad s_8 = -0.22816.$$

Proof. This is Theorem 12.1.   □

Lemma 20.3. *Let F be a set of standard regions bounded by a simple polygon with at most nine edges. Assume that*

$$\sigma_F(D) \leq s_9 \quad and \quad \tau_F(D) \geq t_9,$$

where $s_9 = -0.1972$ and $t_9 = 0.6978$. Then D does not contravene.

Proof. This is Section 12.2.   □

Lemma 20.4. *Let (R, D) be a standard cluster. If R is a triangular region, then*

$$\sigma_R(D) \leq 1\,pt.$$

If R is not a triangular region, then

$$\sigma_R(D) \leq 0.$$

Proof. See Lemma 8.10 and Theorem 8.4. □

Lemma 20.5. $\tau_R(D) \geq 0$, *for all standard clusters R.*

Proof. This is Lemma 10.1. □

Recall that v has *type* (p, q) if every standard region with a vertex at v is a triangle or a quadrilateral, and if there are exactly p triangular faces and q quadrilateral faces that meet at v (see Definition 18.6). We write (p_v, q_v) for the type of v. Define constants $\tau_{\mathrm{LP}}(p, q)/pt$ by Table 20.1. The entries marked with an asterisk will not be needed.

Lemma 20.6. *Let $S_1, \ldots, S_p$ and $R_1, \ldots, R_q$ be the tetrahedra and quad clusters around a vertex of type (p, q). Consider the constants of Table 20.1. Now,*

$$\sum_{}^{p} \tau(S_i) + \sum_{}^{q} \tau(R_i) \geq \tau_{\mathrm{LP}}(p, q).$$

Proof. This is Lemma 10.5. □

Lemma 20.7. *Let $v_1, \ldots, v_k$, for some $k \leq 4$, be distinct vertices of type $(5, 0)$. Let $S_1, \ldots, S_r$ be quasi-regular tetrahedra around the edges $(0, v_i)$, for $i \leq k$. Then*

$$\sum_{i=1}^{r} \tau(S_i) > 0.55k\,pt$$

Table 20.1. $\tau_{\mathrm{LP}}(p, q)/pt.$

p	0	1	2	3	4	5	6
	*	*	15.18	7.135	10.6497	22.27	
1	*	*	6.95	7.135	17.62	32.3	
2	*	8.5	4.756	12.9814	*	*	
3	*	3.6426	8.334	20.9	*	*	
4	4.1396	3.7812	16.11	*	*	*	
5	0.55	11.22	*	*	*	*	
6	6.339	*	*	*	*	*	
7	14.76	*	*	*	*	*	

and

$$\sum_{i=1}^{r} \sigma(S_i) < r \, pt - 0.48k \, pt.$$

Proof.   This is Lemma 10.6.   □

Lemma 20.8.   *Let D be a contravening decomposition star. If the type of the vertex is* (p, q, r) *with* $r = 0$, *then* (p, q) *must be one of the following:*

$$\{(6, 0), (5, 0), (4, 0), (5, 1), (4, 1), (3, 1), (2, 1),$$
$$(3, 2), (2, 2), (1, 2), (2, 3), (1, 3), (0, 3), (0, 4)\}.$$

Proof.   This is Lemma 10.10 and Corollary 12.3.   □

Lemma 20.9.   *A triangular standard region does not contain any enclosed vertices.*

Proof.   This fact is proved in Lemma 3.7 of [Ha6].   □

Lemma 20.10.   *A quadrilateral region does not enclose any vertices of height at most* $2t_0$.

Proof.   This is Lemma 10.13.   □

Lemma 20.11.   *Let F be a union of standard regions. Suppose that the boundary of F consists of four edges. Suppose that the area of F is at most* 2π. *Then there is at most one enclosed vertex over F.*

Proof.   This is Proposition 4.2 of [Ha6].   □

Lemma 20.12.   *Let F be the union of two standard regions, a triangular region and a pentagonal region that meet at a vertex of type* $(1, 0, 1)$ *as shown in Fig. 20.2. Then*

$$\tau_F(D) \geq 11.16 \, pt.$$

Proof.   This is Lemma 14.4.   □

Lemma 20.13.   *Let R be an exceptional standard region. Suppose that R has r different interior angles that are pairwise nonadjacent and such that each is at most 1.32. Then*

$$\tau_R(D) \geq t_n + r(1.47) \, pt.$$

Proof.   This is Remark 14.2.   □

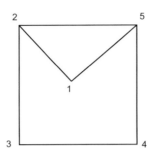

Fig. 20.2. A 4-circuit.

Lemma 20.14. *Every interior angle of every standard region is at least* 0.8638. *Every interior angle of every standard region that is not a triangle is at least* 1.153.

Proof. CALC-208809199 and CALC-853728973-1.   □

Definition 20.15. The *central vertex* of a flat quarter is defined to be the one that does not lie on the triangle formed by the origin and the diagonal.

Lemma 20.16. *If the interior angle at a corner v of a nontriangular standard region is at most* 1.32, *then there is a flat quarter over R whose central vertex is v.*

Proof. This is Lemma 11.30.   □

20.2. *Contravening Plane Graphs Defined*

A plane graph G is attached to every contravening decomposition star as follows. From the decomposition star D, it is possible to determine the coordinates of the set $U(D)$ of vertices at distance at most $2t_0$ from the origin.

If we draw a geodesic arc on the unit sphere at the origin with endpoints at the radial projections of v_1 and v_2 for every pair of vertices $v_1, v_2 \in U(D)$ such that $|v_1|, |v_2|, |v_1 - v_2| \leq 2t_0$, we obtain a plane graph that breaks the unit sphere into standard regions. (The arcs do not meet except at endpoints by Lemma 4.19.)

For a given standard region, we consider the arcs forming its boundary together with the arcs that are internal to the standard region. We consider the points on the unit sphere formed by the endpoints of the arcs, together with the radial projections to the unit sphere of vertices in U whose radial projection lies in the interior of the region.

Remark 20.17. The system of arcs and vertices associated with a standard region in a contravening example must be a polygon, or one of the configurations of Fig. 20.1 (see Lemma 20.1).

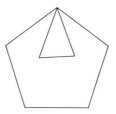

Fig. 20.3. An aggregate forming a pentagon.

Remark 20.18. Observe that one case of Fig. 20.1 is bounded by a triangle and a pentagon, and that the others are bounded by a polygon. Replacing the triangle–pentagon arrangement with the bounding pentagon and replacing the others with the bounding polygon, we obtain a partition of the sphere into simple polygons. Each of these polygons is a single standard region, except in the triangle–pentagon case (Fig. 20.3), which is a union of two standard regions (a triangle and an eight-sided region).

Remark 20.19. To simplify further, if we have an arrangement of six standard regions around a vertex formed from five triangles and one pentagon, we replace it with the bounding octagon (or hexagon). See Fig. 20.4. (It will be shown in Lemma 21.11 that there is at most one such configuration in the standard decomposition of a contravening decomposition star, so we will not worry here about how to treat the case of two overlapping configurations of this sort.)

In summary, we have a plane graph that is approximately that given by the standard regions of the decomposition star, but simplified to a bounding polygon when one of the configurations of Remarks 20.18 and 20.19 occur. We refer to the combination of standard regions into a single face of the graph as *aggregation*. We call it the plane graph $G = G(D)$ attached to a contravening decomposition star D.

Proposition 21.1 will show the vertex set U is nonempty and that the graph $G(D)$ is nonempty.

When we refer to the plane graph in this manner, we mean the combinatorial plane graph as opposed to the embedded metric graph on the unit sphere formed from the system of geodesic arcs. Given a vertex v in $G(D)$, there is a uniquely determined vertex $v(D)$ of $U(D)$ whose radial projection to the unit sphere determines v. We call $v(D)$ the *corner* in $U(D)$ over v.

By construction, the plane graphs associated with a decomposition star do not have loops or multiple joins. In fact, the edges of $G(D)$ are defined by triangles whose sides

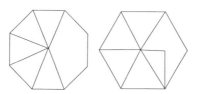

Fig. 20.4. Degree 6 aggregates.

vary between lengths 2 and $2t_0$. The angles of such a triangle are strictly less than π. This implies that the edges of the metric graph on the unit sphere always have an arc-length strictly less than π. In particular, the endpoints are never antipodal. A loop on the combinatorial graph corresponds to an edge on the metric graph that is a closed geodesic. A multiple join on the combinatorial graph corresponds on the metric graph to a pair of points joined by multiple minimal geodesics, that is, a pair of antipodal points on the sphere. By the arc-length constraints on edges in the metric graph, there are no loops or multiple joins in the combinatorial graph $G(D)$.

In Section 18.3 a plane graph satisfying a certain restrictive set of properties is said to be *tame*. If a plane graph $G(D)$ is associated with a contravening decomposition star D, we call $G(D)$ a *contravening plane graph*.

Theorem 20.20. *Let D be a contravening decomposition star. Then its plane graph $G(D)$ is tame.*

This theorem is one of the main steps in the proof of the Kepler conjecture. It is advanced as one of the central claims in Section 3.3. Its proof occupies Sections 21 and 22. In Theorem 19.1 the tame graphs are classified up to isomorphism. As a corollary, we have an explicit list of graphs that contains all contravening plane graphs.

21. Contravention is tame

This section begins the proof of Theorem 20.20 (contravening graphs are tame). To prove Theorem 20.20, it is enough to show that each defining property of tameness is satisfied for every contravening graph. This is the substance of results in the following sections. The proof continues through to the end of Section 22. This section verifies all the properties of tameness, except for the last one (weight assignments).

21.1. First Properties

This section verifies Properties 1, 2, 4, and 8 of tameness. First, we prove the promised nondegeneracy result.

Proposition 21.1. *The construction of Section 20.2 associates a (nonempty) plane graph with at least two faces to every decomposition star D with $\sigma(D) > 0$.*

Proof. First we show that decomposition stars with $\sigma(D) > 0$ have nonempty vertex sets U. (Recall that U is the set of vertices of distance at most $2t_0$ from the center.) The vertices of U are used in Sections 4 and 5 to create all of the structural features of the decomposition star: quasi-regular tetrahedra, quarters, and so forth. If U is empty, the V-cell is a solid containing the ball $B(t_0)$ of radius t_0, and $\sigma(D)$ satisfies

$$\sigma(D) = \mathrm{vor}(D)$$
$$= -4\delta_{\mathrm{oct}}\mathrm{vol}(VC(D)) + 4\pi/3$$
$$< -4\delta_{\mathrm{oct}}\mathrm{vol}(B(t_0)) + 4\pi/3 < 0.$$

By hypothesis, $\sigma(D) > 0$. So U is not empty.

Equation (20.5) shows that the function σ can be expressed as a sum of terms σ_R indexed by the standard regions R. It is proved in Theorem 8.4 that $\sigma_R \leq 0$, unless R is a triangle. Thus, a decomposition star with positive $\sigma(D)$ must have at least one triangle. Its complement contains a second standard region. Even after we form aggregates of distinct standard regions to form the simplified plane graph (Remarks 20.18 and 20.19), there certainly remain at least two faces. $\qquad\square$

Proposition 21.2. *The plane graph of a contravening decomposition star satisfies Property 1 of tameness: The length of each face is at least 3 and at most 8.*

Proof. By the construction of the graph, each face has at least three edges. The upper bound of eight edges is Lemma 20.1. Note that the aggregates of Remarks 20.19 and 20.18 have between five and eight edges. $\qquad\square$

Proposition 21.3. *The plane graph of a contravening decomposition star satisfies Property 2 of tameness: Every 3-circuit is a face or the opposite of a face.*

Proof. The simplifications of the plane graph in Remarks 20.18 and 20.19 do not produce any new 3-circuits. (See the accompanying figures.) The result is Lemma 20.9. $\square$

Proposition 21.4. *Contravening graphs satisfy Property 4 of tameness: The degree of every vertex is at least 2 and at most 6.*

Proof. The statement that degrees are at least 2 trivially follows because each vertex lies on at least one polygon, with two edges at that vertex.

If the type is (p, q), then the impossibility of a vertex of degree 7 or more is found in Lemma 20.8. If the type is (p, q, r), with $r \geq 1$, then Lemma 20.14 shows that the interior angles of the standard regions cannot sum to 2π:

$$6(0.8638) + 1.153 > 2\pi. \qquad\qquad\square$$

Proposition 21.5. *Contravening graphs satisfy Property 8 of tameness: There are never two vertices of type $(4, 0)$ that are adjacent to each other.*

Proof. This is proved in Proposition 4.2 of [Ha6]. $\qquad\square$

21.2. *Computer Calculations and Their Consequences*

This section continues in the proof that all contravening plane graphs are tame. The next few sections verify Properties 6, 5, and then 3 of tameness.

In this section we rely on some inequalities that are not proved in this paper. Recall from Section 8.3 that there is an archive of hundreds of inequalities that have been proved by computer. This full archive appears in [Ha16]. The justification of these inequalities appears in the same archive. (The proofs of these inequalities were executed by computer.)

Each inequality carries a nine digit identifying number. To invoke an inequality, we state it precisely, and give its identifying number, e.g., CALC-123456789.

To use these inequalities systematically, we combine inequalities into linear programs and solve the linear programs on computer. At first, our use of linear programs will be light, but our reliance will become progressively strong as the argument develops.

To start out, we make use of several calculations[132] that give lower bounds on $\tau_R(D)$ when R is a triangle or a quadrilateral. To obtain lower bounds through linear programming, we take a linear relaxation. Specifically, we introduce a linear variable for each function τ_R and a linear variable for each interior angle α_R. We substitute these linear variables for the nonlinear functions $\tau_R(D)$ and nonlinear interior angle function into the given inequalities. Under these substitutions, the inequalities become linear. Given p triangles and q quadrilaterals at a vertex, we have the linear program to minimize the sum of the (linear variables associated with) $\tau_R(D)$ subject to the constraint that the (linear variables associated with the) angles at the vertex sum to at most d. Linear programming yields[133] a lower bound $\tau_{LP}(p, q, d)$ to this minimization problem. This gives a lower bound to the corresponding constrained sum of nonlinear functions τ_R.

Similarly, another group of inequalities[134] yields upper bounds $\sigma_{LP}(p, q, d)$ on the sum of $p + q$ functions σ_R, with p standard regions R that are triangular, and another q that are quadrilateral. These linear programs find their first application in the proof of the following proposition.

21.3. Linear Programs

To continue with the proof that contravening plane graphs are tame, we need to introduce more notation and methods.

If F is a face of $G(D)$, let

$$\sigma_F(D) = \sum \sigma_R(D),$$

where the sum runs over the set of standard regions associated with F. This sum reduces to a single term unless F is an aggregate in the sense of Remarks 20.19 and 20.18.

Lemma 21.6. *The plane graph of a contravening decomposition star satisfies Property 6 of tameness:*

$$\sum_F c(\operatorname{len}(F)) \geq 8.$$

Proof. We will show that

$$c(\operatorname{len}(F))\, pt \geq \sigma_F(D). \tag{21.1}$$

---

[132] The sequence of five inequalities starting with CALC-927432550, Lemma 20.5, and for quads CALC-310151857, CALC-655029773, CALC-73283761, CALC-15141595, CALC-574391221, CALC-396281725.

[133] Although they are closely related, the function τ_{LP} of three arguments introduced here is distinct from the function of two variables of the same name that is introduced in Section 20.1.

[134] CALC-539256862, CALC-864218323, CALC-776305271, and for quads CALC-310151857, CALC-655029773, CALC-73283761, CALC-15141595, CALC-574391221, CALC-396281725.

Assuming this, the result follows for contravening stars D:

$$\sum_F c(\text{len}(F)) \, pt \geq \sum_F \sigma_F(D)$$

$$= \sigma(D) \geq 8 \, pt.$$

We consider three cases for inequality (21.1). In the first case assume that the face F corresponds to exactly one standard region in the decomposition star. In this case inequality (21.1) follows directly from the bounds of Lemma 20.2:

$$\sigma_F(D) \leq s_n \leq c(n) \, pt.$$

In the second case assume the context of a pentagon F formed in Remark 20.18. Then, again by Lemma 20.2, we have

$$\sigma_F(D) \leq s_3 + s_8 \leq (c(3) + c(8)) \, pt \leq c(5) \, pt.$$

(Just examine the constants $c(k)$.)

In the third case we consider the situation of Remark 20.19. The six standard regions give

$$\sigma_F(D) \leq s_5 + \sigma_{\text{LP}}(5, 0, 2\pi - 1.153) < c(8) \, pt.$$

The constant 1.153 comes from Lemma 20.14. $\square$

Proposition 21.7. *Let F be a face of a contravening plane graph $G(D)$. Then*

$$\tau_F(D) \geq d(\text{len}(F)) \, pt.$$

Proof. Similar. $\square$

Lemma 21.8. *If v is a vertex of an exceptional standard region, and if there are six standard regions meeting at v, then the exceptional region is a pentagonal region and the other five standard regions are triangular.*

Proof. There are several cases according to the number k of triangular regions at the vertex.

($k \leq 2$) If there are at least four nontriangular regions at the vertex, then the sum of interior angles around the vertex is at least $4(1.153) + 2(0.8638) > 2\pi$, which is impossible. (See Lemma 20.14.)

($k = 3$) If there are three nontriangular regions at the vertex, then $\tau(D)$ is at least $2t_4 + t_5 + \tau_{\text{LP}}(3, 0, 2\pi - 3(1.153)) > (4\pi\zeta - 8) \, pt.$

($k = 4$) If there are two exceptional regions at the vertex, then $\tau(D)$ is at least $2t_5 + \tau_{\text{LP}}(4, 0, 2\pi - 2(1.153)) > (4\pi\zeta - 8) \, pt.$

If there are two nontriangular regions at the vertex, then $\tau(D)$ is at least $t_5 + \tau_{\text{LP}}(4, 1, 2\pi - 1.153) > (4\pi\zeta - 8) \, pt.$

($k = 5$) We are left with the case of five triangular regions and one exceptional region.

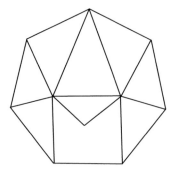

Fig. 21.1. Nonadjacent vertices of degree 6 on a pentagon.

When there is an exceptional standard region at a vertex of degree 6, we claim that the exceptional region must be a pentagon. If the region is a heptagon or more, then $\tau(D)$ is at least $t_7 + \tau_{\mathrm{LP}}(5, 0, 2\pi - 1.153) > (4\pi\zeta - 8)\,pt$.

If the standard region is a hexagon, then $\tau(D)$ is at least $t_6 + \tau_{\mathrm{LP}}(5, 0, 2\pi - 1.153) > t_9$. Also, $s_6 + \sigma_{\mathrm{LP}}(5, 0, 2\pi - 1.153) < s_9$. The aggregate of the six standard regions is nine-sided. Lemma 20.3 gives the bound of $8\,pt$. □

Lemma 21.9. *Consider the standard regions of a contravening star D.*

1. *If a vertex of a pentagonal standard region has degree 6, then the aggregate F of the six faces satisfies*

$$\sigma_F(D) < s_8,$$
$$\tau_F(D) > t_8.$$

2. *An exceptional standard region has at most two vertices of degree 6. If there are two, then they are nonadjacent vertices on a pentagon, as shown in Fig. 21.1.*

Proof. We begin with the first part of the lemma. The sum $\tau_F(D)$ over these six standard regions is at least

$$t_5 + \tau_{\mathrm{LP}}(5, 0, 2\pi - 1.153) > t_8.$$

Similarly,

$$s_5 + \sigma_{\mathrm{LP}}(5, 0, 2\pi - 1.153) < s_8.$$

We note that there can be at most one exceptional region with a vertex of degree 6. Indeed, if there are two, then they must both be vertices of the same pentagon:

$$t_8 + t_5 > (4\pi\zeta - 8)\,pt.$$

Such a second vertex on the octagonal aggregate leads to one of the following constants greater than $(4\pi\zeta - 8)\,pt$. These same constants show that such a second vertex on a hexagonal aggregate must share two triangular faces with the first vertex of degree 6.

$$t_8 + \tau_{LP}(4, 0, 2\pi - 1.32 - 0.8638), \quad \text{or}$$

$$t_8 + 1.47\,pt + \tau_{LP}(4, 0, 2\pi - 1.153 - 0.8638), \quad \text{or}$$

$$t_8 + \tau_{LP}(5, 0, 2\pi - 1.153).$$

(The relevant constants are found at Lemmas 20.13 and 20.14.)   □

21.4. A Noncontravening 4-Circuit

This subsection rules out the existence of a particular 4-circuit on a contravening plane graph. The interior of the circuit consists of two faces: a triangle and a pentagon. The circuit and its enclosed vertex are show in Fig. 20.2 with vertices marked $p_1, \ldots, p_5$. The vertex p_1 is the enclosed vertex, the triangle is (p_1, p_2, p_5), and the pentagon is $(p_1, \ldots, p_5)$. Let $v_1, \ldots, v_4, v_5$ be the corresponding vertices of $U(D)$.

The diagonals $\{v_5, v_3\}$ and $\{v_2, v_4\}$ have length at least $2\sqrt{2}$ by Lemma 4.19. If an interior angle of the quadrilateral is less than 1.32, then, by Lemma 20.16, $|v_1 - v_3| \le \sqrt{8}$. Thus, we assume in the following lemma that all interior angles of the quadrilateral aggregate are at least 1.32.

Lemma 21.10. *A decomposition star that contains this configuration does not contravene.*

Proof. Let P denote the quadrilateral aggregate of these two standard regions. By Lemma 20.12 we have $\tau_P(D) \ge 11.16\,pt$. There are no other exceptional faces, because $11.16\,pt + t_5 > (4\pi\zeta - 8)\,pt$. Every vertex not on P has type $(5, 0)$, by Lemma 20.6. In particular, there are no quadrilateral regions. The interior angles of P are at least 1.32. There are at most four triangles at every vertex of P, because

$$11.16\,pt + \tau_{LP}(5, 0, 2\pi - 1.32) > (4\pi\zeta - 8)\,pt.$$

There are at least three triangles at every vertex of P, otherwise we contradict Lemmas 20.9 or 20.11.

The only triangulation with these properties is obtained by removing one edge from the icosahedron (Exercise). This implies that there are two opposite corners of P each having four quasi-regular tetrahedra. Since the diagonals of P have lengths greater than $2\sqrt{2}$, the results of CALC-325738864 show that the union F of these eight quasi-regular tetrahedra satisfies

$$\tau_F(D) \ge 2(1.5)\,pt.$$

There are two additional vertices of type $(5, 0)$ whose tetrahedra are distinct from these eight quasi-regular tetrahedra. They give an additional $2(0.55)\,pt$. Now $(11.16 + 2(1.5) + 2(0.55))\,pt > (4\pi\zeta - 8)\,pt$ by Lemma 20.7. The result follows.   □

Lemma 21.11. *A contravening plane graph satisfies Property 5 of tameness: If a vertex is contained in an exceptional face, then the degree of the vertex is at most five.*

Proof. An exceptional standard region with a vertex of degree 6 must be pentagonal by Lemma 21.9. If that pentagonal region has two or more such vertices, then by the same lemma it must be the arrangement shown in Fig. 21.1. This arrangement does not appear on a contravening graph by Lemma 21.10. □

Remark 21.12. We have now fully justified the claim made in Remark 20.19: there is at most one vertex on six standard regions, and it is part of an aggregate in such a way that it does not appear as the vertex of $G(D)$.

21.5. *Possible 4-Circuits*

Every 4-circuit divides a plane graph into two aggregates of faces that we may call the interior and exterior. We call vertices of the faces in the aggregate that do not lie on the 4-cycle *enclosed vertices*. Thus, every vertex lies in the 4-cycle, is enclosed over the interior, or is enclosed over the exterior.

Lemma 20.11 asserts that either the interior or the exterior has at most one enclosed vertex. When choosing which aggregate is to be called the interior, we may make our choice so that the interior has area at most 2π, and hence contains at most one vertex. With this choice, we have the following proposition.

Proposition 21.13. *Let D be a contravening plane graph. A 4-circuit surrounds one of the aggregates of faces shown in Property 3 of tameness.*

Proof. If there are no enclosed vertices, then the only possibilities are for it to be a single quadrilateral face or a pair of adjacent triangles.

Assume there is one enclosed vertex v. If v is connected to three or four vertices of the quadrilateral, then that possibility is listed as part of the conclusion.

If v is connected to two opposite vertices in the 4-cycle, then the vertex v has type $(0, 2)$ and the bounds of Lemma 20.6 show that the graph cannot be contravening.

If v is connected to two adjacent vertices in the 4-cycle, then we appeal to Lemma 21.10 to conclude that the graph does not contravene.

If v is connected to at most one vertex, then we appeal to Lemma 20.10. This completes the proof. □

22. Weight Assignments

The purpose of this section is to prove the existence of a good admissible weight assignment for contravening plane graphs. This will complete the proof that all contravening graphs are tame.

Theorem 22.1. *Every contravening plane graph has an admissible weight assignment of total weight less than* tgt $= 14.8$.

Given a contravening decomposition star D, we define a weight assignment w by

$$F \mapsto w(F) = \tau_F(D)/pt.$$

Since D contravenes,

$$\sum_F w(F) = \sum_F \tau_F(D)/pt$$
$$= \tau(D)/pt$$
$$\leq (4\pi\zeta - 8)\,pt/pt$$
$$< \mathrm{tgt} = 14.8.$$

The challenge of the theorem will be to prove that w, when defined by this formula, is admissible.

22.1.   *Admissibility*

The next three lemmas establish that this definition of $w(F)$ for contravening plane graphs satisfies the first three defining properties of an admissible weight assignment.

Lemma 22.2.   *Let F be a face of length n in a contravening plane graph. Define $w(F)$ as above. Then $w(F) \geq d(n)$.*

Proof.   This is Proposition 21.7.   □

Lemma 22.3.   *Let v be a vertex of type (p, q) in a contravening plane graph. Define $w(F)$ as above. Then*

$$\sum_{v \in F} w(F) \geq b(p, q).$$

Proof.   This is Lemma 20.6.   □

Lemma 22.4.   *Let V be any set of vertices of type $(5, 0)$ in a contravening plane graph. Define $w(F)$ as above. If the cardinality of V is $k \leq 4$, then*

$$\sum_{V \cap F \neq \emptyset} w(F) \geq 0.55k.$$

Proof.   This is Lemma 20.7.   □

The following proposition establishes the final property that $w(F)$ must satisfy to make it admissible. *Separated sets* are defined in Section 18.2.

Proposition 22.5. *Let V be any separated set of vertices in a contravening plane graph. Define $w(F)$ as above. Then*

$$\sum_{V \cap F \neq \emptyset} (w(F) - d(\text{len}(F))) \geq \sum_{v \in V} a(\text{tri}(v)),$$

where tri(v) *denotes the number of triangles containing the vertex v.*

The proof occupies the rest of this section. Since the degree of each vertex is 5, and there is at least one face that is not a triangle at the vertex, the only constants tri(v) that arise are

$$\text{tri}(v) \in \{0, \ldots, 4\}.$$

We will prove that in a contravening plane graph Conditions 1 and 4 of a separated set are incompatible with the condition tri(v) ≤ 2, for some $v \in V$. This will allows us to assume that

$$\text{tri}(v) \in \{3, 4\},$$

for all $v \in V$. These cases are treated in Section 22.3.

First we prove the inequality when there are no aggregates involved. Afterwards, we show that the conclusions can be extended to aggregate faces as well.

22.2. Proof that tri(v) > 2

In this subsection D is a contravening decomposition star with associated graph $G(D)$. Let V be a separated set of vertices in $G(D)$. Let v be a vertex in V such that none of its faces is an aggregate in the sense of Remarks 20.18 and 20.19.

Lemma 22.6. *Under these conditions, for every $v \in V$, tri(v) > 1.*

Proof. If there are p triangles, q quadrilaterals, and r other faces, then

$$\tau(D) \geq \sum_{v \in R} \tau_R(D)$$

$$\geq r\, t_5 + \tau_{\text{LP}}(p, q, 2\pi - r(1.153)).$$

If there is a vertex w that is not on any of the faces containing v, then the sum of $\tau_F(D)$ over the faces containing w yields an additional $0.55\, pt$ by Lemma 20.7. We calculate these constants for each (p, q, r) and find that the bound is always greater than $(4\pi \zeta - 8)\, pt$. This implies that D cannot be contravening.

(p, q, r)	Lower bound	Justification
$(0, 5, 0)$	$22.27\, pt$	Lemma 20.6
$(0, q, r \geq 1)$	$t_5 + 4t_4 \approx 14.41\, pt$	
$(1, 4, 0)$	$17.62\, pt$	Lemma 20.6
$(1, 3, 1)$	$t_5 + 12.58\, pt$	(τ_{LP})
$(1, 2, 2)$	$2t_5 + 7.53\, pt$	(τ_{LP})
$(1, q, r \geq 3)$	$3t_5 + t_4$	

□

Lemma 22.7. *Under these same conditions, for every $v \in V$, $\mathrm{tri}(v) > 2$.*

Proof. Assume that $\mathrm{tri}(v) = 2$. We will show that this implies that D does not contravene. Let e be the number of exceptional faces at v. We have $e + \mathrm{tri}(v) \leq 5$.

The constants $0.55\,pt$ and $0.48\,pt$ used throughout the proof come from Lemma 20.7. The constants t_n comes from Lemma 20.2.

($e = 3$) First, assume that there are three exceptional faces around vertex v. They must all be pentagons ($2t_5 + t_6 > (4\pi\zeta - 8)\,pt$). The aggregate of the five faces is an m-gon (some $m \leq 11$). If there is a vertex not on this aggregate, use $3t_5 + 0.55\,pt > (4\pi\zeta - 8)\,pt$. So there are at most nine triangles away from the aggregate, and

$$\sigma(D) \leq 9\,pt + (3s_5 + 2\,pt) < 8\,pt.$$

The argument is the same if there is a quad, a pentagon, or a hexagon ($t_4 + t_6 = 2t_5$, $s_4 + s_6 = 2s_5$).

($e = 2$) Assume next that there are two pentagons and a quadrilateral around the vertex. The aggregate of the two pentagons, quadrilateral, and two triangles is an m-gon (some $m \leq 10$). There must be a vertex not on the aggregate of five faces, for otherwise we have

$$\sigma(D) \leq 8\,pt + (2s_5 + 2\,pt) < 8\,pt.$$

The interior angle of one of the pentagons is at most 1.32. For otherwise, $\tau_{\mathrm{LP}}(2, 1, 2\pi - 2(1.32)) + 2t_5 + 0.55\,pt > (4\pi\zeta - 8)\,pt$.

Lemma 20.13 shows that any pentagon R with an interior angle less than 1.32 yields $\tau_R(D) \geq t_5 + (1.47\,pt)$. If both pentagons have an interior angle < 1.32 the lemma follows easily from this calculation: $2(t_5 + 1.47\,pt)\,pt + \tau_{\mathrm{LP}}(2, 1, 2\pi - 2(1.153)) + 0.55\,pt > (4\pi\zeta - 8)\,pt$. If there is one pentagon with an angle > 1.32, we then have $t_5 + (1.47\,pt) + \tau_{\mathrm{LP}}(2, 1, 2\pi - 1.153 - 1.32) + t_5 + 0.55\,pt > (4\pi\zeta - 8)\,pt$.

($e = 1$) Assume finally that there is one exceptional face at the vertex. If it is a hexagon (or more), we are done: $t_6 + \tau_{\mathrm{LP}}(2, 2, 2\pi - 1.153) > (4\pi\zeta - 8)\,pt$. Assume it is a pentagon. The aggregate of the five faces at the vertex is bounded by an m-circuit (some $m \leq 9$). If there are no more than nine quasi-regular tetrahedra outside the aggregate, then $\sigma(D)$ is at most $(9 - 2(0.48))\,pt + s_5 + \sigma_{\mathrm{LP}}(2, 2, 2\pi - 1.153) < 8\,pt$ (Lemma 20.7). So we may assume that there are at least three vertices not on the aggregate.

If the interior angle of the pentagon is greater than 1.32, then

$$\tau_{\mathrm{LP}}(2, 2, 2\pi - 1.32) + 3(0.55)\,pt + t_5 > (4\pi\zeta - 8)\,pt;$$

if it is less than 1.32, by Lemma 20.13

$$\tau_{\mathrm{LP}}(2, 2, 2\pi - 1.153) + 3(0.55)pt + 1.47\,pt + t_5 > (4\pi\zeta - 8)\,pt. \qquad \square$$

Lemma 22.8. *The bound $\mathrm{tri}(v) > 2$ holds if v is a vertex of an aggregate face.*

Proof. The exceptional region enters into the preceding two proofs in a purely formal way. Pentagons enter through the bounds

$$t_5, \quad s_5, \quad 1.47\,pt$$

and angles 1.153, 1.32. Hexagons enter through the bounds

$$t_6, \ s_6$$

and so forth. These bounds hold for the aggregate faces. Hence the proofs hold for aggregates as well.   □

22.3.   Bounds when $\mathrm{tri}(v) \in \{3, 4\}$

In this subsection D is a contravening decomposition star with associated graph $G(D)$. Let V be a separated set of vertices. For every vertex v in V, we assume that none of its faces is an aggregate in the sense of Remarks 20.18 and 20.19. We assume that there are three or four triangles containing v, for every $v \in V$.

To prove Inequality (4) in the definition of admissible weight assignments, we rely on the following reductions. Define an equivalence relation on exceptional faces by $F \sim F'$ if there is a sequence $F_0 = F, \ldots, F_r = F'$ of exceptional faces such that consecutive faces share a vertex of type $(3, 0, 2)$. (That is, $\mathrm{tri}(v) = 3$.) Let $\mathcal{F}$ be an equivalence class of faces.

Lemma 22.9.   *Let V be a separated set of vertices. For every equivalence class of exceptional faces $\mathcal{F}$, let $V(\mathcal{F})$ be the subset of V whose vertices lie in the union of faces of $\mathcal{F}$. Suppose that for every equivalence class $\mathcal{F}$, Inequality (4) (in the definition of admissible weight assignments) holds for $V(\mathcal{F})$. Then the inequality holds for V.*

Proof.   By construction, each vertex in V lies in some F, for an exceptional face. Moreover, the separating property of V ensures that the triangles and quadrilaterals in the inequality are associated with a well-defined $\mathcal{F}$. Thus, the inequality for V is a sum of the inequalities for each $V(\mathcal{F})$.   □

Lemma 22.10.   *Let v be a vertex in a separated set V at which there are p triangles, q quadrilaterals, and r other faces. Suppose that for some $p' \le p$ and $q' \le q$, we have*

$$\tau_{\mathrm{LP}}(p', q', \alpha) > (p'd(3) + q'd(4) + a(p)) \, pt$$

for some upper bound α on the angle occupied by p' triangles and q' quadrilaterals at v. Suppose further that Inequality 4 (in the definition of admissible weight assignments) holds for the separated set $V' = V \setminus \{v\}$. Then the inequality holds for V.

Proof.   Let $F_1, \ldots, F_m$, $m = p' + q'$, be faces corresponding to the triangles and quadrilaterals in the lemma. The hypotheses of the lemma imply that

$$\sum_1^m (w F_i(D) - d(\mathrm{len}(F_i))) > a(p).$$

Clearly, the inequality for V is the sum of this inequality, the inequality for V', and $d(n) \ge 0$.   □

Recall that the central vertex of a flat quarter is defined to be the one that does not lie on the triangle formed by the origin and the diagonal.

Lemma 22.11. *Let R be an exceptional standard region. Let V be a set of vertices of R. If $v \in V$, let p_v be the number of triangular regions at v and let q_v be the number of quadrilateral regions at v. Assume that V has the following properties:*

1. *The set V is separated.*
2. *If $v \in V$, then there are five standard regions at v.*
3. *If $v \in V$, then the corner over v is a central vertex of a flat quarter in the cone over R.*
4. *If $v \in V$, then $p_v \geq 3$. That is, at least three of the five standard regions at v are triangular.*
5. *If $R' \neq R$ is an exceptional region at v, and if R has an interior angle at least 1.32 at v, then R' also has an interior angle at least 1.32 at v.*

Let F be the union of $\{R\}$ with the set of triangular and quadrilateral regions that have a vertex at some $v \in V$. Then

$$\tau_F(D) > \sum_{v \in V}(p_v d(3) + q_v d(4) + a(p_v)) \, pt.$$

Proof. If $(p_v, q_v) = (3, 1)$ and the internal angle of R at v is at least 1.32, then we use

$$\tau_{LP}(3, 1, 2\pi - 1.32) > 1.4 \, pt + t_4.$$

In this case the inequality of the lemma is a consequence of this inequality and the inequality for $V\setminus\{v\}$. Thus, we may assume without loss of generality that if $(p_v, q_v) = (3, 1)$, then the internal angle of R at v is at most 1.32. The conclusion now follows from Lemma 14.6.   □

Lemma 22.12. *Property 4 of admissibility holds. That is, let V be any separated set of vertices. Then*

$$\sum_{F: \, V \cap F \neq \emptyset} (w(F) - d(\mathrm{len}(F))) \geq \sum_{v \in V} a(\mathrm{tri}(v)).$$

Proof. Let V be a separated set of vertices. The results of Section 22.2 reduce the lemma to the case where $\mathrm{tri}(v) \in \{3, 4\}$ for every vertex $v \in V$.

We say that there is a flat quarter centered at v, if the corner v' over v is the central vertex of a flat quarter and that flat quarter lies in the cone over an exceptional region.

One case is easy to deal with. Assume that there are three triangles, a quadrilateral, and an exceptional face at the vertex. Assume the interior angle on the exceptional region is at least 1.32; then

$$\tau_{LP}(3, 1, 2\pi - 1.32) > 1.4 \, pt + t_4. \tag{22.1}$$

This gives the bound in the sense of Lemma 22.10 at such a vertex. For the rest of the proof, assume that the interior angle on the exceptional region is less than 1.32 at vertices of type $(p, q, r) = (3, 1, 1)$. This implies in particular by Lemma 20.16 that there is a flat quarter centered at each vertex of this type.

Let v be a vertex with no flat quarter centered at v. By Lemma 20.16, the interior angles of the exceptional regions at v are at least 1.32. It follows[135] that

$$\tau_{LP}(p_v, q_v, \alpha) > (p_v d(3) + q_v d(4) + a(p_v)) \, pt. \tag{22.2}$$

Thus, by Lemma 22.10, we reduce to the case where for each $v \in V$, there is a flat quarter centered at v. Assume that V has this property.

Pick a function f from the set V to the set of exceptional standard regions as follows. If there is only one exceptional region at v, then let $f(v)$ be that exceptional region. If there are two exceptional regions at v, then let $f(v)$ be one of these two exceptional regions. Pick it to be an exceptional region with an interior angle at most 1.32 if one of the two exceptional regions has this property. Pick it to have a flat quarter centered at v. Note that by Lemma 20.16, if the exceptional region has an interior angle at most 1.32, then $f(v)$ will have a flat quarter centered at v.

For each exceptional region R, let

$$V_R = \{v \in V : f(v) = R\}.$$

By Lemma 22.11, Property 4 of admissibility is satisfied for each V_R. Since this property is additive in V_R and since V is the disjoint union of the sets V_R, the proof is complete.   □

22.4. Weight Assignments for Aggregates

Lemma 22.13.   *Consider a separated set of vertices V on an aggregated face F as in Remark 20.18. Then inequality 4 holds (in the definition of admissible weight assignments):*

$$\sum_{V \cap F \neq \emptyset} (w(F) - d(\text{len}(F))) \geq \sum_{v \in V} a(\text{tri}(v)).$$

Proof.   We may assume that $\text{tri}(v) \in \{3, 4\}$.

First consider the aggregate of Remark 20.18 of a triangle and an eight-sided region, with pentagonal hull F. There is no other exceptional region in a contravening decomposition star with this aggregate:

$$t_8 + t_5 > (4\pi \zeta - 8) \, pt.$$

A separated set of vertices V on F has cardinality at most 2. This gives the desired bound

$$t_8 > t_5 + 2(1.5) \, pt.$$

---

[135] CALC-551665569, CALC-824762926, and CALC-325738864.

Next, consider the aggregate of a hexagonal hull with an enclosed vertex. Again, there is no other exceptional face. If there are at most $k \leq 2$ vertices in a separated set, then the result follows from

$$t_8 > t_6 + k(1.5) \, pt.$$

There are at most three vertices in V on a hexagon, by the nonadjacency conditions defining V. A vertex v can be removed from V if it is not the central vertex of a flat quarter (Lemma 22.10 and inequalities (22.1) and (22.2)). If there is an enclosed vertex w, it is impossible for there to be three nonadjacent vertices, each the central vertex of a flat quarter:

$$\mathcal{E}(2, 2, 2, \sqrt{8}, \sqrt{8}, \sqrt{8}, 2t_0, 2t_0, 2) > 2t_0.$$

($\mathcal{E}$ is as defined in Definition 4.14.)

Finally consider the aggregate of a pentagonal hull with an enclosed vertex. There are at most $k \leq 2$ vertices in a separated set in F. There is no other exceptional region:

$$t_7 + t_5 > (4\pi\zeta - 8) \, pt.$$

The result follows from

$$t_7 > t_5 + 2(1.5) \, pt. \qquad \square$$

Lemma 22.14. *Consider a separated set of vertices V on an aggregate face of a contravening plane graph as in Remark 20.19. Inequality 4 holds in the definition of admissible weight assignments.*

Proof. There is at most one exceptional face in the plane graph:

$$t_8 + t_5 > (4\pi\zeta - 8) \, pt.$$

Assume first that an aggregate face is an octagon (Fig. 20.4). At each of the vertices of the face that lies on a triangular standard region in the aggregate, we can remove the vertex from V using Lemma 22.10 and the estimate

$$\tau_{LP}(4, 0, 2\pi - 2(0.8638)) > 1.5 \, pt.$$

This leaves at most one vertex in V, and it lies on a vertex of F which is "not aggregated," so that there are five standard regions of the associated decomposition star at that vertex, and one of those regions is pentagonal. The value $a(4) = 1.5 \, pt$ can be estimated at this vertex in the same way it is done for a nonaggregated case in Section 22.3.

Now consider the case of an aggregate face that is a hexagon (Fig. 20.4). The argument is the same: we reduce to V containing a single vertex, and argue that this vertex can be treated as in Section 22.3. (Alternatively, use the fact that the pentagon–triangle combination in this aggregate has been eliminated by Lemma 21.10.) $\qquad \square$

The proof that contravening plane graphs are tame is complete.

23. Linear Program Estimates

We have completed a major portion of the proof of the Kepler conjecture by proving that every contravening plane graph is tame.

The final portion of the proof of the Kepler conjecture consists in showing that tame graphs are not contravening, except for the isomorphism class of graphs isomorphic to G_{fcc} and G_{hcp} associated with the face-centered cubic and hexagonal close packings.

This part of the proof treats all contravening tame graphs except for the three cases G_{fcc}, G_{pent}, and G_{hcp}. The two cases G_{fcc} and G_{hcp} are treated in Theorem 8.1, and the case G_{pent} is treated in Paper V.

The primary tool that will be used is linear programming. The linear programs are obtained as relaxations of the original nonlinear optimization problem of maximizing $\sigma(D)$ over all decomposition stars whose associated graph is a given tame graph G. The upper bounds obtained through relaxation are upper bounds to the nonlinear problem.

To eliminate a tame graph, we must show that it is not contravening. By definition, this means we must show that $\sigma(D) < 8\,pt$. When a single linear program does not yield an upper bound under $8\,pt$, we branch into a sequence of linear programs that collectively imply the upper bound of $8\,pt$. This will call for a sequence of increasingly complex linear programs.

For each of the tame plane graphs produced in Theorem 19.1, we define a linear programming problem whose solution dominates the value of $\sigma(D)$ on the set of decomposition stars associated with the plane graph. A description of the linear programs is presented in this section.

Theorem 23.1.   *If the plane graph of a contravening decomposition star is isomorphic to one in the list [Ha16], then it is isomorphic to one of the following three plane graphs: the plane graph of the pentahedral prism, that of the hexagonal-close packing, or that of the face-centered cubic packing.*

This theorem is one of the central claims described in Section 3.3 that lead to the proof of the Kepler conjecture.

23.1. *Relaxation*

(NLP)   Let $f\colon P \to \mathbb{R}$ be a function on a nonempty set P. Consider the nonlinear maximization problem

$$\max_{p \in P} f(p).$$

(LP)   Consider a linear programming problem

$$\max\ c \cdot x$$

such that $Ax \le b$, where A is a matrix, b and c are vectors of real constants, and x is a vector of variables $x = (x_1, \ldots, x_n)$. We write the linear programming problem as

$$\max(c \cdot x\colon\ Ax \le b).$$

An *interpretation* I of a linear programming problem (LP) is a nonempty set $|I|$, together with an assignment $x_i \mapsto x_i^I$ of functions $x_i^I \colon |I| \to \mathbb{R}$ to variables x_i. We say the constraints $A\,x \le b$ of the linear program are *satisfied* under the interpretation I if for all $p \in |I|$,

$$A\,x^I(p) \le b.$$

The interpretation I is said to be a *relaxation* of the nonlinear program (NLP) if the following three conditions hold:

1. $P = |I|$.
2. The constraints are satisfied under the interpretation.
3. $f(p) \le c \cdot x^I(p)$, for all $p \in |I|$.

Lemma 23.2. *Let* (LP) *be a linear program with relaxation I to* (NLP). *Then* (LP) *has a feasible solution. Moreover, if* (LP) *is bounded above by a constant M, then M is an upper bound on the function $f \colon |I| \to \mathbb{R}$.*

Proof. A feasible solution is $x_i = x_i^I(p)$, for any $p \in |I|$. The rest is clear. □

Remark 23.3. In general, it is to be expected that the interpretations $A\,x^I \le b$ will be nonlinear inequalities on the domain P. In our situation, satisfaction of the constraints will be proved by interval arithmetic. Thus, the construction of an upper bound to (NLP) breaks into two tasks: to solve the linear programs and to prove the nonlinear inequalities required to satisfy the constraints.

There are many nonlinear inequalities entering into our interpretation. These have been proved by interval arithmetic on computer and are listed at [Ha16].

Remark 23.4. There is a second method of establishing the satisfaction of inequalities under an interpretation. Suppose we wish to show that the inequality $e \cdot x \le b'$ is satisfied under the interpretation I. Suppose that we have already established that a system of inequalities $A\,x \le b$ is satisfied under the interpretation I. We solve the linear programming problem $\max(e \cdot x \colon A\,x \le b)$. If this maximum is at most b', then the inequality $e \cdot x \le b'$ is satisfied under the interpretation I. We refer to $e \cdot x \le b'$ as an LP-*derived inequality* (with respect to the system $A\,x \le b$).

23.2. The Linear Programs

Let G be a tame plane graph. Let $DS(G)$ be the space of all decomposition stars whose associated plane graph is isomorphic to G.

Theorem 23.5. *For every tame plane graph G other than G_{fcc}, G_{hcp}, and G_{pent}, there exists a finite sequence of linear programs with the following properties:*

1. *Every linear program has an admissible solution and its solution is strictly less than $8\,pt$.*

2. *For every linear program in this sequence, there is an interpretation I of the linear program that is a relaxation of the nonlinear optimization problem*

$$\sigma: |I| \to \mathbb{R},$$

where $|I|$ is a subset of DS(G).
3. *The union of the subsets $|I|$, as we run over the sequence of linear programs, is DS(G).*

The proof is constructive. For every tame plane graph G a sequence of linear programs is generated by computer and solved. The optimal solutions are all bounded above by $8\,pt$. It will be clear from construction of the sequence that the union of the sets $|I|$ exhausts DS(G). We estimate that nearly 10^5 linear programs are involved in the construction. The rest of this paper outlines the construction of some of these linear programs.

Remark 23.6. Section 3.1.1 of [Ha14] shows how the linear programs that arise in connection with the Kepler conjecture can be formulated in such a way that they always have a feasible solution and so that the optimal solution is bounded. We assume that all our linear programs have been constructed in this way.

Corollary 23.7. *If a tame graph G is not isomorphic to G_{fcc}, G_{hcp}, or G_{pent}, then it is not contravening.*

Proof. This follows immediately from Theorem 23.5 and Lemma 23.2.   □

23.3. Basic Linear Programs

Let G be a tame plane graph. Specifically, G is one of the several thousands of graphs that appear in the explicit classification [Ha16].

To describe the basic linear program, we need the following indexing sets. Let VERTEX be the set of all vertices in G. Let FACE be the set of all faces in G. (Recall that by construction each face F of the graph carries an orientation.) Let ANGLE be the set of all angles in G, defined as the set of pairs (v, F), where the vertex v lies in the face F. Let DIRECTED be the set of directed edges. It consists of all ordered pairs $(v, s(v, F))$, where $s(v, F)$ denotes the successor of the vertex v in the oriented face F. Let TRIANGLES be the subset of FACE consisting of those faces of length 3. Let UNDIRECTED be the set of undirected edges. It consists of all unordered pairs $\{v, s(v, F)\}$, for $v \in F$.

We introduce variables indexed by these sets. Following AMPL notation, we write for instance y{VERTEX} to declare a collection of variables $y[v]$ indexed by vertices v in VERTEX. With this in mind, we declare the variables

$$\alpha\{\text{ANGLE}\}, \quad y\{\text{VERTEX}\}, \quad e\{\text{UNDIRECTED}\},$$
$$\sigma\{\text{FACE}\}, \quad \tau\{\text{FACE}\}, \quad \text{sol}\{\text{FACE}\}.$$

We obtain an interpretation I on the compact space DS(G). First, we define an interpretation at the level of indexing sets. A decomposition star determines the set

$U(D)$ of vertices of height at most $2t_0$ from the origin of D. Each decomposition star $D \in \mathrm{DS}(G)$ determines a (metric) graph with geodesic edges on the surface of the unit sphere, which is isomorphic to G as a (combinatorial) plane graph. There is a map from the vertices of G to $U(D)$ given by $v \mapsto v'$, if the radial projection of v' to the unit sphere at the origin corresponds to v under this isomorphism. Similarly, each face F of G corresponds to a set F' of standard regions. Each edge e of G corresponds to a geodesic edge e' on the unit sphere.

Now we give an interpretation I to the linear-programming variables at a decomposition star D. As usual, we add a superscript I to a variable to indicate its interpretation. Let $\alpha[v, F]'$ be the sum of the interior angles at v' of the metric graph in the standard regions F'. Let $y[v]'$ be the length $|v'|$ of the vertex $v' \in U(D)$ corresponding to v. Let $e[v, w]'$ be the length $|v' - w'|$ of the edge between v' and $w' \in U(D)$. Let

$$\sigma[F]' = \sigma_F(D),$$
$$\mathrm{sol}[F]' = \mathrm{sol}(F'),$$
$$\tau[F]' = \tau_F(D).$$

The objective function for the optimization problems is

$$\max : \quad \sum_{F \in \mathrm{FACE}} \sigma[F].$$

Its interpretation under I is the score $\sigma(D)$.

We can write a number of linear inequalities that will be satisfied under our interpretation. For example, we have the bounds

$$
\begin{aligned}
0 &\leq y[v] \leq 2t_0, & v &\in \mathrm{VERTEX}, \\
0 &\leq e[v, w] \leq 2t_0, & (v, w) &\in \mathrm{EDGE}, \\
0 &\leq \alpha[v, F] \leq 2\pi, & (v, F) &\in \mathrm{ANGLE}, \\
0 &\leq \mathrm{sol}[F] \leq 4\pi & F &\in \mathrm{FACE}.
\end{aligned}
$$

There are other linear relations that are suggested directly by the definitions or the geometry. Here, v belongs to VERTEX:

$$\tau[F] = \mathrm{sol}[F]\zeta\, pt - \sigma[F],$$
$$2\pi = \sum_{F:v \in F} \alpha[v, F],$$
$$\mathrm{sol}[F] = \sum_{v \in F} \alpha[v, F] - (\mathrm{len}(F) - 2)\pi.$$

There are long lists of additional inequalities that come from interval arithmetic verifications. Many are specifically designed to give relations between the variables.

$$\sigma[F], \ \tau[F], \ \alpha[v, F],$$
$$\mathrm{sol}[F], \ y[v], \ e[v, w],$$

whenever F^I is a single standard region having three sides. Similarly, other computer calculations give inequalities for $\sigma[F]$ and related variables, when the length of F is 4. A complete list of inequalities that are used for triangular and quadrilateral faces is found in [Ha16].

For exceptional faces, we have an admissible weight function $w(F)$. According to definitions $w(F) = \tau[F]/pt$, so that the inequalities for the weight function can be expressed in terms of the linear program variables.

When the exceptional face is not an aggregate, then it also satisfies the inequalities of Lemma 20.2.

23.4.   Error Analysis

The variables of the linear programming problem are the dihedral angles, the scores of each of the standard clusters, and their edge lengths.

We subject these variables to a system of linear inequalities. First, the dihedral angles around each vertex sum to 2π. The dihedral angles, solid angles, and score are related by various linear inequalities as described in Section 23.3. The solid-angle variables are linear functions of dihedral angles. We have

$$\sigma(D) = \sigma_{S_1}(D) + \cdots + \sigma_{S_p}(D) + \sigma_{R_1}(D) + \cdots + \sigma_{R_q}(D).$$

Forgetting the origin of the scores, solid angles, and dihedral angles as nonlinear functions of the standard clusters and treating them as formal variables subject only to the given linear inequalities, we obtain a linear programming bound on the score.

Floating-point arithmetic was used freely in obtaining these bounds. The linear programming package *CPLEX* was used (see *www.cplex.com*). However, the results, once obtained, could be checked rigorously as follows.[136]

We present an informal analysis of the floating-point errors. For each quasi-regular tetrahedron S_i we have a nonnegative variable $x_i = pt - \sigma(S_i)$. For each quad cluster R_k, we have a nonnegative variable $x_k = -\sigma(R_k)$. A bound on $\sigma(D)$ is $p\,pt - \sum_{i \in I} x_i$, where p is the number of triangular standard regions, and I indexes the faces of the plane graph. We give error bounds for a linear program involving scores and dihedral angles. Similar estimates can be made if there are edges representing edge lengths. Let the dihedral angles be x_j, for j in some indexing set J. Write the linear constraints as $Ax \leq b$. We wish to maximize $c \cdot x$ subject to these constraints, where $c_i = -1$, for $i \in I$, and $c_j = 0$, for $j \in J$. Let z be an approximate solution to the inequalities $zA \geq c$ and $z \geq 0$ obtained by numerical methods. Replacing the negative entries of z by 0 we may assume that $z \geq 0$ and that $zA_i > c_i - \varepsilon$, for $i \in I \cup J$, and some small error ε. If we obtain the numerical bound $p\,pt + z \cdot b < 7.9999\,pt$, and if $\varepsilon < 10^{-8}$, then $\sigma(D)$ is less than $8\,pt$. In fact, we note that

$$\left(\frac{z}{1+\varepsilon}\right) A_i$$

---

[136] The output from each linear program that has no exceptional regions has been double checked with interval arithmetic. Predictably, the error bounds presented here were satisfactory (1/2002).

is at least c_i for $i \in I$ (since $c_i = -1$), and that it is greater than $c_i - \varepsilon/(1+\varepsilon)$, for $i \in J$ (since $c_i = 0$). Thus, if $N \le 60$ is the number of vertices, and $p \le 2(N-2) \le 116$ is the number of triangular faces,

$$\sigma(D) \le p\,pt + c \cdot x \le p\,pt + \left(\frac{z}{1+\varepsilon}\right) Ax + \frac{\varepsilon}{1+\varepsilon} \sum_{j \in J} x_j$$

$$\le p\,pt + \frac{z \cdot b}{1+\varepsilon} + \frac{\varepsilon}{1+\varepsilon} 2\pi N$$

$$\le \frac{p\,pt + z \cdot b + \varepsilon(p\,pt + 2\pi N)}{(1+\varepsilon)}$$

$$\le \frac{7.9999\,pt + 10^{-8}(116\,pt + 500)}{1 + 10^{-8}} < 8\,pt.$$

In practice, we used $0.4429 < 0.79984\,pt$ as our cutoff, and $N \le 14$ in the interesting cases, so that much tighter error estimates are possible.

24. Elimination of Aggregates

The proof of the following theorem occupies the entire section. It eliminates all the pathological cases that we have had to carry along until now.

Theorem 24.1. *Let D be a contravening decomposition star, and let G be its tame graph. Every face of G corresponds to exactly one standard region of D. No standard region of D has any enclosed vertices from U(D). (That is, a decomposition star with one of the aggregates shown in Fig. 20.1 is not contravening.)*

24.1. Triangle and Quad Branching

Section 25 discusses branch and bound strategies. Branch and bound strategies replace a single linear program with a series of linear program, when a single linear program does not suffice. There is one case of branch and bound that we need before Section 25. This is a branching on triangular and quadrilateral faces.

We divide triangular faces with corners v_1, v_2, v_3 into two cases:

$$e[v_1, v_2] + e[v_2, v_3] + e[v_3, v_1] \le 6.25,$$

$$e[v_1, v_2] + e[v_2, v_3] + e[v_3, v_1] \ge 6.25,$$

whenever sufficiently good bounds are not obtained as a single linear program. We also divide quadrilateral faces into four cases: two flat quarters, two flat quarters with a diagonal running in the other direction, four upright quarters forming a quartered octahedron, and the mixed case. (A mixed case by definition is any case that is not one of the other three.) In general, if there are r_1 triangles and r_2 quadrilaterals, we obtain as many as $2^{r_1+2r_2}$ cases by breaking the various triangles and quadrilaterals into subcases.

We break triangular faces and quadrilaterals into subcases, as needed in the linear programs that follow, without further comment.

24.2.   A Pentagonal Hull with $n = 8$

The next few sections treat the nonpolygonal standard regions described in Remark 20.18. In this subsection there is an aggregate of the octagonal region and a triangle has a pentagonal hull. Let P denote this aggregate.

Lemma 24.2.   *Let G be a contravening plane graph with the aggregate of Remark 20.18. Some vertex on the pentagonal face has type not equal to* $(3, 0, 1)$.

Proof.   If every vertex on the pentagonal face has type $(3, 0, 1)$, then at the vertex of the pentagon meeting the aggregated triangle, the four triangles together with the octagon give

$$t_8 + \sum_{(4)} \tau_{\mathrm{LP}}(4, 0, 2\pi - 2(1.153)) > (4\pi \zeta - 8)\, pt,$$

so that the graph does not contravene.   □

For a general contravening plane graph with this aggregate, we have bounds

$$\sigma_F(D) \ \le \ pt + s_8,$$
$$\tau_F(D) \ \ge \ t_8.$$

We add the inequalities $\tau[F] > t_8$ and $\sigma[F] < pt + s_8$ to the exceptional face. There is no other exceptional face, because $t_8 + t_5 > (4\pi \zeta - 8)\, pt$. We run the linear programs for all tame graphs with the property asserted by Lemma 24.2. Every upper bound is less than $8\, pt$, so that there are no contravening decomposition stars with this configuration.

24.3. $n = 8$, Hexagonal Hull

We treat the two cases from Remark 20.18 that have a hexagonal hull (Fig. 20.1). One can be described as a hexagonal region with an enclosed vertex that has height at most $2t_0$ and distance at least $2t_0$ from each corner over the hexagon. The other is described as a hexagonal region with an enclosed vertex of height at most $2t_0$, but this time with distance less than $2t_0$ from one of the corners over the hexagon.

The argument for the case $n = 8$ with a hexagonal hull is similar to the argument of Section 24.2. Add the inequalities $\tau[R] > t_8$ and $\sigma[R] < s_8$ for each hexagonal region. Run the linear programs for all tame graphs, and check that these additional inequalities yield linear programming bounds under $8\, pt$.

24.4. $n = 7$, Pentagonal Hull

We treat the two cases illustrated in Fig. 20.1 that have a pentagonal hull. These cases require more work. One can be described as a pentagon with an enclosed vertex that has height at most $2t_0$ and distance at least $2t_0$ from each corner of the pentagon. The other is described as a pentagon with an enclosed vertex of height at most $2t_0$, but this time with distance less than $2t_0$ from one of the corners of the pentagon.

In discussing various maps, we let v_i be the corners of the regions, and we set $y_i = |v_i|$ and $y_{ij} = |v_i - v_j|$. The subscript F is dropped, when there is no great danger of ambiguity.

Add the inequalities $\tau[F] > t_7$, $\sigma[F] < s_7$ for the pentagonal face. There is no other exceptional region, because $t_5 + t_7 > (4\pi\zeta - 8)\,pt$. With these changes, of all the tame plane graphs with a pentagonal face and no other exceptional face, all but one of the linear programs give a bound under $8\,pt$.

The plane graph G_0 that remains is easy to describe. It is the plane graph with eleven vertices, obtained by removing from an icosahedron a vertex and all five edges that meet at that vertex.

We treat the case G_0. Let v_{12} be the vertex enclosed over the pentagon. We let $v_1, \ldots, v_5$ be the five corners of $U(D)$ over the pentagon. Break the pentagon into five simplices along $\{0, v_{12}\}$: $S_i = \{0, v_{12}, v_i, v_{i+1}\}$. We have LP-derived bounds (in the sense of Remark 23.4) $y[v_i] \le 2.168$, and $\alpha[v_i, F] \le 2.89$, for $i = 1, 2, 3, 4, 5$. In particular, the pentagonal region is convex, for every contravening star $D \in \mathrm{DS}(G_0)$.

Further LP-derived inequalities are

$$\sigma[F] > -0.2345 \quad \text{and} \quad \tau[F] < 0.644.$$

By using branch and bound arguments on the triangular faces, as described in Section 24.1, we can improve the LP-derived inequality to

$$\tau[F] < 0.6079.$$

Another LP-derived inequality gives a bound on the perimeter:

$$\sum |v_i - v_{i+1}| \le 11.407.$$

Yet another LP-derived inequality states that if v_1, v_2, v_3 are consecutive corners over the pentagonal region, then

$$|v_1 - v_2| + |v_2 - v_3| < 4.804.$$

Lemma 24.3. *Assume that R is a pentagonal standard region with an enclosed vertex v of height at most $2t_0$. Assume further that*

- *$|v_i| \le 2.168$ for each of the five corners.*
- *Each interior angle of the pentagon is at most 2.89.*
- *If v_1, v_2, v_3 are consecutive corners over the pentagonal region, then $|v_1 - v_2| + |v_2 - v_3| < 4.804$.*
- *$\sum_5 |v_i - v_{i+1}| \le 11.407$.*

Then $\sigma_R(D) < -0.2345$ or $\tau_R(D) > 0.6079$.

Proof.   This is Lemma 14.5.   □

Since the bound $\tau_R(D) > 0.6079$ contradicts the LP-derived inequality $\tau[F] < 0.6079$, this case does not occur in a contravening graph.

24.5. *Type* $(p, q, r) = (5, 0, 1)$

We return briefly to the case of six standard regions around a vertex discussed in Remark 20.19. In the plane graph they are aggregated into an octagon. We take each of the remaining cases with an octagon, and replace the octagon with a pentagon and six triangles around a new vertex. There are eight ways of doing this. All eight ways in each of the cases gives an LP bound under $8\,pt$. This completes this case.

The second aggregate shown in Fig. 20.4 contains a pentagon–triangle combination that was ruled out by Lemma 21.10.

24.6. *Summary*

Lemma 24.4. *None of the aggregates of Remarks* 20.19 *and* 20.18 *appears in a contravening star. In particular, all regions are bounded by simple polygons, and each face of the graph* $G(D)$ *corresponds to exactly one standard region.*

Proof. The proof is the main result of this section.   □

25. Branch and Bound Strategies

When a single linear program does not give sufficiently good bounds, we apply branch and bound methods to improve the bound. By branching repeatedly, we are able to show in every case that a given tame graph is not contravening.

By relying to a greater degree on results that appear in unpublished (but publicly available) computer logs, this section is more technical than the others. The purpose of the section is to give a sketch of the various ways that the various decomposition stars are divided into cases according to a branch and bound strategy.

The first branching strategy has already been described in Section 24.1. It divides the decomposition stars with a given graph into subcases according to the structural properties of triangular and quadrilateral standard regions.

We assume the results from Section 24 that eliminate the most unpleasant types of configurations.

25.1. *Review of Internal Structures*

For the past several sections, it has not been necessary to refer to the internal structure of the standard clusters. This section is different. To describe the branching operations, it is necessary to use details about the structure of standard clusters.

Recall that a *quarter* is a set of four vertices with five edges of length at least 2 and at most $2t_0$ and a sixth edge of length at least $2t_0$ and at most $2\sqrt{2}$. The long edge of the quarter is called its diagonal. A set of quarters with pairwise disjoint interiors has been selected. Quarters in this set are said to belong to the Q-system. The Q-system has been constructed in such a way that if one quarter along a diagonal lies in the Q-system, then all quarters along that diagonal lie in the Q-system. An anchor is a vertex of the packing that has distance at least 2 and at most $2t_0$ from both endpoints of a diagonal. Each diagonal has a context (n, k), with $n \geq k$, where n is the number of anchors around the diagonal and $n - k$ is the number of quarters that have that diagonal as an edge. If a diagonal has context (n, k), then k is the number of *gaps* that occur between anchors; that is, spaces that are not filled in by quarters. The context of a quarter is defined to be the context of its diagonal.

Recall that a quarter (or its diagonal) is said to be upright if one endpoint of its diagonal is the origin. A quarter is said to be flat if it is not upright and if some vertex of the quarter is the origin.

There is a process of simplification of the decomposition stars and their scoring functions that eliminates[137] many of the contexts (n, k). (The upright quarters are said to be *erased*.) We assume in the following discussion and lemmas that this procedure has been carried out.

An upright diagonal is said to be a *loop* when there is a reasonable scheme of inserting a simplex into each gap so that the diagonal is completely surrounded by quarters and the inserted simplices. The simplices that are inserted in the gaps are called *anchored simplices*. They are constructed in such a way that every edge of an anchored simplex has length at most 3.2. All simplices in a given loop lie over a single standard region. If the gaps cannot be filled with anchored simplices, the upright diagonal is not a loop. Details of this construction can be found in Section 11.5.

In every case the simplices around a given upright diagonal lie in the cone over a single standard region.

Lemma 25.1. *Consider an upright diagonal that is a loop. Let R be the standard region that contains the upright diagonal and its surrounding simplices. Then the following contexts (n, k) are the only ones possible. Moreover, the constants that appear in the columns marked σ and τ are upper and lower bounds respectively for $\sigma_R(D)$ and $\tau_R(D)$ when R contains one loop of that context.*

Std. region	(n, k)	σ	τ
R quad	$(4, 0)$	-0.0536	0.1362
R pentagon	$(4, 1)$	s_5	0.27385
	$(5, 0)$	-0.157	0.3665
R hexagon	$(4, 1)$	s_6	0.41328
	$(4, 2)$	-0.1999	0.5309
	$(5, 1)$	-0.37595	0.65995
R heptagon	$(4, 1)$	s_7	0.55271
	$(4, 2)$	-0.25694	0.67033
R octagon	$(4, 1)$	s_8	0.60722
	$(4, 2)$	-0.31398	0.72484

Proof. This is Lemma 13.5.   □

---

[137] In detail, we assume that all the contexts that do not carry a penalty have been erased. We leave loops, 3-crowded, 4-crowded, and 3-unconfined upright diagonals unerased at this point.

25.2. 3-Crowded and 4-Crowded Upright Diagonals

Definition 25.2.   Consider an upright diagonal that is not a loop. Let R be the standard region that contains the upright diagonal and its surrounding quarters. Then the contexts $(4, 1)$ and $(5, 1)$ are the only contexts possible. In the context $(4, 1)$, if there does not exist a plane through the upright diagonal such that all three quarters lie in the same half-space bounded by the plane, then we say that the context is 3-*unconfined*. If such a plane exists, then we say that the context is 3-crowded. We call the context $(5, 1)$ a 4-crowded upright diagonal. Thus, every upright diagonal is exactly one of the following: a loop, 3-unconfined, 3-crowded, or 4-crowded. A contravening decomposition star contains at most one upright diagonal that is 3-crowded or 4-crowded. See Section 11.9 for a proof of these facts and for further details.

Lemma 25.3.   *Let R be a standard region that contains an upright diagonal that is 4-crowded. Then*

$$\sigma_R(D) < -0.25 \quad and \quad \tau_R(D) > 0.4.$$

Let R be a standard region that contains an upright diagonal that is 3-crowded. Then

$$\sigma_R(D) < -0.4339 \quad and \quad \tau_R(D) > 0.5606.$$

Proof.   See Lemmas 11.11 and 11.18.   □

Lemma 25.4.   *A contravening decomposition star does not contain any upright diagonals that are 3-crowded.*

Proof.   If we have an upright diagonal that is 3-crowded, then there is only one exceptional region $(0.5606 + t_5 > (4\pi\zeta - 8)\,pt)$. We add the inequalities $\tau > 0.5606$ and $\sigma < -0.4339$ to the exceptional region. All linear programming bounds drop under $8\,pt$ when these changes are made.   □

Upright diagonals that are 4-crowded require more work. We begin with a lemma.

Lemma 25.5.   *Let α be the dihedral angle along the large gap along an upright diagonal that is 4-crowded. Let F be the union of the four upright quarters along the upright diagonal. Let v_1 and v_2 be the anchors of $U(D)$ lying along the large gap. If $|v_1| + |v_2| < 4.6$, then $\alpha > 1.78$ and $\sigma_F(D) < -0.31547$.*

Proof.   The bound $\alpha > 1.78$ comes from the inequality archive.[138] The upper bound on the score is a linear programming calculation involving the inequality $\alpha > 1.78$ and the known inequalities on the score of an upright quarter.   □

---

[138] CALC-161665083.

Lemma 25.6. *A contravening decomposition star does not contain any upright diagonals that are 4-crowded.*

Proof. Add the inequalities $\sigma_R(D) < -0.25$ and $\tau_R(D) > 0.4$ at the exceptional regions. An upright diagonal that is 4-crowded does not appear in a pentagon for purely geometrical reasons. Run the linear programs for all tame plane graphs with an exceptional region that is not a pentagon. If this linear program fails to produce a bound of $8\,pt$, we use the lemma to branch into two cases: either $y[v_1] + y[v_2] \geq 4.6$ or $\sigma[R] < -0.31547$. In every case the bound drops below $8\,pt$.   □

25.3. Five Anchors

Now turn to the decomposition stars with an upright diagonal with five anchors. Five quarters around a common upright diagonal in a pentagonal region can certainly occur. We claim that any other upright diagonal with five anchors leads to a decomposition star that does not contravene. In fact, the only other possible context is $(n, k) = (5, 1)$ (see Lemma 25.1).

Lemma 25.7. *Let D be a contravening decomposition star. Then there are no loops with context $(5, 1)$ in D.*

Proof. By Lemma 25.1, the standard region R that contains the loop must be a hexagon. By the same lemma, we have

$$\tau_R(D) > 0.65995 \quad \text{and} \quad \sigma_R(D) < -0.37595.$$

Add these constraints to the linear program of the tame graphs with a hexagonal face. The LP-bound on $\sigma(D)$ with these additional inequalities is less than $8\,pt$.   □

25.4. Penalties

From now on, we assume that there are no loops with context $(5, 1)$, and no 3-crowded or 4-crowded upright diagonals. This leaves various loops and 3-unconfined upright diagonals.

At times, it is necessary to *erase* certain loops and 3-unconfined upright diagonals. There is a *penalty* for doing so. Let D be a decomposition star with an upright diagonal $\{0, v\}$. Let D' be the decomposition star that is identical in all respects, except that v and all indices in the decomposition star that point to v (in the sense of Section 6.1) have been deleted. Let R be the standard region of D over which v is located, and let R' be the corresponding standard region of D'. We say that the upright diagonal can be *erased* with *penalty* π_R if

$$\sigma_R(D) \leq \sigma_{R'}(D') + \pi_R.$$

Definition 25.8. When we break a single region into smaller regions (by taking the part of the region that meets the cone over a quarter, anchored simplex, and so forth) the smaller regions will be called subregions. An anchored simplex that overlaps a flat quarter is said to *mask* the flat quarter. (Masked flat quarters are not in the Q-system.)

Remark 25.9. A function $\hat{\sigma}$ has been defined in Section 11.10. The details of the definition of this function are not important here. It is proved there that $\hat{\sigma}$ is a good upper bound on the scoring function on flat quarters no matter what the origin of the flat quarter. It gives bounds for flat quarters in the Q-system, masked quarters, isolated quarters, and all the other types of flat quarters. The function $\hat{\tau}$ on the space of flat quarters is defined as

$$\hat{\tau}(Q) = \mathrm{sol}(Q)\zeta\, pt - \hat{\sigma}(Q).$$

Remark 25.10. At times, we work with various upper bounds to $\sigma_R(D)$, say,

$$\sigma_R(D) \le f_R(D).$$

When we have a specific upper bound $f_R(D)$ in view, then we will also say that the upright diagonal can be erased with penalty π_R if

$$f_R(D) \le f_{R'}(D') + \pi_R.$$

In more detail, let $R = \{R_1, \ldots, R_k\}$ be the set of subregions over the anchored simplices in a loop. Let $f_{R_i}(D)$ be the approximations of the score of each anchored simplex. Let $Q_1, \ldots, Q_\ell$ be the flat quarters masked by the anchored simplices in the loop. Let R' be the subregion of points in the union of R that are not in the cone over any Q_i. Then we erase with penalty π_R if

$$\sum_i f_{R_i}(D) \le \sum_\ell \hat{\sigma}(Q_j) = \mathrm{vor}_{R',0}(D) + \pi_R.$$

If the upright diagonal is not a loop, we include in the set R all regions along the "gaps" around the upright diagonal.

Sections 13.4 and 13.6 makes various estimates of the penalties that are involved in erasing various loops and 3-unconfined upright diagonals. Most of the penalties are calculated as integer combinations of the constants $\xi_\Gamma = 0.01561$, $\xi_V = 0.003521$, and 0.008. It is proved[139] in Section 11.7 that ξ_Γ is the penalty for erasing a single upright quarter of compression type, and that ξ_V is the penalty for erasing a single upright quarter of Voronoi type.

Lemma 25.11. *Let $\{0, v\}$ be an upright diagonal.*

- *If the upright diagonal is 3-unconfined, then the upright diagonal can be erased with penalty 0.008.*

---

[139] CALC-751772680 and CALC-310679005.

- *If the upright diagonal is 3-unconfined and it masks a flat quarter, then the upright diagonal can be erased with penalty 0.*
- *If a flat quarter is masked, then its diagonal has length at least 2.6. Also, if the diagonal of a masked flat quarter has length at most 2.7, then the height of its central vertex is at least 2.2.*

Proof. See Section 11.9.   □

25.5. Pent *and* Hex *Branching*

If a single linear program does not yield the bound $\sigma(D) < 8\,pt$, then we divide the set of decomposition stars with graph G into several subsets, according to the arrangements of quarters inside each standard cluster. This section gives a rough classification of possible arrangements of quarters in the cone over pentagonal and hexagonal standard regions.

The possibilities are listed in Figs. 25.1 and 25.2 only up to symmetry by the dihedral group action on the polygon. We do not prove the completeness of the list, but its completeness can be seen by inspection, in view of the comments that follow here and in Section 25.4. Details about the size of the penalties can be found in Section 13.6.

The conventions for generating the possibilities are different for the pentagons and hexagons than for the heptagons and octagons. We describe the pentagons and hexagons first. We erase all 3-unconfined upright diagonals. If there is one loop we leave the loop in the figure. If there are two loops (so that both necessarily have context $(n, k) = (4, 1)$), we erase one and keep the other.

The figures are interpreted as follows. An internal vertex in the polygon represents an upright diagonal. Edges from that vertex are in 1–1 correspondence with the anchors around that upright diagonal. Edges between nonadjacent vertices of the polygon represent the diagonals of flat quarters. We draw all edges from an upright diagonal to its anchors, and all edges of length $[2t_0, 2\sqrt{2}]$ that are not masked by upright quarters. Since the only remaining upright quarters belong to loops, the four simplices around a loop are anchored simplices and the edge opposite the diagonal has length at most 3.2.

Various inequalities in the inequality archive have been designed for subregions of pentagons. Additional inequalities have been designed for subregions in hexagonal regions. Thus, we are able to obtain greatly improved linear programming bounds when

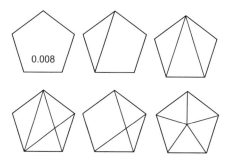

Fig. 25.1. Pentagonal face refinements.

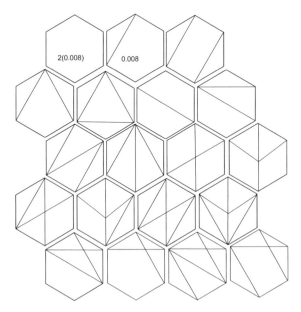

Fig. 25.2. Hexagonal face refinements. The only figures with a penalty are the first two on the top row and those on the bottom row. The first two on the top row have penalties 2(0.008) and 0.008. Those on the bottom row have penalties $3\xi_\Gamma$, $3\xi_\Gamma$, $\xi_\Gamma + 2\xi_V$, and $\xi_\Gamma + 2\xi_V$.

we break each pentagonal region into various cases, according to the list of Figs. 25.1 and 25.2.

25.6. Hept *and* Oct *Branching*

When the figure is a heptagon or octagon, we proceed differently. We erase all 3-unconfined upright diagonals and all loops (either context $(n, k) = (4, 1)$ or $(4, 2)$) and draw only the flat quarters. An undrawn diagonal of the polygon has length at least $2t_0$. Overall, in these cases much less internal structure is represented.

In the cases where 3-confined upright diagonals or loops have been erased, a number indicating a penalty accompanies the diagram (Figs. 25.3 and 25.4). These penalties are derived in Sections 13.6 and 13.4.

Define values

$$Z(3, 1) = 0.00005 \quad \text{and} \quad D(3, 1) = 0.06585.$$

Here are some special arguments that are used for heptagons and octagons.

25.6.1. One Flat Quarter.   Suppose that the standard region breaks into two subregions: the triangular region of a flat quarter Q and one other. Let $n = n(R) \in \{7, 8\}$. We have the inequality

$$\sigma_R(D) < (\hat{\sigma}(Q) - Z(3, 1)) + s_n + \xi_\Gamma + 2\xi_V.$$

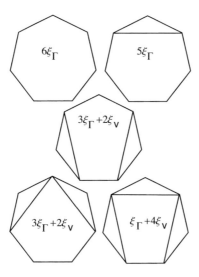

Fig. 25.3. Heptagonal face refinements.

The penalty term $\xi_\Gamma + 2\xi_V$ comes from a possible anchored simplex masking a flat quarter. Let v be the central vertex of the flat quarter Q. Let $\{v_1, v_2\}$ be its diagonal. Masked flat quarters satisfy restrictive edge constraints. It follows from Section 11.10 that we have one of the following three possibilities:

1. $y[v] \geq 2.2$.
2. $e[v_1, v_2] \geq 2.7$.
3. $\sigma_R(D) < (\hat{\sigma}(Q) - Z(3, 1)) + s_{n(R)}$.

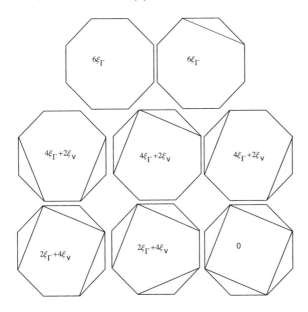

Fig. 25.4. Octagonal face refinements.

25.6.2. *Two Flat Quarters.* We proceed similarly if the standard region R breaks into three subregions: two regions R_1 and R_2 cut out by flat quarters Q_1, Q_2 and one other region made from what remains. Write $\hat{\sigma}_1$ for $\hat{\sigma}(Q_1)$, and so forth. It follows from Section 11.10 that we have one of the following three possibilities:

1. The height of a central vertex is at least 2.2.
2. The diagonal of a flat quarter is at least 2.7.
3.

$$\sigma_R(D) < (\hat{\sigma}_1 - Z(3, 1)) + (\hat{\sigma}_2 - Z(3, 1)) + s_{n(R)},$$
$$\tau_R(D) > (\hat{\tau}_1 - D(3, 1)) + (\hat{\tau}_2 - D(3, 1)) + t_{n(R)}.$$

With heptagons, it is helpful on occasion to use an upper bound on the penalty of $3\xi_\Gamma = 0.04683$. This bound holds if neither flat quarter is masked by a loop. For this, it suffices to show that the first two of the given three cases do not hold.

If there is a loop of context $(n, k) = (4, 2)$, we have the upper bounds of Lemma 25.1. If, on the other hand, there is no loop of context $(n, k) = (4, 2)$, then we have the upper bound

$$\sigma_R(D) \le (\hat{\sigma}(Q_1) - Z(3, 1)) + (\hat{\sigma}(Q_2) - Z(3, 1)) + s_{n(R)} + 2(\xi_\Gamma + 2\xi_V),$$

where $n(R) \in \{7, 8\}$.

25.7. Branching on Upright Diagonals

We divide the upright simplices into two domains depending on the height of the upright diagonal, using $|v| = 2.696$ as the dividing point. We break the upright diagonals (of unerased quarters in the Q-system) into cases:

1. The upright diagonal has height at most 2.696.

2. The upright diagonal $\{0, v\}$ has height at least 2.696, and some anchor w along the flat quarter satisfies $|w| \ge 2.45$ or $|v - w| \ge 2.45$. (There is a separate case here for each anchor w.)

3. The upright diagonal $\{0, v\}$ has height at least 2.696, and every anchor w along the flat quarter satisfies $|w| \le 2.45$ and $|v - w| \le 2.45$.

Many inequalities have been specially designed to hold on these smaller domains. They are included into the linear programming problems as appropriate.

When all the upright quarters can be erased, then the case for upright quarters follows from some other case without the upright quarters. An upright quarter can be erased in the following situations. If the upright quarter Q has compression type (in the sense of Definition 7.8) and the diagonal has height at least 2.696, then[140]

$$\sigma(Q) < \text{s-vor}_0(Q).$$

---

[140] CALC-214637273.

(If there are masked flat quarters, they become scored by $\hat{\sigma}$.) If an upright quarter has Voronoi type and the anchors w satisfy $|w| \leq 2.45$ and $|v - w| \leq 2.45$, then the quarter can be erased[141]

$$\sigma(Q) < \text{s-vor}_0(Q).$$

In general, we only have the weaker inequality[142]

$$\sigma(Q) < \text{s-vor}_0(Q) + 0.003521.$$

In a pentagon or hexagon, consider an upright diagonal with three upright quarters, that is, context $(n, k) = (4, 1)$. If the upright diagonal has height at most 2.696, and if an upright quarter shares both faces along the upright diagonal with other upright quarters, then we may assume that the upright quarter has compression type. For otherwise, there is a face of circumradius at least $\sqrt{2}$, and hence two upright quarters of Voronoi type. The inequality

$$\text{octavor} < \text{octavor}_0 - 0.008, \tag{25.1}$$

if $y_1 \in [2t_0, 2.696]$, and $\eta_{126} \geq \sqrt{2}$, shows that the upright quarters can be erased without penalty because

$$\xi_\Gamma - 0.008 - 0.008 < 0.$$

If erased, the case is treated as part of a different case.

This allows the inequalities[143] to be used that relate specifically to upright quarters of compression type. Furthermore, it can often be concluded that all three upright quarters have compression type. For this, we use various inequalities in the archive which can often be used to show that if the anchored simplex has a face of circumradius at least $\sqrt{2}$, then the linear programming bound on $\sigma(D)$ is less than $8\,pt$.

25.8. Branching on Flat Quarters

We make a few general remarks about flat quarters.

Remark 25.12. Information about the internal structure of an exceptional face gives improvements to the constants $1.4\,pt$ and $1.5\,pt$ of Property 4 in the definition of admissible weight assignments. (The bounds remain fixed at $1.4\,pt$ and $1.5\,pt$, but these arguments allow us to specify more precisely which simplices contribute to these bounds.) These constants contribute to the bound on $\tau(D)$ through the admissible weight assignment. Assume that at the vertex v there are four quasi-regular tetrahedra and an exceptional face, and that the exceptional face has a flat quarter with central vertex v. The calculations of Section 22.3 show that the union F of the four quasi-regular tetrahedra and exceptional region give $\tau_F(D) \geq 1.5\,pt$. If there is no flat quarter with central vertex v, then the union F of four quasi-regular tetrahedra along $\{0, v\}$ give $\tau_F(D) \geq 1.5\,pt$. We can make similar improvements when $\text{tri}(v) = 3$.

---

[141] CALC-378432183.
[142] CALC-310679005.
[143] See, for example, CALC-867513567-*.

Remark 25.13. There are a few other interval-based inequalities that are used in particular cases. The inequalities $y_1 \leq 2.2$, $y_4 \leq 2.7$, η_{234}, $\eta_{456} \leq \sqrt{2}$ imply that the flat quarter has compression type (see Section 7.1). The circumradius is not a linear programming variable, so its upper bound must be deduced from edge-length information.

✱ If all three corners of a flat quarter have height at most 2.14, and if the diagonal has length less than 2.77, then the circumradius of the face containing the origin and diagonal is at most $\eta(2.14, 2.14, 2.77) < \sqrt{2}$. This allows us to branch combine into three cases.

Lemma 25.14. *Let Q be a flat quarter whose corners v_i have height at most 2.14 and whose diagonal is at most 2.77. Then one of the following is true:*

1. *$\sigma(Q) = \Gamma(Q)$.*
2. *The diagonal has length ≤ 2.7, $\eta(y_4, y_5, y_6) \geq \sqrt{2}$, and $\sigma(Q) \leq$ s-vor$_0(Q)$.*
3. *The diagonal has length ≥ 2.7 and $\sigma(Q) \leq$ s-vor$_0(Q)$.*

Proof. Case 1 holds when Q is a quarter of compression type in the Q-system. If Q is in the Q-system but is not of compression type, then $\eta(y_4, y_5, y_6) \geq \sqrt{2}$ and $\sigma(Q) \leq$ s-vor$_0(Q)$. If Q is not in the Q-system, then s-vor$_0(Q)$ is an upper bound Lemma 11.26. If Q is not in the Q-system, then its diagonal has length at least 2.7, or the central vertex has height at most 2.2 (see Lemma 25.11.) In this case we use the upper bound s-vor$_0(Q)$.   □

Various inequalities in the archive have been designed specifically for each of these three cases. Thus, whenever the hypotheses of the lemma are met, we are able to improve on the linear programming bounds by breaking into these three cases.

25.9. Branching on Simplices that Are Not Quarters

Lemma 25.15. *Suppose that a triangular subregion comes from a simplex S with one vertex at the origin and three other vertices of height at most $2t_0$. Suppose that the edge lengths of the fourth, fifth, and sixth edges satisfy y_5, $y_6 \in [2t_0, 2\sqrt{2}]$, $y_4 \in [2, 2t_0]$. Suppose that $\min(y_5, y_6) \leq 2.77$. Then one of the following is true:*

1. *The edges have lengths y_5, $y_6 \in [2t_0, 2.77]$, $\eta_{456} \geq \sqrt{2}$, and $\sigma(S) \leq$ s-vor$_0(S)$.*
2. *y_5, $y_6 \in [2t_0, 2.77]$, and $\sigma(S) \leq$ s-vor(S) (the analytic Voronoi function).*
3. *An edge (say y_6) has length $y_6 \geq 2.77$ and $\sigma(S) \leq$ s-vor$_0(S)$.*

Proof. If we ignore the statements about σ, then the conditions in the lemma concerning edge length are exhaustive. The bounds on σ in each case are given by Section 9.6. □

There are linear programming inequalities that are tailored to each case.

25.10.   Branching on Quadrilateral Subregions

One of the inequalities holds for a quadrilateral subregion, if certain conditions are satisfied. One of the conditions is $y_4 \in [2\sqrt{2}, 3.0]$, where y_4 is a diagonal of the subregion. Since this diagonal is not one of the linear programming variables, these bounds cannot be verified directly from the linear program. Instead we use an inequality which relates the desired bound $y_4 \leq 3$ to the linear programming variables $\alpha[v, F]$, y_2, y_3, y_5, and y_6.

25.11.   Implementation Details for Branching

We now make a detailed examination of the internal structure of exceptional regions.
   A *refinement* $\tilde{F}$ of a face F of a plane graph G is a set $\tilde{F}$ of faces such that:

1. The intersection of the vertex set of G with that of $\tilde{F}$ is the set F.
2. $\tilde{F} \cup \{F^{\mathrm{op}}\}$ is a plane graph.

We use refinements of faces to describe the internal structure of faces.
   We introduce indexing sets FACE-$\tilde{F}$, VERTEX-$\tilde{F}$, ANGLE-$\tilde{F}$, EDGE-$\tilde{F}$, the sets of faces, vertices, angles, and edges in $\tilde{F}$, respectively, analogous to those introduced for G.
   We create variables $\pi[\tilde{F}]$, and indexed variables

$$\text{sol}\{\text{FACE-}\tilde{F}\}, \quad \text{sc}\{\text{FACE-}\tilde{F}\} \qquad \tau\,\text{sc}\{\text{FACE-}\tilde{F}\},$$
$$\alpha\{\text{ANGLE-}\tilde{F}\}, \quad y\{\text{VERTEX-}\tilde{F}\}, \quad e\{\text{EDGE-}\tilde{F}\}.$$

(Variables with names "$y[v]$" and "$e[v, w]$" were already created for some $v, w \in$ VERTEX-$\tilde{F} \cap$ VERTEX. In these cases we use the variables already created.)
   Each vertex v in the refinement will be interpreted either as a vertex $v^l \in U(D)$, or as the endpoint of an upright diagonal lying over the standard region F^l. We interpret the faces of the refinement in terms of the geometry of the decomposition star D variously as flat quarters, upright quarters, anchored simplices, and the other constructs of the fourth paper. This interpretation depends on the context, and is described in greater detail below.
   Once the interpretation of faces is fixed, the interpretations are as before for the variable names introduced already: y, e, α, sol. The lower and upper bounds for α and sol are as before. The lower and upper bounds for $y[v]$ are 2 and $2t_0$ if $v^l \in U(D)$, but if $(0, v^l)$ is an upright diagonal, then the bounds are $[2t_0, 2\sqrt{2}]$. The lower and upper bounds for e will depend on the context.

25.12.   Variables Related to Score

The variables sc are a stand-in for the score σ on a face. We do not call them σ because the sum of these variables will not in general equal the variable $\sigma[F]$, when $\tilde{F}$ is a refinement of F:

$$\left[\sum_{F' \in \tilde{F}} \text{sc}[F'] \neq \sigma[F] \right].$$

We use a weaker relation:

$$\sigma[F] \le \sum_{F' \in \tilde{F}} sc[F'] + \pi[\tilde{F}].$$

The variable $\pi[\tilde{F}]$ is called the penalty associated with the refinement $\tilde{F}$. (Penalties are discussed at length in Sections 13.6 and 13.4.) The interpretations of sc and $\pi[\tilde{F}]$ are rather involved, and are discussed on a case-by-case basis below. The interpretation of τ sc follows from the identity

$$\tau sc[F'] = sol[F']\zeta \, pt - sc[F'], \qquad \forall F' \in \tilde{F}.$$

The interpretation of variables that follows might appear to be hodge-podge at first. However, they are obtained in a systematic way. We analyze the proofs and approximations in Paper IV, and define $sc[F]^l$ as the best penalty-free scoring approximation that is consistent with the given face refinement. Here are the details.

If the subregion is a flat quarter, the interpretation of $sc[F]$ is the function $\hat{\sigma}$, defined in Section 11.10. If the subregion is an upright quarter Q, the interpretation of $sc[F]$ is the function $\sigma(Q)$ from Section 7. If the subregion is an anchored simplex that is not an upright quarter, $sc[F]$ is interpreted as the analytic Voronoi function vor if the simplex has type C or C', and as vor_0 otherwise. (The types A, B, C and C' are defined in Section 9.4.) Whether or not the simplex has type C, the inequality $sc[F] \le 0$ is satisfied. In fact, if vor_0 scoring is used, we note that there are no quoins, and $\varphi(1, t_0) < 0$.

If the subregion is triangular, if no vertex represents an upright diagonal, and if the subregion is not a quarter, then $sc[F]$ is interpreted as vor or vor_0 depending on whether the simplex has type A. In either case, the inequality $sc[F] \le vor_0$ is satisfied.

In most other cases the interpretation of $sc[F]$ is vor_0. However, if R is a heptagon or octagon, and F has four or more sides, then $sc[F]$ is interpreted as vor_0 except on simplices of type A, where it becomes the analytic Voronoi function.

If R is a pentagon or hexagon, and F is a quadrilateral that is not adjacent to a flat quarter, and if there are no penalties in the region, then the interpretation of $sc[F]$ is the actual score of the subregion over the subregion. In this case the score σ_R has a well-defined meaning for the quadrilateral, because it is not possible for an upright quarter in the Q-system to straddle the quadrilateral region and an adjacent region. Consequently, any erasing that is done can be associated with the subregion without ambiguity. By the results of Sections 8.4 and 8.5, we have $sc[F] \le 0$. We also have $sc[F] \le vor_0$.

One other bound that we have not explicitly mentioned is the bound $\sigma_R(D) < s_n$. For heptagons and octagons that are not aggregates, this is a better bound than the one used in the definition of tameness (Property 6). In heptagons and octagons that are not aggregates, if we have a subregion with four or more sides, then $sc[F] < Z(n, k)$ and $\tau sc[F] > D(n, k)$. (See (13.1) and (13.2) in Section 13.5.)

The variables are subject to a number of compatibility relations that are evident from the underlying definitions and geometry.

$$sol[F'] = \sum_{v \in F'} \alpha[v, F'] - (len[F'] - 2)\pi, \qquad \forall F',$$

$$\sum_{F': \, v \in F', F' \in \text{FACE} - \tilde{F}} \alpha[v, F'] = \alpha[v, F], \qquad \forall v.$$

Assume that a face $F_1 \in \tilde{F}$ has been interpreted as a subregion $R = F_1'$ of a standard region. Assume that each vertex of F_1 is interpreted as a vertex in $U(D)$ or as the endpoint of an upright diagonal over F'. One common interpretation of sc is $\mathrm{vor}_{0,F}(U(D))$, the truncated Voronoi function. When this is the interpretation, we introduce further variables:

$$\mathrm{quo}[v, s(v, F_1)], \qquad \forall v \in F_1,$$
$$\mathrm{quo}[s(v, F_1), v], \qquad \forall v \in F_1,$$
$$\mathrm{Adih}[v, F_1], \qquad \forall v \in F_1.$$

We interpret the variables as follows. If $w = s(v)$, and the triangle $(0, v', w')$ has circumradius η at most t_0, then

$$\mathrm{quo}[v, w]' = \mathrm{quo}(R(|v'|/2, \eta, t_0)),$$
$$\mathrm{quo}[w, v]' = \mathrm{quo}(R(|w'|/2, \eta, t_0)).$$

If the circumradius is greater than t_0, we take

$$\mathrm{quo}[v, w]' = \mathrm{quo}[w, v]' = 0.$$

The variable Adih has the following interpretation:

$$\mathrm{Adih}[v, F_1]' = \begin{cases} A(|v'|/2)\alpha(v', F_1'), & |v'| \le 2t_0, \\ 0, & \text{otherwise.} \end{cases}$$

Under these interpretations, the following identity is satisfied:

$$\mathrm{sc}[F_1] = \mathrm{sol}[F_1]\varphi_0 + \sum_{v \in F_1} \mathrm{Adih}[v, F_1] - 4\delta_{\mathrm{oct}} \sum \mathrm{quo}[v, w].$$

The final sum runs over all pairs (v, w), where $v = s(w, F_1)$ or $w = s(v, F_1)$.

For this to be useful, we need good inequalities governing the individual variables. Such inequalities for $\mathrm{Adih}[v, F]$ and $\mathrm{quo}[v, w]$ are found in Calculations CALC-815275408 and CALC-349475742. To make of use these inequalities, it is necessary to have lower and upper bounds on $\alpha[v, F]$ and $y[v]$. We obtain such bounds as LP-derived inequalities in the sense of Remark 23.4.

25.13. Appendix: Hexagonal Inequalities

There are a number of inequalities that have been particularly designed for standard regions that are hexagons. This appendix describes those inequalities. They are generally inequalities involving more than six variables, and because of current technological limitations on interval arithmetic, we were not able to prove these inequalities directly with interval arithmetic.

Instead we give various lemmas that deduce the inequalities from inequalities in a smaller number of variables (small enough to prove by interval arithmetic).

25.13.1. *Statement of Results.* There are a number of inequalities that hold in special situations when there is a hexagonal region. Although these inequalities do not appear in the main text of the proof of the Kepler conjecture, they are used in the linear programs.

After stating all of them, we turn to the proofs.

1. If there are no flat quarters and no upright quarters (so that there is a single subregion F), then

$$\text{vor}_0 < -0.212, \tag{25.2}$$

$$\tau_0 > 0.54525. \tag{25.3}$$

2. If there is one flat quarter and no upright quarters, there is a pentagonal subregion F. It satisfies

$$\text{vor}_0 < -0.221,$$

$$\tau_0 > 0.486.$$

3. If there are two flat quarters and no upright quarters, there is a quadrilateral subregion F. It satisfies

$$\text{vor}_0 < -0.168,$$

$$\tau_0 > 0.352.$$

These are twice the constants appearing in Inequality 11.

4. If there is an edge of length between $2t_0$ and $2\sqrt{2}$ running between two opposite corners of the hexagonal cluster, and if there are no flat or upright quarters on one side, leaving a quadrilateral region F, then F satisfies

$$\text{vor}_0 < -0.075,$$

$$\tau_0 > 0.176.$$

5. If the hexagonal cluster has an upright diagonal with context $(4, 2)$, and if there are no flat quarters (Fig. 25.5), then the hexagonal cluster R satisfies

$$\sigma_R < -0.297,$$

$$\tau_R > 0.504.$$

Fig. 25.5.   A hexagonal cluster with context $(4, 2)$.

Fig. 25.6. A hexagonal cluster with context (4, 2).

6. If the hexagonal cluster has an upright diagonal with context (4, 2), and if there is one unmasked flat quarter (Fig. 25.6), let $\{F\}$ be the set of four subregions around the upright diagonal. (That is, take all subregions except for the flat quarter.) In the following inequality and Inequality 7, let σ_R^+ be defined as σ_R on quarters, and vor_x on other anchored simplices. τ_R^+ is the adapted squander function:

$$\sum_{(4)} \sigma_R^+ < -0.253,$$

$$\sum_{(4)} \tau_R^+ > 0.4686.$$

7. If the hexagonal cluster has an upright diagonal with context (4, 2), and if there are two unmasked flat quarters (Fig. 25.7), let $\{F\}$ be the set of four subregions around the upright diagonal. (That is, take all subregions except for the flat quarters.)

$$\sum_{(4)} \sigma_R^+ < -0.2,$$

$$\sum_{(4)} \tau_R^+ > 0.3992.$$

8. If the hexagonal cluster has an upright diagonal in context (4, 1), and if there are no flat quarters, let $\{F\}$ be the set of four subregions around the upright diagonal. Assume that the edge opposite the upright diagonal on the anchored simplex has length at least $2\sqrt{2}$. (See Fig. 25.8.)

$$\text{vor}_{0,R}(D) + \sum_{(3)} \sigma(Q) < -0.2187,$$

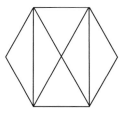

Fig. 25.7. A hexagonal cluster with context (4, 2).

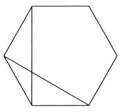

Fig. 25.8. A hexagonal cluster with context (4, 1).

$$\tau_{0,R}(D) + \sum_{(3)} \tau(Q) > 0.518.$$

9. In this same context, let F be the pentagonal subregion along the upright diagonal. It satisfies

$$\text{vor}_0 < -0.137, \tag{25.4}$$

$$\tau_0 > 0.31. \tag{25.5}$$

10. If the hexagonal cluster has an upright diagonal in context (4, 1), and if there is one unmasked flat quarter, let $\{F\}$ be the set of four subregions around the upright diagonal. Assume that the edge opposite the upright diagonal on the anchored simplex has length at least $2\sqrt{2}$. (There are five subregions, shown in Fig. 25.9.)

$$\text{vor}_{0,R}(D) + \sum_{(3)} \sigma(Q) < -0.1657,$$

$$\tau_{0,R}(D) + \sum_{(3)} \tau(Q) > 0.384.$$

11. In this same context, let F be the quadrilateral subregion in Fig. 25.9. It satisfies

$$\text{vor}_0 < -0.084,$$

$$\tau_0 > 0.176.$$

25.13.2. *Proof of the Inequalities*

Proposition 25.16. *Inequalities 1–11 are valid.*

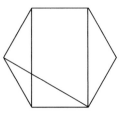

Fig. 25.9. A hexagonal cluster with context (4, 1).

We prove the inequalities in reverse order, 11–1. The bounds[144] $\text{vor}_0 < 0.009$ and $\tau_0 > 0.05925$ for flat quarters with diagonal $\sqrt{8}$ are used repeatedly. Some of the proofs make use of tcc-bounds, which are described in Section 12.9.

Proof of Inequalities 10 and 11.   Break the quadrilateral cluster into two simplices S and S' along the long edge of the anchored simplex S. The anchored simplex S satisfies $\tau(S) \geq 0$, $\sigma(S) \leq 0$. The other simplex satisfies $\tau_0(S') > 0.176$ and $\text{vor}_0(S') < -0.084$ by an interval calculation.[145] This gives Inequality 11. For Inequality 10, we combine these bounds with the linear programming bound on the four anchored simplices around the upright diagonal. From a series of inequalities[146] we find that they score < -0.0817 and squander > 0.208. Adding these to the bounds from Inequality 11, we obtain Inequality 10.   □

Proof of Inequalities 8 and 9.   The pentagon is a union of an anchored simplex and a quadrilateral region. LP-bounds similar to those in the previous paragraph and based on the inequalities of Section 13.12 show that the loop scores at most -0.0817 and squanders at least 0.208. If we show that the quadrilateral satisfies

$$\text{vor}_0 < -0.137, \tag{25.6}$$

$$\tau_0 > 0.31, \tag{25.7}$$

then Inequalities 8 and 9 follow. If by deformations a diagonal of the quadrilateral drops to $2\sqrt{2}$, then the result follows interval calculations.[147] By this we may now assume that the quadrilateral has the form

$$(a_1, 2, a_2, 2, a_3, 2, a_4, b_4), \qquad a_2, a_3 \in \{2, 2t_0\}.$$

If the diagonals drop under 3.2 and $\max(a_2, a_3) = 2t_0$, again the result follows from interval calculations.[148] If the diagonals drop under 3.2 and $a_2 = a_3 = 2$, then the result follows from further interval calculations.[149] So finally we attain by deformations $b_4 = 2\sqrt{2}$ with both diagonals greater than 3.2. However, this does not exist, because

$$\Delta(4, 4, 4, 3.2^2, 4, 8, 3.2^2) < 0. \qquad \square$$

Proof of Inequalities 5–7.   Inequality 7 is derived in Section 13.12. Inequalities 5 and 6 are LP-bounds based on interval calculations.[150]   □

Proof of Inequality 4.   Deform as in Section 12. If at any point a diagonal of the quadrilateral drops to $2\sqrt{2}$, then the result follows from interval calculations[151] and

---

[144] CALC-148776243.
[145] CALC-938091791.
[146] CALC-815492935, CALC-187932932, CALC-485049042, CALC-835344007.
[147] CALC-148776243, CALC-468742136.
[148] calc148776243, CALC-468742136.
[149] CALC-128523606.
[150] CALC-815492935, CALC-187932932, CALC-485049042.
[151] CALC-148776243.

Inequality 11:

$$\text{vor}_0 < 0.009 - 0.084 = -0.075,$$

$$\tau_0 > 0 + 0.176 = 0.176.$$

Continue deformations until the quadrilateral has the form

$$(a_1, 2, a_2, 2, a_3, 2, a_4, b_4), \qquad a_2, a_3 \in \{2, 2t_0\}.$$

There is necessarily a diagonal of length ≤ 3.2, because

$$\Delta(4, 4, 3.2^2, 8, 4, 3.2^2) < 0.$$

Suppose the diagonal between vertices v_2 and v_4 has length at most 3.2. If $a_2 = 2t_0$ or $a_3 = 2t_0$, the result follows from interval calculations[152] and Inequality 11. Take $a_2 = a_3 = 2$. Inequality 4 now follows from interval calculations.[153]   □

Proof of Inequality 3.   We prove that the quadrilateral satisfies

$$\text{vor}_0 < -0.168,$$

$$\tau_0 > 0.352.$$

There are two types of quadrilaterals: (a) two flat quarters whose central vertices are opposite corners of the hexagon, and (b) flat quarters who share a vertex. We consider case (a) first.

Case (a). We deform the quadrilateral as in Section 12.If at any point there is a diagonal of length at most 3.2, the result follows from Inequalities 10 and 11. Otherwise, the deformations give us a quadrilateral:

$$(a_1, 2, a_2, 2t_0, a_3, 2, a_4, 2), \qquad a_i \in \{2, 2t_0\}.$$

The tcc approximation now gives the result (see Section 12.10).

Case (b). Label the vertices of the quadrilateral $v_1, \ldots, v_4$, where (v_1, v_2) and (v_1, v_4) are the diagonals of the flat quarter. Again, we deform the quadrilateral. If at any point of the deformation, we find that $|v_1 - v_3| \leq 3.2$, the result follows from Inequalities 10 and 11. If during the deformation $|v_2 - v_4| \leq 2\sqrt{2}$, the result follows from interval calculations.[154] If the diagonal (v_2, v_4) has length at least 3.2 throughout the deformation, we eventually obtain a quadrilateral of the form

$$(a_1, 2t_0, a_2, 2, a_3, 2, a_4, 2t_0), \qquad a_i \in \{2, 2t_0\}.$$

However, this does not exist:

$$\Delta(4, 4, 3.2^2, (2t_0)^2, (2t_0)^2, 3.2^2) < 0.$$

We may assume that $|v_2 - v_4| \in [2\sqrt{2}, 3.2]$. The result now follows from interval calculations.[155]   □

---

[152] CALC-148776243.
[153] CALC-128523606.
[154] CALC-148776243, CALC-315678695.
[155] CALC-315678695.

Proof of Inequality 2. This case requires more effort. We show that

$$\text{vor}_0 < -0.221,$$

$$\tau_0 > 0.486.$$

Label the corners $(v_1, \ldots, v_5)$ cyclically with (v_1, v_5) the diagonal of the flat quarter in the hexagonal cluster. We use the deformation theory of Section 12. The proof appears in steps (1)–(6).

(1) If during the deformations, $|v_1 - v_4| \le 3.2$ or $|v_2 - v_5| \le 3.2$, the result follows from the Inequalities in Section 25.13.2 and 11. We may assume this does not occur.

(2) If an edge (v_1, v_3), (v_2, v_4), or (v_3, v_5) drops to $2\sqrt{2}$, continue with deformations that do not further decrease this diagonal. If $|v_1 - v_3| = |v_3 - v_5| = 2\sqrt{2}$, then the result follows from interval calculations.[156]

If we have $|v_1 - v_3| = 2\sqrt{2}$, deform the figure to the form

$$(a_1, 2, a_2, 2, a_3, 2, a_4, 2, a_5, 2t_0), \qquad a_2, a_4, a_5 \in \{2, 2t_0\}.$$

Once it is in this form, break the flat quarter $(0, v_1, v_2, v_3)$ from the cluster and deform v_3 until $a_3 \in \{2, 2t_0\}$. The result follows from an interval calculation.[157]

We handle a boundary case of the preceding calculation separately. After breaking the flat quarter off, we have the cluster

$$(a_1, 2\sqrt{2}, a_3, 2, a_4, 2, a_5, 2t_0), \qquad a_3, a_4, a_5 \in \{2, 2t_0\}.$$

If $|v_1 - v_4| = 3.2$, we break the quadrilateral cluster into two pieces along this diagonal and use interval calculations[158] to conclude the result. This completes the analysis of the case $|v_1 - v_3| = 2\sqrt{2}$.

(3) If $|v_2 - v_4| \le 3.2$, then deform until the cluster has the form

$$(a_1, 2, a_2, 2, a_3, 2, a_4, 2, a_5, 2t_0), \qquad a_1, a_3, a_5 \in \{2, 2t_0\}.$$

Then cut along the special simplex to produce a quadrilateral. Disregarding cases already treated by the interval calculations,[159] we can deform it to

$$(a_1, 2, a_2, 2\sqrt{2}, a_4, 2, a_5, 2t_0), \qquad a_i \in \{2, 2t_0\},$$

with diagonals at least 3.2. The result now follows from interval calculations.[160]

In summary of (1)–(3), we find that by disregarding cases already considered, we may deform the cluster into the form

$$(a_1, 2, a_2, 2, a_3, 2, a_4, 2, a_5, 2t_0), \qquad a_i \in \{2, 2t_0\},$$

$|v_1 - v_3| > 2\sqrt{2}$, $|v_3 - v_5| > 2\sqrt{2}$, $|v_2 - v_4| > 3.2$.

---

[156] CALC-148776243, CALC-673399623.

[157] CALC-297256991.

[158] CALC-861511432.

[159] CALC-861511432.

[160] CALC-746445726.

(4) Assume $|v_1 - v_3|, |v_3 - v_5| \le 3.2$. If $\max(a_1, a_3, a_5) = 2t_0$, we invoke interval calculations[161] to prove the inequalities. So we may assume $a_1 = a_3 = a_5 = 2$. The result now follows from interval calculations.[162] This completes the case $|v_1 - v_3|, |v_3 - v_5| \le 3.2$.

(5) Assume $|v_1 - v_3|, |v_3 - v_5| \ge 3.2$. We deform to

$$(a_1, 2, a_2, 2, a_3, 2, a_4, 2, a_5, 2t_0), \qquad a_i \in \{2, 2t_0\}.$$

If $a_2 = 2t_0$ and $a_1 = a_3 = 2$, then the simplex does not exist by Section 13.7. Similarly, $a_4 = 2t_0$, $a_5 = a_3 = 2$ does not exist. The tcc bound gives the result except when $a_2 = a_4 = 2$. The condition $|v_2 - v_4| \ge 3.2$ forces $a_3 = 2$. These remaining cases are treated with interval calculations.[163]

(6) Assume $|v_1 - v_3| \le 3.2$ and $|v_3 - v_5| \ge 3.2$. This case follows from deformations and interval calculations.[164] This completes the proof of Inequality 2.   □

Proof of Inequality 1.   Label the corners of the hexagon $v_1, \ldots, v_6$. The proof to this inequality is similar to the other cases. We deform the cluster by the method of Section 12 until it breaks into pieces that are small enough to be estimated by interval calculations. If a diagonal between opposite corners has length at most 3.2, then the hexagon breaks into two quadrilaterals and the result follows from Inequality 25.13.2.

If a flat quarter is formed during the course of deformation, then the result follows from Inequality 2 and interval calculations.[165] Deform until the hexagon has the form

$$(a_1, 2, a_2, 2, \ldots, a_6, 2), \qquad a_i \in \{2, 2t_0\}.$$

We may also assume that the hexagon is convex (see Section 12.12).

If there are no special simplices, we consider the tcc-bound. The tcc-bound implies Inequality 1, except when $a_i = 2$, for all i. However, if this occurs, the perimeter of the convex spherical polygon is $6 \operatorname{arc}(2, 2, 2) = 2\pi$. Thus, there is a pair of antipodal points on the hexagon. The hexagon degenerates to a lune with vertices at the antipodal points. This means that some of the angles of the hexagon are π. One of the tcc's has the form $C(2, 1.6, \pi)$, in the notation of Section 12.10. With this extra bit of information, the tcc bound implies Inequality 1.

If there is one special simplex, say $|v_5 - v_1| \in [2\sqrt{2}, 3.2]$, we remove it. The score of the special simplex is[166]

$$\begin{array}{lll} \mathrm{vor}_0 < 0, & \tau_0 > 0.05925, & \text{if} \quad \max(|v_1|, |v_5|) = 2t_0, \\ \mathrm{vor}_0 < 0.0461, & \tau_0 > 0, & \text{if} \quad |v_1| = |v_5| = 2, \end{array}$$

The resulting pentagon can be deformed. If by deformations, we obtain $|v_2 - v_5| = 3.2$ or $|v_1 - v_4| = 3.2$, the result follows from Inequalities 25.13.2 and two interval calculations.[167]

---

[161] CALC-148776243, CALC-297256991.
[162] CALC-897046482.
[163] CALC-928952883.
[164] CALC-297256991, CALC-673800906.
[165] CALC-148776243.
[166] CALC-148776243.
[167] CALC-725257062, CALC-977272202.

If $|v_5 - v_1| = 2\sqrt{2}$, we use Inequality 2 and interval calculations[168] unless $|v_1| = |v_5| = 2$. If $|v_1| = |v_5| = 2$, we use interval calculations.[169] If a second special simplex forms during the deformations, the result follows from interval calculations.[170]

The final case of Inequality 1 to consider is that of two special simplices. We divide this into two cases: (a) The central vertices of the specials are v_2 and v_6. (b) The central vertices are opposite v_1 and v_4. In case (a) the result follows by deformations and interval calculations.[171] In case (b) the result follows by deformations and interval calculations.[172] This completes the proof of Inequality 1 and the proof of the proposition. □

25.14. *Conclusion*

By combinations of branching along the lines set forth in the preceding sections, a sequence of linear programs is obtained that establishes that $\sigma(D)$ is less than $8\,pt$. For details of particular cases, the interested reader can consult the log files in [Ha16], which record which branches are followed for any given tame graph. (For most tame graphs, a single linear program suffices.)

This completes the proof of the Kepler conjecture.

References

[Ha6]   T. C. Hales, Sphere packings, I, *Discrete Comput. Geom.* **17** (1997), 1–51.
[Ha14] T. C. Hales, Some algorithms arising in the proof of the Kepler conjecture, in *Discrete and Computational Geometry*, Algorithms and Combinatorics, vol. 25, Springer-Verlag, Berlin, 2003, pp. 489–507.
[Ha15] T. C. Hales, A proof of the Kepler conjecture, *Ann. of Math.* **162** (2005), 1065–1185.
[Ha16] T. C. Hales, Computer Resources for the Kepler Conjecture, http://annals.math.princeton.edu/keplerconjecture/. (The source code, inequalities, and other computer data relating to the solution are also found at http://xxx.lanl.gov/abs/math/ 9811078v1.)

Received November 11, 1998, *and in revised form September* 12, 2003, *and July* 25, 2005.
Online publication February 27, 2006.

---

[168] CALC-148776243.
[169] CALC-377409251.
[170] CALC-586214007.
[171] CALC-89384104.
[172] CALC-859726639.

A Revision to the Proof of the Kepler Conjecture

Introduction to Part III

Part III presents a 2010 follow-up paper of T. C. Hales, J. Harrison, S. McLaughlin, T. Nipkow, S. Obua, and R. Zumkeller, "A Revision of the Proof of the Kepler Conjecture" (*Discrete Comput. Geom.*, **44** (2010), 1–34).

This paper has three objects. First, it describes an ongoing project to give a formal proof of the Kepler Conjecture. A formal proof of the Kepler Conjecture will provide a definitive version of the proof, checkable entirely by computer within a formal language. Second, the initial process of formalization of the original proof uncovered a gap in its detailed logic, and this paper supplies details to fill this gap. Third, this paper lists and corrects errata in the original Hales-Ferguson proof in *Discrete Comput. Geom.*

9

A Revision of the Proof of the Kepler Conjecture, by T. C. Hales, J. Harrison, S. McLaughlin, T. Nipkow, S. Obua, and R. Zumkeller

This paper, completed in 2009, reports on the progress towards obtaining a formal proof of the Kepler Conjecture. It also contains revisions to the Kepler Conjecture proof.

Contents

A Revision of the Proof of the Kepler Conjecture, by T. C. Hales, J. Harrison, S. McLaughlin, T. Nipkow, S. Obua, and R. Zumkeller (Discrete Comput. Geom., **44** (2010), 1–34).

The original version of this chapter was revised. An erratum to this chapter can be found at
http://dx.doi.org/10.1007/978-1-4614-1129-1_12

Discrete Comput Geom (2010) 44: 1–34
DOI 10.1007/s00454-009-9148-4

A Revision of the Proof of the Kepler Conjecture

**Thomas C. Hales · John Harrison ·
Sean McLaughlin · Tobias Nipkow · Steven Obua ·
Roland Zumkeller**

Received: 10 December 2008 / Revised: 31 January 2009 / Accepted: 4 February 2009 /
Published online: 17 March 2009
© The Author(s) 2009

Abstract The Kepler conjecture asserts that no packing of congruent balls in three-dimensional Euclidean space has density greater than that of the face-centered cubic packing. The original proof, announced in 1998 and published in 2006, is long and complex. The process of revision and review did not end with the publication of the proof. This article summarizes the current status of a long-term initiative to reorganize the original proof into a more transparent form and to provide a greater level of certification of the correctness of the computer code and other details of the proof. A final part of this article lists errata in the original proof of the Kepler conjecture.

Keywords Formal proof · Sphere packings · Linear programming · Interval analysis · Higher order logic · Hypermap

Research supported by NSF grant 0804189.

T.C. Hales (✉)
Math Department, University of Pittsburgh, Pittsburgh, PA, USA
e-mail: hales@pitt.edu

J. Harrison
Intel Corporation, JF1-13, 2111 NE 25th Avenue, Hillsboro, OR 97124, USA
e-mail: johnh@ichips.intel.com

S. McLaughlin
Carnegie Mellon University, Pittsburgh, PA, USA
e-mail: seanmcl@gmail.com

T. Nipkow · S. Obua
Department for Informatics, Technische Universität München, Munich, Germany
url: www.in.tum.de/~nipkow

R. Zumkeller
École Polytechnique, Paris, France

Introduction

In 2006, *Discrete and Computational Geometry* devoted an issue to the proof of the Kepler conjecture on sphere packings, which asserts that no packing of congruent balls in three-dimensional Euclidean space can have density greater than that of the face-centered cubic packing [6, 14–16, 23].

The proof is long and complex. The editors' forward to that issue remarks that "the reviewing of these papers was a particularly enormous and daunting task." "The main portion of the reviewing took place in a seminar run at Eötvos University over a 3 year period. Some computer experiments were done in a detailed check. The nature of this proof, consisting in part of a large number of inequalities having little internal structure, and a complicated proof tree, makes it hard for humans to check every step reliably. Detailed checking of specific assertions found them to be essentially correct in every case tested. The reviewing process produced in the reviewers a strong degree of conviction of the essential correctness of the proof approach, and that the reduction method led to nonlinear programming problems of tractable size."

The process of review and revision did not end when the proof was published. This article summarizes the current status of a long-term initiative to reorganize the original proof into a more transparent form and to provide a greater level of certification of the correctness of the computer code and other details of the proof.

The article contains two parts. The first part describes an initiative to give a formal proof of the Kepler conjecture. The second part gives errata in the original proof of the Kepler conjecture. Most of these errata are minor. However, a significant new argument appears in a separate section (Sect. 8). It finishes an incomplete argument in the original proof asserting that there is no loss in generality in assuming (for purposes of the main estimate) that subregions are simple polygons. The incomplete argument was detected during the preparation of the blueprint edition of the proof, which is described in Sect. 2.

In this article, the *original proof* refers to the proof published in [23]. S. Ferguson and T. Hales take full responsibility for every possible error in the original proof of the Kepler conjecture. Over the past decade, many have contributed significantly to making that proof more reliable.

Part 1. Formal Proof Initiative

1 The Flyspeck Project

The purpose of a long-term project, called the Flyspeck project, is to give a formal proof of the Kepler conjecture. This section makes some preliminary remarks about formal proofs and gives a general overview of the current status of this project.

1.1 Formal Proof

A formal proof is a proof in which every logical inference has been checked, all the way back to the foundational axioms of mathematics. No step is skipped no matter how obvious it may be to a mathematician. A formal proof may be less intuitive, and yet is less susceptible to logical errors. Because of the large number of inferences involved, a computer is used to check the steps of a formal proof.

It is a large labor-intensive undertaking to transform a traditional proof into a formal proof. The first stage is to expand the traditional proof in greater detail. This stage fills in steps that a mathematician would regard as obvious, works out arguments that the original proof leaves to the reader, and supplies the assumed background knowledge. In a final stage, the detailed text is transcribed into a computer-readable format inside a computer proof assistant. The proof assistant contains mathematical axioms, logical rules of inference, and a collection of previously proved theorems. It validates each new lemma by stepping through each inference. No other currently available technology is able to provide levels of certification of a complex mathematical proof that is remotely comparable to that available by formal computer verification. A general overview of formal proofs can be found at [8, 22, 28].

Proof assistants differ in detail in the way they treat the formalization of a theorem that is itself a computer verification (such as the proof of the four-color theorem or the proof of the Kepler conjecture). In general, a formal proof of a computer verification can be viewed as a formal proof of the correctness of the computer code used in the verification. That is, the formal proof certifies that the code is a bug-free implementation of its specification.

1.2 Formal Proof of the Kepler Conjecture

As mentioned above, the purpose of the Flyspeck project is to give a formal proof of the Kepler conjecture. (The project name *Flyspeck* comes from the acronym *FPK*, for the *Formal Proof* of the *Kepler* conjecture.) This is the most complex formal proof ever undertaken. We estimate that it may take about twenty work-years to complete this formalization project.

The Flyspeck project is introduced in the article [13]. The project page gives the latest developments [18]. The project is now at an advanced stage; in fact, we estimate that the project is now about half-way complete. One of the main purposes of this article is to present a summary of the current status of this project.

In the original proof of the Kepler conjecture, there was a long mathematical text and three major pieces of computer code. The written part of the proof has been substantially revised with aims of the Flyspeck project in mind. Section 2 compares this revised text with the original. There is now a good match between the mathematical background assumed in the text and the mathematical material that is available in the proof assistant HOL Light. Section 3 describes the current level of support in HOL Light for the formalization of Euclidean space and measure theory. In the years following the publication of the original proof, S. McLaughlin has reworked and largely rewritten the entire body of code in a form that is more transparent and more amenable to formalization. Section 4 points out some difficulties in verifying the computer code in its original form and documents the reimplementation.

There have been three Ph.D. theses written on the Flyspeck project, one devoted to each of the three major pieces of computer code. The first piece of computer code uses interval arithmetic to verify nonlinear inequalities. R. Zumkeller's thesis [49] develops nonlinear inequality proving inside the proof assistant Coq. Section 5 gives an example of this work. The second piece of computer code enumerates all tame graphs. (The definition of tameness is rather intricate; its key property is that the set of tame graphs includes all graphs that give a potential counterexample to the conjecture.) G. Bauer's thesis, together with subsequent work with T. Nipkow, completes the formal proof of the enumeration of tame graphs [38]. Section 6 gives a summary of this formalization project. The third piece of computer code generates and runs some 10^5 linear programs. These linear programs show that none of the potential counterexamples to the Kepler conjecture are actual counterexamples. S. Obua's thesis develops the technology to generate and verify the linear programs inside the proof assistant Isabelle [39]. Section 7 describes this research.

The ultimate aim is to develop a complete formal proof of the Kepler conjecture within a single proof assistant. Because of the scope of the problem and the number of researchers involved, different proof assistants have been used for different parts of the proof: HOL Light for background in Euclidean geometry and the text, Coq for nonlinear inequality verification, and Isabelle/HOL for graph enumeration and linear programming. This raises the issue of how to translate a formal proof automatically from one proof assistant to another. Implementations of automated translation among the proof assistants HOL, Isabelle, and Coq can be found at [4, 33, 40, 47].

2 Blueprint Edition of the Kepler Conjecture

The *blueprint edition* of the proof of the Kepler conjecture is a second-generation proof that contains far more explicit detail than the original proof. The blueprint edition is available at [20, 21]. Many proofs have been significantly simplified and systematized. It has been written in a manner to permit easy formalization. As its name might suggest, this version is intended as a blueprint for the construction of a formal proof. This section compares the blueprint edition with the original.

2.1 Lemmas in Elementary Geometry

A collection of about 200 lemmas that can be expressed in elementary terms has been extracted from the original proof and placed in a separate collection [21]. This has several advantages. First of all, these lemmas, although elementary, are precisely the parts of the original proof that put the greatest burden on the reader's geometrical intuition. (Many of these lemmas deal with the existence or nonexistence of configurations of several points in $\mathbb{R}^3$ subject to various metric constraints.) Also, the lemmas can be stated without reference to the Kepler conjecture and all of the machinery that has been introduced to give a proof. Finally, the proofs of these lemmas rely on similar methods and are best considered together [17]. Section 3 on *Enhanced Automation* gives an approach to proving the lemmas in this collection.

These lemmas can be expressed in the first-order language of the real numbers; that is, they can be expressed in the syntax of first-order logic with equality (allowing quantifiers $\forall x$ and $\exists x$ with variables running over the real numbers), the real constants 0, 1, and ring operations $(+)$, $(-)$, $(\cdot)$ on $\mathbb{R}$. In fact, L. Fejes Tóth's statement of the Kepler conjecture as an optimization problem in a finite number of variables can itself be expressed in the first-order language of the real numbers [5]. (For this, the truncation used by L. Fejes Tóth must be modified slightly so that the truncated Voronoi cells are polyhedra.) Thus, it should come as no surprise that many of the intermediate lemmas in the proof can also be expressed in this manner.

For example, consider the statement asserting the existence of a circumcenter of a triangle: if three points in the plane are not collinear, then there exists a point in the plane that is equidistant from all three. This can be expressed in elementary terms as follows: for every (x_1, y_1), (x_2, y_2), (x_3, y_3), if there do not exist t_1, t_2, and t_3 for which $t_1(x_1, y_1) + t_2(x_2, y_2) + t_3(x_3, y_3) = (0, 0)$ and $t_1 + t_2 + t_3 = 1$, then there exists (x, y) such that

$$(x - x_1)^2 + (y - y_1)^2 = (x - x_2)^2 + (y - y_2)^2 = (x - x_3)^2 + (y - y_3)^2.$$

2.2 Background Material

There is now a close match between what the blueprint edition assumes as background and what the proof assistant *HOL Light* provides, as described in Sect. 3. The blueprint edition develops substantial background material in trigonometry, measure and integration, hypermaps, and fans (a geometric realization of a hypermap). It turns out that only a small number of *primitive* volumes need to be computed for the proof of the Kepler conjecture. These primitives include the volume of a ball, a tetrahedron, and right circular cone. No line or surface integrals are required.

A hypermap consists of three permutations e, n, f (on a finite set D) that compose to the identity $e \circ n \circ f = I$. A hypermap is the combinatorial structure used by Gonthier in his formal proof of the four-color theorem. In 2005, the proof of the Kepler conjecture was rewritten in terms of hypermaps, because it is better suited for formal proofs than planar graphs.

The Jordan curve theorem (JCT) has been formalized as a step in the Flyspeck project [19]. In fact, something weaker than the JCT suffices. The project only uses the JCT for curves on the surface of a unit sphere consisting of a finite number of arcs of great circles; that is, a spherical polygonal version of the JCT. In the proof of the Kepler conjecture, the combinatorial structure of a cluster of spheres is encoded as a hypermap. An Euler characteristic calculation, based on the JCT, shows that these hypermaps are planar. The planarity of these hypermaps is a crucial property that is used in the enumeration of tame graphs (Sect. 6). The background chapter in the blueprint edition contains detailed proofs of these facts.

The blueprint edition contains several introductory essays that introduce the main concepts in the proof, including the algorithms implemented by the computer code.

The formalization of the blueprint text started in 2008 with work of J. Rute (CMU) and the Hanoi Flyspeck group. The current members of this group are Trần Nam Trung, Nguyễn Tất Thắng, Hoàng Lê Trường, Nguyễn Quang Trường, Vũ Khắc Kỷ, Nguyễn Anh Tâm, Nguyễn Tuyên Hoàng, Nguyễn Đức Phương, Vương Anh Quyền, Phan Hoàng Chơn, and managed by Tạ Thị Hoài An. The blueprint formalization is still at an early stage.

3 Formalizing the Ordinary Mathematics

This section describes some of the work on formalizing Euclidean space and measure theory, and the development of further proof automation, which should be useful in this endeavor.

The computer code in Flyspeck has so far received the lion's share of the formal effort. This is entirely reasonable since there are, or at least were, real questions about the feasibility of reproducing these results in a formal way. However, the Flyspeck proof includes a large amount of "ordinary" mathematics, which also needs to be formalized. Here we are on fairly safe ground in principle, because by now we understand the formalization of mainstream mathematics quite well [48]. It is safe to predict that this formalization can be done, and we can even hope for a reasonably accurate estimate of the effort involved. Nevertheless, the formalization is certainly nontrivial and will require considerable work.

3.1 Formalizing Euclidean Space

Much of our work has been devoted to developing a solid general theory of Euclidean space $\mathbb{R}^N$ [26]. For Flyspeck, we invariably just need the special case $\mathbb{R}^3$. While some concepts, e.g., vector cross products, are specific to $\mathbb{R}^3$, most of the theory has been developed for general $\mathbb{R}^N$ so as to be more widely applicable. The theorem prover HOL Light [25] is based on a logic without dependent types, but we can still encode the index N as a type (roughly, an arbitrary indexing type of size N). This means that theorems about specific sizes like 3 really are just type instantiations of theorems for general N stated with polymorphic type variables. The theory contains the following:

- Basic properties of vectors in $\mathbb{R}^N$, linear operators and matrices, dimensions of vector subspaces, and other bits of linear algebra. For example, the following is a formal statement of the theorem that a square matrix A' is a left inverse to another one A iff it is a right inverse. Note that the double use of the same type variable N constrains the theorem to square matrices:

```
|- ∀A:real^N^N A':real^N^N.
        (A ** A' = mat 1) ⇔ (A' ** A = mat 1)
```

- Metric and topological notions like distances, open sets, closure, compactness, and paths. Some of these are very general, others are more specific to Euclidean space. Some results include the Heine–Borel theorem, the Banach fixed-point theorem, and Brouwer's fixed-point theorem. The following is a formal statement that continuous functions preserve connectedness.

```
|- ∀f:real^M->real^N s. f continuous_on s ∧  connected s
        ⇒ connected(IMAGE f s)
```

- Properties of convex sets, convex hulls, cones, etc. Results include Helly's theorem, Carathéodory's theorem, and various classic results about separating and supporting hyperplanes. The following states a simpler but not entirely trivial result that convex hulls preserve compactness.

```
|- ∀s:real^N->bool. compact s ⇒ compact(convex hull s)
```

- Sequences and series of vectors and uniform convergence, Fréchet derivatives and their properties, up to various forms of the inverse function theorem, as well as specific one-dimensional theorems like Rolle's theorem and the Mean Value Theorem. Here is the formal statement of the chain rule for Fréchet derivatives.

```
|- ∀f:real^M->real^N g:real^N->real^P f' g'.
        (f has_derivative f') (at x) ∧
        (g has_derivative g') (at (f x))
        ⇒ ((g o f) has_derivative (g' o f')) (at x)
```

3.2 Formalizing Measure Theory

Although the basic Euclidean theory is an important foundation, and many of the concepts like "convex hull" are used extensively in the Flyspeck mathematics, perhaps the most important thing to formalize is the concept of volume.

We define integrals of general vector-valued functions over subsets of $\mathbb{R}^N$, using the Kurzweil–Henstock gauge integral definition. We develop all the usual properties such as additivity and the key monotone and dominated convergence theorems. We also develop a theory of *absolutely* integrable functions, where both f and $|f|$ are gauge integrable; this is known to coincide with the Lebesgue integral. Here is a formal statement of the simple theorem that integration preserves linear scaling:

```
|- ∀f:real^M->real^N y s h:real^N->real^P.
     (f has_integral y) s ∧  linear h
       ⇒ ((h o f) has_integral h(y)) s
```

Using this integral applied to characteristic functions, we develop a theory of (Lebesgue) measure, which of course gives volume in the three-dimensional case. The specific notion "measure zero" is formalized as negligible, and we also have a general notion of a set having a finite measure, and a function measure to return that measure when it exists. For example, this is the basic additivity theorem:

```
|- ∀s t. measurable s ∧  measurable t ∧  DISJOINT s t
           ⇒ measure(s UNION t) = measure s + measure t
```

We have proved that various "well-behaved" sets such as bounded convex ones and compact ones, or more generally those with negligible frontier (boundary) are measurable, e.g.,

```
|- ∀s:real^N->bool.
            bounded s ∧  negligible(frontier s) ⇒ measurable s
```

The main lack at the moment is a set of results for actually computing the measures of specific sets, as needed for Flyspeck. We can evaluate most basic one-dimensional integrals by appealing to the Fundamental Theorem of Calculus, but we need to enhance the theory of integration with stronger Fubini-type results so that we can evaluate multiple integrals by iterated one-dimensional integrals. This work is in progress at the time of writing.

3.3 Enhanced Automation

Using coordinates, many nontrivial geometric statements in $\mathbb{R}^3$, or other Euclidean spaces of specific finite dimension, can be reduced purely to the elementary theory of reals. This is known to be decidable using quantifier elimination [3, 29, 45]. However, in practice, this is often problematic because quantifier elimination for nonlinear formulas is inefficient. The problem is particularly severe if we want to have any kind of *formal* proof, as we do in Flyspeck, since producing such a proof induces further slowdowns [31, 34]. With this in mind, we have explored a different approach to the case of purely universally quantified formulas [27], based on ideas of Parrilo [41]. This involves reducing the initial problem to semidefinite programming, solving the SDP problem using an external tool and reconstructing a "sum-of-squares" (SOS) certificate that can easily be formally checked.

For example, suppose we wish to verify that if a quadratic equation $ax^2 + bx + c = 0$ has a real root, then $b^2 \geq 4ac$. Using the SDP solver, we find an algebraic certificate $b^2 - 4ac = (2ax + b)^2 - 4a(ax^2 + bx + c)$, from which the required fact follows easily: $(2ax + b)^2 \geq 0$ because it is a square, and $4a(ax^2 + bx + c) = 0$ because x is a root, and so we deduce $b^2 - 4ac \geq 0$. This method seems very useful for automating routine nonlinear reasoning in a way that is easy and quick to formally verify, so that we do not have to rely on the correctness of a complicated program. It is even capable of solving the coordinate forms of some of the simpler Flyspeck inequalities directly, though it seems unlikely to be competitive with customized nonlinear optimization methods as described in Sect. 5. For example, one simple Flyspeck inequality is the following, which after being reduced to a real problem with nine variables (three coordinates for each point) is solved by SOS in a second:

$$\|u - v\| \geq 2 \wedge \|u - w\| \geq 2 \wedge \|v - w\| \geq 2 \wedge \|u - v\| < \sqrt{8}$$
$$\Rightarrow \quad \|w - (u + v)/2\| > \|u - v\|/2.$$

A quite different approach to geometric theorem proving is to work in the setting of a general real vector space or normed real vector space. In this case, other decision methods are available [44]. In particular, one of these decision procedures that we have implemented in HOL Light can sometimes handle simple forms of spatial reasoning in a purely "linear" way, and so be much more efficient than the direct reduction to coordinates, even if we do in fact have a specific dimension in mind. One real example from formalizing complex analysis is the following in $\mathbb{R}^2$:

$$|\|w - z\| - r| = d \wedge \|u - w\| < d/2 \wedge \|x - z\| = r \quad \Rightarrow \quad d/2 \leq \|x - u\|.$$

4 Standard ML Reimplementation of Code

This section describes a reimplementation of the computer code used in the proof of the Kepler conjecture. The code has been substantially redesigned to avoid various difficulties with the original implementation.

4.1 Code

The original proof of the Kepler conjecture relies significantly on computation. Computer code is used extensively and is central both to the correctness of the result and to a thorough understanding of the proof.

There are four major difficulties with understanding and verifying the original code base. The first and most glaring difficulty is simply the amount of code. At the website [11] that posts the code for the original proof, there are well over 50,000 lines of programs in Java, C++, and Mathematica (among others). This represents only the calculations that Hales did himself. Samuel Ferguson also completed many of the calculations with an entirely different code base[1] of 137,000 lines of C. By contrast, the proof of the four-color theorem by Robinson et al. [43] is less than 3,000 lines of C.

The second difficulty is in the organization of the code. The calculations were done over the space of four years and involved thousands of executions of a multitude of independent programs. Section 1.2 identifies three main computational tasks: tame graph generation, linear program bounding, and nonlinear inequality verification. Each of these main tasks consists of several subtasks. For example, verifying the inequalities required dozens of relatively complicated preprocessing phases where second derivatives of the relevant functions were bounded over fixed domains. As another example, many linear programs were solved only after a branch and bound period which were recorded in voluminous log files. In an attempt to organize the complex web of calculations, Hales devised a labeling scheme to uniquely identify the calculations. However, even now some computations relied upon by the proof are difficult to find in the original source code. To locate, for instance, computation "CALC-821707685" from the original proof [23, p. 159], one can search on the website [11] in vain. While records of the computations were made, it is not always an easy matter to find them without guidance from the authors.

The third difficulty lies in the complexity of the implementation. For instance, the software developed to prove the inequalities upon which the original proof rests is relatively complicated. Processing power at the time (1994–1998) was just barely capable of completing the computations requested. To keep the length of execution to days or weeks instead of months or years, the code is extensively optimized. The optimizations were often implemented without comment in the source and in some cases are difficult to understand.

The fourth difficulty is that the original code uses C and C++ to carry out *interval arithmetic* calculations based on *floating point arithmetic*. In the process, it explicitly sets the IEEE 754 [30] rounding modes on the processor's floating point unit. While floating point is desirable for its speed, there are difficulties with using floating point for software that requires a very high level of rigor such as that supporting mathematical proof. The first is that reasoning about floating point instructions requires a relatively deep understanding of the machine architecture [36]. For instance, setting the rounding mode changes the state of the processor itself. Such an instruction has a

---

[1]There is a large amount of code copying in Ferguson's code, resulting in a much larger code base. The number of distinct lines is difficult to measure.

global effect on all subsequent floating point computations. In the original code base, the rounding modes are explicitly changed at least 400 times. Moreover, compilers, libraries, and even processors are notorious for unsound implementations of the 754 standard.

4.2 Reimplementation

In 2004, we decided to reimplement the original code base. We decided that the new implementation should not require floating point numbers and rounding modes. Though speed was important, we wanted the code to be independent of any particular interval arithmetic implementation. This meant we could use a fast floating point implementation of interval arithmetic for our daily work, but could use a slower but more trustworthy implementation such as MPFI [42] to double check important computations. We also wanted to organize the new implementation such that any of the many computations upon which the proof relies could be evaluated from a single interface. This would allow Flyspeck developers to find and easily check the text of the proof during the formalization process. Finally, we wished to bring the computational aspects of the Kepler conjecture closer to the level of simplicity and clarity necessary for formalization by a proof assistant. We began this work in the spirit of Robinson et al. [43], which simplified the original code of Appel and Haken [1] and was used by Gonthier to construct the fully formal proof [8] in the Coq proof assistant.

We chose Standard ML for the reimplementation for a number of reasons. It has a formal definition [35], and thus programs have a meaning apart from the particular compiler used. It has an efficient compiler, named MLton [46]. (Our reimplementation runs between 50% and 200% the speed of the original implementation compiled with GCC.) MLton has the ability to use external libraries written in languages other than SML with relative ease. This allowed us, from one programming environment, to control multiple linear programming solvers, interval arithmetic implementations, and nonlinear optimization packages. SML has an expressive module system, and thus it was simple to write our code with respect to an abstract type of interval arithmetic. Thus we could use multiple independent implementations with ease. As of 2008, most of the code has been completely rewritten in SML and is executable from a single command-line program. The code is freely available at the project website [32].

In the original implementation, the myriad computations were done with many different programs written in a half dozen programming languages. The results of these computations are not always easy to find or interpret. Now all the computations are executed from the same source, with organized output. In addition to giving us added confidence that the original computations were sound, we have a fairly complete suite of software support for the Flyspeck project. We are now in the process of organizing and reevaluating the thousands of computations upon which the proof depends.

5 Proving Nonlinear Inequalities with Bernstein Bases

The hardest computational part of the original proof of the Kepler conjecture is the verification of a list of about a thousand nonlinear inequalities. This section presents a technique aimed at proving them, based on polynomial approximation and Bernstein bases. We feel that this approach better fits the requirements of formal proof, as outlined in Sect. 4.2. We hope to refine the method to cover all Flyspeck inequalities.

We exhibit the method on a single inequality CALC-586468779. The original proof contains the following definitions [10]:

$$
\mathrm{pt} := -\frac{\pi}{3} + 4 \arctan \frac{\sqrt{2}}{5},
$$

$$
\delta_{\mathrm{oct}} := \frac{\pi - 4 \arctan \frac{\sqrt{2}}{5}}{2\sqrt{2}},
$$

$$
\Delta(y) := \frac{1}{2}
\begin{vmatrix}
0 & 1 & 1 & 1 & 1 \\
1 & 0 & y_3^2 & y_2^2 & y_1^2 \\
1 & y_3^2 & 0 & y_4^2 & y_5^2 \\
1 & y_2^2 & y_4^2 & 0 & y_6^2 \\
1 & y_1^2 & y_5^2 & y_6^2 & 0
\end{vmatrix},
$$

$$
a_0(y) := y_1 y_2 y_3 + \frac{1}{2}\left(y_1^2 y_2 + y_1 y_2^2 + y_1^2 y_3 + y_2^2 y_3 + y_1 y_3^2 \right.
$$

$$
\left. + y_2 y_3^2 - y_1 y_4^2 - y_2 y_5^2 - y_3 y_6^2 \right),
$$

$$
a_1(y) := a_0(y_1, y_5, y_6, y_4, y_2, y_3),
$$

$$
a_2(y) := a_0(y_2, y_4, y_6, y_5, y_1, y_3),
$$

$$
a_3(y) := a_0(y_4, y_5, y_3, y_1, y_2, y_6),
$$

$$
\gamma(y) := -\frac{\delta_{\mathrm{oct}}}{6}\sqrt{\Delta(y)} + \frac{2}{3}\sum_{i=0}^{3} \arctan \frac{\sqrt{\Delta(y)}}{a_i(y)}.
$$

The statement of the inequality is:

$$
\forall y \in [2, 2.51]^6 \quad \gamma(y) \le \mathrm{pt}. \tag{1}
$$

Define the difference of two intervals by $[a_1, b_1] - [a_2, b_2] = [a_1 - b_2, b_1 - a_2]$. Interval arithmetic is used to prove the inequalities in the original proof. It suffers from the *dependency problem*: the minimum and maximum of the formula $x - x$ are overestimated because $[a, b] - [a, b] = [a - b, b - a]$, although $x - x$ is clearly 0. Subdividing $[a, b]$ into $[a, \frac{a+b}{2}]$ and $[\frac{a+b}{2}, b]$, and then re-evaluating the formula yields an improved result. However, depending on the problem, the number of required subdivisions can be very large. This is why checking some inequalities takes a very long time.

5.1 From Geometrical Functions to Polynomials

Fortunately, better methods than interval arithmetic are available if the function under consideration is polynomial. A quick look at γ tells us that (1) is not polynomial, since it has occurrences of $\sqrt{\cdot}$, $1/\cdot$, and arctan. Can it nevertheless be reduced to a polynomial problem? Two strategies come to mind:

First, algebraic laws such as $\sqrt{a} \leq b \Leftrightarrow a \leq b^2$ (if $b \geq 0$) and $\frac{a}{b} \leq c \Leftrightarrow a \leq bc$ (if $b > 0$) can often be used to eliminate occurrences of $\sqrt{\cdot}$ and $1/\cdot$. The list of trigonometric identities is endless. For our example, Vega's rule $\arctan a + \arctan b = \arctan \frac{a+b}{1-ab}$ seems useful. Unfortunately, this technique quite often yields huge expressions that are difficult to deal with by virtue of their sheer size. Also, an algebraic transformation to a polynomial problem may simply be impossible (we suspect that this is the case for (1)).

A second technique is based on replacing γ with a polynomial g that dominates it but is still smaller than pt. Clearly, if there exists a g such that

$$\forall y \in [2, 2.51]^6 \quad \gamma(y) \leq g(y) \tag{2}$$

and

$$\forall y \in [2, 2.51]^6 \quad g(y) \leq \text{pt}, \tag{3}$$

then by transitivity (1) holds.

Such a polynomial g can be obtained by replacing $\sqrt{\cdot}$, $1/\cdot$, and arctan with polynomial approximations. We only need to ensure that we use upper approximations for positive occurrences and lower approximations for negative ones. Only occurrences whose arguments contain variables need to be replaced, since, e.g., $\sqrt{2}$ is a (constant) polynomial itself.

In the definition of γ, the function arctan occurs positively, so it is replaced by an upper approximation $\overline{\arctan}$. The term $\frac{\sqrt{\Delta(y)}}{a_i(y)}$ is first unfolded to $\sqrt{\Delta(y)} \cdot \frac{1}{a_i(y)}$. Both the square root and reciprocal occur positively again here, so they can be replaced by upper approximations $\overline{\text{sqrt}}$ and $\overline{\text{rcp}}$, respectively. This yields $\overline{\arctan}(\overline{\text{sqrt}}(\Delta(y)) \cdot \overline{\text{rcp}}(a_i(y)))$ in all four summands. There remains only $\sqrt{\cdot}$ occurring negatively after $-\frac{\delta_{\text{oct}}}{6}$, which is to be replaced by a lower approximation $\underline{\text{sqrt}}$.

We choose the following approximations:

$$\overline{\arctan}(t) := \arctan \frac{\sqrt{2}}{5} + \frac{25}{27}\left(t - \frac{\sqrt{2}}{5}\right),$$

$$\overline{\text{rcp}}(t) := \frac{1}{4} - \frac{37t}{1600} + \frac{t^2}{1000} - \frac{13t^3}{640000} + \frac{t^4}{6400000},$$

$$\underline{\text{sqrt}}(t) := 8\sqrt{2} + \frac{3}{64(\pi - 4\arctan\frac{\sqrt{2}}{5})}(t - 128),$$

$$\overline{\text{sqrt}}(t) := 8\sqrt{2} + \frac{1}{16\sqrt{2}}(t - 128).$$

These approximations are valid on the domain (1). For example,

$$\forall t \in \Delta\left([2, 2.51]^6\right) \quad \sqrt{t} \leq \overline{\text{sqrt}}(t).$$

This can be established by elementary means, knowing that $\Delta([2, 2.51]^6) \subseteq$ [128; 501]. The latter can be shown automatically by the method outlined in the next subsection.

In summary, we arrive at the following definition of g:

$$g(y) := -\frac{\delta_{oct}}{6}\underline{sqrt}(\Delta(y)) + \frac{2}{3}\sum_{i=0}^{3}\overline{arctan}\big(\underline{sqrt}(\Delta(y)) \cdot \overline{rcp}(a_i(y))\big).$$

In form, it is almost identical to the definition of γ. Our construction of g therefore ensures (2). Moreover, the approximations were chosen (using polynomial interpolation) in a way such that

$$\gamma(2, 2, 2, 2, 2, 2) = g(2, 2, 2, 2, 2, 2) = \text{pt}. \tag{4}$$

This is important, because otherwise (3) cannot hold.

5.2 Bounding Polynomials

In order to prove (1), it remains to be shown that $g(y) \le \text{pt}$. This can be done with the help of Bernstein polynomials. We briefly outline the case of a single variable x here.

The ith Bernstein basis polynomial of order k is defined as

$$B_i^k(x) := \binom{k}{i}x^i(1 - x)^{k-i}.$$

For a polynomial p and a vector $b \in \mathbb{R}^k$, if

$$p(x) = \sum_{i=0}^{k}b_i \cdot B_i^k(x),$$

then b is called the *Bernstein representation* of p. In this case,

$$\forall x \in [0; 1] \quad p(x) \le \max_i b_i.$$

This property is tremendously useful: it gives us an upper bound on p, namely the largest coefficient of p's Bernstein representation. By a change of variable we can reduce any interval to [0, 1]. The generalization to the multivariate case is straightforward [7, 49].

In order to bound a polynomial, it thus suffices to convert it into Bernstein representation. This can be done by a matrix multiplication (the Bernstein basis of order k forms a basis of the vector space of all polynomials of degree up to k). For practical purposes, it is however crucial to use a more efficient algorithm (cf. [7, 49]).

Note that g contains irrational coefficients. This is a consequence of requirement (4) and cannot be avoided. However, we were able to choose the approximation polynomials in a way such that the transcendental parts can be factored out (hence the occurrence of $\pi - 4\arctan\frac{\sqrt{2}}{5}$ in the definition of $\underline{\text{sqrt}}$). As can be easily checked with symbolic algebra software, the polynomial $p(y) := \sqrt{2}(g(y) - \text{pt})$ has rational coefficients! It can thus be converted to a Bernstein representation without rounding, using the algorithm presented in [49]. With this method the only divisions are by powers of 2, which can be efficiently represented using dyadic numbers.

The polynomial p consists of 12945 monomials and has total degree 18. A prototype implementation in Haskell returns 0 as the maximum for p in about half a minute. Thus $\sqrt{2}(g(y) - \text{pt}) \leq 0$ and $g(y) \leq \text{pt}$.

6 Tame Graph Enumeration

Tame graphs are particular plane graphs that represent potential counterexamples to the Kepler conjecture. The *Archive* is a list of over 5000 plane graphs. The original proof generates the Archive with the help of a Java program that enumerates all plane graphs. Tameness is defined in Sect. 18, and the enumeration is sketched in Sect. 19 of [23]. This section sketches the formally machine-checked proof of Claim 3.13 and Theorem 19.1 in the original proof [23]:

Theorem 1 *Any tame plane graph is isomorphic to a graph in the Archive.*

There are two potential reasons why an error in the original proof of this theorem might have gone undetected: the publications only sketch the details of the enumeration, and the referees only made a passing glance at the implementation, consisting of more than 2000 lines of Java.

We recast the Java program for the enumeration of all tame graphs in logic, proved its completeness with the help of an interactive theorem prover, ran it, and compared the output to the Archive. It turns out that the original proof was right, the Archive is complete, although redundant (there are at most 2771 tame graphs). Doing all this inside a logic and a theorem prover requires two things:

- The logic must contain a programming language. We used Church's *higher-order logic* (HOL) based on λ-calculus, the foundation of functional programming. Programs in HOL are simply sets of recursion equations, i.e., pure logic.
- The programming language contained in the logic must be efficiently executable, and such executions must count as proofs. The theorem prover that we used, Isabelle/HOL [37], fulfills this criterion. If all functions that appear in a term are either data, e.g., numbers, or functions defined by recursion equations, Isabelle/HOL offers the possibility to evaluate this term t in one big and relatively efficient step to a value v, giving rise to the theorem $t = v$.

The enumeration of all tame graphs generates 23 million plane graphs—hence the need to perform massive computations in reasonable time.

Now we give a top-level overview of the formalization and proof of completeness of the enumeration of tame graphs in HOL. For details, see [38]. The complete machine-checked proof, over 17000 lines, is available online in the Archive of Formal Proofs at afp.sf.net [2].

6.1 Plane Graphs

Following the original proof, we represent finite, undirected, plane graphs as lists (= finite sets) of faces and faces as lists of vertices. Note that by representing faces as lists they have an orientation. The enumeration of plane graphs requires an additional distinction between *final* and *nonfinal* faces. Hence a face is really a pair of a list of vertices and a Boolean. A plane graph is *final* iff each of its faces is. In final graphs, we can ignore the Boolean component of the faces.

6.2 Enumeration of Plane Graphs

The original proof characterizes plane graphs by an executable enumeration and sketches a proof of completeness of this enumeration. We have followed the original proof and taken this enumeration as the definition of planarity. The enumeration of plane graphs in the original proof proceeds inductively: you start with a seed graph with two faces, the final outer one and the (reverse) nonfinal inner one. If a graph contains a nonfinal face, it can be subdivided into a final face and any number of nonfinal ones. Because a face can be subdivided in many ways, this process defines a tree of graphs. By construction the leaves must be final graphs, and they are the plane graphs we are interested in: any plane graph of n faces can be generated in $n - 1$ steps by this process, adding one (final) face at a time. For details, see [23] or [38].

The enumeration is parameterized by a natural number p which controls the maximal size of final faces in the generated graphs. The seed graph $Seed_p$ contains two $(p + 3)$-gons, and the final face created in each step may at most be a $(p + 3)$-gon. As a result, different parameters lead to disjoint sets of graphs.

The HOL formalization defines an executable function $next\text{-}plane_p$ that maps a graph to a list of graphs, the successor graphs reachable by subdividing one nonfinal face. The plane graphs are the final graphs reachable from $Seed_p$ via $next\text{-}plane_p$ for some p.

6.3 Enumeration of Tame Graphs

The definition of tameness in the original proof is already quite close to a direct logical formulation. Hence the HOL formalization is very close to this. Of course pictures of graphs had to be translated into formulae, taking implicit symmetries in pictures into account. We found one simplification: in the definition of an admissible weight assignment, one can drop condition 3 (a condition on the 4-circuits in graphs) without changing the set of tame graphs. What facilitated our work considerably was that a number of the eight tameness conditions in the original proof are directly executable. The details are described elsewhere [38].

The enumeration of tame graphs is a modified enumeration of plane graphs where we remove final graphs that are definitely not tame, and prune the search tree at nonfinal graphs that cannot lead to tame graphs anymore. The published description [23] is deliberately sketchy, and the precise formulation of the pruning criteria is based on the original Java programs. This is the most delicate part of the proof because we need to balance effectiveness of pruning with simplicity of the completeness proof:

weak pruning criteria are easy to justify but lead to unacceptable run times of the enumeration, sophisticated pruning techniques are difficult to justify formally. Since computer-assisted proofs are still very laborious, simplifying those proofs was of prime importance. In the end, the HOL formalization defines a function *next-tame*$_p$ from a graph to a graph list. It computes the list of plane successor graphs *next-plane*$_p$ g and post-processes it as follows:

(1) Remove all graphs from the list that cannot lead to tame graphs because of lower bound estimates for the total admissible weight of the final graph.
(2) Finalize all triangles in all of the graphs in the list (because every 3-cycle in a tame graph must be a face).
(3) Remove final graphs that are not tame from the list.

A necessary but possibly not sufficient check for tameness is used in the last step. Hence the enumeration may actually produce nontame graphs. This is unproblematic: in the worst case, a fake counterexample to the Kepler conjecture is produced, but we do not miss any real ones.

Although we have roughly followed the procedure of the original proof, we have simplified it in many places. In particular, we removed the special treatment of *Seed*$_0$ and *Seed*$_1$, which is a fairly intricate optimization that turned out to be unnecessary.

The following completeness theorem is the key result:

Theorem 2 *If a tame and final graph g is reachable from Seed*$_p$ *via next-plane*$_p$, *then g is also reachable from Seed*$_p$ *via next-tame*$_p$.

Each step *next-tame*$_p$ is executable, and an exhaustive enumeration of all graphs reachable from a seed graph is easily defined on top of it. We call this function *tameEnum*$_p$. By definition, tame graphs may contain only triangles up to octagons, which corresponds to the parameters $p = 0, \ldots, 5$.

6.4 Archive

In order to build on the above enumeration of all tame graphs without having to rerun the enumeration, the results of running *tameEnum*$_p$ with $p = 0, \ldots, 5$ are put into an Archive, and isomorphic graphs are eliminated. This results in 2771 graphs, as opposed to 5128 in the original proof. The reasons are twofold: there are many isomorphic copies of graphs in the Archive, and it contains a number of nontame graphs, partly because, for efficiency reasons, the original proof did not enforce all tameness conditions in its Java program. The new reduced Archive is also available online [2].

Finally we can prove Theorem 1: if g is tame plane graph, Theorem 2 and the definition of *tameEnum* tell us that g must be contained in *tameEnum*$_p$ for some $p = 0, \ldots, 5$. Hence it suffices to enumerate *tameEnum*$_p$, $p = 0, \ldots, 5$, and check that, modulo graph isomorphism, the result is the same as the Archive. This is a proposition that can be proved by executing it (because the HOL formalization also includes a verified executable test for graph isomorphism which we do not discuss).

7 Verifying Linear Programs

This section reports on the current state of the formal verification of the linear programming part of the proof of the Kepler conjecture. The results of the linear programming in the original proof are recorded in several gigabytes of log files. This section presents the formalization of the generation and bounding of these linear programs in the mechanical proof assistant Isabelle [37]. A more detailed version of the material presented here can be found in S. Obua's thesis [39].

This formalization relies on the archive of tame graphs. Each tame graph (except the graphs associated with the face-centered cubic and hexagonal close packings) represents a potential counterexample to the Kepler conjecture. Each potential counterexample obeys certain constraints. The original proof refutes each potential counterexample by building a linear program from the constraints and showing that these linear programs are infeasible.

A structure, which we call a *graph system*, makes this precise in a formal environment. An instance of a graph system is a tame graph which obeys the constraints listed in the definition of a graph system. The current formalization does not use all of the constraints of the original proof but only those that do not require branch and bound strategies. Our current notion of graph system can be viewed as a detailed formalization of what is called *basic linear programs* in [23, Sect. 23.3]. Because the current formalization does not capture not all constraints of the original proof, we cannot hope to refute all potential counterexamples. Nevertheless we manage to refute most of them.

We represent tame graphs as *hypermaps* [8, 20]. As mentioned in Sect. 2, a hypermap is just a finite set D of *darts* together with three permutations on D: the edge, the face, and the node permutation that compose to the identity $e \circ n \circ f = I$. See Fig. 1. This representation greatly simplifies the axiomatization of a graph system. For a detailed description of how we represent hypermaps and for a complete list of all of the axioms of a graph system, see [39, Sect. 4].

Figure 2 summarizes how we generate and solve the linear programs. We apply the axioms of a graph system to each tame graph. This results in a large Isabelle theorem which is a conjunction of linear equalities and inequalities. We then normalize this

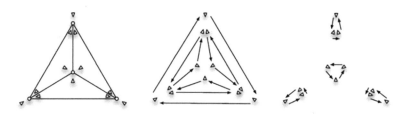

Fig. 1 A hypermap is a combinatorial structure attached to a planar graph. A dart, represented as a *small triangle*, is place at each face angle of the graph. In the second frame, the face permutation f cycles through the darts in each face. In the third frame, the node permutation n cycles through the darts at each node. The edge permutation e (not shown) is an involution exchanging darts from opposite ends of the edge of a graph. (This figure has been reproduced from [24].)

Fig. 2 Refuting a potential
counterexample to the Kepler
conjecture

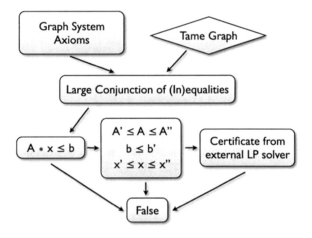

conjunction to bring it into the form of a matrix inequality

$$Ax \leq b. \tag{5}$$

The entries of A and b are symbolic expressions which contain various real constants
(such as π) that are needed to define the axioms of a graph system. In order to apply
linear programming, we need to replace A and b with numerical approximations.
We achieve this via a formalization of interval arithmetic in Isabelle and arrive at
numerical matrices A', A'', and b' for which we have the formally proven Isabelle
theorems

$$A' \leq A \leq A'', \qquad b \leq b'. \tag{6}$$

We then apply a simple preprocessing step which gives us formally proven a priori
bounds x' and x'' for x such that

$$x' \leq x \leq x''. \tag{7}$$

It is then possible to obtain a certificate from an external linear programming solver
like GLPK (Gnu Linear Programming Kit) that allows us to formally reach a contradiction from $Ax \leq b$. The beauty of a certificate is that we can use results obtained
from an untrusted source (in our case, this untrusted source is a heavily optimized
linear program solver programmed in C) in a trusted and completely mechanically
verifiable way.

   In this way, we proved the inconsistency of 2565 of the graph systems but failed
to prove the inconsistency of the remaining 206. This yields a success rate of about
92.5%. Future work will extend the notion of graph system and generate linear programs that take all the constraints of the original proof into account.

Part 2. Addendum to and Errata in the Original Proof

8 Biconnected Graphs

This section gives further detail to the argument of [15, Sect. 12.7, p. 131]. There it is claimed that the proof of the main estimate [15, Theorem 12.1] can be reduced to the case of polygonal standard regions. This claim is correct. However, the justification of this claim is not complete in the original proof. This section gives complete justification of the claim. The main result is Theorem 3 stated below.

This section patches the original proof. As such, it should be imagined this section to be inserted as an addendum, directly following Sect. 12.13 in the original proof, as a new Sect. 12.14. (In fact, Sect. 12.13 is itself an addendum to the 1998 preprint, making this section the second addendum to Sect. 12.) If the actual insertion were to be carried out, the corresponding changes to the numbering of the results in this section would be made: Theorem 3 below would be inserted as a new Theorem 12.21, and so forth.

In the original proof, the boundaries of standard regions may fail to be simple polygons. In fact, the failure of the boundaries to be polygons may be quite severe: the boundaries may contain multiple components and articulation vertices. An *articulation* vertex in a graph is a vertex whose removal increases the number of connected components of the graph. A connected graph is *biconnected* if it contains no articulation vertex.

In the original proof, each standard region is first prepared by a process of erasure (described in [15, Sect. 11.1]), then deformed in a way to transform its boundary into a simple polygon. The deformation is made in such a way that the values of two key functions $\text{vor}_{0,R}$ and $\tau_{0,R}$ are left unchanged. These are the functions that enter into the main estimate [15, Theorem 12.1].

What the original proof fails to consider is a particular hypothetical situation where the deformation might fail. This situation is illustrated in Fig. 3, where the rigid movement of set of vertices (such as the illustrated triangle with vertex w) is blocked by a nearby vertex v that is not visible from w, when the distance between v and w drops to the minimum value 2. This section presents a proof that this hypothetical situation does not occur.

The strategy of the proof in this section is to use a deformation argument. A general decomposition star is deformed until the graph attached to it becomes biconnected. In a biconnected planar graph with at least three vertices, each face is a simple polygon. Thus, by producing a biconnected graph, we achieve our objective. In the original proof of the Kepler conjecture, biconnected graphs are not mentioned by name. Nevertheless, most of the graphs that occur in the late stages of the original proof are biconnected.

In more detail, the proof in this section will produce a sequence of admissible deformations of a decomposition star D in such a way that the individual standard regions R are preserved in number and identity (but not in shape) by the deformations. The deformations will preserve the values of $\text{vor}_{0,R}(D)$ and $\tau_{0,R}(D)$ for each standard region R. The deformations will change the combinatorial structure of the graph formed by all the subregions of the decomposition star. At the conclusion of

Fig. 3 In this figure (not to scale), segments represent geodesic arcs on the unit sphere. The *shaded region* is a subregion whose boundary is not a simple polygon. We wish to rigidly slide the triangle containing the vertex w until a new visible distinguished edge forms (say from w to u). We will show that this rigid motion does not decrease the distance between two vertices (say w and v) to the minimum distance 2

the deformation, the decomposition star D will have a graph (formed by all the sub-regions of the decomposition star) that is biconnected. As in the original proof, the decomposition star is first simplified by a process of erasure, before the deformations begin.

In the remainder of this section, we adopt notation, definitions, and conventions without further comment from [15].

8.1 Context

We work in the following restrictive context for Theorem 3 and its proof. We fix a packing centered at a vertex at the origin. As described in Sect. 12.6 of the original proof, we assume that all upright quarters are erased, except loops (that is, those surrounded by anchored simplices). Let U be the set of (non-null) vertices of height at most 2.51. As usual, we say two edges $\{u_1, u_2\}$ and $\{u'_1, u'_2\}$ *cross* if the interiors of the triangles formed by $\{0, u_1, u_2\}$ and $\{0, u'_1, u'_2\}$ intersect.

We form the set of edges E' between vertices in U, consisting of

- All standard edges; that is, $\{v, w\} \subseteq U$ such that $0 < \|v - w\| \le 2.51$.
- All edges $\{v, w\} \subseteq U$ of an anchored simplex, whenever the upright diagonal of the anchored simplex is an unerased loop.
- All edges $\{v, w\} \subseteq U$ such that $0 < \|v - w\| \le \sqrt{8}$, where $\{v, w\}$ does not cross any other edge in previous two items. (If two of these edges cross, pick only one of them. This can only happen with conflicting diagonals of a quad cluster.)

These edges do not cross. A *special simplex* $\{0, u, v, w\}$ has one edge $\{v, w\}$ of length at least $\sqrt{8}$, called the *special edge*. A special edge has length at most 3.2. The other vertex u is called a *special vertex* (or *special corner*). Let $E = E' \setminus S$, where S is the set of special edges (that is, the edges of special simplices shared with an anchored simplex). The projection of the line segments formed by E to the unit sphere is a planar graph. The complement of this graph in the unit sphere is a disjoint union of connected components. The closures of these connected components are called *subregions*.

We call a *loop subregion* one that contains an unerased loop $\{0, v\}$. If R is a loop subregion, then there are no enclosed vertices of height ≤ 2.51 over the subregion. The corners of R are the anchors of the upright diagonal together with the special corners of the subregion. The subregion R is star convex with center point at the projection of v to the unit sphere. It follows that the boundary of R is a simple polygon.

The graph Γ with vertices U and edges E is not necessarily connected. The aim is to deform U (and D) to create a biconnected graph (without changing the values of $\text{vor}_{0,R}(D)$ and $\tau_{0,R}(D)$ for *standard regions* R; the values for individual subregions will change). Once the graph Γ is biconnected, the subregions are simple polygons as desired.

The deformation of U is *admissible* if it satisfies the following three conditions.

- If the graph Γ is not connected, the deformation acts by a rotation about the origin on the vertices of a single chosen connected component of Γ, leaving all other vertices of U fixed. (For example, in Fig. 3, the entire triangle with vertex w may be rotated.) In particular, $\|v\|$, for $v \in U$, is constant.
- If the graph Γ is connected, but not biconnected, with a chosen articulation vertex a, then the deformation acts by a rotation about the axis $\{0, a\}$ on the vertices of a single component of $\Gamma \setminus \{a\}$, leaving all other vertices of U fixed.
- The distance between v, w remains at least 2 for all $v, w \in U$.

The deformation stops when either of the following *halting conditions* are met:

(H1) $\|v - w\| \leq \sqrt{8}$, where the edge $\{v, w\} \notin E'$, with $\{v, w\} \subseteq U$, does not cross any edge in E', or

(H2) $\|v - w\|$ decreases to 2 for some $v, w \in U$, and $\{v, w\}$ crosses an edge in E'.

We warn that the deformation is allowed to assume configurations in which the distance between some $u \in U$ and the upper endpoint of an unerased upright diagonal d is less than 2. This is not a problem because the vertex u does not end up as a corner of the loop subregion to which the upright diagonal d belongs. (In particular, the halting condition (H1) prevents a new special vertex from being added to a loop subregion.) Thus, when a single subregion is considered in isolation, all distances between pairs of vertices in that subregion are at least 2.

In the restrictive context that has been described in this section, we have the following result.

Theorem 3 *If the graph $\Gamma = (U, E)$ is not biconnected, then a nontrivial admissible deformation of U exists. The halting condition* (H2) *never holds. The admissible deformation can always be continued until the halting condition* (H1) *holds.*

We break the proof of Theorem 3 into cases (a), (b), (c), (d) according to the number and combinatorial structure of the edges $\{u_1, u_2\}$ that pass through the triangle $\{0, v, w\}$. See Fig. 4. Lemma 4 below gives Case (a): we cannot have two such edges $\{u_1, u_2\}$ and $\{u'_1, u'_2\}$ with distinct endpoints. It follows that there is an endpoint u_2 shared by every edge that passes through the triangle $\{0, v, w\}$. Lemma 9 below gives Case (b): there cannot be three or more edges. Finally, Cases (c) and (d) of one or two edges $\{u_1, u_2\}$ are treated in further subsections. Each subsection is organized

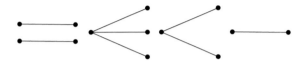

Fig. 4 Theorem 3 breaks into various cases: (**a**) two edges with distinct endpoints, (**b**) three or more edges, (**c**) two edges, or (**d**) a single edge passing through the triangle $\{0, v, w\}$

around one of the main cases (with introductory Sect. 8.2): Cases (a), (b), (c), and (d) are treated in the four Sects. 8.3, 8.4, 8.5, 8.6, respectively.

When U is deformed until the halting condition (H1) holds, a new edge $\{v, w\}$ can be added to E. The theorem is repeated to add further edges, until a biconnected graph is obtained. The remainder of this section explains why the halting condition (H2) never holds. For this, we assume, on the contrary, that $\|v - w\| = 2$, where $\{v, w\}$ belong to different connected components of Γ (if Γ is not connected) and different connected components of $\Gamma \setminus \{a\}$ (if a is an articulation vertex of a connected graph Γ). When this situation occurs, we say that v, w belong to different bicomponents. Without loss of generality, we may also assume that $\{v, w\}$ crosses some edge of E'.

8.2 Preliminary Lemmas

Lemma 1 below shows that a vertex does not cross over an edge during a deformation; the halting conditions for the vertex are met before it reaches the edge. Recall that the term *geometric considerations* refers to a specific collection of methods introduced in [23, Sect. 4.2] to prove the existence and nonexistence of various simple configurations of points in $\mathbb{R}^3$.

Lemma 1 *Let $S = \{0, v, u_1, u_2\}$ be a set of four distinct points in $\mathbb{R}^3$ whose pairwise distances are at least 2. Suppose that $\|u\| \leq 2.51$ for all $u \in S$. Suppose that $\|u_1 - u_2\| < 3.2$. If the segment $\{0, v\}$ meets the segment $\{u_1, u_2\}$, then $\|v - u_1\|$, $\|v - u_2\| \leq 2.51$.*

Proof This follows by geometric considerations. $\square$

The next two lemmas give conditions under which the halting condition (H2) cannot hold.

Lemma 2 *Let $S = \{v, w, u_1, u_2, v_0\}$ be a set of five distinct points in $\mathbb{R}^3$ whose pairwise distances are at least 2. Suppose that the segment $\{v, w\}$ passes through $\{v_0, u_1, u_2\}$. Assume*

$$\|u_1 - u_2\| \leq 3.2,$$

$$\|v_0 - u\| \leq 2.51 \quad \text{for } u \in S.$$

Then $\|v - w\| > 2$.

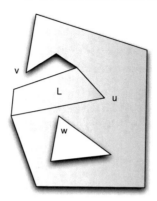

Fig. 5 In this figure (not to scale), segments represent geodesic arcs on the unit sphere. The *two shaded regions* represent subregions. There are possibly other subregions partitioning the unshaded portions of the figure. These other subregions are not represented in this figure. The *shaded region* marked L represents a loop subregion. The *other shaded region* represents a subregion with two boundary components that is undergoing a deformation, by the motion of the triangle with vertex v. The situation of Fig. 3 does not exist (as will be shown). The intervening loop subregion L forces v and w to be more than the minimum distance apart

Proof This follows by geometric considerations.   □

Lemma 3 *Let* $S = \{v, w, u_1, u_2, v_0\}$ *be a set of five distinct points in* $\mathbb{R}^3$ *whose pairwise distances are at least 2. Suppose that the segment* $\{u_1, u_2\}$ *passes through* $\{v_0, v, w\}$. *Assume*

$$\|u_1 - u_2\| \le 2.91,$$

$$\|v_0 - u\| \le 2.51 \quad \text{for } u \in S.$$

Then $\|v - w\| > 2$.

Proof This follows by geometric considerations.   □

By Lemmas 2 and 3, we may assume that for each edge $\{u_1, u_2\}$ that $\{v, w\}$ crosses, we have that $\{u_1, u_2\}$ passes through $\{0, v, w\}$ and that $\|u_1 - u_2\| > 2.91$. This means that the edge $\{u_1, u_2\}$ is an edge of an anchored simplex, so that the edge is special or bounds a loop subregion. This loop subregion will provide the key to the proof, because we will show that it prevents the distance between v and w from becoming 2, as assumed (Fig. 5).

Remark 1 As an aside, we mention that this issue of biconnected graphs is a major issue only in the proof of the Kepler conjecture and not in the proof of the dodecahedral conjecture [24]. In the proof of the dodecahedral conjecture, there are no loops of anchored simplices, and without loops there are no difficulties: edges of length at most $\sqrt{8}$ can be used instead of the set E'. Lemmas 2 and 3 suffice.

8.3 Case (a): Edge Crossings with Distinct Endpoints

Lemma 4 *There does not exist a set* $S = \{0, v, w, u_1, u_2, u_1', u_2'\}$ *of seven distinct points in* $\mathbb{R}^3$ *whose pairwise distances are at least 2 and that satisfies the following conditions.*

- *The edges* $\{u_1, u_2\}$ *and* $\{u_1', u_2'\}$ *do not cross.*
- *The edges* $\{u_1, u_2\}$ *and* $\{u_1', u_2'\}$ *both pass through* $\{0, v, w\}$.
- $\|u\| \leq 2.51$ *for all* $u \in S$.
- $\|v - w\| = 2$.
- $\|u_1 - u_2\|, \|u_1' - u_2'\| \leq 3.2$.
- u_1 *and* u_1' *lie in the same half-space bounded by the plane* $\{0, v, w\}$.
- *The directed segment from* v *to* w *crosses the segment* $\{u_1', u_2'\}$ *before the segment* $\{u_1, u_2\}$.

Note that the last two conditions can always be achieved by suitable labels on the points $\{u_1, u_2, u_1', u_2'\}$.

Proof This will follow as a direct consequence of Lemmas 6 and 8 below. □

Lemma 5 *Let* $S = \{u, v, w\}$ *be a triangle such that each side has length at least 2, and such that* $\|u - v\| \leq 2.51$, $\|u - w\| \leq 2.51$, $\|v - w\| = 2$. *Let* X *be the set of points in the convex hull of* S *that have distance at least 1.2 from each vertex of* S. *Then the diameter of* X *is less than 1.044.*

Proof Let $x = \|u - v\|$ and $y = \|u - w\|$. Assume $x \geq y$. As u moves away from v, along a fixed line through v and an initial position u_0, the region X expands. Thus, we may assume that $x = 2.51$. The boundary of X is a polygonal curve consisting of line segments and concave arcs of circles. The diameter is realized by the distance between two vertices $p_i(y)$ of the polygonal curve. We consider two cases according to $y \leq 2.4$ and $y \geq 2.4$ because the structure of the polygonal curve changes at $y = 2.4$. We calculate $\|p_i(y) - p_j(y)\|$ directly, checking for each (i, j) that the distances are less than 1.044. □

Lemma 6 *Let* $S = \{0, v, w, u_1, u_2, u_1', u_2'\} \subseteq \mathbb{R}^3$ *be a configuration of seven points that satisfies the conditions of Lemma 4. Then there exist a point* p *on the segment* $\{u_1, u_2\}$ *and a point* p' *on the segment* $\{u_1', u_2'\}$ *such that* $\|p - p'\| < 1.044$.

Proof Let $u \in \{0, v, w\}$. By the metric constraints, the distance from u to the segment $\{u_1, u_2\}$ (resp. $\{u_1', u_2'\}$) is at least

$$\sqrt{2^2 - (3.2/2)^2} = 1.2.$$

Let p (resp. p') be the point of intersection of the segment $\{u_1, u_2\}$ (resp. $\{u_1', u_2'\}$) with the convex hull of $\{0, v, w\}$. By Lemma 5, we have $\|p - p'\| < 1.044$. □

Lemma 7 *Let* $S = \{u_1, u_1', u_2, u_2'\} \subseteq \mathbb{R}^3$ *be a set of four distinct points such that the distance between each pair of points is at least 2. Assume that*

$$\|u_1 - u_2\| \le 3.2, \qquad \|u_1' - u_2'\| \le 3.2.$$

If any of the following conditions hold:

(A) $\|u_1 - u_2'\|, \|u_1' - u_2\| \ge 2.91$
(B) $\|u_1 - u_1'\|, \|u_2 - u_2'\| \ge 2.85, \|u_1 - u_2'\| \ge 2.91$; *or*
(C) $\|u_1 - u_1'\| \ge 3.64, \|u_1 - u_2'\| \ge 2.91$

then every point on the segment $\{u_1, u_2\}$ *has distance greater than* 1.044 *from every point on the segment* $\{u_1', u_2'\}$.

Proof Assume for a contradiction that the assumptions hold and that the conclusion is false for some configuration. The metric constraints can be used to show that the segments $\{u_1, u_2\}$ and $\{u_1', u_2'\}$ can be stretched along their axes without decreasing any edge length. Thus we may assume without loss of generality that $\|u_1 - u_2\| = \|u_1' - u_2'\| = 3.2$. Decreasing one dihedral angle of the simplex S at a time, we may move the segments closer together, until all four edges $\{u_i, u_j'\}$ attain their minimum length. Then we have three rigidly determined simplices (A), (B), (C) (with equality constraints). An explicit coordinate calculation of the distance between the two segments shows that the distance is greater than 1.044 in each case. ☐

Lemma 8 *Let* $S = \{0, v, w, u_1, u_2, u_1', u_2'\} \subseteq \mathbb{R}^3$ *be a configuration of seven points that satisfies the conditions of Lemma 4. Then there do not exist a point* p *on the segment* $\{u_1, u_2\}$ *and a point* p' *on the segment* $\{u_1', u_2'\}$ *such that* $\|p - p'\| < 1.044$.

Proof Assume for a contradiction that such p, p' exist. In Lemma 7, we may assume that none of the conditions A, B, C hold. By obvious symmetry, without loss of generality, we may assume by case A that $\|u_2 - u_1'\| < 2.91$. By Lemma 3, this implies that $\{u_1', u_2\}$ does not pass through $\{0, v, w\}$. By the conditions of the lemma, the segment $\{u_1, u_1'\}$ does not meet the plane $\{0, v, w\}$, so that $\{u_1, u_1'\}$ does not pass through $\{0, v, w\}$. This means that the triangle $\{u_1, u_1', u_2\}$ is linked around $\{0, v, w\}$, and some edge of $\{0, v, w\}$ passes through $\{u_1, u_1', u_2\}$. Up to symmetry there are two cases: (1) $\{0, v\}$ passes through $\{u_1, u_1', u_2\}$, or (2) $\{v, w\}$ passes through $\{u_1, u_1', u_2\}$.

In the first case, recall that $\{u_1', u_2'\}$ does not cross $\{u_1, u_2\}$. So u_2' is enclosed over $(0, \{u_1, u_1', u_2\})$. Then $\{u_2', u_1\}$ and $\{u_2', u_1'\}$ pass through $\{0, v, w\}$. By Lemma 3, we have $\|u_2' - u_1\|, \|u_2' - u_1'\| \ge 2.91$. As we are assuming that C does not hold, we have $\|u_1 - u_1'\| \le 3.64$. We claim that $\{0, u_2'\}$ passes through $\{u_1, u_1', u_2\}$. Otherwise, u_2' lies in the convex hull of $\{0, u_1, u_1', u_2\}$ and a coordinate calculation shows that the upper bounds on the edges of the simplex $\{0, u_1, u_1', u_2\}$ are inconsistent with the lower bounds on the distances from u_2' to the vertices of the simplex. Since $\{0, u_2'\}$ passes through $\{u_1, u_1', u_2\}$, we may use geometric considerations to show that $\|u_2'\| > 2.51$. This is contrary to the hypotheses of Lemma 4.

In the second case, $\{v, w\}$ passes through both $\{u_1, u_1', u_2\}$ and $\{u_1', u_2', u_2\}$. Geometric considerations give $\|u_1 - u_1'\| > 2.85$ and $\|u_2 - u_2'\| > 2.85$. Assuming that B does not hold gives $\|u_1 - u_2'\| < 2.91$. The first paragraph of the proof now gives that $\{u_2, u_2', u_1\}$ links around the triangle $\{0, v, w\}$. The edge $\{v, w\}$ does not pass through $\{u_2, u_2', u_1\}$. (This can be seen by drawing the relative positions of $p(u)$ for $u \in S$ in the projection p of the points to a plane orthogonal to $\{v, w\}$.) Thus, we are in the first case, which has already been treated.   □

8.4 Case (b): Triple Edge Crossings

By Lemma 4, there is a common endpoint u_2 such that every edge of E' that passes through $\{0, v, w\}$ has u_2 as an endpoint. Next we show that there cannot be three such edges.

Lemma 9 *There does not exist a set of seven distinct points*

$$S = \{0, v, w, u_1, u_1', u_1'', u_2\}$$

in $\mathbb{R}^3$ that satisfies the following conditions.

- *The distance between each pair of distinct points in S is at least 2.*
- *The edges $\{u_1, u_2\}$, $\{u_1', u_2\}$, and $\{u_1'', u_2\}$ pass through $\{0, v, w\}$.*
- $\|u\| \le 2.51$ *for all $u \in S$.*
- $\|v - w\| = 2$.
- $\|u - u_2\| \le 3.2$ *for $u = u_1, u_1', u_1''$.*

Proof Assume for a contradiction that S exists. We may pivot w around the axis $\{0, v\}$ until $\|w - u_2\| \le 2.51$. (The metric constraints on edge lengths show that the condition $\Delta > 0$ is preserved for the simplices $\{u, u_2, v, w\}$ and $\{u, u_2, w, 0\}$ for $u = u_1, u_1', u_1''$ throughout this pivot.) A similar pivot of v gives $\|v - u_2\| \le 2.51$. We may order the vertices in cyclic order around $\{0, u_2\}$ as

$$(w_1, w_2, w_3, w_4, w_5) = (w, u_1, u_1', u_1'', v),$$

so that setting $d(w_i, w_j) = \mathrm{dih}(0, u_2, w_i, w_j)$, we have

$$d(w_1, w_5) = \sum_{i=1}^{4} d(w_i, w_{i+1}) \ge d(w_2, w_3) + d(w_3, w_4).$$

Interval calculations[2] give $d(w_2, w_3), d(w_3, w_4) \ge 0.7$ and $d(w_1, w_5) < 1.4$. We obtain an immediate contradiction:

$$1.4 > d(w_1, w_5) \ge 0.7 + 0.7.$$

□

---

[2]CALC-2799256461, CALC-5470795818.

Fig. 6 The upright diagonal u cannot be placed over any of the regions A, B, C, D. The *lines* (not to scale) represent geodesic arcs on the sphere passing through the pairs of points in $\{p(u_1), p(u_2), p(v)\}$, where p denotes projection to the unit sphere

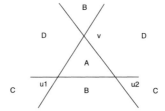

8.5 Case (c): Double Edge Crossings

This subsection treats the case of two edges crossings in the proof of Theorem 3. We continue to assume the general context of Theorem 3. As usual, the edge $\{u_1, u_2\} \in E'$ crosses $\{v, w\}$.

Lemma 10 *Let $\{u_1, u_2, w, v\}$ be a set of four distinct points in $\mathbb{R}^3$ (in the given context). Assume that $\|v - u_i\| \le 2.51$ for $i = 1, 2$. Assume that no edge of E' crosses $\{v, w\}$ in the open half-space A containing v bounded by the plane $\{0, u_1, u_2\}$. (That is, $\{u_1, u_2\}$ is the first edge to cross $\{v, w\}$, moving from v toward w.) Assume that there is a loop subregion L along $\{u_1, u_2\}$ on the A-side of $\{u_1, u_2\}$. Then $\{u_1, u_2\}$ is a special edge of E' with corner v. In particular, $\{u_1, u_2\} \notin E$, so that it is not a bounding edge of a subregion.*

Proof Assume for a contradiction that $\{u_1, u_2\}$ is not special. Note that loop subregions have simple polygonal boundaries and remain rigid under all the deformations. In particular, the upright diagonal, special corners, and so forth remain rigidly positioned with respect to the corners of the subregion.

Since there are no further edges crossing $\{v, w\}$, the subregion L extends to v. Hence v is a corner of the subregion L. It is either an anchor or a special corner (with respect to L). However, it cannot be a special corner, by the assumption that $\{u_1, u_2\}$ is not a special edge. Hence it is an anchor. Also, u_1 and u_2 are anchors.

To reach a contradiction, we consider possible locations of the upright diagonal $\{0, u\}$ and show that it has nowhere to go (Fig. 6). Since $\{u_1, u_2\}$ is an edge, the points u_1, u_2 are consecutive anchors. This prevents u from lying over the region B. Also, the upright diagonal of any unerased loop has at least four anchors (say v, u_1, u_2, w'). Moreover, some anchored simplex around the upright diagonal is not an upright quarter (because of the edge $\{u_1, u_2\}$). Geometric considerations based on these constraints (say $\|w' - u_1\| \ge 2.51$) show that the fourth anchor w' is not in A. This prevents u from being located over the region A. A vertex u_j cannot be enclosed over an upright quarter $\{0, u, v, u_i\}$. This excludes region C. Finally, an edge $\{v, u_i\}$ of length at most 2.51 cannot pass through a triangle $\{0, u, u_j\}$ of sides at most $2.51, 2.51, \sqrt{8}$ (excluding D). □

Lemma 11 *Let $\{u_1, u_2, w, v\}$ be a set of four distinct points in $\mathbb{R}^3$ (in the given context). Assume that $\|v - u_i\| \le 2.51$ for $i = 1, 2$. Assume that no edge of E crosses $\{v, w\}$ in the open half-space A containing v bounded by the plane $\{0, u_1, u_2\}$. (That*

is, $\{u_1, u_2\}$ *is the first edge to cross* $\{v, w\}$, *moving from* v *toward* w.) *Then both edges* $\{u_1, v\}$ *and* $\{u_2, v\}$ *belong to* E. *In particular, there is a circuit of the graph* Γ *through* v, u_1, u_2.

Proof If there is a loop subregion L along $\{u_1, u_2\}$ on the A-side of $\{u_1, u_2\}$, Lemma 10 implies that $\{u_1, u_2\}$ is a special edge of E' with corner v. In particular, $\{u_1, v\}$ and $\{u_2, v\}$ are edges of E. The conclusion follows in this case.

Now assume that there is no loop subregion along $\{u_1, u_2\}$ on the A-side of $\{u_1, u_2\}$.

Let S be the finite set of points of U enclosed over the simplex $\{0, u_1, u_2, v\}$. We show by contradiction that S is empty. The plane $\{0, v, w\}$ separates S into a disjoint union $S = S_1 \cup S_2$, according to those in the same half-space as u_i, $i = 1, 2$. We form the convex hull of the projection p to the unit sphere of the points $S_i \cup \{u_i, v\}$. As in [15, Sect. 12.13], form a sequence of geodesic arcs on the unit sphere from $p(u_i)$ to $p(v)$. Let $p(w_i)$, for $w_i \in S_i$, be the other endpoint of the arc starting at $p(u_i)$ (or set $w_i = v$ if $S_i = \emptyset$. For some $w' \in \{w_1, w_2\} \setminus \{v\}$, the edges $\{w', u_i\}$ do not cross any edges of E. Furthermore, geometric considerations show that $\|w' - u_i\| \le 2.51$ for $i = 1, 2$. By the criteria for forming edges of E, we must have $\{w', u_i\} \in E$ for $i = 1, 2$. This contradicts the assumption that $\{v, w\}$ does not cross any edges of E over A. Hence $S = \emptyset$.

Since $S = \emptyset$, the edges $\{v, u_i\}$ do not cross any edges of E. By the criteria for forming edges of E, they belong to E. This completes the proof. $\square$

We are ready to prove the next major case of Theorem 3. We continue to work in the general context of that theorem, with v and w in different bicomponents of the graph Γ.

Lemma 12 *In this context, no set of six points* $\{0, v, w, u_1, u_1', u_2\}$ *exists, where* $\{u_1, u_2\}$ *and* $\{u_1', u_2\}$ *pass through* $\{0, v, w\}$ *and such that*

$$\|u_1 - u_2\| \le 3.2, \qquad \|u_1' - u_2\| \le 3.2.$$

Proof We argue by contradiction. We may assume that $\{u_1, u_1'\}$ are ordered so that the cyclic order around $\{0, u_2\}$ is $(w_1, w_2, w_3, w_4) = (v, u_1', u_1, w)$. By the previous results, there are at most two edges that pass through $\{0, v, w\}$ in this manner. In particular, the part of the line segment $\{v, w\}$ between the crossings of $\{u_1, u_2\}$ and $\{u_1', u_2\}$ lies in a single subregion L.

We claim that we do not have $\|v - u_1'\|, \|v - u_2\|, \|w - u_1\|, \|w - u_2\| \le 2.51$. Otherwise, we break into two cases and derive a contradiction as follows. Either (1) L is a loop subregion, or (2) L is not a loop subregion, but both regions adjacent to L (along edges $\{u_1, u_2\}$, $\{u_1', u_2\}$) are loop subregions. In case (1), $\{u_1, u_2, u_1'\}$ are corners of the loop subregion L. Hence, they lie on a circuit in Γ formed by the corners of that loop subregion. If $\|v - u_2\|, \|v - u_1'\| \le 2.51$, then by Lemma 11, the points $\{v, u_1', u_2\}$ lie on a circuit in Γ. A similar conclusion holds if corresponding inequalities hold $\|w - u_2\|, \|w - u_1\| \le 2.51$. If all four inequalities hold, then these circuits put v, w in the same bicomponent of Γ, which is contrary to hypothesis.

In case (2), then by Lemma 11, $\{v, u_2\}, \{v, u'_1\} \in E$, so $\{u'_1, u_2\}$ is a special edge and L is a loop subregion. This is contrary to the assumption of case (2). Hence we may assume by symmetry and without loss of generality that $\|v - u'_1\| \geq 2.51$ or $\|v - u_2\| \geq 2.51$.

We may stretch along the edges $\{u_2, u_1\}, \{u_2, u'_1\}$, moving u_1, u'_1, until $\|u_2 - u_1\| = \|u_2 - u'_1\| = 3.2$. We may add the inequality

$$\|u_2\| \leq 2.23,$$

for otherwise by geometric considerations $\|u_2 - u'_1\| > 3.2$. Similarly, $\|u'_1\| \leq 2.23$. If $\|u_2 - v\| \geq 2.51$, we may pivot v toward u_2 around the axis $\{0, w\}$ until $\|u_2 - v\| \leq 2.51$. Similarly, we may assume that $\|u_2 - w\| \leq 2.51$. Set $d(i, j) = \text{dih}(0, u_2, w_i, w_j)$. Then interval arithmetic calculations[3] give the contradiction:

$$1.3 > d(1, 4) = d(1, 2) + d(2, 3) + d(3, 4) > 0.5 + 0.8 + 0 = 1.3. \qquad \square$$

8.6 Case (d): Single Edge Crossings

This subsection treats the proof of Theorem 3 in the case of a single edge crossing $\{u_1, u_2\}$. This is the final case of the proof. We continue to assume the notation and general context of that theorem. In particular, v and w lie in different bicomponents of the graph Γ.

Lemma 13 *Let $\{0, u_1, u_2, v, w\}$ be a set of five distinct points such that $\{u_1, u_2\}$ is the only edge of E' that crosses $\{v, w\}$. Then $\|v - w\| > 2$.*

Proof We assume for a contradiction that $\|v - w\| = 2$. We consider four cases depending on lengths.

Case 1: $\|u - u_i\| \leq 2.51$ for $i = 1, 2$ and $u = v, w$. By Lemma 11, there are circuits running through $\{u, u_1, u_2\}$ for $u = v, w$. This is contrary to the assumption that v, w lie in different bicomponents of the graph Γ. (In the remaining cases, there is no loss in generality to assume $\|w - u_2\| \geq 2.51$.)

Case 2: $\|w - u_2\| \geq 2.51$, $\|v - u_1\| \geq 2.51$. Geometric considerations give the contradiction $\|u_1 - u_2\| > 3.2$.

Case 3: $\|w - u_2\| \geq 2.51$, $\|v - u_2\| \geq 2.51$. Geometric considerations gives the contradiction $\|u_1 - u_2\| > 3.2$.

Case 4: $\|w - u_2\| \geq 2.51$, $\|v - u_i\| \leq 2.51$ for $i = 1, 2$. The edge $\{u_1, u_2\}$ cannot be a special edge of E'. Otherwise, v, w are corners of the same loop subregion. This contradicts the running assumption that these two vertices belong to separate bicomponents of the graph Γ. By Lemma 10, there is no loop subregion along $\{u_1, u_2\}$ on the v-side. Since $\{u_1, u_2\}$ has length greater than $\sqrt{8}$, there is a loop subregion L bounded by the edge, and it must then lie on the w-side. Thus, w is a corner of L, and the circuit of Γ described by the boundary of L passes through w, u_1, u_2. By Lemma 11, there is a circuit of Γ through v, u_1, u_2. Hence, v, w lie in the same biconnected component, which is contrary to the running assumption. $\qquad \square$

---

[3]CALC-7431506800, CALC-5568465464, CALC-4741571261, CALC-6915275259.

9 Errata Listing

The abridged version of the Kepler conjecture in the Annals [12] was generated by the same TeX files as the unabridged version in [23]. For this reason, it seems that every correction to the abridged version should also be a correction to the unabridged version. We list errata in the unabridged version. The same list applies to corresponding passages in the abridged version.

Each correction gives its location in [23]. The location $\ell + n$ counts down from the top of the page, or if a section or lemma number is provided, it counts from the top of that organizational unit. The location $\ell - n$ counts up from the bottom of the page. Footnotes are not included in the count from the bottom. Every line containing text of any sort is included in the count, including displayed equations, section headings, and so forth. The material to the left of $\rightsquigarrow$ indicates original text, and material to the right of the arrow gives replacement text. The original text and replacement text appear in italic. Comments about the corrections appear in roman.

In addition to the corrections to the text mentioned below, there have been some corrections to the computer code, including some typos in the listings of nonlinear inequalities. They are described in detail in [9].

9.1 Listing

[p. 47, Lemma 5.16] $Q \rightsquigarrow F$

[p. 49, $\ell + 2$] *supposed* $\rightsquigarrow$ *suppose*

[p. 63, Lemma 7.10] S-*system* $\rightsquigarrow Q$-*system*

[p. 75, Remark 8.11] *show* $\rightsquigarrow$ *shows*

[p. 78, $\ell - 7$] *constraints* $\rightsquigarrow$ *constraint*

[p. 86, $\ell + 14$] *Let* $\{0, v\}$ *be the diagonal of an upright quarter in the* Q_0 $\rightsquigarrow$ *Let v be a vertex with* $2t_0 < |v| < \sqrt{8}$. Section 9 assumes that the diagonal belongs to a quarter in the Q-system, but Lemma 10.14 uses these results when $\{0, v\}$ has 0 or 1 anchors. To make this coherent, we should assume throughout Sect. 9 that we have the weaker condition that whenever $\{0, v\}$ has two or more anchors, it is a diagonal of a quarter in the Q-system. The proofs of Sect. 9 all go through in this context. (Lemma 9.7 is all that is relevant here.)

[p. 87, Definition 9.3] In definition of $\Delta(v, W^e)$, we can have some Q (as in Fig 9.1) with negative orientation. In this case, $E_v \cap E_i$ can clip the other side. We want the object without clipping. $\Delta(v, W^e)$ should be understood as the unclipped object.

[p. 88, Definition 9.6] The definition is poorly worded. First of all, it requires that the subscript to ϵ to be a vertex, but then in the displayed equation, it makes $w/2$ the subscript, which is not a vertex. To define ϵ', move from $w/2$ along the ray through x' until an edge of the Voronoi cell is encountered. If v, w, u are the three vertices defining that edge, then set $\epsilon'_v(\Lambda, x) = u$. Degenerate cases, such as when two different edges are encountered at the same time, can be resolved in any consistent fashion. In [21], these degeneracies are avoided altogether, by replacing functions ϵ, ϵ' with sets Φ, Φ'.

[p. 88, Lemma 9.7, $\ell + 2$] w *and* $v \rightsquigarrow w$ *and* u

[p. 88, L. 9.7, Claim 1] *with* $|w - w'| \le 2t_0$, *and* $\rightsquigarrow$ *with*

[p. 88, L. 9.7, $\ell + 5$] *Then:* $\rightsquigarrow$ *Let* $R'_w = \{x \in R_w \cap (0, \{u, w\}) \mid \epsilon_0(x, \{u, w\}) = u\}$. *Assume that R'_w is not empty. Then:*

[p. 88, L. 9.7, Claim 3] $R_w \rightsquigarrow R'_w$

[p. 89, $\ell + 2$] $\{w, v\} \rightsquigarrow \{w, u\}$

[p. 92, $\ell + 16, \ell + 21$] $\max_j u_j \rightsquigarrow \max_j |u_j|$

[p. 93, $\ell - 4$] *obstructed from w* $\rightsquigarrow$ *obstructed from w'*
[p. 93, $\ell - 2$] *from some* $\rightsquigarrow$ *for some*
[p. 99, $\ell + 1$] *start* $\rightsquigarrow$ *star*
[p. 105, Lemma 10.14] In the proof of the cases involving 0 or 1 anchor, a combination of the decompositions from Sects. 8.4 and 9 are used. These decompositions have not been shown to be compatible. Instead, it is better to combine $\Delta(v, W)$ with t_0-truncation on the rest of the quad-cluster. With a t_0 truncation, we no longer have the nonpositivity results from Sect. 8. (The quoins give a positive contribution.) However, a routine calculation shows that the estimate on $\Delta(v, W)$ is sufficiently small so that we still obtain a constant less than -1.04 pt.
[p. 116] [p. 121] Definition 11.7 allows masked flat in definition of 3-unconfined. Definition 11.24 requires no masked flats in the same definition. Use Definition 11.24 (no masked flats), rather than 11.7. Where masked flats occur, treat them with Lemma 11.23, parts (1) and (2).
[p. 116, $\ell + 1$] *Lemma* 4.16 $\rightsquigarrow$ *Lemma* 4.17
[p. 117, before Lemma 11.9] *two others* $\rightsquigarrow$ *three others*
[p. 117, Definition 11.8] $y1 \rightsquigarrow y_1$
[p. 119, Definition 11.5] By definition, we require a masked flat quarter to be a strict quarter.
[p. 121] See p. 116.
[p. 121, $\ell - 5$] $0.2274 \rightsquigarrow 0.02274$
[p. 123, flat case (2)] It is missing isolated quarters cut from the side. To fix this, in condition 2(f), $\eta_{456} \geq \sqrt{2} \rightsquigarrow \eta_{456} \geq \sqrt{2}$ *or* $\eta_{234} \geq \sqrt{2}$
[p. 124, Lemma 11.27] *The bound of 0 is established in Theorem 8.4* $\rightsquigarrow$ *The bound of 0 for upright quarters appears in Lemma 8.12. The bound of 0 for the other anchored simplex appears in Lemma 8.7 or 8.13, depending on the circumradius.*
[p. 126] Theorem 12.1 should include $\sigma_R(D) \leq s_n$ with $s_3 = 1$ pt and $s_4 = 0$, and $\tau_R(D) \geq t_3 = 0$.
[p. 131] Section 8 gives the deformation arguments that produce a biconnected graph.
[p. 139, Lemma 12.18, proof, $\ell + 3$] $C_0(|v|, \pi) \rightsquigarrow C_0^u(|v|, \pi)$
[p. 139, Lemma 12.18, proof, $\ell + 6$] $\tau_0(C_0^u(2t_0, \pi)) - \pi_{\max} \rightsquigarrow \tau_0(C_0^u(2.2, \pi)) - \pi_{\max}$
[p. 144, $\ell + 11, \ell + 17$] $2t_0^2 \rightsquigarrow (2t_0)^2$
[p. 146] $S_n^{\pm} \rightsquigarrow$ *of 3-crowded, 3-undefined, and 4-crowded combinations*
[p. 148, Sect. 13.6] This entire section is misplaced. It belongs with Sects. 25.5 and 25.6.
[p. 149, before 13.7] *the diagrams* $\rightsquigarrow$ *Figs.* 25.1–25.4
[p. 149, p. 156] The definition of δ_{loop} was accidentally dropped from the published version. Set $\delta_{\text{loop}}(4, 2) = 0.12034$, $\delta_{\text{loop}}(5, 1) = 0.24939$. These constants and their properties appear in the earlier 2002 arXiv preprint of the proof *The Kepler conjecture (Sphere Packings VI)*.
[p. 156, Lemma 13.5, $\ell + 4$] *respectively for* $\tau_R(D) \rightsquigarrow$ *respectively, for* $\sigma_R(D)$ *and* $\tau_R(D)$
[p. 164, $\ell - 1$] *This shows... occur.* $\rightsquigarrow$ *This completes the proof.*
[p. 173, $\ell + 4$] Insert the subscript on b, as in Proposition 15.5, starting on p. 173: $b \rightsquigarrow b_q$.
[p. 182, Lemma 16.7] The bound of 0 has not been shown to hold on each half. This is not a direct consequence of Theorem 8.4 as claimed. This can be fixed as follows. Let v_1, v_3 be the corners giving the endpoints of the long edge of the acute triangle at 0, and let v_2, v_4 be the other two corners. If either vertex v_1, v_3 has height greater than 2.3, show that the voro_0-scored quad cluster scores less than -1.04 pt. For this, we may use the deformations of Lemmas 12.10. The length of the diagonal along the acute face remains fixed and at least $\sqrt{8}$. We claim that these deformations produce a diagonal of length less than $\sqrt{8}$ between opposite corners of the quadrilateral. (If not, the deformations produce a rhombus with side 2 and diagonals both greater than $\sqrt{8}$, which is a geometric impossibility.) We cut the quad cluster along the diagonal of length $\sqrt{8}$ and continue with deformations until the top edges on each simplex are $(y_4, y_5, y_6) = (2, 2, \sqrt{8})$. We may apply CALC-474496219 and CALC-

8990938295 to the two separate simplices to obtain the result. Now we may assume that the heights of v_1, v_3 are at most 2.3. If either height is at least 2.1, the result follows from CALC-5127197465, which gives the bound of 0 on each half.

Finally, we have the case where both heights are at most 2.1. We may apply dimension-reduction techniques so that the each of the two remaining corners $v \neq v_1, v_3$ of the quad cluster either has height 2 or has distance 2 from v_1 or v_3. We then reprove Lemma 16.8 without using the bound of 0 and Lemma 16.9 for tetrahedra without the bound -1.04 pt. This appears in CALC-1551562505 and CALC-3013446042.

If the dimension reduction drops the cross-diagonal $\{v_2, v_4\}$ all the way to $\sqrt{8}$, then we may swap diagonals and continue, until both diagonals are exactly $\sqrt{8}$. In this case, by the cases already considered, we may assume that each corner has height at most 2.1. Also, geometric considerations give that the other edges are at most 2.02:

$$\mathcal{E}(2, 2, 2, \sqrt{8}, 2.1, 2.1, 2, 2, 2.02) > \sqrt{8}.$$

The result follows in this case by CALC-4723770703.

[p. 241] *Mixed* is defined so as to include the pure analytic case. In earlier articles, "mixed" excludes the pure analytic. *mixed ⤳ mixed or pure*

[p. 243, $\ell + 13$, $\ell + 14$, $\ell + 15$] Delete three sentences: '*Let v_{12} be … We let … Break the pentagon …*'

[p. 248, last displayed formula] $=\ ⤳\ +$ so that it reads

$$\sum_i f_{R_i}(D) \leq \hat{\sigma}(Q_i) + \mathrm{vor}_{R',0}(D) + \pi_R.$$

[p. 252, Sect. 25.7, Cases 2 and 3] "The flat quarter" is mentioned, but there are no flat quarters that have been introduced into the context. This passage has been displaced from its original context.

[p. 254, $\ell + 7$] *to branch combine ⤳ to combine*

References

1. Appel, K., Haken, W.: The four color proof suffices. Math. Intell. **8**(1), 10–20 (1986)
2. Bauer, G., Nipkow, T.: Flyspeck I: Tame graphs. In: Klein, G., Nipkow, T., Paulson, L. (eds.) The Archive of Formal Proofs. http://afp.sf.net/entries/Flyspeck-Tame.shtml, May 2006
3. Collins, G.E.: Quantifier elimination for real closed fields by cylindrical algebraic decomposition. In: Automata Theory and Formal Languages, Second GI Conf., Kaiserslautern, 1975. Lecture Notes in Comput. Sci., vol. 33, pp. 134–183. Springer, Berlin (1975)
4. Denney, E.: A prototype proof translator from HOL to Coq. In: TPHOLs'00: Proceedings of the 13th International Conference on Theorem Proving in Higher Order Logics, London, UK, 2000, pp. 108–125. Springer, Berlin (2000)
5. Fejes Tóth, L.: Lagerungen in der Ebene auf der Kugel und im Raum, 2nd edn. Springer, Berlin (1972)
6. Ferguson, S.P.: Sphere packings V. Pentahedral prisms. Discrete Comput. Geom. **36**(1), 167–204 (2006)
7. Garloff, J.: Convergent bounds for the range of multivariate polynomials. In: Proceedings of the International Symposium on Interval Mathematics on Interval Mathematics 1985, London, UK, 1985, pp. 37–56. Springer, Berlin (1985)
8. Gonthier, G.: Formal proof—the four-colour theorem. Not. Am. Math. Soc. **55**(11), 1382–1393 (2008)
9. Hales, T.C.: Errata and revisions to the proof of the Kepler conjecture. http://code.google.com/p/flyspeck/
10. Hales, T.C.: Sphere packings. I. Discrete Comput. Geom. **17**, 1–51 (1997)

11. Hales, T.C.: Kepler conjecture source code (1998). http://www.math.pitt.edu/~thales/kepler98/
12. Hales, T.C.: A proof of the Kepler conjecture. Ann. Math. **162**, 1065–1185 (2005)
13. Hales, T.C.: Introduction to the Flyspeck project. In: Coquand, T., Lombardi, H., Roy, M.-F. (eds.) Mathematics, Algorithms, Proofs, number 05021 in Dagstuhl Seminar Proceedings, Dagstuhl, Germany, 2006. Internationales Begegnungs- und Forschungszentrum für Informatik (IBFI), Schloss Dagstuhl, Germany. http://drops.dagstuhl.de/opus/volltexte/2006/432
14. Hales, T.C.: Sphere packings. III. Extremal cases. Discrete Comput. Geom. **36**(1), 71–110 (2006)
15. Hales, T.C.: Sphere packings. IV. Detailed bounds. Discrete Comput. Geom. **36**(1), 111–166 (2006)
16. Hales, T.C.: Sphere packings. VI. Tame graphs and linear programs. Discrete Comput. Geom. **36**(1), 205–265 (2006)
17. Hales, T.C.: Some methods of problem solving in elementary geometry. In: LICS '07: Proceedings of the 22nd Annual IEEE Symposium on Logic in Computer Science, Washington, DC, USA, 2007, pp. 35–40. IEEE Comput. Soc., Los Alamitos (2007)
18. Hales, T.C.: The Flyspeck Project (2007). http://code.google.com/p/flyspeck
19. Hales, T.C.: The Jordan curve theorem, formally and informally. Am. Math. Mon. **114**(10), 882–894 (2007)
20. Hales, T.C.: Flyspeck: A blueprint for the formal proof of the Kepler conjecture (2008). Source files at http://code.google.com/p/flyspeck/source/browse/trunk/
21. Hales, T.C.: Lemmas in elementary geometry (2008). Source files at http://code.google.com/p/flyspeck/source/browse/trunk/
22. Hales, T.C.: Formal proof. Not. Am. Math. Soc. **55**(11), 1370–1380 (2008)
23. Hales, T.C., Ferguson, S.P.: The Kepler conjecture. Discrete Comput. Geom. **36**(1), 1–269 (2006)
24. Hales, T.C., McLaughlin, S.: A proof of the Dodecahedral conjecture. J. Am. Math. Soc. (2009, to appear). math/9811079
25. Harrison, J.: HOL Light: A tutorial introduction. In: Srivas, M., Camilleri, A. (eds.) Proceedings of the First International Conference on Formal Methods in Computer-Aided Design (FMCAD'96). Lecture Notes in Computer Science, vol. 1166, pp. 265–269. Springer, Berlin (1996)
26. Harrison, J.: A HOL theory of Euclidean space. In: Theorem Proving in Higher Order Logics. Lecture Notes in Comput. Sci., vol. 3603, pp. 114–129. Springer, Berlin (2005)
27. Harrison, J.: Verifying nonlinear real formulas via sums of squares. In: Schneider, K., Brandt, J. (eds.) Proceedings of the 20th International Conference on Theorem Proving in Higher Order Logics, TPHOLs 2007. Lecture Notes in Computer Science, vol. 4732, pp. 102–118. Springer, Kaiserslautern (2007)
28. Harrison, J.: Formal proof—theory and practice. Not. AMS **55**(11), 1395–1406 (2008)
29. Hörmander, L.: The Analysis of Linear Partial Differential Operators. II. Classics in Mathematics. Springer, Berlin (2005). Differential Operators with Constant Coefficients; reprint of the 1983 original
30. IEEE Standards Committee 754. IEEE Standard for binary floating-point arithmetic, ANSI/IEEE Standard 754-1985. Institute of Electrical and Electronics Engineers, New York (1985)
31. Mahboubi, A., Pottier, L.: Elimination des quantificateurs sur les réels en Coq. In: Journées Francophones des Langages Applicatifs (JFLA) (2002). Available on the Web from http://www.lix.polytechnique.fr/~assia/Publi/jfla02.ps
32. McLaughlin, S.: KeplerCode: computer resources for the Kepler and Dodecahedral Conjectures. http://code.google.com/p/kepler-code/
33. McLaughlin, S.: An interpretation of Isabelle/HOL in HOL Light. In: Furbach, U., Shankar, N. (eds.) IJCAR. Lecture Notes in Computer Science, vol. 4130, pp. 192–204. Springer, Berlin (2006)
34. McLaughlin, S., Harrison, J.: A proof-producing decision procedure for real arithmetic. In: Automated deduction—CADE-20. Lecture Notes in Comput. Sci., vol. 3632, pp. 295–314. Springer, Berlin (2005)
35. Milner, R., Tofte, M., Harper, R.: The Definition of Standard ML. MIT Press, Cambridge (1990)
36. Monniaux, D.: The pitfalls of verifying floating-point computations. TOPLAS **30**(3), 12 (2008)
37. Nipkow, T., Paulson, L., Wenzel, M.: In: Isabelle/HOL: A Proof Assistant for Higher-Order Logic. Lect. Notes in Comp. Sci., vol. 2283. Springer, Berlin (2002). http://www.in.tum.de/~nipkow/LNCS2283/
38. Nipkow, T., Bauer, G., Schultz, P.: Flyspeck I: Tame graphs. In: Furbach, U., Shankar, N. (eds.) Automated Reasoning (IJCAR 2006). Lect. Notes in Comp. Sci., vol. 4130, pp. 21–35. Springer, Berlin (2006)
39. Obua, S.: Flyspeck II: The basic linear programs. PhD thesis, Technische Universität München (2008)
40. Obua, S., Skalberg, S.: Importing HOL into Isabelle/HOL. In: Automated Reasoning. Lecture Notes in Computer Science, vol. 4130, pp. 298–302. Springer, Berlin (2006)

41. Parrilo, P.A.: Semidefinite programming relaxations for semialgebraic problems. Math. Program., Ser. B **96**(2), 293–320 (2003). Algebraic and geometric methods in discrete optimization
42. Revol, N., Rouillier, F.: Motivations for an arbitrary precision interval arithmetic and the MPFI library. Reliab. Comput. **11**(4), 275–290 (2005)
43. Robertson, N., Sanders, D., Seymour, P., Thomas, R.: The four-colour theorem. J. Comb. Theory, Ser. B **70**, 2–44 (1997)
44. Solovay, R.M., Arthan, R.D., Harrison, J.: Some new results on decidability for elementary algebra and geometry. APAL (2009, submitted)
45. Tarski, A.: A Decision Method for Elementary Algebra and Geometry, 2nd edn. University of California Press, Berkeley (1951)
46. Weekes, S.: MLton. http://mlton.org
47. Wiedijk, F.: Encoding the HOL Light logic in Coq. http://www.cs.ru.nl/~freek/notes/holl2coq.pdf
48. Wiedijk, F. (eds.): The Seventeen Provers of the World. Lecture Notes in Computer Science, vol. 3600. Springer, Berlin (2006). Foreword by Dana S. Scott, Lecture Notes in Artificial Intelligence
49. Zumkeller, R.: Global optimization in type theory. PhD thesis, École Polytechnique Paris (2008)

Initial Papers of the Hales Program

Introduction to Part IV

Part IV presents two initial papers of T. C. Hales published in 1997, establishing his approach for proving the Kepler Conjecture. These are T. C. Hales, "Sphere Packings I" (*Discrete Comput. Geom.*, **17** (1997), 1–51) and T. C. Hales, "Sphere Packings II" (*Discrete Comput. Geom.*, **18** (1997), 135–149).

The general approach outlined in these papers had to be modified in the eventual proof, as explained in the first of the six papers in Part II. The final proof was then rewritten to be logically independent of these papers. These papers contain crucial ideas suggesting the form of the final proof.

10

Sphere Packings I, by T. C. Hales

This 1997 paper is Hales's first paper on the Kepler Conjecture project, outlining his initial proof approach.

Contents

Sphere Packings I, by T. C. Hales (Discrete Comput. Geom., **17** (1997), 1–51).

The original version of this chapter was revised. An erratum to this chapter can be found at
http://dx.doi.org/10.1007/978-1-4614-1129-1_12

Discrete Comput Geom 17:1–51 (1997)

Discrete & Computational

Geometry

© 1997 Springer-Verlag New York Inc.

Sphere Packings, I

T. C. Hales

Department of Mathematics, University of Michigan,
Ann Arbor, MI 48109, USA
hales@umich.edu

Abstract. We describe a program to prove the Kepler conjecture on sphere packings. We then carry out the first step of this program. Each packing determines a decomposition of space into Delaunay simplices, which are grouped together into finite configurations called Delaunay stars. A score, which is related to the density of packings, is assigned to each Delaunay star. We conjecture that the score of every Delaunay star is at most the score of the stars in the face-centered cubic and hexagonal close packings. This conjecture implies the Kepler conjecture. To complete the first step of the program, we show that every Delaunay star that satisfies a certain regularity condition satisfies the conjecture.

1. Introduction

The Kepler conjecture asserts that no packing of spheres in three dimensions has density exceeding that of the face-centered cubic lattice packing. This density is $\pi/\sqrt{18} \approx 0.74048$. In an earlier paper [H2] we showed how to reduce the Kepler conjecture to a finite calculation. That paper also gave numerical evidence in support of the method and conjecture. This finite calculation is a series of optimization problems involving up to 53 spheres in an explicit compact region of Euclidean space. Computers have little difficulty in treating problems of this size numerically, but a naive attempt to make a thorough study of the possible arrangements of these spheres would quickly exhaust the world's computer resources.

The first purpose of this paper is to describe a program designed to give a rigorous proof of the Kepler conjecture. A sketch of a related program appears in [H2]. Although the approach of [H2] is based on substantial numerical evidence, some of the constructions of that paper are needlessly complicated. This paper streamlines some of those constructions and replaces others with constructions that are more amenable to rigorous methods. For this program to succeed, the original optimization problem must be partitioned into a series of much smaller problems that may be treated by current computer technology or hand calculation.

The second purpose of this paper is to carry out the first step of the proposed program. A statement of the result is contained in Theorem 1 below.

Background to another approach to this problem is found in [H3]. To add more detail to the proposed program, we recall some constructions from earlier papers [H1], [H2]. Begin with a packing of nonoverlapping spheres of radius 1 in Euclidean three-space. The *density* of a packing is defined in [H1]. It is defined as a limit of the ratio of the volume of the unit balls in a large region of space to the volume of the large region. The density of the packing may be improved by adding spheres until there is no further room to do so. The resulting packing is said to be *saturated*. It has the property that no point in space has distance greater than 2 from the center of some sphere.

Every saturated packing gives rise to a decomposition of space into simplices called the *Delaunay decomposition*. The vertices of each Delaunay simplex are centers of spheres of the packing. None of the centers of the spheres of the packing lie in the interior of the circumscribing sphere of any Delaunay simplex. In fact, this property is enough to determine the Delaunay decomposition completely except for certain degenerate packings. A degeneracy occurs, for instance, when two Delaunay simplices have the same circumscribing sphere. In practice, these degeneracies are important, because they occur in the face-centered cubic and hexagonal close packings. The paper [H2] shows how to resolve the degeneracies by taking a small perturbation of the packing. In general, the Delaunay decomposition will depend on this perturbation. We refer to the centers of the packing as *vertices*, since the structure of the simplicial decomposition of space is our primary concern. For a proof that the Delaunay decomposition is a dissection of space into simplices, we refer the reader to [R].

The Delaunay decomposition is dual to the well-known Voronoi decomposition. If the vertices of the Delaunay simplices are in nondegenerate position, two vertices are joined by an edge exactly when the two corresponding Voronoi cells share a face, three vertices form a face exactly when the three Voronoi cells share an edge, and four vertices form a simplex exactly when the four corresponding Voronoi cells share a vertex. In other words, two vertices are joined by an edge if they lie on a sphere that does not contain any other of the vertices, and so forth (again assuming the vertices to be in nondegenerate position). The collection of all simplices that share a given vertex is called a *Delaunay star*. (This is a provisional definition: it is refined below.)

Every Delaunay simplex has edges between 2 and 4 in length and, because of the saturation of the packing, a circumradius of at most 2. We assume that every simplex S in this paper comes with a fixed order on its edges, $1, \ldots, 6$. The order on the edges is to be arranged so that the first, second, and third edges meet at a vertex. We may also assume that the edges numbered i and $i + 3$ are opposite edges for $i = 1, 2, 3$. We define $S(y_1, \ldots, y_6)$ to be the (ordered) simplex whose ith edge has length y_i. If S is a Delaunay simplex in a fixed Delaunay star, then it has a distinguished vertex, the vertex common to all simplices in the star. In this situation we assume that the edges are numbered so that the first, second, and third edges meet at the distinguished vertex.

A function, known as the *compression* $\Gamma(S)$, is defined on the space of all Delaunay simplices. Let $\delta_{oct} = (-3\pi + 12 \arccos(1/\sqrt{3}))/\sqrt{8} \approx 0.720903$ be the density of a regular octahedron with edges of length 2. That is, place a unit ball at each vertex of the octahedron, and let δ_{oct} be the ratio of the volume of the part of the balls in the octahedron to the volume of the octahedron. Let S be a Delaunay simplex. Let B be the union of four unit balls placed at each of the vertices of S. Define the compression as

$$\Gamma(S) = -\delta_{oct}\,\text{vol}(S) + \text{vol}(S \cap B).$$

We extend the definition of compression to Delaunay stars D^* by setting $\Gamma(D^*) = \sum \Gamma(S)$, with the sum running over all the Delaunay simplices in the star.

In this and subsequent work, we single out for special treatment the edges of length between 2 and 2.51. The constant 2.51 was determined experimentally to have a number of desirable properties. This constant appears throughout the paper. We call vertices that come within 2.51 of each other *close neighbors*.

We say that the convex hull of four vertices is a *quasi-regular tetrahedron* (or simply a *tetrahedron*) if all four vertices are close neighbors of one another. Suppose that we have a configuration of six vertices in bijection with the vertices of an octahedron with the property that two vertices are close neighbors if and only if the corresponding vertices of the octahedron are adjacent. Suppose further that exactly one of the three diagonals has length at most $2\sqrt{2}$. In this case we call the convex hull of the six vertices a *quasi-regular octahedron* (or simply an *octahedron*).

The compatibility of quasi-regular tetrahedra and octahedra with the Delaunay decomposition is established in Section 3. We think of Euclidean space as the union of quasi-regular tetrahedra, octahedra, and various less-interesting Delaunay simplices. From now on, a Delaunay star is to be the collection of all quasi-regular tetrahedra, octahedra, and Delaunay simplices that share a common vertex v. This collection of Delaunay simplices and quasi-regular solids is often, but not always, the same as the objects called Delaunay stars in [H2]. We warn the reader of this shift in terminology.

It is convenient to measure the compression in multiples of the compression of the regular simplex of edge length 2. We define a *point* (abbreviated pt) to be $\Gamma(S(2,2,2,2,2,2))$. We have $pt = 11\pi/3 - 12\arccos(1/\sqrt{3}) \approx 0.0553736$.

One of the main purposes of this paper and its sequel is to replace the compression by a function (called the *score*) that has better properties than the compression. Further details on the definition of score appear in Section 2. We are now able to state the main theorem of this paper.

Theorem 1. *If a Delaunay star is composed entirely of quasi-regular tetrahedra, then its score is less than 8 pt.*

The idea of the proof is the following. Consider the unit sphere whose center is the center of the Delaunay star D^*. The intersection of a simplex in D^* with this unit sphere is a spherical triangle. For example, a regular tetrahedron with edges of length 2 gives a triangle on the unit sphere of arc length $\pi/3$. The star D^* gives a triangulation of the unit sphere. The restriction on the lengths of the edges of a quasi-regular tetrahedron constrains the triangles in the triangulation. We classify all triangulations that potentially come from a star scoring more than 8 pt. Section 5 develops a long list of properties that must be satisfied by the triangulation of a high-scoring star.

It is then necessary to classify all the triangulations that possess the properties on this list. The original classification was carried out by a computer program, which generated all potential triangulations and checked them against the list. D. J. Muder has made a significant improvement in the argument by giving a direct, computer-free classification. His result appears in the Appendix.

As it turns out, there is only one triangulation that satisfies all of the properties on the list. Section 7 proves that Delaunay stars with this triangulation score less than 8 *pt*. This will complete the main thread of the argument.

There are a number of estimates in this paper that are established by computer. These estimates are used throughout the paper, even though their proofs are not discussed until Sections 8 and 9. These sections may be viewed as a series of technical appendices giving explicit formulas for the compression, dihedral angles, solid angles, volumes, and other quantities that must be estimated. The final section states the inequalities and gives details about the computerized verification. There is no vicious circle here: the results of Sections 8 and 9 do not rely on anything from Sections 2–7.

There are several functions of a Delaunay simplex that are used throughout this paper. The compression $\Gamma(S)$ has been defined above. The *dihedral angle* dih(S) is defined to be the dihedral angle of the simplex S along the first edge (with respect to the fixed order on the edges of S). Set $\text{dih}_{min} = \text{dih}(S(2, 2, 2.51, 2, 2, 2.51)) \approx 0.8639$; $\text{dih}_{max} = \text{dih}(S(2.51, 2, 2, 2.51, 2, 2)) = \arccos(-29003/96999) \approx 1.874444$. We will see that dih_{min} and dih_{max} are lower and upper bounds on the dihedral angles of quasi-regular tetrahedra. The *solid angle* (measured in steradians) at the vertex joining the first, second, and third edges is denoted sol(S). The intersection of S with the ball of unit radius centered at this vertex has volume sol$(S)/3$ (see [H1, 2.1]). For example, $\text{sol}(S(2, 2, 2, 2, 2, 2)) \approx 0.55$. Let rad$(S)$ be the circumradius of the simplex S. In Section 2 we define two other functions: vor(S), which is related to the volume of Voronoi cells, and the score $\sigma(S)$. Finally, let $\eta(a, b, c)$ denote the circumradius of a triangle with edges a, b, c. Explicit formulas for all these functions appear in Section 8.

2.   The Program

By proving Theorem 1, the main purpose of this paper will be achieved. Nevertheless, it might be helpful to give a series of comments about how Theorem 1 may be viewed as the solution to the first of a handful of optimization problems that would collectively provide a solution to the Kepler conjecture.

We begin with some notation and terminology. We fix a Delaunay star D^* about a vertex v_0, which we take to be the origin, and we consider the unit sphere at v_0. Let v_1 and v_2 be vertices of D^* such that v_0, v_1, and v_2 are all close neighbors of one another. We take the radial projections p_i of v_i to the unit sphere with center at the origin and connect the points p_1 and p_2 by a geodesic arc on the sphere. We mark all such arcs on the unit sphere. Lemma 3.10 shows that the arcs meet only at their endpoints. The closures of the connected components of the complement of these arcs are regions on the unit sphere, called the *standard regions*. We may remove the arcs that do not bound one of the regions. The resulting system of edges and regions is referred to as the *standard decomposition* of the unit sphere.

Let C be the cone with vertex v_0 over one of the standard regions. The collection of the Delaunay simplices, quasi-regular tetrahedra, and quasi-regular octahedra of D^* in C (together with the distinguished vertex v_0) is called a *standard cluster*. Each Delaunay simplex in D^* belongs to a unique standard cluster. Each triangle in the standard decomposition of the unit sphere is associated with a unique quasi-regular tetrahedron, and each tetrahedron determines a triangle in the standard decomposition (Lemma 3.7). We may identify quasi-regular tetrahedra with clusters over triangular regions.

We assign a score to each standard cluster in [H4, 3]. In this section we define the score of a quasi-regular tetrahedron and describe the properties that the score should have in general.

Let S be a quasi-regular tetrahedron. It is a standard cluster in a Delaunay star with center v_0. If the circumradius of S is at most 1.41, then we define the score to be $\Gamma(S)$.

If the circumradius is greater than 1.41, then embed the simplex S in Euclidean three-space. Partition Euclidean space into four infinite regions (infinite Voronoi cells) by associating with each vertex of S the points of space closest to that vertex. By intersecting S with each of the four regions, we partition S into four pieces $\hat{S}_0$, $\hat{S}_1$, $\hat{S}_2$, and $\hat{S}_3$, corresponding to its four vertices v_0, v_1, v_2, and v_3. Let sol_i be the solid angle at the vertex v_i of the simplex. The expression $-4\delta_{oct} \mathrm{vol}(\hat{S}_0) + 4\,\mathrm{sol}_0/3$ is an analytic function of the lengths of the edges for simplices S that contain their circumcenters. (Explicit formulas appear in Section 8.) This function may be analytically continued to a function of the lengths of the edges for simplices S that do not necessarily contain their circumcenters. Let $\mathrm{vor}(S)$ be defined as the analytic continuation of $-4\delta_{oct} \mathrm{vol}(\hat{S}_0) + 4\,\mathrm{sol}_0/3$.

In this case define the score of S to be $\mathrm{vor}(S)$. Write $\sigma(S)$ for the score of a quasi-regular tetrahedron. In summary, the score is

$$\sigma(S) = \begin{cases} \Gamma(S), & \text{if the circumradius of } S \text{ is at most } 1.41, \\ \mathrm{vor}(S) & \text{otherwise.} \end{cases}$$

The quasi-regular tetrahedron S appears in four Delaunay stars. In the other three Delaunay stars, the distinguished vertices will be v_1, v_2, and v_3, so that S, viewed as a standard cluster in the other Delaunay stars, will have scores $-\delta_{oct} \mathrm{vol}(\hat{S}_i) + \mathrm{sol}_i/3$ (or their analytic continuations), for $i = 1, 2, 3$. By definition,

$$\Gamma(S) = \sum_{i=0}^{3} (-\delta_{oct} \mathrm{vol}(\hat{S}_i) + \mathrm{sol}_i/3).$$

The sum of the scores of S, for each of the four vertices of S, is $4\Gamma(S)$. This is the same total that is obtained by summing the compression of S at each of its vertices. This is the property we need to relate the score to the density of the packing. It means that although the score reapportions the compression among neighboring Delaunay stars, the average of the compression over a large region of space equals the average of the score over the same region, up to a negligible boundary term.

The analytic continuation in the definition of $\mathrm{vor}(S)$ has the following geometric interpretation. If the circumcenter of a quasi-regular tetrahedron S is not contained in S, a small tip of the infinite Voronoi cell at v_0 (or some other vertex) will protrude through the opposite face of the Delaunay simplex. The volume of this small protruding tip is not counted in $-4\delta_{oct} \mathrm{vol}(\hat{S}_0) + 4\,\mathrm{sol}_0/3$, but it is counted in the analytic continuation. The analytic continuations of the scores of S for each of the other three vertices acquire a term representing the negative volume of a part of the tip. The three parts together constitute the entire tip, so that the negative volumes exactly offset the volume of the tip, and the sum of the four scores of S is $4\Gamma(S)$. Details appear in [H4].

The general definition of the score will have similar properties. To each standard cluster of a Delaunay star D^* a score will be assigned. The rough idea is to let the score of a simplex in a cluster be the compression $\Gamma(S)$ is the circumradius of every face of S is small, and otherwise to let the score be defined by Voronoi cells (in a way that generalizes the definition for quasi-regular tetrahedra).

The score $\sigma(D^*)$ of a Delaunay star is defined as the sum of the scores of its standard clusters. The score has the following properties [H4, 3.1 and 3.5]:

1. The score of a standard cluster depends only on the cluster, and not on the way it sits in a Delaunay star or in the Delaunay decomposition of space.
2. The Delaunay stars of the face-centered cubic and hexagonal close packings score exactly 8 pt.
3. The score is asymptotic to the compression over large regions of space. We make this more precise. Let Λ denote the vertices of a saturated packing. Let Λ_N denote the vertices inside the ball of radius N. (Fix a center.) Let $D^*(v)$ denote the Delaunay star at $v \in \Lambda$. Then the score satisfies (in Landau's notation)

$$\sum_{\Lambda_N} \sigma(D^*(v)) = \sum_{\Lambda_N} \Gamma(D^*(v)) + O(N^2).$$

Lemma 2.1. *If the score of every Delaunay star in a saturated packing is at most $s < 16\pi/3$, then the density of the packing is at most $16\pi\,\delta_{oct}/(16\pi - 3s)$. If the score of every Delaunay star in a packing is at most 8 pt, then the density of the packing is at most $\pi/\sqrt{18}$.*

Proof. The second claim is the special case $s = 8\,pt$. The proof relies on property 3. The number of vertices such that $D^*(v)$ meets the boundary of the ball B_N of radius N has order $O(N^2)$. Since the Delaunay stars give a fourfold cover of $\mathbb{R}^3$, we have

$$4\left(-\delta_{oct}\,\mathrm{vol}(B_N) + |\Lambda_N|\frac{4\pi}{3}\right) = \sum_{\Lambda_N}(-\delta_{oct}\,\mathrm{vol}(D^*(v)) + \frac{4\,\mathrm{sol}(D^*(v))}{3} + O(N^2)$$

$$= \sum_{\Lambda_N} \Gamma(D^*(v)) + O(N^2)$$

$$= \sum_{\Lambda_N} \sigma(D^*(v)) + O(N^2)$$

$$\leq s|\Lambda_N| + O(N^2).$$

Rearranging, we get

$$\frac{4\pi|\Lambda_N|}{3\,\mathrm{vol}(B_N)} \leq \frac{\delta_{oct}}{(1 - 3s/16\pi)} + \frac{O(N^2)}{\mathrm{vol}(B_N)}.$$

In the limit the left-hand side is the density and the right-hand side is the bound. Similar arguments can be found in [H1] and [H2]. $\qquad\square$

The following conjecture is fundamental. By the lemma, this conjecture implies the Kepler conjecture. The lemma also shows that weaker bounds than 8 *pt* on the score might be used to give new upper bounds on the density of sphere packings.

Conjecture 2.2.   *The score of every Delaunay star is at most 8 pt.*

The basic philosophy behind the approach of this paper is that quasi-regular tetrahedra are the only clusters that give a positive score, standard clusters over quadrilateral regions should be the only other clusters that may give a score of zero, and every other standard cluster should give a negative score. Moreover, we will prove that no quasi-regular tetrahedron gives more than 1 *pt*.

Thus, heuristically, we try to obtain a high score by including as many triangular regions as possible. If we allow any other shape, preference should be given to quadrilaterals. Any other shape of region should be avoided if possible. If these other regions occur, they should be accompanied by additional triangular regions to compensate for the negative score of the region. We will see later in this paper that even triangular regions tend to give a low score unless they are arranged to give five triangles around each vertex.

The main steps in a proof of the Kepler conjecture are

1. *A proof that even if all regions are triangular the total score is less than 8 pt.*
2. *A proof that standard clusters in regions of more than three sides score at most 0 pt.*
3. *A proof that if all of the standard regions are triangles or quadrilaterals, then the total score is less than 8 pt (excluding the case of pentagonal prisms).*
4. *A proof that if some standard region has more than four sides, then the star scores less than 8 pt.*
5. *A proof that pentagonal prisms score less than 8 pt.*

The division of the problem into these steps is quite arbitrary. They were originally intended to be steps of roughly equal magnitude, although is has turned out that a construction in [H4] has made the second step substantially easier than the long calculations of the third step.

This paper carries out the first step. The second step of the program is also complete [H4]. Partial results are known for the third step [H5]. In the fourth step it will be necessary to argue that these regions take up too much space, give too little in return, and have such strongly incompatible shapes that they cannot be part of a winning strategy.

To make step 5 precise, we define *pentagonal prisms* to be Delaunay stars whose standard decomposition has ten triangles and five quadrilaterals, with the five quadrilaterals in a band around the equator, capped on both ends by five triangles (Diagram 2.3). The conjecture in this section asserts, in particular, that pentagonal prisms, which created such difficulties in [H2], score less than 8 *pt*. The final step has been separated from the third step, because the estimates are expected to be more delicate for pentagonal prisms than for a general Delaunay star in the third step.

One of the main shortcomings of the compression is that pentagonal prisms have compression greater than 8 *pt* (see [H2]). Numerical evidence suggests that the upper bound on the compression is attained by a pentagonal prism, denoted D^*_{ppdp} in [H2], at

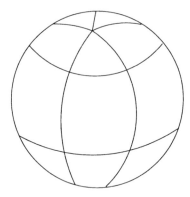

Diagram 2.3

about 8.156 *pt*, and this means that the link between the compression and the Kepler conjecture is indirect. The score appears to correct this shortcoming.

What evidence is there for the conjecture and program? They come as the result of extensive computer experimentation. I have checked the conjecture against much of the data obtained in the numerical studies of [H2, 9.3]. The data suggest that the score tends to give a dramatic improvement over the compression, often improving the bound by more than a point. The score of the particularly troublesome pentagonal prism D^*_{ppdp} drops safely under 8 *pt*. I have checked a broad assortment of other pentagonal prisms and have found them all to score less than 8 *pt*.

The second step shows that no serious pathologies can arise. The only way to form a Delaunay star with a positive score is by arranging a number of quasi-regular tetrahedra around a vertex (together with other standard clusters than can only lower the score). There must be at least eight tetrahedra to score 8 *pt*, and if there are any distortions in these tetrahedra, there must be at least nine. However, as this paper shows, too many quasi-regular tetrahedra in any star are also harmful. Future papers will impose additional limits on the structure of the optimal Delaunay star.

3. Quasi-Regular Tetrahedra

This section studies the compatibility of the Delaunay simplices and the quasi-regular solids. Fix three vertices v_1, v_2, and v_3 that are close neighbors to one another. Let T be the triangle with vertices v_i. It does not follow that T is the face of a Delaunay simplex. However, as we will see, when T is not the face of a simplex, the arrangement of the surrounding simplices is almost completely determined.

If T is not the face of a Delaunay simplex, then we will show that there are two additional vertices v_0 and v'_0, where v_0 and v'_0 are close neighbors to v_1, v_2, and v_3. This means that there are two quasi-regular tetrahedra S_1 and S_2 with vertices (v_0, v_1, v_2, v_3) and (v'_0, v_1, v_2, v_3), respectively, that have the common face T (see Diagram 3.1(a)). We see that $S_1 \cup S_2$ is the union of three Delaunay simplices with vertices (v_0, v'_0, v_1, v_2), (v_0, v'_0, v_2, v_3), and (v_0, v'_0, v_3, v_1) in Diagram 3.1(b). This section establishes that this is the only situation in which quasi-regular tetrahedra are not Delaunay simplices: they must come in pairs and their union must be three Delaunay simplices joined along a common edge. The decomposition of this paper is obtained by taking each such triple of Delaunay simplices (3.1(b)) and replacing the triple by a pair of quasi-regular tetrahedra (3.1(a)).

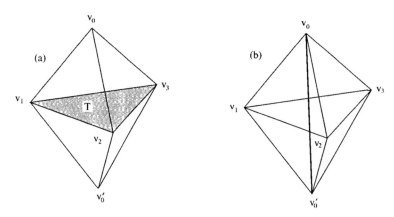

Diagram 3.1

From the dual perspective of Voronoi cells, the Voronoi cell at v_0 (or v_0') would have a small tip protruding from S_1 through T, if the vertex v_0' were not present. The vertex v_0' slices off this protruding tip so that the Voronoi cells at v_0 and v_0' have a small face in common.

Lemma 3.2. *Suppose that the circumradius of the triangle T is less that $\sqrt{2}$. Then T is the face of a Delaunay simplex.*

Proof. Let $r < \sqrt{2}$ be the radius of the circle that circumscribes T, and let c be the center of the circle. The sphere of radius r at c does not contain any vertices of D^* other than v_1, v_2, and v_3. By the definition of the Delaunay decomposition (as described in the Introduction), this implies that T is the face of a simplex. □

Remark 3.3. We have several constraints on the edge lengths, if T is not a face of a Delaunay simplex. Consider the circumradius $\eta(a, b, c)$ of a triangle whose edges have lengths a, b, c between 2 and 2.51. Since $2.51^2 < 2^2 + 2^2$, we see that the triangle is acute, so that $\eta(a, b, c,)$ is monotonically increasing in a, b, and c. This gives simple estimates relating the circumradius to a, b, and c. The circumradius is at most $\eta(2.51, 2.51, 2.51) = 2.51/\sqrt{3} \approx 1.449$. If the circumradius is at least $\sqrt{2} \approx 1.41421$, then a, b, and c are greater than $2.3(\eta(2.3, 2.51, 2.51) \approx 1.41191 < \sqrt{2})$. Under the same hypothesis, two of a, b, and c are greater than $2.41(\eta(2.41, 2.41, 2.51) < \sqrt{2})$. Finally, at least one edge has length greater than $2.44(\eta(2.44, 2.44, 2.44) < \sqrt{2})$.

Lemma 3.4. *Let T be the triangle with vertices v_i. Assume that the vertices v_i are close neighbors of one another. Suppose there is a vertex v_0 that lies closer to the circumcenter* *of T than the vertices of T do. Then the vertex v_0 satisfies $2 \leq |v_0 - v_i| < 2.15$, for $i = 1, 2, 3$. In particular, the convex hull of $v_0, \ldots, v_3$ is a quasi-regular tetrahedron S with face T.*

Another way of stating the hypothesis on the circumcenter is to say that the plane of T separates v_0 from the circumcenter of S. Because of the constraints on the edge lengths in Remark 3.3, the three other faces of S are faces of Delaunay simplices.

Proof. We defer the proof to Section 8.2.5. □

Lemma 3.5. *Let v, v_1, v_2, v_3, and v_4 be distinct vertices with pairwise distances at least 2. Suppose that the pairs (v_i, v_j) are close neighbors for $\{i, j\} \neq \{1, 4\}$. Then v does not lie in the convex hull of (v_1, v_2, v_3, v_4).*

Proof. For a contradiction, suppose v lies in the convex hull. Since $|v - v_i| \geq 2$ and for $\{i, j\} \neq \{1, 4\}$, v_i and v_j are close neighbors, the angle formed by v_i and v_j at the vertex v is at most $\theta_0 = \arccos(1 - 2.51^2/8) \approx 1.357$. Each such pair of vertices gives a geodesic arc of length at most θ_0 radians on the unit sphere centered at v. We obtain in this way a triangulation of the unit sphere by four triangles, two with edges of length at most $\theta_0 = 2\arcsin(2.51/4)$ radians, and two others that fit together to form a quadrilateral with edges of at most θ_0 radians. By the spherical law of cosines, the area formula for a spherical triangle, and [H2, 6.1], each of the first two triangles has area at most $3\arccos(\cos\theta_0/(1 + \cos\theta_0)) - \pi \approx 1.04$. By the same lemma, the quadrilateral has area at most the area of a regular quadrilateral of side θ_0, or about 2.8. Since the combined area of the two triangles and quadrilateral is less than 4π, they cannot give the desired triangulation. To see that v cannot lie on the boundary, it is enough to check that a triangle having two edges of lengths between 2 and 2.51 cannot contain a point that has distance 2 or more from each vertex. We leave this as an exercise. □

Corollary 3.5.1. *No vertex of the packing is ever an interior point of a quasi-regular tetrahedron or octahedron.*

Proof. The corollary is immediate for a tetrahedron. For an octahedron, draw the distinguished diagonal and apply the lemma to each of the four resulting simplices. □

Let T be a triangle of circumradius between $\sqrt{2}$ and $2.51/\sqrt{3}$, with edges of length between 2 and 2.51. Consider the line through the circumcenter of T, perpendicular to the plane of T. Let s be the finite segment in this line whose endpoints are the circumcenters of the two simplices with face T formed by placing an additional vertex at distance 2 from the three vertices of T on either side of the plane through T.

Lemma 3.6. *Let S be a simplex formed by the vertices of T and a fourth vertex v_0'. Suppose that the circumcenter of S lies on the segment s. Assume that v_0' has distance at least 2 from each of the vertices of T. Then v_0' has distance less than 2.3 from each of the vertices of T.*

Proof. Let v_1, v_2, v_3 be the vertices of T. For a contradiction, assume that v_0' has distance at least 2.3 from a vertex v_1 of T. Lemma 3.4 shows that the plane through T does not separate v_0' from the circumcenter of S. If the circumcenter lies in s (and is not separated from v_0' by the plane through T), then by moving v_0' to decrease the circumradius, the circumcenter remains in s.

Let S be the simplex with vertices v_0', v_1, v_2, and v_3. The circumcenter of S lies in the interior of S. We omit the proof, because it is established by methods similar to (but longer than) the proof of Lemma 3.4. Thus, the circumradius is increasing in the lengths of $|v_0' - v_i|$, for $i = 1, 2, 3$ (see paragraph 8.2.4).

Let R be the circumradius of a simplex with face T and center an endpoint of s. We will prove that the circumradius of S is greater than R, contrary to our hypothesis. Moving v_0' to decrease the circumradius, we may take the distances to v_i to be precisely 2.3, 2, and 2. We may move v_1, v_2, and v_3 along their fixed circumscribing circle until $|v_2 - v_3| = 2.51$ and $|v_1 - v_2| = |v_1 - v_3|$ in a way that does not decrease any of the distances from v_0' to v_i. Repeating the previous step, we may retain our assumption that v_0' has distances exactly 2.3, 2, and 2 from the vertices v_i as before. We have reduced the problem to a one-dimensional family of tetrahedra parametrized by the radius r of the circumscribing circle of T. Set $x(r) = |v_1 - v_3| = |v_2 - v_3|$. To obtain our desired contradiction, we must show that the circumradius $R'(r)$ of the simplex $S(2.3, 2, 2, 2.51, x(r), x(r))$ satisfies $R'(r) > R(r)$. Since both R' and R are increasing in r, for $r \in [\sqrt{2}, 2.51/\sqrt{3}]$, the desired inequality follows if we evaluate the 200 constants

$$R'(r_i) - R(r_{i+1}) \qquad \text{for} \quad r_i = \sqrt{2}\left(\frac{2.51}{\sqrt{3}} - \sqrt{2}\right)\frac{i}{200},$$

for $i = 0, \ldots, 199$, and check that they are all positive. (The smallest is about 0.00005799, which occurs for $i = 199$.) ☐

Let T be a triangle made up of three close neighbors. Suppose that T is not the face of a Delaunay simplex. There exists a vertex v_0 whose distance to the circumcenter of T is less than the circumradius of T. Let S be the quasi-regular tetrahedron formed by v_0 and the vertices of T. It is not a Delaunay simplex, so there exists a vertex v_0' that is less than the circumradius of S from the circumcenter of S. Let S' be the simplex formed by v_0' and the vertices of T. It is not a Delaunay simplex either. The circumcenter of S lies in the segment s of Lemma 3.4, so the circumcenter of S' does too. The lengths of the edges of S and S' other than T are constrained by Lemmas 3.4 and 3.6. In particular, in light of Remark 3.3, the faces other than T of S and S' are faces of Delaunay simplices.

If v_0 and v_0' lie on the same side of the plane through T, then either v_0' lies in S, v_0 lies in S', or the faces of S and S' intersect nonsimplicially. None of these situations can occur because a nondegenerate Delaunay decomposition is a Euclidean simplicial complex and because of Lemma 3.5. We conclude that v_0 and v_0' lie on opposite sides of the plane through T.

$S \cup S'$ is bounded by Delaunay faces, so $S \cup S'$ is a union of Delaunay simplices. The fourth vertex of the Delaunay simplex in $S \cup S'$ with face (v_0, v_1, v_2) cannot be v_3 (S is not a Delaunay simplex), so it must be v_0'. Similarly, (v_0, v_0', v_2, v_3) and (v_0, v_0', v_3, v_1) are Delaunay simplices. These three Delaunay simplices cannot be quasi-regular tetrahedra by Lemma 8.3.2.

The assumption that T is not a face has completely determined the surrounding geometry: there are two quasi-regular tetrahedra S and S' along T such that $S \cup S'$ is a union of three Delaunay simplices.

Lemma 3.7. *Let L be a union of standard regions. Suppose that the boundary of L consists of three edges. Then either L or its complement is a single triangle.*

For example, the interior of L cannot have the form of Diagram 3.8. Lemma 3.5 (proof) shows that if all regions are triangles, then there are at least 12 triangles, so that the exterior of L cannot have the form of Diagram 3.8 either.

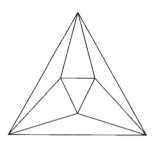

Diagram 3.8

Proof. Replacing L by its complement if necessary, we may assume that the area of L is less than its complement. The triangular boundary corresponds to four vertices v_0 (the origin), v_1, v_2, and v_3. The close-neighbor constraints on the lengths show that the convex hull of $v_0, \ldots, v_3$ is a quasi-regular tetrahedron. By construction each quasi-regular tetrahedron is a single cluster. $\square$

We say that a point $v \in \mathbb{R}^3$ is *enclosed* by a region on the unit sphere is the interior of the cone (with vertex v_0) over that region contains v. For example, in Diagram 3.9, the point v is enclosed by the given spherical triangle.

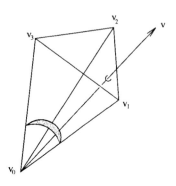

Diagram 3.9

The following lemma was used in Section 2 to define the standard decomposition.

Lemma 3.10. *Fix a Delaunay star D^* with center v_0. Draw geodesic arcs on the unit sphere at v_0 for every triple of close neighbors v_0, v_1, v_2 (as in Section 2). The resulting system of arcs do not meet except at endpoints.*

Proof. Our proof is based on the fact that a nondegenerate Delaunay decomposition is a Euclidean simplicial complex. Let T_1 and T_2 be two triangles made from two such triples of close neighbors. We have $T_i \subset S_i \cup S_i'$, where S_i and S_i are the Delaunay simplices with face T_i if T_i is the face of a Delaunay simplex, and they are the two quasi-regular tetrahedra with face T_i constructed above, otherwise. Since a nondegenerate Delaunay decomposition is a Euclidean simplicial complex, $S_1 \cup S_1'$ meets $S_2 \cup S_2'$ simplicially. By the restrictions on the lengths of the edges in Lemmas 3.4 and 3.6, this forces T_1 to intersect T_2 simplicially. The result follows.   □

4. Quadrilaterals

Fix a Delaunay star composed entirely of quasi-regular tetrahedra and consider the associated triangulation of the unit sphere. Let L be a region of the sphere bounded by four edges of the triangulation. L will be the union of two or more triangles. Replacing L by its complement in the unit sphere if necessary, we assume that the area of L is less than that of its complement.

We claim that, in this context, L is the union of either two or four triangles, as illustrated in Diagram 4.1. In particular, L encloses at most one vertex. If a diagonal to the quadrilateral L is an edge of the triangulation, the region L is divided into two triangles each associated with a quasi-regular tetrahedron. In particular, there is no enclosed vertex. (Lemma 3.7 precludes any subtriangulation of a triangular region.) If, however, there is a single enclosed vertex and neither diagonal is an edge of the triangulation, then the only possible triangulation of L is the one of the diagram. Proposition 4.2 completes the proof of the claim.

Diagram 4.1

Proposition 4.2. *A union of regions (of area less than 2π) bounded by exactly four edges cannot enclose two vertices of distance at most 2.51 from the origin.*

This argument is somewhat delicate: if our parameter 2.51 had been set at 2.541, for instance, such an arrangement would exist. First we prove a useful reduction.

Lemma 4.3. *Assume a figure exists with vectors $v_1, \ldots, v_4$, v, and v' subject to the constraints*

$$2 \le |v_i| \le 2.51,$$
$$2 \le |v_i - v_{i+1}| \le k_i,$$
$$2 \le |v_i - v_{i+2}|,$$
$$2 \le |v - v'|,$$
$$h_i \le |w - v_i|,$$
$$2 \le |w| \le \ell \quad \text{for} \quad w = v, v' \quad \text{and} \quad i = 1, \ldots, 4 \pmod 4,$$

where ℓ, h_i, and k_i are fixed constants that satisfy $\ell \in [2.51, 2\sqrt{2}]$, $h_i \in [2, 2\sqrt{2})$, $k_i \in [2, 2.51]$. Let L be the quadrilateral on the unit sphere with vertices $v_i/|v_i|$ and edges running between consecutive vertices. Assume that v and v' lie in the cone at the origin obtained by scaling L. Then another figure exists made of a (new) collection of vectors $v_1, \ldots, v_4$, v, and v' subject to the constraints above together with the additional constraints

$$|v_i - v_{i+1}| = k_i,$$
$$|v_i| = 2 \quad \text{for} \quad i = 1, \ldots, 4,$$
$$|v| = |v'| = \ell.$$

Moreover, the quadrilateral L may be assumed to be convex.

Proof. By rescaling v and v', we may assume that $|v| = |v'| = \ell$. (Moving v or v' away from the origin increases its distance from the other vertices of the configuration.)

The diagonals satisfy $|v_2 - v_4| > 2.1$ and $|v_1 - v_3| > 2.1$. Otherwise, if say $|v_1 - v_3| \le 2.1$, then the faces with vertices (v_1, v_2, v_3) and (v_1, v_4, v_3) have circumradius less than $\sqrt{2}$. The edge from 0 to v has length at most $\sqrt{2}$, so this edge cannot intersect these faces by the Euclidean simplicial complex argument used before Lemma 3.7 and in Lemma 3.10. By Lemma 3.5, v cannot lie in the convex hull of $(0, v_1, v_3, v_i)$. This leaves v nowhere to go, and a figure with $|v_1 - v_3| \le 2.1$ does not exist.

Next, we claim that we may assume that the quadrilateral L is convex (in the sense that it contains the geodesic arcs between any two points in the region). To see this, suppose the vertex v_i lies in the cone over the convex hull of the other three vertices v_j. Consider the plane P through the origin, v_{i-1}, and v_{i+1}. The reflection v_i' of v_i through P is no closer to v, v', or v_{i+2} and has the same distance to the origin, v_{i-1}, and v_{i+1}. Thus, replacing v_i with v_i' if necessary, we may assume that L is convex.

Most of the remaining deformations are described as pivots. We fix an axis and rotate a vertex around a circle centered along and perpendicular to the given axis. If, for example, $|v_3 - v_4| < k_3$, we pivot the vertex v_4 around the axis through 0 and v_1 until $|v_3 - v_4| = k_3$. It follows from the choice of axes that the distances from v_4 to the origin and v_1 are left unchanged, and it follows from the convexity of L that $|w - v_4|$ increases for $w = v_2, v_3, v$, and v'. Similarly, we may pivot vertices v_i along the axis through the origin and v_{i+1} until $|v_i - v_{i-1}| = k_i$, for all i.

Fix an axis through two opposite vertices (say v_1 and v_3) and pivot another vertex (say v_2) around the axis toward the origin. We wish to continue by picking different axes and pivoting until $|v_i| = 2$, for $i = 1, 2, 3, 4$. However, this process appears to break down in the event that a vertex v_i has distance h_i from one of the enclosed vertices and the pivot toward the origin moves v_i closer to that enclosed vector. We must check that this situation can be avoided.

Interchanging the roles of (v_1, v_3) with (v_2, v_4) as necessary, we continue to pivot until $|v_1| = |v_3| = 2$ or $|v_2| = |v_4| = 2$ (say the former). We claim that either v_2 has distance greater than h_2 from v or that pivoting v_2 around the axis through v_1 and v_3 moves v_2 away from v. If not, we find that $|v_1| = |v_3| = 2$, $|v - v_2| = h_2$ and that v lies in the cone $C = C(v_2)$ with vertex v_2 spanned by the vectors from v_2 to the origin, v_1, and v_3. (This relies on the convexity of the region L.)

To complete the proof, we show that this figure made from $(0, v_1, v_2, v_3, v)$ cannot exist. Contract the edge (v, v_2) as much as possible keeping the triangle $(0, v_1, v_3)$ fixed, subject to the constraints that $v \in C(v_2)$ and $|w - w'| \geq 2$, for $w = v, v_2$ and $w' = 0$, v_1, v_3. This contraction gives $|v - v_2| \leq h_2 < 2\sqrt{2}$.

Case 1: v lies in the plane of $(0, v_2, v_3)$. This gives an impossibility: crossing edges (v, v_2), $(0, v_3)$ of length less than $2\sqrt{2}$. Similarly, v cannot lie in the plane of $(0, v_1, v_2)$.

Case 2: v lies in the interior of the cone $C(v_2)$. The contraction gives $|v - w'| = 2$ for $w = v, v_2$ and $w' = 0$, v_1, v_3. The edge (v, v_2) divides the convex hull of $(0, v_1, v_2, v_3, v)$ into three simplices. Consider the dihedral angles of these simplices along this edge. The dihedral angle of the simplex (v_1, v_2, v_3, v) is less than π. The dihedral angles of the other two are less than $\mathrm{dih}(S(2\sqrt{2}, 2, 2, 2, 2, 2)) = \pi/2$. Hence, the dihedral angles along the diagonal cannot sum to 2π and the figure does not exist.

Case 3: v lies in the plane of (v_2, v_1, v_3). Let r be the radius of the circle in this plane passing through v_1 and v_3 obtained by intersecting the plane with a sphere of radius 2 at the origin. We have $2r \geq |v_1 - v_3| > 2\sqrt{2}$ because otherwise we have the impossible situation of crossing edges (v_1, v_3) and (v_2, v) of length less than $2\sqrt{2}$. Let H be the perpendicular bisector of the segment (v_1, v_3). By reflecting v through H if necessary we may assume that v and v_2 lie on the same side of H, say the side of v_3. Furthermore, by contracting (v, v_2), we may assume without loss of generality that $|v_3 - v| = |v_2 - v_3| = 2$. Let $f(r) = |v - v_2|$, as a function of r. f is increasing in r. The inequalities $2\sqrt{2} > h_2 \geq |v - v_2| \geq f(\sqrt{2}) = 2\sqrt{2}$ give the desired contradiction. $\qquad\square$

Proof of Proposition 4.2.   Assume for a contradiction that v and v' are vertices enclosed by L. Let the center of the Delaunay star be at the origin, and let $v_1, \ldots, v_4$, indexed consecutively, be the four vertices of the Delaunay star that determine the extreme points of L.

We describe a sequence of deformations of the configuration (formed by the vertices $v_1, \ldots, v_4, v, v'$) that transform the original configuration of vertices into particular rigid arrangements below. We show that these rigid arrangements cannot exist, and from this it follows that the original configuration does not exist either. These deformations preserve the constraints of the problem. To be explicit, we assume that $2 \leq |w| \leq 2.51$, that $2 \leq |w - v_i|$ if $w \neq v_i$, that $2 \leq |v - v'|$, and that $|v_i - v_{i+1}| \leq 2.51$, for $i = 1, 2, 3, 4$ and $w = v_1, \ldots, v_4, v, v'$. Here and elsewhere we take our subscripts modulo 4, so that

$v_1 = v_5$, and so forth. The deformations also keep v and v' in the cone at the origin that is determined by the vertices v_i.

We consider some deformations that increase $|v - v'|$. By Lemma 4.3, we may assume that $|v_i| = 2$, $|v| = |v'| = |v_i - v_{i+1}| = 2.51$, for $i = 1, 2, 3, 4$. If, for some i, we have $|v_i - v| > 2$ and $|v_i - v'| > 2$, then we fix v_{i-1} and v_{i+2} and pivot v_{i+1} around the axis through the origin and v_{i+2} away from v and v'. The constraints $|v_{i+1} - v_i| = |v_i - v_{i-1}| = 2.51$ force us to drag v_i to a new position on the sphere of radius 2. By making this pivot sufficiently small, we may assume that $|v_i - w|$ and $|v_{i+1} - w|$, for $w = v, v'$, are greater than 2.

The vertices v and v' cannot both have distance 2 from both v_{i+2} and v_{i-1}, for then we would have $v = v'$. So one of them, say v, has distance exactly 2 from at most one of v_{i+2} and v_{i-1} (say v_{i+2}). Thus, v may be pivoted around the axis through the origin and v_{i+2} away from v'. In this way, we increase $|v - v'|$ until $|v_i - v| = 2$ or $|v_i - v'| = 2$, for $i = 1, 2, 3, 4$.

Suppose one of v, v' (say v) has distance 2 from v_i, v_{i+1}, and v_{i+2}. The configuration is completely rigid. By symmetry, the vertices v_{i-1} and v' must be the reflections of v_{i+1} and v, respectively, through the plane through 0, v_i, and v_{i+2}. In particular, v' has distance 2 from v_i, v_{i-1}, and v_{i+2}. We pick coordinates and evaluate the length $|v - v'|$. We find that $|v - v'| \approx 1.746$, contrary to the hypothesis that the centers of the spheres of our packing are separated by distances of at least 2. Thus, the hypothesis that v has distance 2 from three other vertices is incorrect.

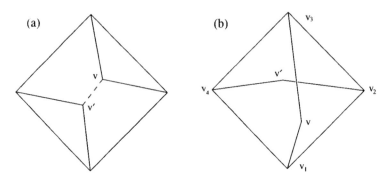

Diagram 4.4

We are left with one of the configurations of Diagram 4.4. An edge is drawn in the diagram, when the distance between the two endpoints has the smallest possible value (that is, 2.51 for the four edges of the quadrilateral, and 2 for the remaining edges). Deform the figure of case (a) along the one remaining degree of freedom until $|v - v'| = 2$. In case (a), referring to the notation established by Diagram 4.5, we have a quadrilateral on the unit sphere of edges $t_1 = 2 \arcsin(2.51/4) \approx 1.357$ radians, $t_2 = \arccos(2.51/4)$ radians, and $t_3 = 2 \arcsin(1/2.51)$ radians. The form of this quadrilateral is determined by the angle α, and it is clear that the angle $\beta(\alpha)$ is decreasing in α. We have

$$\theta = \mathrm{dih}_{\max} = \mathrm{dih}(S(2.51, 2, 2, 2.51, 2, 2)) \approx 1.874444.$$

The figure exists if and only if there exists α such that $\beta(\alpha) + \beta(2\pi - \theta - \alpha) = 2\pi - \theta$.

By symmetry, we may assume that $\alpha \leq \pi - \theta/2 \approx 2.20437$. The condition $|v - v_1| \geq 2$ implies that

$$\alpha \geq \arccos\left(\frac{\cos t_2 - \cos t_2 \cos t_3}{\sin t_2 \sin t_3}\right) > 1.21.$$

However, by monotonicity,

$$\beta(\alpha) + \beta(2\pi - \theta - \alpha) < \beta(\alpha_i) + \beta(2\pi - \theta - \alpha_i - 0.1) < 2\pi - \theta,$$

for $\alpha_i \leq \alpha \leq 0.1 + \alpha_i$, with $\alpha_i = 1.21 + 0.1i$, and $i = 0, 1, \ldots, 9$, as a direct calculation of the constants $\beta(\alpha_i) + \beta(2\pi - \theta - \alpha_i - 0.1)$ will reveal. (The largest constant, which is about $2\pi - \theta - 0.113$, occurs for $i = 0$.) Hence, the figure of Diagram 4.4(a) does not exist.

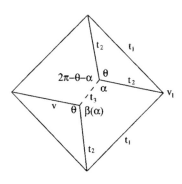

Diagram 4.5

To rule out Diagram 4.4(b), we reflect v, if necessary, to its image through the plane P through $(0, v_1, v_3)$, so that P separates v and v'. The vertex v can then be pivoted away from v' along the axis through v_1 and v_3. This decreases $|v|$, but we may rescale so that $|v| = 2.51$. Eventually $|v - v_2| = 2$ or $|v - v_4| = 2$. This is the previously considered case in which v has distance 2 from three of the vertices v_i. This completes the proof that the original arrangement of two enclosed vertices does not exist.   □

5.   Restrictions

If a Delaunay star D^* is composed entirely of tetrahedra, then we obtain a triangulation of the unit sphere. As explained in Section 2, we wish to prove that no matter what the triangulation is, we always obtain a score less than 8 *pt*. In this section we make a long list of properties that a configuration must have if it is to have a score of 8 *pt* or more. The next section and the Appendix show that only one triangulation satisfies all of these properties. Additional arguments will show that this triangulation scores less than 8 *pt*. This will complete the proof of Theorem 1.

In this section the term *vertices* refers to the vertices of the triangulation of the unit sphere. The edges of the triangulation give a planar graph. We adopt the standard terminology of graph theory to describe the triangulation. We speak of the degree of a vertex, adjacent vertices, and so forth. The n triangles around a vertex are referred to as an n-gon. We also refer to the corresponding n tetrahedra that give the triangles of the polygon. We say that a triangulation contains a *pattern* $(a_1, \ldots, a_n)$, for $a_i \in \mathbb{N}$, if there are distinct vertices v_i of degrees a_i that are pairwise nonadjacent, for $i = 1, \ldots, n$. Let N be the number of vertices in the triangulation, and let N_i be the number of vertices of degree i. We have $N = \sum N_i$.

In this section we start to use our various inequalities from Section 9 related to the score. Since the score $\sigma(S)$ may be either $\Gamma(S)$ or vor(S) depending on the circumradius of S, there are two cases to consider for every inequality. In general, the inequalities for $\Gamma(S)$ are more difficult to establish. In the following sections we only cite the inequalities pertaining to $\Gamma(S)$. Section 9 shows how all the same inequalities hold for vor(S).

Proposition 5.1. *Consider a Delaunay star D^* that is composed entirely of quasi-regular tetrahedra. Suppose that $\sigma(D^*) \geq 8$ pt. Then the following restrictions hold on the triangulation of the unit sphere given by the standard decomposition:*

1. *$13 \leq N \leq 15$.*
2. *$N = N_4 + N_5 + N_6$.*
3. *A region bounded by three edges is either a single triangle or the complement of a single triangle.*
4. *Two degree 4 vertices cannot be adjacent.*
5. *$N_4 \leq 2$.*
6. *Patterns $(6, 6, 6)$ and $(6, 6, 4)$ do not exist.*
7. *The pattern $(6, 6)$ or $(6, 4, 4)$ implies that $N \geq 14$.*
8. *If there are two adjacent degree 6 vertices, and a third degree 6 vertex adjacent to neither of the first two, then $N = 15$ and all other vertices are adjacent to at least one of these three.*
9. *The triangulation is made of geodesic arcs on the sphere whose radian lengths are between 0.8 and 1.36.*

Lemma 5.2. *Consider a vertex of degree n, for some $4 \leq n \leq 7$. Let $S_1, \ldots, S_n$ be the tetrahedra that give the n triangles. Then $\sum_{i=1}^{n} \sigma(S_i)$ is less than z_n, where $z_4 = 0.33$ pt, $z_5 = 4.52$ pt, $z_6 = -1.52$ pt, and $z_7 = -8.9$ pt. Suppose that $n \geq 6$. Let $S_1, \ldots, S_4$ be any four of the n tetrahedra around the vertex. Then $\sum_{i=1}^{4} \sigma(S_i) < 1.5$ pt.*

Proof. When $n = 4$, this is Lemma 9.5. When $n = 5$, this is Lemma 9.6. When $n = 6$ or $n = 7$, we have, by Calculation 9.4,

$$\sum_{i=1}^{n} \sigma(S_i) < \sum_{i=1}^{n} (0.378979 \, \text{dih}(S_i) - 0.410894)$$
$$= 2\pi(0.378979) - 0.410894n.$$

The right-hand side evaluates to about -1.520014 pt and -8.940403 pt, respectively, when $n = 6$ and $n = 7$.

Assume that $n \geq 6$, and select any four $S_1, \ldots, S_4$ of the n tetrahedra around the vertex. Let S_5 and S_6 be two other tetrahedra around the vertex. The dihedral angles of

S_5 and S_6 are at least $\text{dih}_{\min}$, by Calculation 9.3. Each of the four triangles (associated with $S_1, \ldots, S_4$) must then, on average, have an angle at most $(2\pi - 2\,\text{dih}_{\min})/4$ at v. By Calculation 9.4,

$$\sum_{i=1}^{4} \sigma(S_i) < \sum_{i=1}^{4}(0.378979\,\text{dih}(S_i) - 0.410894)$$
$$\leq (2\pi - 2\,\text{dih}_{\min})0.378979 + 4(-0.410894) < 1.5\ pt. \qquad \square$$

Proof of Proposition 5.1. Let $t = 2(N-2)$ be the number of triangles, an even number. By Euler's theorem on polyhedra,

$$3N_3 + 2N_4 + N_5 + 0N_6 - N_7 \cdots = 12.$$

Let $S_1, \ldots, S_t$ denote the tetrahedra of D^*. Let $\sigma_i = \sigma(S_i)$ be the corresponding score. Let sol_i denote the solid angle cut out by S_i at the origin. We have $\sum \text{sol}_i = 4\pi$. Often, without warning, we rearrange the indices i so that the tetrahedra that give the triangles around a given vertex are numbered consecutively. When a vertex v of the triangulation has been fixed, we let α_i denote the angles of the triangles at v, so that $\sum \alpha_i = 2\pi$. The angle α_i of a triangle is equal to the corresponding dihedral angle of the simplex S_i. We abbreviate certain sums over n elements to $\sum_{(n)}$, when the context makes the indexing set clear. Throughout the argument, we use Calculation 9.1, which asserts that $\sigma_i \leq 1\ pt$, for all i. The proofs show that if a triangulation fails to have any of properties 1–9, then the total score must be less than $8\ pt$.

(Proof of 5.1.1.) Assume that $t \geq 28$. By Calculation 9.9,

$$\sum_{(t)} \sigma_i < \sum_{(t)}(0.446634\,\text{sol}_i - 0.190249) \leq 4\pi(0.446634) + t(-0.190249)$$
$$\leq 4\pi(0.446634) + 28(-0.190249) \leq 8\ pt.$$

Suppose that $t \leq 18$. By Calculation 9.8,

$$\sum_{(t)} \sigma_i < \sum_{(t)}(-0.37642101\,\text{sol}_i + 0.287389) = 4\pi(-0.37642101) + t(0.287389)$$
$$\leq 4\pi(-0.37642101) + 18(0.287389) \leq 8\ pt.$$

This proves that $12 \leq N \leq 15$. The case $N = 12$ is excluded after 5.1.2.

(Proof of 5.1.2.) By Lemma 8.3.2 and Calculation 9.3, each angle α of each triangle satisfies $\text{dih}_{\min} \leq \alpha \leq \text{dih}_{\max}$. In particular, $\alpha > \pi/4$, so that each vertex has degree less than 8, and $\alpha < 2\pi/3$, so that each vertex has degree greater than 3.

Assume for a contradiction that there is a vertex of degree 7. Consider the seven tetrahedra around the given vertex. By Lemma 5.2, the tetrahedra satisfy

$$\sum_{(7)} \sigma_i < -8.9\ pt.$$

Suppose $t \leq 24$. For each vertex v, set $\zeta_v = \sum_v \sigma_i$, the sum running over the tetrahedra around v. Clearly, $\sum_{(t)} \sigma_i = (\sum_v \zeta_v)/3$. Pick a vertex v that is not a vertex of any of the seven triangles of the heptagon. By Lemma 5.2, we see that $\zeta_v < 0.33$ *pt* if v has degree 4, $\zeta_v < 4.52$ *pt* if v has degree 5, $\zeta_v < -1.52$ *pt* if v has degree 6, and $\zeta_v < -8.9$ *pt* if v has degree 7. In particular, if v has degree n, then ζ_v falls short of n points by at least 0.48 *pt*. Thus,

$$\sum_{(t)} \sigma_i < \sum_{(7)} \sigma_i + \sum_{(t-7)} pt - 0.48\ pt \le (-8.9 + (24 - 7) - 0.48)\ pt < 8\ pt.$$

Finally, we assume that $t = 26$ and $N = 15$. If $N_6 = 0$ and $N_7 = 1$, then Euler's theorem gives the incompatible conditions $2N_4 + N_5 - 1 = 12$ and $N_4 + N_5 + 1 = 15$. Thus, $N_6 > 0$ or $N_7 > 1$. This gives a second k-gon ($k = 6$ or 7) around a vertex v. This second polygon shares at most two triangles with the original heptagon. This leaves at least four triangles of a second polygon exterior to the first. Thus, by Lemma 5.2,

$$\sum_{(26)} \sigma_i \le \sum_{(7)} \sigma_i + \sum_{(4)} \sigma_i + \sum_{(t-11)} pt < -8.9\ pt + 1.5\ pt + (26 - 11)\ pt < 8\ pt.$$

(Proof of 5.1.1, continued) Assume $N = 12$. We define three classes of quasi-regular tetrahedra. In the first class all the edges have lengths between 2 and 2.1. In the second class the fourth, fifth, and sixth edges have lengths greater than 2.1. The third class is everything else.

Set $\varepsilon = 0.001$, $a = -0.419351$, $b = 0.2856354$. The following are established by Calculations 9.10–9.12:

$$\sigma(S) \le a\ \text{sol}(S) + b + \varepsilon \qquad \text{for } S \text{ in the first class,}$$
$$\sigma(S) \le a\ \text{sol}(S) + b \qquad \text{for } S \text{ in the second class,}$$
$$\sigma(S) \le a\ \text{sol}(S) + b - 5\varepsilon \qquad \text{for } S \text{ in the third class.}$$

Consider a vertex v of degree $n = 4, 5$, or 6 and the surrounding tetrahedra $S_1, \ldots, S_n$. We claim that

(5.1.1.1) $$\sum_{i=1}^{n} \sigma(S_i) \le \sum_{i=1}^{n} (a\ \text{sol}(S_i) + b).$$

This follows directly from the stated inequalities if none of these tetrahedra are in the first class. It is also obvious if at least one of these tetrahedra is in the third class, because then the inequality is violated by at most $-5\varepsilon + (n - 1)\varepsilon \le 0$. So assume that all of the tetrahedra are in the first two classes with at least one in the first class. By the restrictions on the lengths of the edges, a tetrahedron in the first class cannot be adjacent to one in the second class. We conclude that the tetrahedra are all in the first class. By Lemma 5.2, $z_n \le n(0.904)$ *pt*. If $\sum_{(n)} \text{sol}(S_i) \le n(0.56176)$, then

$$\sum_{(n)} \sigma(S_i) \le n(0.904)\ pt \le \sum_{(n)} (a(0.56176) + b) \le \sum_{(n)} (a\ \text{sol}(S_i) + b).$$

So we may assume that $\sum_{(n)} \text{sol}(S_i) \ge n(0.56176)$. By Calculation 9.13,

$$\sum_{(n)} \sigma(S_i) \le \sum_{(n)} (-0.65557\ \text{sol}(S_i) + 0.418)$$

$$\leq \sum_{(n)}(a\,\mathrm{sol}(S_i)+b)+\sum_{(n)}(0.132365-0.236219(0.56176))$$

$$< \sum_{(n)}(a\,\mathrm{sol}(S_i)+b).$$

This establishes inequality (5.1.1.1). By averaging over every vertex, we see that the average of the scores $\sigma(S)$ is less than the average of $a\,\mathrm{sol}(S)+b$. So

$$\sigma(D^*)=\sum_{(t)}\sigma(S)\leq\sum_{(t)}(a\,\mathrm{sol}(S)+b)=4\pi a+20b\approx 7.99998\,pt<8\,pt.$$

(Proof of 5.1.3.) This is Lemma 3.7. (If we had not introduced quasi-regular tetrahedra, then this result would no longer hold.)

(Proof of 5.1.4.) If two degree 4 vertices are adjacent, we then have the arrangement of Diagram 5.3. This is a quadrilateral enclosing two vertices, contrary to Proposition 4.2.

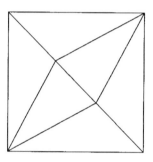

Diagram 5.3

(Proof of 5.1.5.) As in the proof of 5.1.2, for each vertex v, set $\zeta_v=\sum_v\sigma_i$, the sum running over the tetrahedra around the vertex v. We use the estimates of ζ_v that appear in Lemma 5.2.

We have found that $N_i=0$, if $i\neq 4,5,6$. By Euler's theorem, $N_5=12-2N_4$ and $N_6=N_4+N-12$. Assume that $N\geq 13$ and that $N_4\geq 3$. Then

$$\sum_{(t)}\sigma_i=\tfrac{1}{3}\sum_{(N)}\zeta_v<\tfrac{1}{3}(0.33N_4+4.52(12-2N_4)-1.52(N_4+N-12))\,pt$$

$$\leq \tfrac{1}{3}(0.33(3)+4.52(12-6)-1.52(3+13-12))\,pt<8\,pt.$$

(Proof of 5.1.6.) Suppose that we have the pattern $(6,6,6)$ Then reordering indices according to the polygons in the pattern, we have, by the estimates of Lemma 5.2,

$$\sum_{(t)}\sigma_i\leq\sum_{(6)}\sigma_i+\sum_{(6)}\sigma_i+\sum_{(6)}\sigma_i+\sum_{(t-18)}\sigma_i$$

$$<-1.52\,pt-1.52\,pt-1.52\,pt+(26-18)\,pt<3.5\,pt.$$

Similarly, if we have the pattern $(6,6,4)$, then we find

$$\sum_{(t)}\sigma_i\leq-1.52\,pt-1.52\,pt+0.33\,pt+(26-16)\,pt<8\,pt.$$

(Proof of 5.1.7.) We use the same method as in the proof of 5.1.6. If we have the pattern (6, 6), and if $t \leq 22$, then

$$\sum_{(t)} \sigma_i \leq \sum_{(6)} \sigma_i + \sum_{(6)} \sigma_i + \sum_{(t-12)} \sigma_i \leq -1.52\,pt - 1.52\,pt + (22 - 12)\,pt < 8\,pt.$$

Similarly, if we have the pattern (6, 4, 4), then $\sum_{(t)} \sigma_i$ is less than $-1.52\,pt + 0.33\,pt + 0.33\,pt + 8\,pt < 8\,pt$.

(Proof of 5.1.8.) Let the two adjacent degree 6 vertices be v_1 and v_2. Let the third be v_3. The six triangles in the hexagon around v_3 give less than $-1.52\,pt$. The ten triangles in the hexagons around v_1 and v_2 give at most

$$\sum_{(4)} \sigma_i \leq \sum_{(6)} \sigma_i < 1.5\,pt - 1.52\,pt < 0\,pt,$$

by the argument described in the case $t = 26$ of 5.1.2 (see Lemma 5.2).

Suppose that $t \leq 24$. There remain at most $24 - 16 = 8$ triangles, and they give a combined score of at most $8\,pt$. The total score is then less than $(-1.52 + 8)\,pt$, as desired.

Now assume that $t = 26$. Suppose there is a vertex v that is not adjacent to any of v_1, v_2, or v_3. As in the proof of 5.1.8, the ten triangles in the two overlapping hexagons give less than $0\,pt$. The other hexagon gives less than $-1.52\,pt$. By Lemma 5.2, the n triangles around v fall short of n points by at least $(5 - 4.52)\,pt = 0.48\,pt$. Each of the remaining triangles gives at most $1\,pt$. The score is then less than $(0 - 1.52 + (26 - 16) - 0.48)\,pt = 8\,pt$, as desired.

(Proof of 5.1.9.) This follows directly from the construction of the triangulation and the close-neighbor restrictions on the lengths of the edges of a quasi-regular tetrahedron. The lengths are between $0.8 < 2\arcsin(1/2.51)$ and $2\arcsin(2.51/4) < 1.36$. This completes the proof of the proposition.   □

6. Combinatorics

Theorem 6.1. *Suppose that a triangulation satisfies Proposition 4.2 and properties 1–9 of Proposition 5.1. Then it must be the triangulation of Diagram 6.2 with 14 vertices and 24 triangles.*

Proof. Fix a polygon centered at a vertex u_0, such as the hexagon in Diagram 6.3. The six vertices $v_1, \ldots, v_6$ of the hexagon are distinct, for otherwise two distinct geodesic

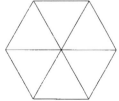

Diagram 6.2

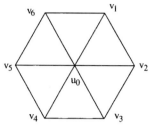

Diagram 6.3

arcs on the sphere would run between u_0 and v_i for some i. This is impossible, because u_0 and v_i are not antipodal by Property 5.1.9. Similarly, the vertices of every other polygon of the triangulation are distinct.

We then extend the polygon to a second layer of triangles. Each of the vertices v_i has degree 4, 5, or 6, and two degree 4 vertices cannot be adjacent. The new vertices are denoted $w_1, \ldots, w_k$. One example is shown in Diagram 6.4.

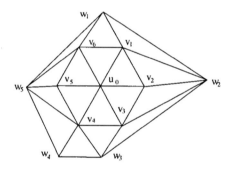

Diagram 6.4

There can be no identification of a vertex v_i with a vertex w_j, for otherwise there is a triangle (say with vertices $w_j, v_k, u_0, v_i = w_j$) that is subtriangulated, contrary to Property 5.1.3. Similarly, there is no identification of two vertices w_i and w_j, for otherwise it can be checked that there is a quadrilateral (with vertices $w_i, v_k, u_0, v_\ell, w_j = w_i$) that encloses more than one vertex, which is impossible by Proposition 4.2. A purely combinatorial problem remains. It is solved in the Appendix. □

Proposition 6.5. *The triangulation of Theorem 6.1 scores less than 8 pt.*

Proof. Our initial bound on the score comes by viewing the triangulation as made up of two hexagons and twelve additional triangles. By Lemma 5.2 and Calculation 9.1,

$$(6.6) \quad \sum_{(24)} \sigma_i \le \sum_{(6)} \sigma_i + \sum_{(6)} \sigma_i + \sum_{(12)} pt < -1.52\, pt - 1.52\, pt + 12\, pt = 8.96\, pt.$$

A refinement is required in order to lower the upper bound to 8 *pt*.

If a vertex v has height $|v| \geq 2.2$, then, by Calculation 9.2, we find $\sigma_i < 0.5 \, pt$ for the tetrahedra at v. Thus, in inequality (6.6), if the vertex v of some pentagon has height $|v| \geq 2.2$, then the term $\sum_{(12)} pt$ may be replaced by $\sum_{(9)} pt + \sum_{(3)} 0.5 \, pt$ (there are three triangles in the pentagon that do not belong to the hexagon), and the upper bound on the score falls to 7.46 pt.

If a vertex of degree 6 has height $|v| \geq 2.05$, then we claim that $\sum_{(6)} \sigma_i < -3.04 \, pt$. In fact, by Calculation 9.7, the hexagon gives

$$\sum_{(6)} \sigma_i < \sum_{(6)} (0.389195 \, \text{dih}(S_i) - 0.435643)$$
$$= 2\pi(0.389195) - 6(0.435643) < -3.04 \, pt.$$

Estimate (6.6) is improved to

$$\sum_{(24)} \sigma_i < -3.04 \, pt - 1.52 \, pt + 12 \, pt = 7.44 \, pt.$$

If the twelve tetrahedra have combined solid angle less than 6.48, then, by Calculation 9.9,

$$\sum_{(12)} \sigma_i < \sum_{(12)} (0.446634 \, \text{sol}(S_i) - 0.190249)$$
$$\leq 6.48(0.446634) + 12(-0.190249) < 11.039 \, pt.$$

Then estimate (6.6) is improved to the bound $-1.52 \, pt - 1.52 \, pt + 11.039 \, pt = 7.999 \, pt$.

Now assume, on the other hand, that the combined solid angle of the two hexagons is at most $4\pi - 6.48$. Set $K = (4\pi - 6.48)/12$ and define $\sigma'(S) := \sigma(S) + (K - \text{sol}(S))/3$. The solid angle of one of the two hexagons is at most $6K$. For that hexagon, we have

$$\sum_{(6)} \sigma'(S_i) = \sum_{(6)} \sigma_i + \frac{6K - \sum_{(6)} \text{sol}(S_i)}{3} \geq \sum_{(6)} \sigma_i.$$

By our previous estimates, we now assume without loss of generality that the heights $|v|$ of the vertices of triangles in the hexagon are at most 2.05, 2.2, and 2.2, the bound of 2.05 occurring at the center of the hexagon. By Calculation 9.15,

$$\sum_{(6)} \sigma'(S_i) \leq \sum_{(6)} 0.564978 \, \text{dih}_i - 0.614725$$
$$= 2\pi(0.564978) + 6(-0.614725) < -2.5 \, pt.$$

Estimate (6.6) becomes

$$\sum_{(24)} \sigma_i \leq \sum_{(6)} \sigma'(S_i) + \sum_{(6)} \sigma_i + \sum_{(12)} \sigma_i < (-2.5 - 1.52 + 12) \, pt = 7.98 \, pt.$$

This completes the proof of the proposition.   □

7. The Method of Subdivision

The rest of this paper is devoted to the verification of the inequalities that have been used in Sections 5 and 6. In this section we describe the method used to obtain several of our bounds. We call it the method of subdivision. Let $p(x) = \sum_I c_I x^I$ be a polynomial with real coefficients c_I, where $I = (i_1, \ldots, i_n)$, $x = (x_1, \ldots, x_n) \in \mathbb{R}^n$, and $x^I = x_1^{i_1} \ldots x_n^{i_n}$. It is clear that if C is the product of intervals $[a_1, b_1] \times \cdots \times [a_n, b_n] \subset \mathbb{R}^n$ in the positive orthant ($a_i > 0$, for $i = 1, \ldots, n$), then

$$\forall x \in C, \qquad p_{\min}(C) \le p(x) \le p_{\max}(C),$$

where

$$p_{\min}(C) = \sum_{c_I > 0} c_I a^I + \sum_{c_I < 0} c_I b^I \quad \text{and} \quad p_{\max}(C) = \sum_{c_I > 0} c_I b^I + \sum_{c_I < 0} c_I a^I.$$

Another bound comes from the Taylor polynomial $p(x) = \sum d_I (x - a)^I$ at a:

$$(7.1) \qquad d_0 + \sum_{d_I < 0, I \neq 0} d_I (b - a)^I \le p(x) \le d_0 + \sum_{d_I > 0, I \neq 0} d_I (b - a)^I.$$

If $r(x) = p(x)/q(x)$ is a rational function, and if $q_{\min}(C) > 0$, then

$$\forall x \in C, \qquad \frac{p_{\min}(C)}{q_1(C, p)} \le r(x) \le \frac{p_{\max}(C)}{q_2(C, p)},$$

where $q_1(C, p)$ (resp. $q_2(C, p)$) is defined as $q_{\max}(C)$ whenever $p_{\min}(C) \ge 0$ (resp. $p_{\max}(C) < 0$) and as $q_{\min}(C)$ otherwise.

We define a *cell* to be a product of intervals in the positive orthant of $\mathbb{R}^n$. By covering a region with a sufficiently fine collection of cells, various inequalities of rational functions are easily established. To prove an inequality of rational functions with positive denominators (say $r_1(x) < r_2(x)$, for all $x \in C$), we cover C with a finite number of cells and compare the upper bound of $r_1(x)$ with the lower bound of $r_2(x)$ on each cell. If it turns out that some of the cells give too coarse a bound, then we subdivide each of the delinquent cells into a number of smaller cells and repeat the process. If at some stage we succeed in covering the original region C with cells on which the upper bound of $r_1(x)$ is less than the lower bound of $r_2(x)$, the inequality is established.

A refinement of this approach applies the method to the partial derivatives. If, for instance, we establish by the method of subdivision that, for some i,

$$\frac{\partial p}{\partial x_i}(x) \ge 0, \qquad \forall x \in C,$$

then we may compute an upper bound of p by applying the method of subdivision to the polynomial obtained from p by the specialization $x_i = b_i$, where b_i is the upper bound of x_i on C. Thus, we obtain an upper bound on a polynomial by fixing all the variables that are known to have partial derivatives of fixed sign and then applying the method subdivision to the resulting polynomial. Similar considerations apply to lower bounds and to rational functions with positive denominators.

It is a fortunate circumstance that many of the polynomials we encounter in sphere packings are quadratic in each variable with negative leading coefficient (u, ρ, Δ, χ in the next section). In this case the lower bound is attained at a corner of the cell. Of course, the maximum of a quadratic function with negative leading coefficient is also elementary: $-\alpha(x - x_0)^2 + \beta \leq \beta$, if $\alpha \geq 0$.

8. Explicit Formulas for Compression, Volume, and Angle

Many of the formulas in this section are classical. They can typically be found in 19th century primers on solid geometry. The formula for solid angles, for example, is due to Euler and Lagrange. For anyone equipped with symbolic algebra software, the verifications are elementary, so we omit many of the details. All formulas in this section are valid for Delaunay simplices and for quasi-regular tetrahedra, unless otherwise noted.

8.1. The Volume of a Simplex

As in the previous section, a *cell* in $\mathbb{R}^n$ is a product of intervals in $\mathbb{R}^n$. We define a function $\Delta: [4, 16]^6 \subset \mathbb{R}^6 \to \mathbb{R}$ by

$$
\begin{aligned}
(8.1.1) \qquad \Delta(x_1, \ldots, x_6) = {} & x_1 x_4(-x_1 + x_2 + x_3 - x_4 + x_5 + x_6) \\
& + x_2 x_5(x_1 - x_2 + x_3 + x_4 - x_5 + x_6) \\
& + x_3 x_6(x_1 + x_2 - x_3 + x_4 + x_5 - x_6) \\
& - x_2 x_3 x_4 - x_1 x_3 x_5 - x_1 x_2 x_6 - x_4 x_5 x_6.
\end{aligned}
$$

We set $y_i = \sqrt{x_i}$, for $i = 1, \ldots, 6$. This relationship between x_i and y_i remains in force to the end of the paper. Index the edges of a simplex as in Diagram 8.1.2.

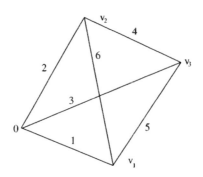

Diagram 8.1.2

We also define $u: [4, 16]^3 \to \mathbb{R}$ by

$$
\begin{aligned}
(8.1.3) \qquad u(x_1, x_2, x_6) = {} & (y_1 + y_2 + y_6)(y_1 + y_2 - y_6)(y_1 - y_2 + y_6) \\
& \times (-y_1 + y_2 + y_6) \\
= {} & -x_1^2 - x_2^2 - x_6^2 + 2x_1 x_6 + 2x_1 x_2 + 2x_2 x_6.
\end{aligned}
$$

Lemma 8.1.4. *There exists a simplex of positive volume with edges of length $y_1, \ldots, y_6$ if and only if $\Delta(x_1, \ldots, x_6) > 0$. If these conditions hold, then the simplex has volume $\Delta(x_1, \ldots, x_6)^{1/2}/12$.*

Proof. The function $u = u(x_1, x_2, x_6)$ is quadratic in each variable, with negative leading coefficient, so the minimum of u, which is 0, is attained at a vertex of the cube $[4, 16]^3$. At the vertices where the minimum is attained, $\Delta(x_1, \ldots, x_6) \leq 0$.

Assume $\Delta > 0$. Then $u > 0$, and a simplex exists with vertices 0, $X = (y_1, 0, 0)$, $Y = \frac{1}{2}(*, u^{1/2}/y_1, 0)$, $Z = (*, *, (\Delta/u)^{1/2})$. Conversely, if the simplex exists, then it must be of the given form, up to an orthogonal transformation, so $u > 0$ (otherwise O, X, and Y are colinear) and $\Delta > 0$. The volume is $|\det(X, Y, Z)|/6 = \Delta^{1/2}/12$. □

Let C be a cell contained in $[4, 16]^6$. The minimum of Δ on C is attained at a vertex of C: this is clear because Δ is quadratic in each variable with negative leading coefficient. To obtain an upper bound on Δ on a cell, we may use the method of Section 7. The restriction of Δ to the zero set of $\partial \Delta / \partial x_1$ is

$$\frac{u(x_4, x_5, x_6) \cdot u(x_2, x_3, x_4)}{4x_4}.$$

This is an upper bound on $\Delta(x_1, \ldots, x_6)$.

8.2. The Circumradius of a Simplex

The circumradius of a face with edges y_4, y_5, and y_6 is

$$\eta(y_4, y_5, y_6) = \frac{y_4 y_5 y_6}{u(y_4^2, y_5^2, y_6^2)^{1/2}}.$$

Define $\rho: [4, 16]^6 \to \mathbb{R}$ by

(8.2.1)
$$\rho(x_1, \ldots, x_6) = -x_1^2 x_4^2 - x_2^2 x_5^2 - x_3^2 x_6^2 + 2x_1 x_2 x_4 x_5 + 2x_1 x_3 x_4 x_6 + 2x_2 x_3 x_5 x_6.$$

Lemma 8.2.2. *Suppose $\Delta > 0$; then $\rho > 0$.*

Proof. If $\Delta > 0$, then the simplex of Lemma 8.1.4 exists. Let the coordinates of the circumcenter of the simplex be (x, y, z). Direct calculation shows that $0 < x^2 + y^2 + z^2 = \rho/(4\Delta)$. □

Corollary 8.2.3. *Let S be a Delaunay simplex of positive volume with edges of lengths $y_i = \sqrt{x_i}$. The circumradius is $\frac{1}{2}(\rho/\Delta)^{1/2}$.*

Set

$$\chi(x_1, x_2, x_3, x_4, x_5, x_6) = x_1 x_4 x_5 + x_1 x_6 x_4 + x_2 x_6 x_5 + x_2 x_4 x_5 + x_5 x_3 x_6$$
$$+ x_3 x_4 x_6 - 2x_5 x_6 x_4 - x_1 x_4^2 - x_2 x_5^2 - x_3 x_6^2.$$

We have

$$\frac{\rho}{4\Delta} - \eta(y_4, y_5, y_6)^2 = \frac{\chi(x_1, \ldots, x_6)^2}{4u(x_4, x_5, x_6)\Delta}.$$

The vanishing of χ is the condition for the circumcenter of the simplex to lie in the plane through the face T bounded by the fourth, fifth, and sixth edges. χ is positive if the circumcenter of S and the vertex of S opposite T lie on the same side of the plane through T and negative if they lie on opposite sides of the plane. We say that T has *positive orientation* when $\chi > 0$.

8.2.4. The function ρ is quadratic in each of the variables $x_1, \ldots, x_6$ with negative leading coefficient. Thus, the minimum of ρ is attained at a vertex of a given cell C. The derivative is $\Delta^2\partial(\rho/\Delta)/\partial x_1 = \chi(x_5, x_6, x_1, x_2, x_3, x_4)\chi(x_1, x_2, x_3, x_4, x_5, x_6)$. This leads to bounds on the circumradius by the method of Section 7.

The circumradius of a quasi-regular tetrahedron is increasing in $x_1, \ldots, x_6$ if the orientation of each face is positive. When a face fails to have a positive orientation, it satisfies the constraints of Section 3 (for example, $y_1, y_2, y_3 \in [2, 2.15]$, $y_4, y_5, y_6 \in [2.3, 2.51]$, $\eta(y_4, y_5, y_6) \geq \sqrt{2}$, if it is the face opposite the origin). This allows us to determine bounds on the circumradius for most of the quasi-regular tetrahedra we encounter in this paper by inspection. For example, in Section 9 we study the simplices constrained by $y_i \in [2, 2.1]$. An upper bound on the circumradius is $\mathrm{rad}(S(2.1, 2.1, 2.1, 2.1, 2.1, 2.1))$.

8.2.5. *Proof of Lemma* 3.4. In the notation of Section 3.4 let T be the given face, and let v_0 be a vertex satisfying the conditions of the lemma. Let S by the simplex at the origin $v_0 = 0$, whose first, second, and third edges abut at v_i, for $i = 1, 2, 3$. Suppose for a contradiction that $|v_0 - v_3|^2 = x_3 \geq 2.15^2$. In light of the results of this section, a contradiction follows if $\chi(x_1, \ldots, x_6) > 0$, for $x_i \geq 4$ for $i = 1, 2, x_3 \geq 2.15^2$, $2.3^2 \leq x_i \leq 2.51^2$ for $i = 4, 5, 6$. (The lower bound of 2.3 comes from Remarks 3.2 and 3.3: the circumradius of T is at least $\sqrt{2}$). Since T is acute,

$$\frac{\partial\chi}{\partial x_1} = x_4(-x_4 + x_5 + x_6) > 0.$$

Similarly, χ is increasing in x_2 and x_3. Thus, to minimize χ, we take $x_1 = x_2 = 4$, $x_3 = 2.15^2$. However, χ is quadratic in each variable x_4, x_5, and x_6 with negative leading coefficient, so the minimum occurs at a corner point of $[2.3^2, 2.51^2]^3$. We find $\chi \geq \chi(2^2, 2^2, 2.15^2, 2.51^2, 2.51^2, 2.51^2) \approx 0.885 > 0$. □

For the set of Delaunay stars to be a compact topological space, simplices of zero volume must be included. Since the circumradius remains bounded, the degenerate Delaunay simplices of zero volume are planar quadrilaterals that possess a circumscribing circle.

8.3. Dihedral Angles

Let $\mathrm{dih}(S)$ be the dihedral angle of a simplex S along the first edge. It is the (interior) angle formed by faces with edges $(2, 1, 6)$ and $(1, 3, 5)$.

Lemma 8.3.1.

$$\cos \text{dih}(S) = \frac{\partial \Delta / \partial x_4}{(u(x_1, x_2, x_6)u(x_1, x_3, x_5))^{1/2}},$$

where u is the function defined by (8.1.3).

The partial derivative of $\cos \text{dih}(S)$ with respect to x_3 is

$$\frac{2x_1(\partial \Delta / \partial x_2)}{u(x_1, x_2, x_6)^{1/2}u(x_1, x_3, x_5)^{3/2}},$$

so that the sign of the partial derivative is determined by the sign of $\partial \Delta / \partial x_2$. Similar considerations apply to the partial derivatives with respect to x_2, x_5, and x_6.

Lemma 8.3.2. *Let S be a quasi-regular tetrahedron. The the dihedral angle of S along any edge is at most* $\arccos(-29003/96999) = \text{dih}_{max} \approx 1.874444$.

Proof. Suppose that $\text{dih}(S) > \pi/2$, so that the numerator $n = \partial \Delta / \partial x_4$ in Lemma 8.3.1 is negative. To bound $\cos \text{dih}(S)$ from below, we minimize $u(x_1, x_2, x_6)$, $u(x_1, x_3, x_5)$, and the numerator n over the variables x_2, x_3, x_5, and x_6.

We have $\partial_i u = x_j + x_k - x_i > 0$ on the indicated domain, for $i = 2, 3, 5$, and 6, so that

$$u(x_1, x_2, x_6)u(x_1, x_3, x_5) \geq u(x_1, 2^2, 2^2)^2 = (-x_1^2 + 16x_1)^2.$$

Similarly, $\partial_i n = x_1 + x_j - x_k > 0$, for $i = 2, 3, 5$, and 6, so that

$$n(x_1, \ldots, x_6) > n(x_1, 2^2, 2^2, x_4, 2^2, 2^2) = x_1(16 - x_1 - 2x_4).$$

Thus,

$$0 \geq \cos \text{dih}(S) \geq \frac{16 - x_1 - 2x_4}{16 - x_1} = 1 - \frac{2x_4}{16 - x_1}$$

$$\geq 1 - \frac{2(2.51)^2}{16 - 2.51^2} = \frac{-29003}{96999}. \qquad \square$$

8.4.   *The Solid Angles of a Delaunay Simplex*

Let S be a Delaunay simplex with vertices $v_0, \ldots, v_3$. By the solid angle $\text{sol}_i(S)$ of the simplex at the vertex i, we mean $3 \text{vol}(S \cap B_i)$, where B_i is a unit ball centered at v_i. For simplicity, suppose that v_0 is located at the origin, and that, in the notation of Diagram 8.1.2, the vertices v_1, v_2, v_3 are the edges of lengths y_1, y_2, and y_3. We write sol for sol_0. Set

$$(8.4.1) \quad a(y_1, y_2, \ldots, y_6) = y_1 y_2 y_3 + \tfrac{1}{2} y_1(y_2^2 + y_3^2 - y_4^2)$$
$$+ \tfrac{1}{2} y_2(y_1^2 + y_3^2 - y_5^2) + \tfrac{1}{2} y_3(y_1^2 + y_2^2 - y_6^2).$$

Lemma 8.4.2.

$$\text{sol}(S) = 2\arccot\left(\frac{2a}{\Delta^{1/2}}\right).$$

Proof. (See [H2, p. 64].) We use the branch of arccot taking values in $[0, \pi]$. ☐

The function a is increasing in y_1, y_2, and y_3 on $[2, 4]$ and is decreasing in the variables y_4, y_5, and y_6 on the same interval.

We claim that $a > 0$ for any Delaunay simplex. We prove this in the case $\Delta > 0$, leaving the degenerate case $\Delta = 0$ to the reader. The area of the face T opposite the origin is at most $4\sqrt{3}$ (since its edges are all at most 4). There exists a plane that is tangent to the unit sphere between this face and the unit sphere [H1, 2.1]. The area of the radial projection of T to this plane is less than the area of a disk D of radius 1.5 on the plane centered at the point of tangency. The solid angle (that is, the area of the radial projection of T to the unit sphere) is less than the area of the radial projection of D to the unit sphere). This area is $2\pi(1 - \cos(\arctan(1.5))) < \pi$. Lemma 8.4.2 now gives the result. This allows us to use Lemma 8.4.2 in the form $\text{sol}(S) = 2\arctan(\Delta^{1/2}/(2a))$.

8.4.3. The solid angle is the area of a spherical triangle. Let x, y, and z be the cosines of the radian lengths of the edges of the triangle. By the spherical law of cosines, the solid angle, expressed as a function of x, y, and z is

$$c(x, y, z) + c(y, z, x) + c(z, x, y) - \pi, \qquad c(x, y, z) := \arccos\left(\frac{x - yz}{\sqrt{(1 - y^2)(1 - z^2)}}\right).$$

The partial derivative with respect to x of this expression for the solid angle is $(-1 - x + y + z)/((x + 1)\sqrt{t})$, where $t = 1 - x^2 - y^2 - z^2 + 2xyz$. The second derivative of this expression evaluated at $-1 - x + y + z = 0$ is $-2(1 - y)(1 - z)t^{-3/2} \leq 0$. So the unique critical point is always a local maximum. If neither edge at a vertex of a spherical triangle is constrained, then the triangle can be contracted or expanded by moving the vertex. If the lengths of the edges are constrained to lie in a product of intervals, then the minimum area occurs when two of the edges are as short as possible and the third is at one of the extremes. The maximum area is attained when two of the edges are as long as possible and the third is at one of the extremes or at a critical point: $x = y + z - 1$ (or it symmetries, $y = x + z - 1$, $z = x + y - 1$).

8.5. The Compression of a Delaunay Simplex

The compression $\Gamma(S)$ of a Delaunay simplex S is defined as

$$(8.5.1) \qquad \Gamma(S) = -\delta_{oct}\,\text{vol}(S) + \frac{\sum_{i=1}^{4}\text{sol}_i(S)}{3},$$

where

$$(8.5.2) \qquad \delta_{oct} = \frac{-3\pi + 12\arccos(1/\sqrt{3})}{2\sqrt{2}} \approx 0.72.$$

We let $a_0 = a(y_1, y_2, y_3, y_4, y_5, y_6)$, where a is the function defined in (8.4.1). We also let a_1, a_2, and a_3 be the functions a for the vertices denoted v_1, v_2, and v_3 in Diagram 8.1.2. For example, $a_1 = a(y_1, y_5, y_6, y_4, y_2, y_3)$. Set $t = \sqrt{\Delta}/2$. By the results of Sections 8.1 and 8.4, we find that $\Gamma(S) = -\delta_{oct}t/6 + \sum_0^3 \frac{2}{3}\arctan(t/a_i)$. Recall that $a_i > 0$ on a cell C. We wish to give an elementary upper bound of Γ on C. Set, for $t \geq 0$ and $a = (a_0, a_1, a_2, a_3) \in \mathbb{R}_+^4$,

$$\gamma(t, a) := \frac{-\delta_{oct}t}{6} + \sum_{i=0}^{3} \frac{2}{3}\arctan\left(\frac{t}{a_i}\right).$$

The partial derivative of γ with respect to a_i is $-2t/(3(t^2 + a_i^2)) \leq 0$, so $\gamma(t, a) \leq \gamma(t, a^-)$, where $a^- = (a_0^-, a_1^-, a_2^-, a_3^-)$ is a lower bound of a on C, as determined by Section 8.4. We study $\gamma(t) = \gamma(t, a^-)$ as a function of t to obtain an upper bound on $\Gamma(S)$. We note that

$$\gamma'(t) = \frac{-\delta_{oct}}{6} + \frac{2}{3}\sum_{i=0}^{3}\frac{a_i}{(t^2 + a_i^2)} \quad \text{and} \quad \gamma''(t) = -\frac{4}{3}\sum\frac{a_i t}{(t^2 + a_i^2)^2} \leq 0.$$

Upper and lower bounds on t are known from Section 8.1. An upper bound on γ for t in $[t_{\min}, t_{\max}]$ is the maximum of $\ell(t_{\min})$ and $\ell(t_{\max})$, where ℓ is the tangent to γ at any point in $[t_{\min}, t_{\max}]$.

8.6. Voronoi cells

We assume in this section that the simplex S has the property that the circumcenter of each of the three faces with vertex at the origin lies in the cone at the origin over the face. This condition is automatically satisfied if these three faces are acute triangles. In particular, it is satisfied for a quasi-regular tetrahedron.

When the cone over a quasi-regular tetrahedron S contains the circumcenter of S, Section 2 sets $\text{vor}(S) = -4\delta_{oct}\,\text{vol}(\hat{S}_0) + 4\,\text{sol}(S)/3$, where $\hat{S}_0$ is the intersection of S with a Voronoi cell at the origin. Otherwise, $\text{vor}(S)$ is defined as an analytic continuation. This section gives formulas for $\text{vol}(\hat{S}_0)$.

As usual, set $S = S(y_1, \ldots, y_6)$ and $x_i = y_i^2$. Suppose at first that the circumcenter of S is contained in the cone over S. The polyhedron $\hat{S}_0$ breaks into six pieces, called the *Rogers simplices*. A Rogers simplex is the convex hull of the origin, the midpoint of an edge (the first, second, or third edge), the circumcenter of a face along the given edge, and the circumcenter of S. Each Rogers simplex has the form

$$R = R(a, b, c) := S(a, b, c, (c^2 - b^2)^{1/2}, (c^2 - a^2)^{1/2}, (b^2 - a^2)^{1/2})$$

for some $1 \leq a \leq b \leq c$. Here a is the half-length of an edge, b is the circumradius of a face, and c is the circumradius of the Delaunay simplex.

The volume is $\text{vol}(R(a, b, c)) = a(b^2 - a^2)^{1/2}(c^2 - b^2)^{1/2}/6$. The density $\delta(a, b, c)$ of $R = R(a, b, c)$ is defined as the ratio of the volume of the intersection of R with a unit ball at the origin to the volume of R. It follows from the definitions that

(8.6.1) $$\text{vor}(S) = \sum 4\,\text{vol}(R(a, b, c))(-\delta_{oct} + \delta(a, b, c)),$$

where c is the circumradius $\mathrm{rad}(S)$, and (a, b) runs over the six pairs

$$\left(\frac{y_1}{2}, \eta(y_1, y_2, y_6)\right), \qquad \left(\frac{y_2}{2}, \eta(y_1, y_2, y_6)\right),$$

$$\left(\frac{y_2}{2}, \eta(y_2, y_3, y_4)\right), \qquad \left(\frac{y_3}{2}, \eta(y_2, y_3, y_4)\right),$$

$$\left(\frac{y_3}{2}, \eta(y_3, y_1, y_5)\right), \qquad \left(\frac{y_1}{2}, \eta(y_3, y_1, y_5)\right).$$

Upper and lower bounds on $\mathrm{vol}(R(a, b, c))$ follow without difficult from upper and lower bounds on a, b, and c. Thus, an upper bound on $\delta(a, b, c)$ leads to an upper bound on $\mathrm{vor}(S)$. The next lemma, which is due to Rogers, gives a good upper bound [R].

Lemma 8.6.2. *The density $\delta(a, b, c)$ is monotonically decreasing in each variable for $1 < a < b < c$.*

Proof. Let $1 \le a_1 \le b_1 \le c_1$, $1 \le a_2 \le b_2 \le c_2$, $a_2 \le a_1$, $b_2 \le b_1$, $c_2 \le c_1$. The points of $R(a_1, b_1, c_1)$ are realized geometrically by linear combinations

$$\mathbf{s}_1 = \lambda_1(a_1, 0, 0) + \lambda_2(a_1, (b_1^2 - a_1^2)^{1/2}, 0) + \lambda_3(a_1, (b_1^2 - a_1^2)^{1/2}, (c_1^2 - b_1^2)^{1/2}),$$

where $\lambda_1, \lambda_2, \lambda_3 \ge 0$ and $\lambda_1 + \lambda_2 + \lambda_3 \le 1$. The points of $R(a_2, b_2, c_2)$ are realized geometrically by linear combinations

$$\mathbf{s}_2 = \lambda_1(a_2, 0, 0) + \lambda_2(a_2, (b_2^2 - a_2^2)^{1/2}, 0) + \lambda_3(a_2, (b_2^2 - a_2^2)^{1/2}, (c_2^2 - b_2^2)^{1/2}),$$

with the same restrictions on λ_i. Then

$$|\mathbf{s}_1|^2 - |\mathbf{s}_2|^2 = \lambda_1(\lambda_1 + 2\lambda_2 + 2\lambda_3)(a_1^2 - a_2^2) + \lambda_2(\lambda_2 + 2\lambda_3)(b_1^2 - b_2^2) + \lambda_3^2(c_1^2 - c_2^2).$$

So $|\mathbf{s}_1|^2 \ge |\mathbf{s}_2|^2$. This means that the linear transformation $\mathbf{s}_1 \mapsto \mathbf{s}_2$ that carries the simplex S_1 to S_2 moves points of the simplex S_1 closer to the origin. In particular, the linear transformation carries the part in S_1 of the unit ball at the origin into the unit ball. This means that the density of $R(a_1, b_1, c_1)$ is at most that of $R(a_2, b_2, c_2)$. □

If the circumcenter of S is not in the cone over S, then the analytic continuation gives

$$\mathrm{vor}(S) = \sum_R 4\varepsilon_R \, \mathrm{vol}(R(a, b, c))(-\delta_{oct} + \delta(a, b, c)),$$

where $\varepsilon_R = 1$ if the face of the Delaunay simplex S corresponding to R has positive orientation, and $\varepsilon_R = -1$ otherwise. (The face of S "corresponding" to $R(a, b, c)$ is the one used to compute the circumradius b.)

8.6.3. A calculation based on the explicit coordinates of S and its circumcenter given in Lemma 8.1.4 shows that

$$\varepsilon_R \, \mathrm{vol}\left(R\left(\frac{y_1}{2}, \eta(y_1, y_2, y_6), \mathrm{rad}(S)\right)\right) = \frac{x_1(x_2 + x_6 - x_1)\chi(x_4, x_5, x_3, x_1, x_2, x_6)}{48 u(x_1, x_2, x_6)\Delta(x_1, \ldots, x_6)^{1/2}}.$$

If the circumcenter of S is not contained in S, then the same formula holds by analytic continuation. By definition, $\varepsilon_R = -1$ exactly when the function χ is negative. Although this formula is more explicit than the earlier formula, it tends to give weaker estimates of $\mathrm{vor}(S)$ and was not used in the calculations in Section 9.

8.6.4. There is another approximation to $\mathrm{vor}(S)$ that will be useful. Set $S_y = S(2, 2, 2, y, y, y)$. (We hope there is no confusion with the previous notation S_i.) For $1 \leq a \leq b \leq c$, let $\mathrm{vol}(R(a, b, c)) = a((b^2 - a^2)(c^2 - b^2))^{1/2}/6$ be as above. Set $r(a) = \mathrm{vol}(R(a, \eta(2, 2, 2a), 1.41))$.

Lemma 8.6.5. *Assume that the circumradius of a quasi-regular tetrahedron S is at least 1.41, and that $6 \leq y_1 + y_2 + y_3 \leq 6.3$. Set $a = (y_1 + y_2 + y_3 - 4)/2$. Pick y to satisfy $\mathrm{sol}(S_y) = \mathrm{sol}(S)$. Then*

$$\mathrm{vor}(S) \leq \mathrm{vor}(S_y) - 8\delta_{oct}\left(1 - \frac{1}{a^3}\right)r(a)$$

Proof. If S is any quasi-regular tetrahedron, let S_{tan} be the simplex defining the "tangent" Voronoi cell, that is, S_{tan} is the simplex with the same origin that cuts out the same spherical triangle as S on the unit sphere, but that satisfies $y_1 = y_2 = y_3 = 2$. The lengths of the fourth, fifth, and sixth edges of S_{tan} are between $\sqrt{8 - 2(2.3)} = \sqrt{3.4}$ and 2.51. The faces of S_{tan} are acute triangles. A calculation similar to the proof in paragraph 8.2.5, based on $\chi(3.4, 3.4, 4, 4, 4, 2.51^2) > 0$, shows that the circumcenter of S_{tan} is contained in the cone over S_{tan}.

Since S_{tan} is obtained by "truncating" S, we observe that $\mathrm{vor}(S_{\mathrm{tan}}) - 4\delta_{oct}\,\mathrm{vol}(\hat{S} \backslash \hat{S}_{\mathrm{tan}})$, where $\hat{S}$ and $\hat{S}_{\mathrm{tan}}$ are the pieces of Voronoi cells denoted $\hat{S}_0$ in Section 2 for S and S_{tan}, respectively. By a convexity result of Fejes Tóth, $\mathrm{vor}(S_{\mathrm{tan}}) \leq \mathrm{vor}(S_y)$ [FT, p. 125]. (This inequality relies on the fact that the circumcenter of S_{tan} is contained in the cone over S_{tan}.)

Let a_1 be the half-length of the first, second, or third edge. Since $\mathrm{vol}(R(a_1, b, c))$ is increasing in c, we obtain a lower bound on $\mathrm{vol}(R(a_1, b, c))$ for $c = 1.41$, the lower bound on the circumradius of S. The function $\mathrm{vol}(R(a_1, b, 1.41))$, considered as a function of b, has at most one critical point in $[a_1, 1.41]$ and it is always a local maximum (by a second derivative test). Thus,

$(8.6.6)\quad \mathrm{vol}(R(a_1, b, c)) \geq \min(\mathrm{vol}(R(a_1, b_{\min}, 1.41)), \mathrm{vol}(R(a_1, b_{\max}, 1.41))),$

where $b_{\min}$ and $b_{\max}$ are upper and lower bounds on $b \in [a_1, 1.41]$. A lower bound on b is $\eta(2, 2, 2a_1)$, which means that $\mathrm{vol}(R(a_1, b_{\min}, 1.41))$ may be replaced with $r(a_1)$ in inequality (8.6.6). By Heron's formula, the function $\eta(a, b, c)$, for acute triangles, is convex in pairs of variables:

$$\eta_{aa}\eta_{bb} - \eta_{ab}^2 = \frac{\eta^6(a^2 + b^2 - c^2)(a^2 - b^2 + c^2)(-a^2 + b^2 + c^2)(a^2 + b^2 + c^2)}{a^6 b^6 c^4} > 0.$$

Thus, an upper bound on $b_{\max}$ is $\eta(2, 2.51, 2a)$, where $a = (y_1 + y_2 + y_3 - 4)/2$. This means that

$$\mathrm{vol}(R((a_1, b_{\max}, 1.41)))$$

may be replaced with the function $\text{vol}(R(a_1, \eta(2, 2.51, 2a), 1.41))$ in inequality (8.6.6). Now $1 \leq a_1 \leq a$, and $\text{vol}(R(a_1, b, c))$ is decreasing in the first variable (for $1 \leq a_1$ and $b \leq \sqrt{2}$), so we may use the lower bound

$$\text{vol}(R(a, \eta(2, 2.51, 2a), 1.41))$$

instead. By Calculations 9.20.2 and 9.20.3, we conclude that $\text{vol}(R(a_1, b, c)) \geq r(a)$. This lower bound is valid for each Rogers simplex. The volume of $\hat{S} \backslash \hat{S}_{\text{tan}}$ is then at least

$$\left(\left(1 - \frac{8}{y_1^3} \right) + \left(1 - \frac{8}{y_2^3} \right) + \left(1 - \frac{8}{y_3^3} \right) \right) 2r(a).$$

The concavity of $1 - 8/y^3$ gives $\sum_{i=1}^{3} (1 - 8/y_i^3) \geq 1 - 1/a^3$. We have established that

$$\text{vol}(\hat{S} \backslash \hat{S}_{\text{tan}}) \geq \left(1 - \frac{1}{a^3} \right) 2r(a).$$

The result follows.   □

8.6.7. We conclude our discussion of Voronoi cells with a few additional comments about the case in which analytic continuation is used to define $\text{vor}(S)$, with S a quasi-regular tetrahedron. Assume the circumcenter c of S lies outside S and that the face T of S with negative orientation is the one bounded by the first, second, and sixth edges. It follows from Section 3 that $y_1, y_2, y_6 \in [2.3, 2.51]$ and $y_3, y_4, y_5 \in [2, 2.15]$.

Let p_1 (resp. p_2) be the point on T equidistant from the origin, v_3, and v_1 (resp. v_2).

$$|p_1|^2 = \frac{x_1}{4} + \frac{x_1 u(x_1, x_2, x_6)(-x_1 + x_3 + x_5)^2}{4(\partial \Delta / \partial x_4)^2}.$$

Let p_0 be the circumcenter of T (see Diagram 8.6.8).

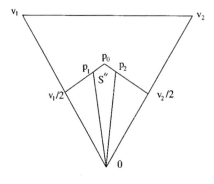

Diagram 8.6.8

$\varepsilon_R = -1$ for the Rogers simplex with vertices the origin, p_0, c, and $v_1/2$. It lies outside S. The other Rogers simplex along the first edge has $\varepsilon_R = 1$, so that the part common to both of these Rogers simplices cancels in the definition of vor(S). This means that vor(S) becomes the sum of the usual contributions from the two Rogers simplices along the third edge,

$$4 \operatorname{vol}(R(a, b, c))(-\delta_{oct} + \delta(a, b, c))$$

for $(a, b, c) = (y_1/2, \eta(y_1, y_3, y_5), |p_1|)$ and $(y_2/2, \eta(y_2, y_3, y_4), |p_2|)$, and

$$v(S'') := 4\delta_{oct} \operatorname{vol}(S'') - \frac{4 \operatorname{sol}(S'')}{3},$$

where S'' is the convex hull of 0, p_0, p_1, p_2, and c.

We claim that in any cell satisfying the constraints given above, $|p_1|$ is minimized by making y_1, y_2, y_6 as large as possible and y_3, y_4, y_5 as small as possible. To show this, the derivatives have to be written explicitly from the formula for $|p_1|^2$ given above. The partial derivatives with respect to $x_2, \ldots, x_6$ factor into products of the polynomials $\partial \Delta / \partial x_i$, for $i = 3, 4, 5$, x_1, $(x_3 + x_5 - x_1)$, $u(x_1, x_2, x_6)$, $(-x_1 x_4 + x_2 x_5 + x_1 x_6 - x_5 x_6)$, and $(x_1 x_2 - x_2 x_3 - x_1 x_4 + x_3 x_6)$. The signs of these polynomials are easily determined. The partial with respect to x_1 is complicated (the numerator has 88 terms), and we had to resort to the method of subdivision to determine its sign. We omit the details.

To complete our estimate, we describe an upper bound on $v(S'')$. The ratio $\operatorname{sol}(S'')/(3 \operatorname{vol}(S''))$ is at least $1/\operatorname{rad}(S)^3$, because S'' is contained in a sphere of radius $\operatorname{rad}(S)$, centered at the origin. We have

$$\frac{\operatorname{vol}(S'')}{2} \leq \tfrac{1}{6}|p_0 - p_1||c - p_0||p_0|,$$

because the convex hull of 0, p_0, p_1, and c, which is one side of S'', is a pyramid with base the right triangle (p_1, p_0, c) and height at most $|p_0| = \eta(y_1, y_2, y_6)$. Of course, $|c - p_0|^2 = \operatorname{rad}(S)^2 - \eta(y_1, y_2, y_6)^2$. We now have a bound on $v(S'')$ in terms of quantities that have been studied in Section 8, if we rely on the bound $|p_0 - p_1| \leq 0.1381$. Write $p_1 = p_1(y_1, \ldots, y_6)$. This bound is obtained from the following inequalities:

$$
\begin{aligned}
|p_0 - p_1| &= \left(\eta(y_1, y_2, y_6)^2 - \frac{x_1}{4} \right)^{1/2} - \left(|p_1|^2 - \frac{x_1}{4} \right)^{1/2} \\
&\leq \left(\eta(y_1, 2.51, 2.51)^2 - \frac{x_1}{4} \right)^{1/2} - \left(|p_1(y_1, 2.51, 2, 2, 2, 2.51)|^2 - \frac{x_1}{4} \right)^{1/2} \\
&\leq \left(\eta(2.51, 2.51, 2.51)^2 - \frac{2.51^2}{4} \right)^{1/2} \\
&\quad - \left(|p_1(2.51, 2.51, 2, 2, 2, 2.51)|^2 - \frac{2.51^2}{4} \right)^{1/2} \\
&< 0.1381.
\end{aligned}
$$

The inequality that replaces y_1 with 2.51 results from Calculation 9.21.

8.7. A Final Reduction

Let S be a Delaunay simplex. Suppose that the lengths of the edges y_1, y_5, and y_6 are greater than 2. Let S' be a simplex formed by contracting the vertex joining edges 1, 5, and 6 along the first edge by a small amount. We assume that the lengths y_1', y_5', and y_6' of the new edges are still at least 2 and that the circumradius of S' is at most 2, so that S' is a Delaunay simplex.

Proposition 8.7.1. $\Gamma(S') > \Gamma(S)$.

We write sol_i, for $i = 1, 2, 3$, for the solid angles at the three vertices p_1, p_2, and p_3 of S terminating the edges 1, 2, and 3. Let p_1' be the vertex terminating edge 1 of S'. Similarly, we write sol_i', for $i = 1, 2, 3$, for the solid angles at the corresponding vertices of S'. We set $\mathrm{vol}(V) = \mathrm{vol}(S) - \mathrm{vol}(S')$ and $w_i = \mathrm{sol}_i - \mathrm{sol}_i'$. It follows directly from the construction of S' that w_2 and w_3 are positive. The dihedral angle α along the first edge is the same for S and S'. The angle β_i of the triangle $(0, p_1, p_i)$ at p_1 is less than the angle β_i' of the triangle $(0, p_1', p_i)$ at p_1', for $i = 2, 3$. It follows that w_1 is negative, since $-w_1$ is the area of the quadrilateral region of Diagram 8.7.2 on the unit sphere.

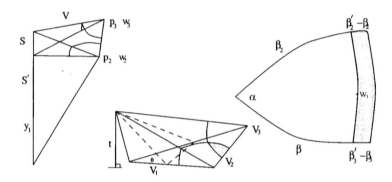

Diagram 8.7.2

Lemma 8.7.3. $\delta_{oct}\,\mathrm{vol}(V) > w_2/3 + w_3/3$.

The lemma immediately implies the proposition because $w_1 < 0$ and

$$\Gamma(S') - \Gamma(S) = \frac{-w_1}{3} - \frac{w_2}{3} - \frac{w_3}{3} + \delta_{oct}\,\mathrm{vol}(V).$$

Proof.   Let T' be the face of S' with vertices p_1', p_2, and p_3. We consider S' as a function of t, where t is the distance from p_1 to the plane containing T'. (See Diagram 8.7.2.) It is enough to establish the lemma for t infinitesimal.

As shown in Diagram 8.7.2, let V_1 be the pyramid formed by intersecting V with the plane through p_1 that meets the fifth edge at distance $t_0 = 1.15$ from p_3 and the sixth edge at distance t_0 from p_2. Also, let the intersection of V with a ball of radius t_0 centered at p_i be denoted V_i, for $i = 2, 3$. For t sufficiently small, the region V_1 is (essentially) disjoint from V_2 and V_3.

We claim that $\mathrm{vol}(V_1) > \mathrm{vol}(V_2 \cap V_3)$. Let θ be the angle of T' subtended by the fifth and sixth edges of S'. Then $\mathrm{vol}(V_1) = Bt/3$, where B is the area of the intersection of

V_1 and T':

(8.7.4)
$$B = \frac{\sin\theta(y'_5 - t_0)(y'_6 - t_0)}{2}.$$

Since S' is a Delaunay simplex, the estimates $\pi/6 \leq \theta \leq 2\pi/3$ from [H1, 2.3] hold. In particular, $\sin\theta \geq 0.5$, so $B \geq 0.25(2 - 1.15)^2$, and $\text{vol}(V_1) > 0.06t$.

If $\text{vol}(V_2 \cap V_3)$ is nonempty, the fourth edge of S must have length less than $2t_0$. The dihedral angle α' of V along the fourth edge is then less than the tangent of the angle, which is at most $t + O(t^2)$, in Landau's notation. As in [H1, 5], we obtain the estimate

$$\text{vol}(V_2 \cap V_3) \leq \alpha' \int_1^{t_0} t_0^2 - t^2 \, dt = \frac{(t_0 - 1)^2(2t_0 + 1)\alpha'}{3} < 0.025t + O(t^2).$$

This establishes the claim.

Thus, for t sufficiently small

$$\delta_{oct}\,\text{vol}(V) \geq \delta_{oct}(\text{vol}(V_1) + \text{vol}(V_2) + \text{vol}(V_3) - \text{vol}(V_2 \cap V_3))$$
$$> (\delta_{oct}t_0^3)\left(\frac{w_2}{3} + \frac{w_3}{3}\right) > 1.09\left(\frac{w_2}{3} + \frac{w_3}{3}\right). \qquad \square$$

9. Floating-Point Calculations

This section describes various inequalities that have been established by the method of subdivision on SUN workstations. The full source code (in C++) for these calculations is available [H6].

Floating-point operations on computers are subject to round-off errors, making many machine computations unreliable. Methods of interval arithmetic give users control over round-off errors [AH]. These methods may be reliably implemented on machines that allow arithmetic with directed rounding, for example, those conforming to the IEEE/ANSI standard 754 [W], [IEEE], [P].

Interval arithmetic produces an interval in the real line that is guaranteed to contain the result of an arithmetic operation. As the round-off errors accumulate, the interval grows wider, and the correct answer remains trapped in the interval. Apart from the risk of compiler errors and defective hardware, a bound established by interval arithmetic is as reliable as a result established by integer arithmetic on a computer. We have used interval arithmetic wherever computer precision is a potential issue (Calculations 9.1–9.19, in particular).

Every inequality of this section has been reduced to a finite number of inequalities of the form $r(x_0) < 0$, where r is a rational function of $x \in \mathbb{R}^n$ and x_0 is a given element in $\mathbb{R}^n$. To evaluate each rational expression, interval arithmetic is used to obtain an interval Y containing $r(x_0)$. The stronger inequality, $y < 0$ for all $y \in Y$, which may be verified by computer, implies that $r(x_0) < 0$.

To reduce the calculations to rational expressions $r(x_0)$, rational approximations to the functions $\sqrt{x}$, $\arctan(x)$, and $\arccos(x)$ with explicit error bounds are required. These were obtained from [H7]. Reliable approximations to various constants (such as π, $\sqrt{2}$, and δ_{oct}) with explicit error bounds are also required. These were obtained in *Mathematica* and were double checked against *Maple*.

Let $S = S(y_1, \ldots, y_6)$ be a quasi-regular tetrahedron. We label the indices as in Diagram 8.1.2. Let $\Gamma = \Gamma(S(y_1, \ldots, y_6))$ be the compression. We also let dih $=$ dih$(S(y_1, \ldots, y_6))$ be the dihedral angle along the first edge, and let sol $=$ sol$(S(y_1, \ldots, y_6))$ be the solid angle at the origin, that is, the solid angle formed by the first, second, and third edges (y_1, y_2, y_3) of S.

All of the following inequalities are to be considered as inequalities of analytic functions of $y_1, \ldots, y_6$. Although each of the calculations is expressed as an inequality between functions of six variables, Proposition 8.7.1 has been invoked repeatedly to reduce the number of variables to three or four. For instance, suppose that we wish to establish $I(\text{sol}, \Gamma) < 0$, where $I(\text{sol}, \Gamma)$ is an expression in $\Gamma(S(y_1, \ldots, y_6))$ and sol$(S(y_1, \ldots, y_6))$. Invoking Proposition 8.7.1 three times, we may assume that the vertices marked v_1, v_2, and v_3 in Diagram 8.1.2 each terminate an edge of minimal length. It is then sufficient to establish the inequality in seven situations of smaller dimension; that is, we may assume the edges $i \in I$ have minimal length, where I is one of

$$\{1, 4\}, \{2, 5\}, \{3, 6\}, \{4, 5\}, \{4, 6\}, \{5, 6\}, \{1, 2, 3\}.$$

Similarly, for an inequality in Γ and dihedral angle, we reduce to the seven cases

$$I = \{1, 4\}, \{2, 5\}, \{3, 6\}, \{1, 2, 3\}, \{1, 5, 6\}, \{3, 4, 5\}, \{2, 4, 6\}.$$

The first two calculations are inequalities of the compression Γ of a quasi-regular tetrahedron.

Calculation 9.1. $\Gamma \leq 1$ *pt.*

This first inequality and Calculation 9.3 are the only ones that are not strict inequalities. Set $S_0 = S(2, 2, 2, 2, 2, 2)$. By definition, $\Gamma(S_0) = 1$ *pt.* We must give a direct proof that S_0 gives the maximum in an explicit neighborhood. Then we use the method of subdivision to bound Γ away from 1 *pt* outside the given neighborhood. An infinitesimal version of the following result is proved in [H1].

Lemma 9.1.1. *If $y_i \in [2, 2.06]$, for $i = 1, \ldots, 6$, then $\Gamma(S(y_1, \ldots, y_6)) \leq \Gamma(S_0)$, with equality if and only if $S(y_1, \ldots, y_6) = S_0$.*

Set $a_{00} = a(2, 2, 2, 2, 2, 2) = 20$, $\Delta_0 = \Delta(2^2, 2^2, 2^2, 2^2, 2^2, 2^2)$, $t_0 = \sqrt{\Delta_0}/2 = 4\sqrt{2}$, and $b_0 = \frac{2}{3}(1 + t_0^2/a_{00}^2)^{-1} = 50/81$. Set $f = \max_i(y_i - 2) \leq 0.06$ and $a^- = a(2, 2, 2, 2 + f, 2 + f, 2 + f) = 20 - 12f - 3f^2$. As in Section 8, we set $t = \sqrt{\Delta}/2$. We have $\arctan(x) \leq \arctan(x_0) + (x - x_0)/(1 + x_0^2)$, if $x, x_0 \geq 0$. We verify below that $\Delta \geq \Delta_0$ under the restrictions given above. Then

$$(9.1.2) \quad \Gamma(S) = -\frac{\delta_{oct} t}{6} + \frac{2}{3} \sum_{i=0}^{3} \arctan\left(\frac{t}{a_i}\right)$$

$$\leq \Gamma(S_0) - \frac{\delta_{oct}(t - t_0)}{6} + b_0 \sum_{i=0}^{3} \left(\frac{t}{a_i} - \frac{t_0}{a_{00}}\right)$$

$$= \Gamma(S_0) - \frac{\delta_{oct}(\Delta - \Delta_0)}{24(t + t_0)} + b_0 \sum_{i=0}^{3} \frac{t - t_0}{a_i}$$

$$+ \frac{b_0 t_0}{a_{00}^2} \sum_{i=0}^{3} (a_{00} - a_i) + \frac{b_0 t_0}{a_0^2} \sum_{i=0}^{3} \frac{(a_{00} - a_i)^2}{a_i}$$

$$\leq \Gamma(S_0) + \frac{(\Delta - \Delta_0)}{t + t_0} \left(-\frac{\delta_{oct}}{24} - \frac{b_0}{a^-} \right) + \frac{b_0 t_0}{a_{00}^2} \sum_{i=0}^{3} (a_{00} - a_i)$$

$$+ \frac{b_0 t_0}{a_0^2 a^-} \sum_{i=0}^{3} (a_{00} - a_i)^2.$$

Set $c_0 = -\delta_{oct}/24 + b_0/a^- > 0$. Write $y_i = 2 + f_i$, with $0 \leq f_i \leq f$. Set $x_i = 4 + e_i = y_i^2 = 4 + 4f_i + f_i^2$. Then $f_i \leq e_i/4$, for $i = 1, \ldots, 6$. Set $e = 4f + f^2$. Set $\tilde{\Delta}(e_1, \ldots, e_6) = \Delta(x_1, \ldots, x_6)$. We find that $\partial^2 \tilde{\Delta}/\partial e_2 \, \partial e_1 = 4 + e_4 + e_5 - e_6 > 0$ and similarly that $\partial^2 \tilde{\Delta}/\partial e_i \partial e_1 > 0$, for $i = 3, 5, 6$. So $\partial \tilde{\Delta}/\partial e_1$ is minimized by taking $e_2 = e_3 = e_5 = e_6 = 0$, and this partial derivative is at least

$$\frac{\partial \tilde{\Delta}}{\partial e_1}(e_1, 0, 0, e_4, 0, 0) = 16 - 8e_1 - 2e_1 e_4 - e_4^2 \geq 16 - 8e - 3e^2 > 0.$$

Thus $\Delta \geq \Delta_0$. The partial derivative $\partial \tilde{\Delta}/\partial e_1$ is at most

$$\frac{\partial \tilde{\Delta}}{\partial e_1}(e_1, e, e, e_4, e, e) = 16 + 16e - 8e_1 + 4ee_4 - 2e_1 e_4 - e_4^2 \leq 16 + 16e + 4ee_4 - e_4^2$$

$$\leq 16 + 16e + 3e^2.$$

So $\Delta - \Delta_0 \leq (16 + 16e + 3e^2)(e_1 + \cdots + e_6)$.

We expand $\sum_{i=0}^{3}(a_i - a_{00}) = 5(e_1 + \cdots + e_6) + h_1 + h_2 + h_3$ as a sum of homogeneous polynomials h_i of degree i in $f_1, \ldots, f_6$. A calculation shows that $h_1 = 0$. Also h_2 is quadratic in each variable f_i with negative leading coefficient, so h_2 attains its minimum at an extreme point of the cube $[0, f]^6$. A calculation then shows that $h_2 \geq -14f^2$ on $[0, f]^6$. By discarding all the positive terms of h_3, we find that

$$h_3 \geq -f_1^2 f_4 - f_1 f_4^2 - f_2^2 f_5 - f_2 f_5^2 - f_3^2 f_6 - f_3 f_6^2 \geq -6f^3.$$

Thus, $\sum_{i=0}^{3}(a_i - a_{00}) \geq 5(e_1 + \cdots + e_6) - 14f^2 - 6f^3$. Since $e = \max(e_i)$, we find that $f \leq e/4 \leq (e_1 + \cdots + e_6)/4$. This gives

$$\sum_{i=0}^{3}(a_{00} - a_i) \leq (e_1 + \cdots + e_6) \left(-5 + \frac{7f}{2} + \frac{3f^2}{2} \right).$$

Similarly, we expand $\sum_{i=0}^{3}(a_{00} - a_i)^2$ as a polynomial in $f_1, \ldots, f_6$. To obtain an upper bound, we discard all the negative terms of the polynomial and evaluate all the positive terms at $(f_1, \ldots, f_6) = (f, \ldots, f)$. This gives

$$\sum_{i=0}^{3}(a_{00} - a_i)^2 \leq 4944 f^2 + 7296 f^3 + 3684 f^4 + 828 f^5 + 73 f^6$$

$$\leq \tfrac{1}{4}(e_1 + \cdots + e_6)(4944 f + 7296 f^2 + 3684 f^3 + 828 f^4 + 73 f^5).$$

We now insert these estimates back into inequality (9.1.2). This gives

$$(9.1.3) \quad \frac{\Gamma(S) - \Gamma(S_0)}{(e_1 + \cdots + e_6)} \leq \frac{c_0}{2t_0}(16 + 16e + 3e^2) + \frac{b_0 t_0}{a_{00}^2}(-5 + 3.5f + 1.5f^2)$$

$$+ \frac{b_0 t_0}{4a_{00}^2 a^-}(4944f + 7296f^2 + 3684f^3 + 828f^4 + 73f^5).$$

The right-hand side of this inequality is a rational function of f. (Both a^- and e depend on f.) Each of the three terms on the right-hand side is increasing in f. Therefore, the right-hand side reaches its maximum at $f = 0.06$. Direct evaluation at $f = 0.06$ gives $\Gamma(S) - \Gamma(S_0) \leq -0.00156(e_1 + \cdots + e_6)$.

To verify Calculation 9.14, the computer examined only 7 cells. Calculation 9.1 required 1899 cells. Calculation 9.6.1 required over 2 million cells. The number of cells required in the verification of the other inequalities falls between these extremes.

Calculation 9.2. $\Gamma < 0.5 \, pt$, if $y_1 \in [2.2, 2.51]$.

The next several calculations are concerned with the relationship between the dihedral angle and the compression.

Calculation 9.3. $\mathrm{dih}(S(y_1, \ldots, y_6)) \geq \mathrm{dih}_{\min} := \mathrm{dih}(S(2, 2.51, 2, 2, 2.51, 2)) \approx 0.8639$.

Since this bound is realized by a simplex, we must carry out the appropriate local analysis in a neighborhood of $S(2, 2.51, 2, 2, 2.51, 2)$.

Lemma 9.3.1. *Suppose that* $2 \leq y_i \leq 2.2$, *for* $i = 1, 3, 6$, *and that* $2 \leq y_i = 2.51$, *for* $i = 2, 4, 5$. *Then* $\mathrm{dih}(S(y_1, \ldots, y_6)) \geq \mathrm{dih}_{\min}$.

Proof. This is an application of Lemma 8.3.1. By that lemma, $\mathrm{dih}(S)$ is increasing in x_4, so we fix $x_4 = 4$. The sign of $\partial \cos \mathrm{dih}/\partial x_2$ is the sign of $\partial \Delta/\partial x_3$, and a simple estimate based on the explicit formulas of Section 8 shows that $\partial \Delta/\partial x_3 > 0$ under the given constraints. Thus, we minimize $\mathrm{dih}(S)$ by setting $x_2 = 2.51^2$. By symmetry, we set $x_5 = 2.51^2$.

Now consider $\mathrm{dih}(S)$ as a function of x_1, x_3, and x_6. The sign of $\partial \cos \mathrm{dih}/\partial x_3$ is the sign of $\partial \Delta(x_1, 2.51^2, x_3, 2^2, 2.51^2, x_6)/\partial x_2$. The maximum of this partial (about -2.39593) is attained when x_1, x_3, and x_6 are as large as possible: $x_1 = x_3 = x_6 = 2.2^2$. So $\mathrm{dih}(S)$ is increasing in x_3. We take $x_3 = 4$, and by symmetry $x_6 = 4$.

We have

$$\frac{\partial \cos \mathrm{dih}(S)}{\partial x_1} = \frac{2t_1(x_1)}{u(x_1, x_2, x_6)^{3/2} u(x_1, x_3, x_5)^{3/2}},$$

where $t_1(x_1) \approx 464.622 - 1865.14x_1 + 326.954x_1^2 - 92.9817x_1^3 + 4x_1^4$. An estimate of the derivative of $t_1(x)$ shows that $t_1(x)$ attains its maximum at $x_1 = 4$, and $t_1(4) < -6691 < 0$. Thus $\mathrm{dih}(S)$ is minimized when $x_1 = 4$. $\qquad\square$

Calculation 9.4. $\Gamma < 0.378979 \, \mathrm{dih} - 0.410894$.

Lemma 9.5. *If S_1, S_2, S_3, and S_4 are any four tetrahedra such that* $\text{dih}(S_1) + \text{dih}(S_2) + \text{dih}(S_3) + \text{dih}(S_4) \geq 2\pi$, *then* $\Gamma(S_1) + \cdots + \Gamma(S_4) < 0.33\ pt$.

This lemma is a consequence of the following three calculations, each established by interval arithmetic and the method of subdivision.

Calculation 9.5.1.
$$\Gamma < -0.19145\,\text{dih} + 0.2910494,$$

provided dih $\in [\text{dih}(S(2, 2, 2, 2, 2, 2)),\ 1.42068]$.

Calculation 9.5.2.
$$\Gamma < -0.0965385\,\text{dih} + 0.1562106,$$

provided dih $\in [1.42068,\ \text{dih}(S(2, 2, 2, 2.51, 2, 2))]$.

Calculation 9.5.3.
$$\Gamma < -0.19145\,\text{dih} + 0.31004,$$

provided dih $\geq \text{dih}(S(2, 2, 2, 2.51, 2, 2))$.

To deduce Lemma 9.5, we consider the piecewise linear bound ℓ on Γ obtained from these estimates. The linear pieces are $\ell_1(x) \leq 1\ pt$, ℓ_2, ℓ_3, and ℓ_4 on $[\text{dih}_{\min}, d_1]$, $[d_1, d_2]$, $[d_2, d_3]$, and $[d_3, \text{dih}_{\max}]$, where $d_1 = \text{dih}(S(2, 2, 2, 2, 2, 2))$, $d_2 = 1.42068$, and $d_3 = \text{dih}(S(2, 2, 2, 2.51, 2, 2))$. (See Calculation 9.1 and Lemma 8.3.2.) Diagram 9.5.4 illustrates these linear bounds. (There are small discontinuities at d_1, d_2, and d_3 that may be eliminated by replacing ℓ_i by $\ell_i + \varepsilon_i$, for some $\varepsilon_i > 0$, for $i = 1, 2$, and 4.) We then ask for the maximum of $\ell(t_1) + \ell(t_2) + \ell(t_3) + \ell(t_4)$ under the constraint $(t_1 + t_2 + t_3 + t_4)/4 \geq \pi/2 \in [d_2, d_3]$. Since $\ell(x)$ is constant on $(\text{dih}_{\min}, d_1)$ and decreasing on $[d_1, \text{dih}_{\max}]$, we may assume that $t_i \geq d_1$, for all i. Since the slope of ℓ_2 is equal to the slope of ℓ_4, we may assume that $t_i \geq d_2$, for all i, or that $t_i \leq d_3$, for all i. If $t_i \leq d_3$, for all i, then we find that $t_i \geq 2\pi - 3d_3 > d_2$. So in either case, $t_i \geq d_2$, for all i. By convexity, an upper bound is $4\ell_3(\pi/2) < 0.33\ pt$.

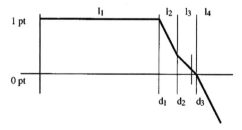

Diagram 9.5.4

Lemma 9.6. *If $S_1, \ldots, S_5$ are any five tetrahedra such that* $\text{dih}(S_1) + \cdots + \text{dih}(S_5) \geq 2\pi$, *then*
$$\Gamma(S_1) + \cdots + \Gamma(S_5) < 4.52\ pt.$$

This is a consequence of two other calculations.

Calculation 9.6.1. $\Gamma < -0.207045 \dim + 0.31023815$, provided dih $\in [d_0, 2\pi - 4d_0]$, where $d_0 = \dim(S(2, 2, 2, 2, 2, 2))$.

Calculation 9.6.2. $\Gamma < 0.028792018$, if dih $> 2\pi - 4d_0$.

Calculations 9.1, 9.6.1, and 9.6.2 give a piecewise linear bound ℓ on Γ as a function of dihedral angle. See Diagram 9.6.3. (Again, there are minute discontinuities that may be eliminated in the same manner as before.) We claim that $4.52 \ pt > \ell(t_1) + \cdots + \ell(t_5)$ whenever $t_1 + \cdots + t_5 \geq 2\pi$. As in Lemma 9.5, we may assume that $t_i \geq d_0$. Since ℓ is decreasing on $[d_0, \dim_{\max}]$, we may assume that $t_1 + \cdots + t_5 = 2\pi$. Thus, only the interval $[d_0, 2\pi - 4d_0]$ is relevant for the optimization. On this interval, the bound is linear, so $\ell(t_1) + \cdots + \ell(t_5) \leq 5\ell(2\pi/5) < 4.52 \ pt$.

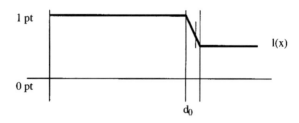

Diagram 9.6.3

Calculation 9.7. $\Gamma < 0.389195 \dim - 0.435643$, if $y_1 \geq 2.05$.

The next inequalities relate the solid angles to the compression.

Calculation 9.8. $\Gamma < -0.37642101 \ \mathrm{sol} + 0.287389$.

Calculation 9.9. $\Gamma < 0.446634 \ \mathrm{sol} - 0.190249$.

Calculation 9.10. $\Gamma < -0.419351 \ \mathrm{sol} + 0.2856354 + 0.001$, if $y_i \in [2, 2.1]$, for $i = 1, \ldots, 6$.

The following calculation involves the circumradius. We leave it to the reader to check that the dimension-reduction techniques of Lemma 8.7.1 may still be applied.

Calculation 9.11. $\Gamma < -0.419351 \ \mathrm{sol} + 0.2856354$, provided that $y_4, y_5, y_6 \geq 2.1$ and that the circumradius of S is at most 1.41.

Calculation 9.12. $\Gamma < -0.419351 \ \mathrm{sol} + 0.2856354 - 5(0.001)$, if $y_i > 2.1$ for some i, and $y_4 \in [2, 21]$.

Calculation 9.13. $\Gamma < -0.65557 \ \mathrm{sol} + 0.418$, if $y_i \in [2, 2.1]$, for $i = 1, \ldots, 6$.

Calculation 9.14. $\text{sol}(S) > 0.21$.

Calculation 9.15. $\Gamma + (K - \text{sol})/3 < 0.564978\,\text{dih} - 0.614725$, where $K = (4\pi - 6.48)/12$, provided $y_1 \in [2, 2.05]$, and $y_2, y_3 \in [2, 2.2]$.

Calculation 9.16. $\text{dih} > 0.98$ and $\text{sol} > 0.45$, provided $y_1 \in [2, 2.05]$, and $y_2, y_3 \in [2, 2.2]$.

Let $\Gamma(S)$ be replaced by $\text{vor}(S)$ in each of the Calculations 9.1–9.16 to obtain a new list of inequalities 9.1′–9.16′. (In 9.11′ we drop the constraint on the circumradius of S.) We claim that all of the inequalities 9.*′ hold whenever S is a quasi-regular tetrahedron of circumradius at least 1.41. In fact, inequalities 9.1′, 9.2′, 9.4′, 9.5.1′, 9.5.2′, 9.5.3′, 9.6.1′, 9.6.2′, and 9.7′ follow directly from 9.17 and the inequalities $\text{dih}_{\min} \leq \text{dih} \leq \text{dih}_{\max}$. Calculations 9.3, 9.14, and 9.16 are independent of Γ, and so do not require modification. Inequalities 9.8′, 9.11′, and 9.12′ also rely on 9.17, 9.18, and 9.19, inequality 9.9′ on 9.14, and inequality 9.15′ on 9.16. Inequalities 9.10′ and 9.13′ are vacuous by the comments of paragraph 8.2.4.

Write $S = S(y_1, y_2, y_3, y_4, y_5, y_6)$, $\text{vor} = \text{vor}(S)$, and let $\text{rad} = \text{rad}(S)$ be the circumradius of S.

Lemma 9.17. *If the circumradius is at least* 1.41, *then* $\text{vor} < -1.8\,pt$.

Proof. If $\text{sol} \geq 0.91882$, then the lemma is a consequence of Lemma 9.18. (The proof of Lemma 9.18, under the restriction $\text{sol} \geq 0.91882$, is independent of the proof of this lemma.) Assume that $\text{sol} < 0.91882$.

If S' and S and Delaunay stars, related as in Section 8.7, then $\text{vor}(S') > \text{vor}(S)$ because $\hat{S}'_0$ is obtained by slicing a slab from $\hat{S}_0$. This means that we may apply the dimension-reduction techniques described at the beginning of this section, unless the deformation decreases the circumradius to 1.41. An interval calculation establishes the result when $y_1 + y_2 + y_3 \geq 6.3$ and the circumradius constraint is met (Calculation 9.17.1). When the dimension-reduction techniques apply, the dimension of the search space may be reduced to four, and an interval calculation similar to the others in this section gives the result (Calculation 9.17.2).

We lacked the computer resources to perform the interval analysis directly when the dimension-reduction techniques fail and found it necessary to break the problem up into smaller pieces when $y_1 + y + 2 + y_3 \leq 6.3$. We make use of the following calculations.

Calculation 9.17.1. $\text{vor}(S) < -1.8\,pt$ provided $y_1 + y_2 + y_3 \geq 6.3$, $\text{rad}(S) = 1.41$, and $\text{sol}(S) \leq 0.91882$.

Calculation 9.17.2. $\text{vor}(S) < -1.8\,pt$ provided that $\text{rad}(S) > 1.41$, $y_1 + y_2 + y_3 \geq 6.3$, $\text{sol}(S) \leq 0.912882$, and S lies in one of the seven subspaces of smaller dimension associated with $I(\text{sol}, \Gamma)$ at the beginning of the section.

Calculation 9.17.3.1. Assume that $y_1 + y_2 + y_3 \le 6.3$ and that rad ≥ 1.41. Then sol ≥ 0.767.

Calculation 9.17.3.2. Assume that $y_1 + y_2 + y_3 \le 6.192$ and that rad ≥ 1.41. Then sol ≥ 0.83.

Calculation 9.17.3.3. Assume that $y_1 + y_2 + y_3 \le 6.106$ and that rad ≥ 1.41. Then sol ≥ 0.87.

Calculation 9.17.3.4. Assume that $y_1 + y_2 + y_3 \le 6.064$ and that rad ≥ 1.41. Then sol ≥ 0.9.

Calculation 9.17.3.5. Assume that $y_1 + y_2 + y_3 \le 6.032$ and that rad ≥ 1.41. Then sol ≥ 0.91882.

In the interval arithmetic verification of Calculations 9.17.3, we may assume that rad $= 1.41$ and that $y_1 + y_2 + y_3$ is equal to the given upper bound 6.3, 6.192, etc. To see this we note that the circumradius constraint is preserved by a deformation of S that increases y_1, y_2, or y_3 while keeping fixed the spherical triangle on the unit sphere at the origin cut out by S. We increase y_1, y_2, and y_3 in this way until the sum equals the given upper bound. Then fixing y_1, y_2, and y_3, and one of y_4, y_5, and y_6, we decrease the other two edges in such a way as to decrease the solid angle and circumradius until rad $= 1.41$.

This deformation argument would break down if we encountered a configuration in which two of y_4, y_5, and y_6 equal 2, but this cannot happen when rad$(S) \ge 1.41$ because this constraint on the edges would lead to the contradiction

$$1.41 \le \text{rad}(S) \le \text{rad}(S(2.3, 2.3, 2.3, 2, 2, 2.51)) < 1.39$$

It is possible to reduce Calculations 9.17.3 further to the four-dimensional situation where two of y_4, y_5 and y_6 are either 2 or 2.51. Consider a simplex S with vertices 0, v_1, v_2, and v_3. Let p_i be the corresponding vertices of the spherical triangle cut out by S on the unit sphere at the origin. Fix the origin, v_2, and v_3, and vary the vertex v_1. The locus on the unit sphere described as p_1 traces out spherical triangles of fixed area is an arc of a Lexell circle $\mathcal{C}$. Define the "interior" of $\mathcal{C}$ to be the points on the side of $\mathcal{C}$ corresponding to spherical triangles of smaller area. The locus traced by v_1 on the circumsphere (of S) with $|v_1|$ constant is a circle. Let $\mathcal{C}'$, also a circle, be radial projection of this locus to the unit sphere. Define the "interior" of $\mathcal{C}'$ to be the points coming from larger $|v_1|$. The two circles $\mathcal{C}$ and $\mathcal{C}'$ meet at p_1, either tangentially or transversely. The interior of $\mathcal{C}$ cannot be contained in the interior of $\mathcal{C}'$ because the Lexell arc contains p_2^*, the point on the unit sphere antipodal to p_2 [FT, p. 23], but the interior of of $\mathcal{C}'$ does not. Furthermore, if $|v_1| = 2$ and $|v_2| + |v_3| > 4$, then v_2 or v_3 lies in the interior of $\mathcal{C}$ and $\mathcal{C}'$, so that the circles have interior points in common. This means that v_1 can always be moved in such a way that the solid angle is decreasing, the circumradius is constant, and the length $|v_1|$ is decreasing (or constant if $|v_1| = 2$). If any two of y_4, y_5, and y_6 are not at an extreme point, this argument can be applied to v_1, v_2, or v_3 to decrease the solid angle. This proves the reduction.

We are now in a position to prove the lemma for simplices satisfying $y_1+y_2+y_3 \leq 6.3$. Calculation 9.17.3.1 allows us to assume sol ≥ 0.767. As in Section 8.6, let $S_y = S(2, 2, 2, y, y, y)$. We rely on the fact that vor(S_y) is decreasing in y, for $y \in [2.26, 2.41]$. In fact, the results of Section 8.6 specialize to the formula

$$(9.17.4) \quad \text{vor}(S_y) = \frac{-8\delta_{oct}y^2}{(12-y^2)^{1/2}(16-y^2)} + \frac{8}{3}\arctan\left(\frac{(12-y^2)^{1/2}y^2}{64-6y^2}\right),$$

and the sign of its derivative is determined by a routine *Mathematica* calculation. It is clear that sol(S_y) is continuous and increasing in y. Since sol($S_{2.26}$) < 0.767 and sol($S_{2.41}$) > 0.91882, our conditions imply that sol(S) $=$ sol(S_y) for some $y \in [2.26, 2.41]$. Let $r(a)$ and vol($R(a, b, c)$) be the functions introduced in Section 8.6.4.

This suggests the following procedure. Pick y so that sol(S) $\geq$ sol(S_y). Calculate the smallest (or at least a reasonably small) a for which

$$\xi(y, a) := \text{vor}(S_y) - 4\delta_{oct}\left(1 - \frac{1}{a^3}\right)2r(a)$$

is less than -1.8 *pt*. Monotonicity (Calculation 9.20.1) and Lemma 8.6.5 imply that vor(S) < -1.8 *pt*, if $y_1+y_2+y_3 \geq 2(2+a)$. To treat the case that remains ($y_1+y_2+y_3 < 2(2+a)$), use Calculations 9.17.3 to obtain a new lower bound on sol(S), and hence a new value for y. The procedure is repeated until $a = 0.016$. Calculation 9.17.3.5 completes the argument by covering the case $y_1 + y_2 + y_3 \leq 6.032$. We leave it to the reader to check that

$$\xi(2.2626, 1.096), \xi(2.326, 1.053), \xi(2.364, 1.032), \xi(2.391, 1.016)$$

are less than -1.8 *pt* and that

$$\text{sol}(S_{2.2626}) \leq 0.767, \text{sol}(S_{2.326}) \leq 0.83, \text{sol}(S_{2.364}) \leq 0.87, \text{sol}(S_{2.391}) \leq 0.9.$$

This completes the proof of Lemma 9.17.   □

Lemma 9.18. *If* rad(S) ≥ 1.41, *then* vor < -0.419351 sol $+0.2856354$.

Proof. We adopt the notation and techniques of Lemma 9.17. If sol ≤ 0.918819, then the result follows from Lemma 9.17. (The proof of Lemma 9.17 is independent of the argument that follows under that restriction on solid angles.) Let $f(S) = -0.419351$ sol(S) $+ 0.2856354 - $ vor(S). We show that $f(S)$ is positive. We use the inequality $f(S) \geq f(S_{tan}) + 4\delta_{oct}$ vol($\hat{S}\backslash\hat{S}_{tan}$) of Lemma 8.6.4. A routine calculation based on formula (9.17.4) shows that $f(S_y)$ is increasing, for $y \in [2.4085, 2.51]$.

If sol ≥ 0.951385, then we appeal to Fejes Tóth's convexity argument described in Section 8.6. To justify the use of his argument, we must verify that the cone over S_{tan} contains the circumcenter of S. The first, second, and third edges have length 2, and the fourth, fifth, and sixth edges are between 2.21 and 2.51 (Calculation 9.18.2). These are stronger restrictions on the edges than in the proof of Lemma 8.6.5, so the justification there applies here as well. We observe that $\text{sol}(S_{2.4366}) < 0.951385$ so that

$$f(S) \geq f(S_{\tan}) \geq f(S_y) \geq f(S_{2.4366}) \approx 0.000024 > 0,$$

where y satisfies $\text{sol}(S_y) = \text{sol}(S_{\tan})$. We may assume that $0.918819 \leq \text{sol} \leq 0.951385$.

Lemma 9.18.1. *The combined volume of the two Rogers simplices along a common edge of a quasi-regular tetrahedron S is at least* 0.01.

Proof. The combined volume is at least that of the right-circular cone of height a and base a wedge of radius $\sqrt{b^2 - a^2}$ and dihedral angle $\text{dih}_{\min}$, where a is the half-length of an edge and b is a lower bound on the circumradius of a face with an edge $2a$. This gives the lower bound of

$$\frac{\pi}{3}(b^2 - a^2)a\frac{\text{dih}_{\min}}{2\pi} > 0.1439(b^2 - a^2).$$

We minimize b by setting $b^2 = \eta(2, 2, 2a)^2 = 4/(4 - a^2)$. Then $b^2 - a^2 = (a^2 - 2)^2/(4 - a^2)$, which is decreasing in $a \in [1, 2.51/2]$. Thus, we obtain a lower bound on $b^2 - a^2$ by setting $a = 2.51/2$, and this gives the estimate of the lemma. $\square$

We remark that $\text{sol}(S_{2.4085}) < 0.918819$. If $1.15 \leq (y_1 + y_2 + y_3 - 4)/2$, then Lemma 9.18.1 and Section 8.6 give

$$f(S) \geq f(S_{2.4085}) + 48_{oct}\left(1 - \frac{1}{1.15^3}\right)0.01 > 0.$$

(Analytic continuation is not required here, because of the constraints on the edges in Calculation 9.18.2.) We may now assume that $y_1 + y_2 + y_3 \leq 6.3$. To continue, we need a few more calculations.

Calculation 9.18.2. If $\text{sol} \geq 0.918$, then $y_4, y_5, y_6 \geq 2.21$.

Calculation 9.18.3.1. If $y_1 + y_2 + y_3 \leq 6.02$ and $\text{rad} \geq 1.41$, then $\text{sol} \geq 0.928$.

Calculation 9.18.3.2. If $y_1 + y_2 + y_3 \leq 6.0084$ and $\text{rad} \geq 1.41$, then $\text{sol} \geq 0.933$.

Calculation 9.18.3.3. If $y_1 + y_2 + y_3 \leq 6.00644$ and $\text{rad} \geq 1.41$, then $\text{sol} \geq 0.942$.

In the verification of these calculations, we make the same reductions as in Calculations 9.17.3.

Adapting the procedure of Lemmas 9.17 and 8.6.5, we fine that a lower bound on the solid angle leads to an estimate of a constant a with the property that $f(S) > 0$ whenever $y_1 + y_2 + y_3 \geq 4 + 2a$. That is, we pick a so that $\xi_1(y, a) := f(S_y) + 88_{oct}(1 - 1/a^2)r(a)$

is positive, where y is chosen so that $\mathrm{sol}(S_y)$ is a lower bound on the solid angle. The values

$$\xi_1(2.4085, 1.01), \xi_1(2.4165, 1.0042), \xi_1(2.42086, 1.00322), \xi_1(2.4286, 1.0017)$$

are all positive. This yields the bound $y_1 + y_2 + y_3 \le 6.0034$.

Assume that S satisfies $y_1 + y_2 + y_3 \le 6.0034$ and $\mathrm{rad}(S) \ge 1.41$. Then by Calculation 9.18.3.3, $\mathrm{sol}(S) = \mathrm{sol}(S_{\tan}) \ge 0.942$. Also $\mathrm{rad}(S_{\tan}) > \mathrm{rad}(S)/1.0017 \ge 1.41/1.0017 > 1.4076$, because rescaling $S_{\tan}$ by a factor of 1.0017 gives a simplex containing S. This means that $S_{\tan}$ satisfies the hypotheses of the following calculation. Calculation 9.18.4 completes the proof of Lemma 9.18.   □

Calculation 9.18.4. If $\mathrm{rad}(S_{\tan}) \ge 1.4076$ and $0.933 \le \mathrm{sol}(S_{\tan}) \le 0.951385$, then $f(S_{\tan}) > 0$.

In this verification we may assume that the fourth, fifth, and sixth edges of $S_{\tan}$ are at least 2.27, for otherwise the circumradius is at most

$$\mathrm{rad}(S(2, 2, 2, 2.27, 2.51, 2.51)) < 1.4076.$$

In the verification of Calculation 9.18.4, we also rely on the fact that

$$f_1(x_4, x_5, x_6) := f(S(2, 2, 2, \sqrt{x_4}, \sqrt{x_5}, \sqrt{x_6}))$$

is increasing in $(x_4, x_5, x_6) \in [2.27^2, 2.51^2]$. Here is a sketch justifying this fact. The details were carried out in *Mathematica* with high-precision arithmetic. The explicit formulas of Section 8.6 lead to an expression for $\partial f_1/\partial x_4$ as

$$\frac{W(x_4, x_5, x_6)(-x_4 + x_5 + x_6)}{(-16 + x_4)^2 \Delta^{3/2}},$$

where W is a polynomial in x_4, x_5, and x_6 (with 13 terms). To show that W is positive, expand it in a Taylor polynomial about $(x_4, x_5, x_6) = (2.27^2, 2.27^2, 2.27^2)$ and check that the lower bound of inequality (7.1) is positive.

Calculation 9.19. If $y_4 \in [2, 2.1]$, then $\mathrm{sol} < 0.906$.

For $a \le b \le c$, we have $\mathrm{vol}(R(a, b, c)) = a((b^2 - a^2)(c^2 - b^2))^{1/2}/6$. The final four calculations are particularly simple (to the extent that any of these calculations are simple), since they involve a single variable. They were verified in *Mathematica* with rational arithmetic.

Calculation 9.20.1. The function $(1 - 1/a^2)\,\mathrm{vol}(R(a, \eta(2, 2, 2a), 1.41))$ is increasing on $[1, 1.15]$.

Calculation 9.20.2. The function $\mathrm{vol}(R(a, \eta(2, 2, 2a), 1.41))$ is decreasing for $a \in [1, 1.15]$.

Calculation 9.20.3.   For $a \in [1, 1.15]$ we have

$$\text{vol}(R(a, \eta(2, 2, 2a), 1.41)) < \text{vol}(R(a, \eta(2, 2.51, 2a), 1.41)).$$

Calculation 9.21.   Let $p_1 = p_1(y_1, \ldots, y_6)$ be the point in Euclidean space introduced in Section 8.6.7. For $y \in [2.3, 2.51]$,

$$\frac{d}{dy}\left(\eta(2.51, 2.51, y) - \frac{y^2}{4}\right)^{1/2} > -0.75 > \frac{d}{dy}\left(|p_1(y, 2.51, 2, 2, 2, 2.51)|^2 - \frac{y^2}{4}\right)^{1/2}.$$

Acknowledgments

I would like to thank D. J. Muder for the Appendix and the referees for suggesting other substantial improvements.

Appendix.   Proof of Theorem 6.1

D. J. Muder

Notation and Observations.   Let P be a point of degree d. If we consider P to be the center of the configuration, then the *first rim* of points will be the set of d points adjacent to P, and the *second rim* will be those at distance 2 from P. Let $\delta(P)$ be the sum of the degrees of the first rim points. If $d = 6$, it is easy to see that the number of points on the second rim is $s = \delta(P) - 24$, and the total number of points of distance at most 2 from P is $\delta(P) - 17$. The second rim is thus an s-gon. This s-gon and the triangulation of the P-side of the s-gon are referred to as the *inner graph*. The inner-graph degree of a second rim point is called its *inner degree*. The number of second-rim points with inner degree 5 is equal to the number of degree 4 points in the first rim, and the number of second-rim points with inner degree 3 is equal to the number of degree 6 points on the first rim. All other second-rim points have inner degree 4. Points beyond the second rim are called *extra points*. If there are no extra points, then there are at least two second-rim points whose degrees are equal to their inner degrees. Let $N_\Delta^i(P)$ be the number of points of degree Δ in rim i around P. Notice that if $d = 6$, then

$$\delta(P) = 30 - N_4^1(P) + N_6^1(P).$$

Further, Euler's formula gives $N - 12 = N_6 - N_4$. Putting this together with 5.1.1 and 5.1.5, we see

$$13 \le 12 + N_6 - N_4 \le 15,$$
$$1 \le 1 + N_4 \le N_6 \le 3 + N_4 \le 5.$$

Lemma 1.   *There is no triangle whose vertices all have degree 6.*

Lemma 2. *Suppose $N_6 \geq 3$. Let $\mathbf{G}_6$ be the subgraph of points of degree 6. Either $\mathbf{G}_6$ is three points in two components, or it is a single (open or closed) path, with no other edges.*

With these lemmas and this notation, we consider the different possible values of N_6.

$N_6 = 1$. This forces $N_4 = 0$. Let P be the degree 6 point. Then $\delta(P) = 30$, all thirteen points are in the inner graph, and all six second-rim points have inner degree 4. However, at least two of these second-rim points have inner degree equal to degree. Contradiction.

$N_6 = 2$. If the two degree 6 points are adjacent, let P be either one. Now $\delta(P) = 31 - N_4^1$, and we see that there are no extra points, and all second-rim points have degree 5. All but N_4^1 second-rim points have inner degree less than 5, and at least two of them will have inner degree equal to degree. So $N_4 \geq N_4^1 = 2$, and $N \leq 12$. Contradiction.

If the degree 6 points are not adjacent, then the $(6, 6)$ forces $N \geq 14$, so $N_4 = 0$. If $\delta(P)$ is either of the two degree 6 points, $\delta(P) = 30$, so the second rim has six points, all of inner degree 4, and there is one extra point. If the extra point has degree 5, it is adjacent to all but one of the second-rim points. This unique second-rim point is then part of a quadrilateral, which can only be triangulated by the diagonal edge that does not include the extra point. This creates two second-rim points of degree 6, in violation of our assumptions. Therefore the extra point has degree 6, and we have Diagram 6.2.

$N_6 = 3$. Either $\mathbf{G}_6$ is an open path or it has two components. In either case there is a $(6, 6)$, so $N \geq 14$ and $N_4 = 0$ or 1.

In the first case let P be the center of the path. Now $\delta(P) = 32 - N_4^1$. We see that there are no extra points and any degree 4 point is in the first rim. The second rim has at most one point of inner degree greater than 4, and two points whose inner degrees are equal to their degrees. So there must be a point of degree 4 on the second rim. Contradiction.

In the second case 5.1.8 forces $N = 15$, hence $N_4 = 0$. If P is one of the points in the two-point component of $\mathbf{G}_6$, then $\delta(P) = 31$, so there is one extra point and seven second-rim points. Either one or two of the second rim points is not connected to the extra point. In either case at least one of these two has degree = inner degree, which is at most 4. Contradiction.

$N_6 = 4$. Now $\mathbf{G}_6$ is either an open or closed path of length 4. If open, then, by 5.1.8, $N = 15$ and $N_4 = 1$. If P is either of the interior degree 6 points, then $\delta(P) = 32 - N_4^1$, the second rim is an $(8 - N_4^1)$-gon with N_4^1 points of inner degree at least 5 and N_4^1 extra points. We cannot have $N_4^1 = 0$, or else there would be at least two second-rim points of degree 4, so $N_4^1 = 1$. This holds no matter which interior degree 6 point we started with, so the degree 4 point must be adjacent to both. It must also be adjacent to one of the other two, or else there is a $(6, 6, 4)$. Therefore there is a degree 6 point whose first-rim degree sequence is $6/4/6/5/5/5$, producing a second-rim inner degree sequence of $4/3/5/3/4/4/4$. The point of inner degree 5 is the fourth neighbor of the degree 4 point, and cannot be degree 6 without making $\mathbf{G}_6$ closed. Therefore there is an edge connecting the two points of inner degree 3. Adding this edge to the inner graph creates a hexagon with all points of "inner" degree 4, and the extra point in its interior. The extra point is connected to five of these points, and the sixth has degree 4. Contradiction.

If $\mathbf{G}_6$ is a closed path, then a degree 4 point Q must triangulate this quadrilateral. Using Q as the center, there are four first-rim points and eight second-rim points. The second rim can be drawn in a square with the four points of inner degree 3 on the corners and the four points of inner degree 4 at the midpoints. There is at most one additional degree 4 point (other than Q). If it is anywhere but at one of the second-rim midpoints, a $(6, 6, 4)$ exists. The four midpoints cannot be adjacent to any additional second-rim points without creating an illegal triangle or quadrilateral. Since there can be at most one more degree 4 point, at least three of the four second-rim midpoints must be joined to the extra point(s). Two midpoints of consecutive sides cannot be joined to the same extra point without forcing the corresponding corner to be degree 4, which is impossible. So there must be two extra points, and thus no degree 4 points other than Q. One extra point must be joined to the midpoints of each pair of parallel sides of the square. However, all these edges cannot be drawn without intersecting.

$N_6 = 5$. Now $\mathbf{G}_6$ is a path of length 5, closed so that a $(6, 6, 6)$ does not exist. Any point in this pentagon is nonadjacent to two others, so the situation of 5.1.8 applies in five different ways. Let $P, Q, R \in \mathbf{G}_6$ with P and Q adjacent, but neither adjacent to R. Then there are eight points adjacent to either P or Q, and six points adjacent to R, so there are exactly two common points in these two sets. In our case these two points are precisely the other two points of degree 6. This means that no point inside or outside the pentagon can be connected to more than two points of the pentagon. Thus the only possible configuration for the 15 points is to form three concentric pentagons, with $\mathbf{G}_6$ as the middle one. Prior to triangulating the inner and outer pentagons, all of their points have degree 4. Triangulating either creates more degree 6 points. Contradiction.

Proof of Lemma 1. Let P_1, P_2, and P_3 be the vertices of such a triangle. If P_j has no neighbors of degree 4, then the only possibility is $\delta(P_j) = 32$, which forces $N_5^1 = 4$, $N_6^1 = 2$, and $N = 15$. All second-rim points have inner degree at most 4, and there are no extra points. Therefore two second-rim points have degree and inner degree 4. Call them Q_1 and Q_2. Therefore, $N_4 \geq 2$, $N_6 \geq 5$, and $N_6^2 \geq 2$. Let R_1 and R_2 be second-rim points of degree 6. Each Q_k and R_ℓ must be adjacent, or else (P_j, R_ℓ, Q_k) is a $(6, 6, 4)$. However, all edges from Q_j are inner-graph edges. So $Q_1 R_1 Q_2 R_2$ must be a second-rim quadrilateral, which is impossible. Thus $N_4^1(P_j) \geq 1$ for each P_j.

This forces $N_4 = 2$. At least one of the degree 4 points is adjacent to two of the P_j. One of these two, say P_1, is not adjacent to the other degree 4 point. Since $\delta(P) \leq 32$, we are left with four possible first-rim degree sequences for P_1: (1) $6/6/4/5/5/5$; (2) $6/6/4/6/5/5$; (3) $6/6/4/5/6/5$: or (4) $6/6/4/5/5/6$.

The last three possibilities are easily dealt with. Sequence (3) contains a $(6, 6, 4)$. Sequence (4) creates two 6-6-6 triangles joined at an edge. Let P_1 and P_2 be the vertices of the edge, and let P_3 and P_4 be the other vertices of the triangles. The conditions $N_4^1(P_j) \geq 1$ and $N_4 = 2$ force P_3 and P_4 to have sequence (1). In (2) we can invoke 5.1.8 with the three degree 6 points in the first rim. Call them R_1, R_2, and T, with R_1 adjacent to R_2. In order for the numbers to work out, there can be precisely two points which are adjacent to T as well as one of the R_j. However, in (2) the center, the first-rim degree 4 point, and the second-rim neighbor of the first-rim degree 4 point must all fit this description.

Sequence (1) is more difficult to eliminate. Let $R_1, \ldots, R_6$ be the first-rim points listed in the order of (1). Let $S_1 \ldots S_7$ be the second-rim points, with S_1 being adjacent to both R_1 and R_2, and S_7 having only R_1 as a first-rim neighbor. There is at least one degree 6 point in the second rim—if it is at S_1, then $\delta(R_1) = 32$; if it is anywhere else we can invoke 5.1.8. In either case, $N = 15$ and there are two degree 6 points outside the first rim. These must lie in $\{S_6, S_7, S_1, S_3\}$ to avoid making a $(6, 6, 4)$ with R_1 and R_3. At most one of these points can be adjacent to R_1 (since $\delta(R_1) < 33$), so S_3 must have degree 6. Now S_3 must connect to both the degree 6 and degree 4 points in $\{S_6, S_7, S_1\}$ to avoid making a $(6, 6, 6)$ or $(6, 6, 4)$ with the center point. However, S_3 has inner degree 5 and so can connect to at most one of those points. Contradiction.

Proof of Lemma 2.   If G_6 has at least three points and no triangles, then there exists a $(6, 6)$ forcing $N \geq 14$. Any point in G_6 can be adjacent to at most two others, or else either a triangle or a $(6, 6, 6)$ is created. Therefore each component is a path. There cannot be three or more components without producing a $(6, 6, 6)$. If G_6 has two components, each of them must be a complete graph for the same reason. Since there are no triangles, both of the components must have at most two points. If there are two components with two points each, then 5.1.8 forces $N = 15$. Each component is adjacent to eight non-G_6 points of the original graph, and not adjacent to three. Now 5.1.8 forces each point in the other component to be adjacent to these three. However, two points can have only two common neighbors. So the only two-component G_6 is the one described in the lemma. □

References

[AH]   G. Alefeld and J. Herzeberger, *Introduction to Interval Computations*, Academic Press, New York, 1983.

[FT]   L. Fejes Tóth, *Lagerungen in der Ebene, auf der Kugel und im Raum*, Springer-Verlag, Berlin, 1953.

[H1]   T. C. Hales, Remarks on the density of sphere packings in three dimensions, *Combinatorica*, **13**(2) (1993), 181–197.

[H2]   T. C. Hales, The sphere packing problem, *J. Comput. Appl. Math*, **44** (1992), 41–76.

[H3]   T. C. Hales, The status of the Kepler conjecture, *Math. Intelligencer*, **16**(3) (1994), 47–58.

[H4]   T. C. Hales, Sphere packings, II, *Discrete Comput. Geom.*, to appear.

[H5]   T. C. Hales, Sphere packings, III$_\alpha$, in preparation.

[H6]   T. C. Hales, Packings, http://www-personal.math.lsa.umich.edu/~hales/packings.html

[H7]   J. F. Hart *et al.*, *Computer Approximations*, Wiley, New York, 1968.

[IEEE]   IEEE Standard for Binary Floating-Point Arithmetic, ANSI/IEEE Std. 754–1985, IEEE, New York.

[P]   W. H. Press *et al.*, Numerical recipes in C, *Less-Numerical Algorithms*, second edition, Cambridge University Press, Cambridge, 1992. Chapter 20.

[R]   C. A. Rogers, The packing of equal spheres, *Proc. London Math. Soc.* (3) **8** (1958), 609–620.

[W]   What every computer scientist should know about floating-point arithmetic, *Comput. Surveys*, **23**(1) (1991), 5–48.

Received May 12, 1994, and in revised form April 24, 1995, and April 11, 1996.

Sphere Packings II, by T. C. Hales

This 1997 paper is Hales's second initial paper on the Kepler Conjecture project. It proves some basic inequalities useful to his proof approach.

Contents

The original version of this chapter was revised. An erratum to this chapter can be found at
http://dx.doi.org/10.1007/978-1-4614-1129-1_12

Discrete Comput Geom 18:135–149 (1997)

Discrete & Computational

Geometry
© 1997 Springer-Verlag New York Inc.

Sphere Packings, II[*]

T. C. Hales

Department of Mathematics, University of Michigan,
Ann Arbor, MI 48109, USA

Abstract. An earlier paper describes a program to prove the Kepler conjecture on sphere packings. This paper carries out the second step of that program. A sphere packing leads to a decomposition of $\mathbb{R}^3$ into polyhedra. The polyhedra are divided into two classes. The first class of polyhedra, called quasi-regular tetrahedra, have density at most that of a regular tetrahedron. The polyhedra in the remaining class have density at most that of a regular octahedron (about 0.7209).

1. Introduction

This paper is a continuation of the first part of this series [4]. The terminology and notation of this paper are consistent with this earlier paper, and we refer to results from that paper by prefixing the relevant section numbers with "I."

We review some definitions from [4]. Begin with a packing of nonoverlapping spheres of radius 1 in Euclidean three-space. The *density* of a packing is defined in [2]. It is defined as a limit of the ratio of the volume of the unit balls in a large region of space to the volume of the large region. The density of the packing may be improved by adding spheres until there is no further room to do so. The resulting packing is said to be *saturated*.

Every saturated packing gives rise to a decomposition of space into simplices called the *Delaunay decomposition* [8]. The vertices of each Delaunay simplex are centers of spheres of the packing. By the definition of the decomposition, none of the centers of the spheres of the packing lie in the interior of the circumscribing sphere of any Delaunay simplex. We refer to the centers of the packing as *vertices*. Vertices that come within 2.51 of each other are called *close neighbors*.

The Delaunay decomposition is dual to the well-known Voronoi decomposition. If the vertices of the Delaunay simplices are in nondegenerate position, two vertices are joined

---

[*] This research was supported by the National Science Foundation.

by an edge exactly when the two corresponding Voronoi cells share a face, three vertices form a face exactly when the three Voronoi cells share an edge, and four vertices form a simplex exactly when the four corresponding Voronoi cells share a vertex. In other words, two vertices are joined by an edge if they lie on a sphere that does not contain any other of the vertices, and so forth (again assuming the vertices to be in nondegenerate position).

We say that the convex hull of four vertices is a *quasi-regular tetrahedron* (or simply a *tetrahedron*) if all four vertices are close neighbors of one another. If the largest circumradius of the faces of a Delaunay simplex is at most $\sqrt{2}$, we say that the simplex is *small*. Suppose that we have a configuration of six vertices in bijection with the vertices of an octahedron with the property that two vertices are close neighbors if and only if the corresponding vertices of the octahedron are adjacent. Suppose further that there is a unique diagonal of length at most $2\sqrt{2}$. In this case we call the convex hull of the six vertices a *quasi-regular octahedron* (or simply an *octahedron*). A *Delaunay star* is defined as the collection of all quasi-regular tetrahedra, octahedra, and Delaunay simplices that share a common vertex v.

We assume that every simplex S in this paper comes with a fixed order on its edges, $1, \ldots, 6$. The order on the edges is to be arranged so that the first, second, and third edges meet at a vertex. We may also assume that the edges numbered i and $i + 3$ are opposite edges for $i = 1, 2, 3$. We define $S(y_1, \ldots, y_6)$ to be the (ordered) simplex whose ith edge has length y_i. If S is a Delaunay simplex in a fixed Delaunay star, then it has a distinguished vertex, the vertex common to all simplices in the star. In this situation, we assume that the edges are numbered so that the first, second, and third edges meet at the distinguished vertex.

A function, known as the *compression* $\Gamma(S)$, is define on the space of all Delaunay simplices. Set $\delta_{oct} = (-3\pi + 12 \arccos(1/\sqrt{3}))/\sqrt{8} \approx 0.720903$. Let S be a Delaunay simplex. Let B be the union of four unit balls placed at each of the vertices of S. Define the compression as

$$\Gamma(S) = -\delta_{oct} \operatorname{vol}(S) + \operatorname{vol}(S \cap B).$$

We extend the definition of compression to Delaunay stars D^* by setting $\Gamma(D^*) = \sum \Gamma(S)$, with the sum running over all the Delaunay simplices in the star. We define a *point* (abbreviated *pt*) to be $\Gamma(S(2, 2, 2, 2, 2, 2)) \approx 0.0553736$. The compression is often expressed as a multiple of *pt*.

There are several other functions of a Delaunay simplex that will be used. The *dihedral angle* dih(S) is defined to be the dihedral angle of the simplex S along the first edge (with respect to the fixed order on the edges of S). The *solid angle* (measured in steradians) at the vertex joining the first, second, and third edges is denoted sol(S). Let rad(S) be the circumradius of the simplex S. More generally, let rad(F) denote the circumradius of the face of a simplex. Let $\eta(a, b, c)$ denote the circumradius of a triangle with edges a, b, c. Explicit formulas for all these functions appear in Section I.8.

Fix a Delaunay star D^* about a vertex v_0, which we take to be the origin, and we consider the unit sphere at v_0. Let v_1 and v_2 be vertices of D^* such that v_0, v_1, and v_2 are all close neighbors of one another. We take the radial projections p_i of v_i to the unit sphere with center at the origin and connect the points p_1 and p_2 by a geodesic arc on the sphere. We mark all such arcs on the unit sphere. The closures of the connected components of the complement of these arcs are regions on the unit sphere, called the *standard regions.* We may remove the arcs that do not bound one of the regions. The resulting system of edges and regions will be referred to as the *standard decomposition* of the unit sphere.

Let C be the cone with vertex v_0 over one of the standard regions. The collection of the Delaunay simplices, quasi-regular tetrahedra, and quasi-regular octahedra of D^* in C (together with the distinguished vertex v_0) will be called a *standard cluster.* Each Delaunay simplex in D^* belongs to a unique standard cluster.

A real number, called the *score*, will be attached to each cluster. Each star receives a score by summing the scores for the clusters in the star.

The steps of the Kepler conjecture, as outlined in Part I, are:

1. A proof that even if all standard regions are triangular, the total score is less than 8 *pt.*
2. A proof that the standard clusters with more than three sides score at most 0 *pt.*
3. A proof that if all of the standard regions are triangles of quadrilaterals, then the total score is less than 8 *pt* (excluding the case of pentagonal prisms).
4. A proof that if some standard region has more than four sides, then the star scores less than 8 *pt.*
5. A proof that pentagonal prisms score less than 8 *pt.*

The proof of the first step is complete. The other steps are briefly discussed in Part I. This paper establishes Step 2. Partial results have been obtained for Step 3 [5]. C. A. Rogers has shown that the density of a regular tetrahedron is a bound on the density of packings in $\mathbb{R}^3$ [8]. The main result of this paper may be interpreted as saying that the density ($\delta_{oct} \approx 0.7209$) of a regular octahedron is a bound on the density of the complement in $\mathbb{R}^3$ of the quasi-regular tetrahedra in the packing.

The score of a Delaunay star is obtained by mixing Delaunay stars with the dual Voronoi cells. Delaunay stars D^* and the associated function Γ behave much better than estimates of density by Voronoi cells, provided each Delaunay simplex in the Delaunay star has a small circumradius. Unfortunately, $\Gamma(S)$ gives an increasingly poor bound on the density as the circumradius of the Delaunay simplex S increases. When the circumradius of S is greater than about 1.8, it becomes extremely difficult to prove anything about sphere packings with the function $\Gamma(S)$. The score is introduced to regularize the irregular behavior of $\Gamma(S)$.

Voronoi cells also present enormous difficulties. The dodecahedron shows that a single Voronoi cell cannot lead to a bound on the density of packings better than about 0.75. This led L. Fejes Tóth to propose an approach to the Kepler conjecture in which two layers of Voronoi cells are considered: one central Voronoi cell and a number of surrounding ones. Wu-Yi Hsiang has made some progress in this direction, but there remain many technical difficulties [3], [7].

The method of scoring in this paper seeks to combine the best aspects of both approaches. When the circumradius of a simplex is small, we proceed as in Part I. However, when the circumradius of a simplex is large, we switch to Voronoi cells. Remarkably, these two approaches may be coherently combined to give a meaningful score to Delaunay stars and, by extension, a bound on the density of a packing. The calculations of this paper suggest that this hybrid approach to packings retains the best features of both methods with no (foreseeable) negative consequences.

2. Some Polyhedra

Sometimes the tip of a Voronoi cell protrudes beyond the face of a corresponding De-launay simplex (see Diagram 2.1(a)). This section describes a construction that amounts to slicing off the protruding tip of a Voronoi cell and reapportioning it among the neigh-boring cells (see Diagram 2.1(b)).

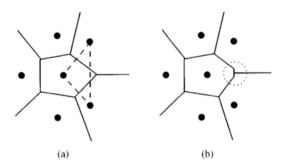

(a)   (b)

Diagram 2.1. (a) Voronoi cells and (b) reapportioned.

Let D^* be a Delaunay star with center $v_0 = 0$. Let V be the Voronoi cell around v_0, obtained by duality from D^*. As a matter of convenience, we may assume that each point in $\mathbb{R}^3$ belongs to a unique Voronoi cell by making an arbitrary choice for each point on the boundary of a cell. If R is a standard cluster (possibly a single quasi-regular tetrahedron) in D^*, let $C(R)$ denote its cone over v_0:

$$C(R) = \{tx : t \geq 0, x \in R\}.$$

In general, $V \cap C(R)$ depends on more vertices than just those in the cluster R. It is convenient to consider the slightly larger polyhedron V_R^0 defined by just the vertices of D^* that are in R. That is, let V_R^0 be the intersection of $C(R)$ with the half-spaces $\{x : x \cdot v_i \leq v_i \cdot v_i/2, \forall i \neq 0\}$, where $\{v_i\}_i$ are the vertices (other than v_0) of the simplices and quasi-regular solids in the cluster R. The faces of V_R^0 at v_0 are contained in the triangular faces bounding the standard region of R. The other faces of V_R^0 are contained in planes through the faces of the Voronoi cell V. We refer to these as Voronoi faces. If R is not a quasi-regular tetrahedron, set $V_R = V_R^0$. If R is a quasi-regular tetrahedron, we take the slightly smaller polyhedron V_R obtained by intersecting V_R^0 with the half-space (containing v_0) bounded by the hyperplane through the face of R opposite the origin v_0. (This may cut a tip from the Voronoi cell.) By construction, V_R depends only on the simplices in R. The polyhedron V_R is based at the center of some Delaunay star, giving it a distinguished vertex v. We write $V_R = V_R(v)$ when we wish to make this dependence explicit.

By construction, $V_R^0 \supset V \cap C(R)$. It is often true that $V_R = V \cap C(R)$. Let us study the conditions under which this can fail. We say that a vertex w *clips* a standard cluster R (based at v_1) if $w \neq v_1$ and some point of $V_R^0(v_1)$ belongs to the Voronoi cell at w. Part I makes a thorough investigation of the geometry when a vertex w clips a quasi-regular tetrahedron. (The vertex w must belong to a second quasi-regular tetrahedron that shares a face with S. The shared face must have circumradius greater than $\sqrt{2}$, and so forth.)

Lemma 2.2. *Let R, based at a vertex v_0, be a standard cluster other than a quasi-regular tetrahedron. Suppose it is clipped by a vertex w. Then there is a face (v_0, v_1, v_2) of R such that (w, v_0, v_1, v_2) is a quasi-regular tetrahedron. Furthermore, (v_0, v_1, v_2) is the unique face of the quasi-regular tetrahedron of circumradius at least $\sqrt{2}$.*

Proof. Consider a point p in $V_R \setminus V$. Then there exists a vertex $w \notin C(R)$ of D^* such that $p \cdot w > w \cdot w/2$. The line segment from p to w intersects the cone $C(F)$ of some triangular face F that bounds the standard region of R and has v_0 as a vertex. Let v_1 and v_2 be the other vertices of F. By the construction of the faces bounding a standard region, the edges of F have lengths between 2 and 2.51.

Consider the region X containing p and bounded by the planes $H_1 = \text{span}(v_1, w)$, $H_2 = \text{span}(v_2, w)$, $H_3 = \text{span}(v_1, v_2)$, $H_4 = \{x : x \cdot v_1 = v_1 \cdot v_1/2\}$, and $H_5 = \{x : x \cdot v_2 = v_2 \cdot v_2/2\}$. The planes H_4 and H_5 contain the faces of the Voronoi cell at v_0 defined by the vertices v_1 and v_2. The plane H_3 contains the face F. The planes H_1 and H_2 bound the region containing points, such as p, that can be connected to w by a segment that passes through $C(F)$.

Let $P = \{x : x \cdot w > w \cdot w/2\}$. The choice of w implies that $X \cap P$ is nonempty. We leave it as an exercise to check that $X \cap P$ is bounded. If the intersection of a bounded polyhedron with a half-space is nonempty, then some vertex of the polyhedron lies in the half-space. So some vertex of X lies in P.

We claim that the vertex of X lying in P cannot lie on H_1. To see this, pick coordinates (x_1, x_2) on the plane H_1 with origin $v_0 = 0$ so that $v_1 = (0, z)$ (with $z > 0$) and $X \cap H_1 \subset X' := \{(x_1, x_2) : x_1 \geq 0, x_2 \leq z/2\}$. See Diagram 2.3. If X' meets P, then the point $v_1/2$ lies in P. This is impossible, because every point between v_0 and v_1 lies in the Voronoi cell at v_0 or v_1, and not in the Voronoi cell of w. (Recall that $|v_1 - v_0| < 2.51 < 2\sqrt{2}$.)

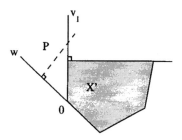

Diagram 2.3

Similarly, the vertex of X in P cannot lie on H_2. Thus, the vertex must be the unique vertex of X that is not on H_1 or H_2, namely, the point of intersection of H_3, H_4, and H_5. This point is the circumcenter c of the face F. We conclude that the polyhedron $X_0 := X \cap P$ contains c. Since $c \in X_0$, the hypotheses of Lemma I.3.4 are met for $T = F$, and the vertices v_0, v_1, v_2, and w are the vertices of a quasi-regular tetrahedron S. By Lemma I.3.4, $|w' - v_i| < 2.3$, for $i = 0, 1, 2$. The circumradius of the face F is between $\sqrt{2}$ and $2.51/\sqrt{3} \approx 1.449$. $\qquad\square$

In the same context, if w and w' both clip R, then the regions they cut from $V_R(v_0)$ are disjoint. For otherwise, a common point would belong to both $V_S^0(w)$ and $V_S^0(w')$, where S and S' are the two quasi-regular tetrahedra constructed by the lemma. Section I.3 shows that S and S' share their unique face of circumradius greater than $\sqrt{2}$. This is impossible, because the lemma states that this face is shared with R.

Although the polyhedron X_0 belongs to the Voronoi cell at w, it is included in the polyhedron V_R. Similarly, by repeating the construction at v_1 and v_2, we find that there are small regions X_1, X_2 (with vertex c) in polyhedra V_{R_1} and V_{R_2} at v_1 and v_2, respectively, that belong to the Voronoi cell at w.

Call the union $X_0 \cup X_1 \cup X_2$ the *tip* protruding form the quasi-regular tetrahedron S. Associated with a quasi-regular tetrahedron is at most one such tip. (The tip must protrude from the face of S with circumradius greater than $\sqrt{2}$.) By construction, the tip is the set of points

$$\{x : |x - w| \leq |x - v_i|, \text{ for } i = 0, 1, 2; \det(x, v_1, v_2) \det(w, v_1, v_2) \leq 0\}.$$

This is $V_S^0(w) \backslash V_S(w)$.

The tip is a subset of the Voronoi cell at w. Section I.3 explains the conditions under which this can fail to hold. There must be another vertex $u \neq w$ with the property that $|u - v_i| < 2.3$, for $i = 0, 1, 2$. Then u, v_0, v_1, and v_2 are the vertices of a second quasi-regular tetrahedron S' with face F, and this contrary to our assumption that R is not a quasi-regular tetrahedron.

Corollary 2.4. *The polyhedra V_R cover $\mathbb{R}^3$ evenly as we range over all the standard clusters of all the Delaunay stars of the packing.*

Proof. The preceding analysis shows that the polyhedra V_R are obtained from the Voronoi cells by taking each protruding tip, breaking it into three pieces X_0, X_1, X_2, and attaching the piece X_i to the Voronoi cell at v_i. The Voronoi cells cover $\mathbb{R}^3$ evenly. As a result of this analysis, we see that the polyhedra V_R cover $\mathbb{R}^3$ evenly. □

To give one example of the size of the tip, we consider the extreme case of the tetrahedron $S = S(2, 2, 2, 2.51, 2.51, 2.51)$. Diagram 2.5 shows a correctly scaled drawing of a tip protruding from the largest face of S.

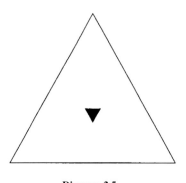

Diagram 2.5

3.   The Score Attached to a Delaunay Star

This section gives some rules for computing the score. They were developed as a result of computer experimentation suggesting when it is advantageous to use Voronoi cells over Delaunay simplices. This section actually gives an entire family of scoring systems. This extra bit of flexibility will be useful as we encounter new examples in the remaining steps of the program. We expect the score to satisfy Conjecture I.2.2, which asserts that the score of a Delaunay star is at most 8 *pt*, for all the scoring systems satisfying Properties 1–4 below. The Kepler conjecture is true if Conjecture I.2.2 holds for any one such scoring system. We have found through experimentation that small or seemingly innocent changes in the score can lead to enormous changes in the complexity of the optimization problem.

3.1.   This paper proves the second step of the program for all of the scoring systems presented below. Write $\sigma(S)$ for the score of S.

1. Suppose that the standard cluster R is a single quasi-regular tetrahedron: $R = S$. When the circumcenter of S is contained in S,

$$-4\delta_{oct} \, \mathrm{vol}(V_R \cap C(S)) + \frac{4 \, \mathrm{sol}(S)}{3}$$

is an analytic function of the lengths of the edges. This expression has an analytic continuation, denoted $\mathrm{vor}(S, V_R)$, to simplices S that do not necessarily contain their circumcenter.

If $\mathrm{rad}(S) > 1.41$, then define the score to be $\mathrm{vor}(S, V_R)$. If $\mathrm{rad}(S) \leq 1.41$, then define the score to be the compression $\Gamma(S)$. (This rule agrees with the definition of $\mathrm{vor}(S)$ given in Section I.2.)

2. Let S be a small simplex that is not a quasi-regular tetrahedron. The score of S will be either $\mathrm{vor}(S)$ or $\Gamma(S)$ depending on criteria to be determined by future research.[1] These criteria may depend on whether S belongs to a quasi-regular octahedron, but not on the position of any vertices of the packing outside S. It is essential for the scoring at all four vertices to have the same type (Voronoi or compression). The only constraint imposed by the second step of the Kepler conjecture will be $\sigma(S) \leq 0$, if S is small. This leads to the following mild restrictions on the use of Voronoi scoring.

If one of the first three edges is the long edges (say the first), compression scoring is to be used if the second, third, and fourth edges have length at most 2.06, and the fifth and sixth edges have length at most 2.08.

If one of the last three edges (say the fourth) is the long edge, compression scoring is to be used if (a), (b), (c), and (d) hold.

(a) The first edge has length at most 2.06.

---

[1] More generally, we might add a small constant c to the score of S at one of its vertices and subtract the same constant from another vertex.

(b) The second and third edges have length at most 2.08.
(c) The fifth and sixth edges have length at most 2.2.
(d) The fourth edge has length at most 2.58, or the fifth and sixth edges have lengths at most 2.12.

3. Suppose that R is any standard cluster other than a quasi-regular tetrahedron. The cluster is a union of Delaunay simplices $S_1, \ldots, S_r$. Index the simplices so that $S_1, \ldots, S_p$, for some $p \leq r$ are the small simplices in the cluster. We define the score of the cluster R to be

$$\sum_{1 \leq i \leq p} \sigma(S_i) + \sum_{p < i \leq r} \text{vor}(S_i, V_R),$$

where $\text{vor}(S, V_R) = 4(-\delta_{oct} \text{vol}(V_R \cap C(S)) + \text{sol}(S)/3)$.
4. If D^* is a Delaunay star, then its total score $\sigma(D^*)$ is a sum of the scores of the standard clusters of D^*.

Consider the quasi-regular tetrahedron S of Section 2 with vertices v_0, v_1, v_2, and w that has a protruding tip $X_0 \cup X_1 \cup X_2$. Let $\text{sol}_v(S)$ denote the solid angle of S at the vertex v. The analytic continuation $\text{vor}(S, V_R)$ has the following geometric interpretation.

$$\text{vor}(S, V_R(v)) = -4\delta_{oct}(\text{vol}(S, V_R(v)) + A(v)) + \frac{4 \, \text{sol}_v(S)}{3},$$

with the correction term $A(v_i) = -\text{vol}(X_i)$, for $i = 0, 1, 2$, and

$$A(w) = \sum_{i=1}^{3} \text{vol}(X_i).$$

The only pieces that are compression scored are small simplices, everything else is Voronoi scored. The small simplices that are compression scored will be called simplices of *compression type*. The Voronoi-scored small simplices will be called simplices of *Voronoi type*. We define the *restricted cell* of a cluster R to be the complement in V_R of the small simplices in the packing.

Lemma 3.2.

(1) *The score of a cluster depends only on the cluster, and not on the way it sits in a Delaunay star or in the Delaunay decomposition of space.*
(2) *Let Λ denote the vertices of a saturated packing. Let Λ_N denote the vertices inside the ball of radius N. (Fix any center for the ball.) Let $D^*(v)$ denote the Delaunay star at $v \in \Lambda$. Then the score satisfies (in Landau's notation)*

$$\sum_{\Lambda_N} \sigma(D^*(v)) = \sum_{\Lambda_N} \Gamma(D^*(v)) + O(N^2).$$

Proof. Statement (1) holds by construction.

(2) The score reapportions the compression of a given star among surrounding stars. The second part of the lemma follows from the claim that everything is accounted for, if we ignore the boundary effects caused by the truncation N. Space is partitioned into regions each counted $-4\delta_{oct}$ times by the compression of some star. Each point in a sphere of the packing is counted four times by the compression of some star. To verify Lemma 3.2(2), we must check that the same holds of the score.

We switch from Voronoi to compression scoring on certain small simplices. The faces F of a small simplex S satisfy rad$(F) \leq \sqrt{2}$, so no point on a face F of S can be closer to another vertex in the packing than it is to the closest vertex of F. This has two implications. First, the only polyhedra V_R meeting a small simplex S are the four based at the vertices of S. Second, let R be a standard cluster. Let S be a small simplex in R. Then $V_R \cap C(S) = V_R \cap S$. (In other words, tips cannot protrude from a small simplex.) This means that the restricted cells and small simplices cover space evenly. This decomposition is compatible with the standard decomposition of a Delaunay star.

Consider the rules defining the score. In counting the part of the volume of a sphere contained in a simplex S, we see that it appears four times with weight 1 for a total weight of 4, when S is a small simplex of compression type. It appears once with weight 4 for a total weight of 4, when S is of Voronoi type.

The result is now clear.   □

Remark 3.3. It is useful to summarize the proof from a slightly different point of view. If S is a quasi-regular tetrahedron or a small Delaunay simplex, then the sum of its four scores, for each of its four vertices, is $4\Gamma(S)$. This follows directly from the definitions (and the proof of Lemma 3.2) if the circumcenter of S is contained in S (which is always the case for small simplices), and it follows by analytic continuation in general. Any other point in space belongs to a unique Voronoi cell centered at some vertex v. If the point is not in a tip protruding from a quasi-regular tetrahedron, it is counted in the score at v. If, however, the point belongs to a protruding tip, it is counted in the score at exactly one of the three vertices, other than v, of the quasi-regular tetrahedron. In this way, every point in $\mathbb{R}^3$ is accounted for.

Remark 3.4. The choice of the parameter $\mu = 1.41$ in Section 3.1(1) is somewhat arbitrary. The choice is based on the comparison of the functions

$$f_1(x) = \text{vor}(S(2, 2, 2, 2.51, 2.51, x), V_S) \quad \text{and} \quad f_2(x) = \Gamma(S(2, 2, 2, 2.51, 2.51, x)).$$

The difference $f_1(x) - f_2(x)$ has a zero for some $x \in [2.2603, 2.2604]$. This gives a crude estimate of when it is advantageous to switch from $\Gamma(S)$ to vor(S, V_S). The constant 1.41 is a little more than rad$(S(2, 2, 2, 2.51, 2.51, 2.604)) \approx 1.405656$.

Proposition 3.5. *The Delaunay stars in the face-centered cubic and hexagonal-close packings score 8 pt.*

Proof. The eight regular tetrahedra each score 1 *pt*, and each regular octahedron scores 0 *pt*, because it has density δ_{oct}, for a total of 8 *pt*.   □

We will see in Proposition 4.6 that the regular octahedron can be broken into smaller pieces that score 0 *pt*.

4. The Main Theorem

Theorem 4.1.

(a) *The score of any small quasi-regular tetrahedron is at most 1 pt.*
(b) *The score of any other standard cluster is at most 0 pt.*

Proof. Statement (a) is a special case of Calculation I.9.1. A quasi-regular tetrahedron of Voronoi type scores less than 0 *pt* by Lemma I.9.17. In the remainder of the proof, we actually prove a much stronger statement. We explicitly decompose each cluster (other than a quasi-regular tetrahedron) into a number of pieces and show that the density of each piece is at most δ_{oct}. Since vor(S, V_R) and $\Gamma(S)$ are zero precisely when the corresponding densities are δ_{oct} (or when the volumes are zero), the theorem will follow. The relevant pieces will be congruent to one of the following *types*:

1. A small simplex that is not a quasi-regular tetrahedron.
2. A set $\{tx : 0 \le t \le 1, x \in P_2\} \subset \mathbb{R}^3$, where P_2 is a measurable set and every point of P_2 has distance at least 1.18 from the origin (Diagram 4.2(a)).
3. A set $\{tx : 0 \le t \le 1, x \in P_3\} \subset \mathbb{R}^3$, where P_3 is a wedge of a disk of the form

$$P_3 = \{(x_1, x_2, x_3) : x_3 = z_0, x_1^2 + x_2^2 \le 2, 0 \le x_2 \le \alpha x_1\},$$

for some $\alpha > 0$ and some $1 \le z_0 \le 1.18$ (Diagram 4.2(b)).
4. A Rogers simplex $R(a, b, \sqrt{2})$ where $1 \le a \le 1.18$ and $\frac{4}{3} \le b^2 \le 2$ (see Section I.8.6 and Diagram 4.2(c)).

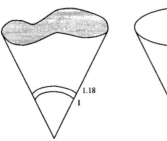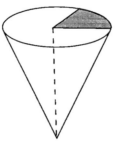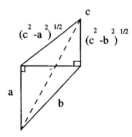

Diagram 4.2

In the first type, a unit ball is placed at each vertex of the simplex S, and the density is the ratio of the volume of the part of the balls in S to the volume of S. In the second, third, and fourth types, a unit ball is placed at the origin, and the density is the ratio of the volume of the part of the ball in the region to the volume of the region.

We decompose all of $\mathbb{R}^3$ into these four types and quasi-regular tetrahedra. Set all the quasi-regular tetrahedra aside. Classify all the small simplices, including those contained in a quasi-regular octahedron, as regions of the first type. There remain the restricted cells. Now fix a Delaunay star D^*, with center at the origin, and consider the restricted cell of one of the clusters in the star. We may assume that the restricted cell does not lie in a quasi-regular tetrahedron. Break the restricted cell up further by taking its intersection with the cones over each of its Voronoi faces F. Let X be one such intersection. If the face F has distance more than 1.18 from the origin, classify X as a region of the second type. Now assume the face F has distance h at most 1.18 from the center. Because $h < \sqrt{2}$, the point in the plane of F closest to the origin lies on the face F. The set of points P_2 on the face F at distance greater than $\sqrt{2}$ from the origin gives rise to a region of the second type. To study what remains, we may truncate F by intersecting it with a ball of radius $\sqrt{2}$. Let $F' \subset F$ be the truncated face.

By Voronoi–Delaunay duality, the face F' lies in the bisecting plane between 0 and some vertex v of the Delaunay star. Consider the collection of triangles formed by 0, v, and another vertex of the Delaunay star D^*, with the property that either the triangle has circumradius at most $\sqrt{2}$ or all three edges of the triangle have lengths between 2 and 2.51. Consider the half-planes (bounded by the line through 0 and v) containing the various triangles in this collection. This fan of half-planes partitions the face F' into a collection of wedge-shaped pieces. Consider one of them F''. We claim that is has the form of Diagram 4.3.

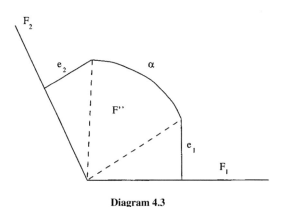

Diagram 4.3

More precisely, F'' is bounded by two triangular faces F_1 and F_2 (in this collection of triangles), two edges e_1 and e_2 of the Voronoi cell dual to the triangles, and an arc α obtained from the truncation. The two edges e_1 and e_2 are perpendicular to the faces F_1 and F_2, respectively, by the definition of Voronoi–Delaunay duality. The edges e_1 and e_2 meet the faces F_1 and F_2, respectively, by the construction of restricted cells. The edges e_1 and e_2 cannot intersect at any point less than $\sqrt{2}$ from the origin, because the point of intersection would be a point equidistant from the four vertices of a simplex formed by the vertices of F_1 and F_2. The simplex would have circumradius less than $\sqrt{2}$. Its faces would then also have circumradius less than $\sqrt{2}$, so that the Delaunay simplex is small. This is impossible, since all small simplices have already been classified as regions of the first type.

Lemma 4.4. *In this context, assume that the faces F_1 and F_2 form an acute angle, and let p be the point at which the line through e_1 meets the plane through F_2. Let w_1 and w_2 be the third vertices of the faces F_1 and F_2, respectively (that is, those other than 0 and v). If the distance from 0 to p is at most $\sqrt{2}$, then the simplex $(0, v, w_1, w_2)$ is small or a quasi-regular tetrahedron.*

Proof. Suppose that the distance from p to 0 is at most $\sqrt{2}$. Let c_1 be the circumcenter of F_1. Let S be the simplex $(0, v, w_1, w_2)$.

We claim that p lies in the interior of the triangle F_2. The bisecting line ℓ between 0 and v in the plane of F_2 contains p. The line ℓ intersects two edges of F_2, once at $v/2$ and once at some other point p'. If p is (strictly) outside F_2, then $|p| > |p'|$. This leads to a contradiction, once we show $|p'| \geq \sqrt{2}$. If p' lies on the edge between v and w_2, this is clear, because an elementary exercise shows that every point on the line passing through w_2 and v has distance at least $\sqrt{2}$ from the origin. Assume p' lies between 0 and w_2, and consider p' as a function of w_2 and v. Its length $|p'|$ attains its minimum when F_2 is the right triangle $|v| = 2$, $|v - w_2| = 2$, $|w_2| = 2\sqrt{2}$. Thus, $|p'| \geq \sqrt{2}$.

So p lies in the interior of F_2. The vertex w_1 has distance at most $\sqrt{2}$ from p. No vertex w_1 can come within $\sqrt{2}$ of an interior point of F_2 unless the circumradius of F_2 is at least $\sqrt{2}$. If the circumradius of F_2 is $\sqrt{2}$, then p is the circumcenter of S, so that the circumradius of S is $\sqrt{2}$, making S a small simplex. If the circumradius is greater than $\sqrt{2}$, then since $|p - w_1| \leq \sqrt{2}$, p lies in the Voronoi cell at w_1. Thus, w_1 clips (possibly degenerately) a standard region across the faces F_2 from w_1, based at 0, v, or w_2. By Lemma 2.2 $(w_1, 0, v, w_2)$ is a quasi-regular tetrahedron.   □

We continue with our description of the figure in Diagram 4.3. The arc α cannot be interrupted by a further (Voronoi) edge of F'. Such an edge would be dual to a (Delaunay) face with vertices 0, v, and some v'. The circumradius of the triangle with these three vertices would be less than $\sqrt{2}$ (because every edge in F' comes within distance $\sqrt{2}$ of the origin). This contradicts the construction of F'' with half-planes given above. This completes our discussion of the figure in Diagram 4.3. We emphasize, however, that the edges e_1 or e_2 may degenerate to length 0, and the circular arc α may degenerate to a point.

This Voronoi face-wedge can be broken into three convex pieces: the convex hull of 0, $v/2$, and the circular arc, and the convex hulls of 0, $v/2$, and the edge e_i, for $i = 1, 2$. The first piece has the third type, the others have the fourth type. The boundary condition $\frac{4}{3} \leq b^2$ expresses the fact that the circumradius of a triangle with sides of length at least 2 cannot be less than $2\sqrt{2}/3$. This completes the reduction to the four given types.

Now we must show that each of the given types has density at most δ_{oct}.

Type 1. A small simplex that is not a quasi-regular tetrahedron. Let S be a small simplex of Voronoi type with at least one edge longer than 2.51. By the monotonicity properties of the circumradius, we know that the circumradius of S is at least $\text{rad}(S(2, 2, 2, 2.51, 2, 2)) > 1.3045$. Let $\delta(a, b, c)$ denote the density of the Rogers simplex $(R(a, b, c)$ (see Lemma I.8.6). By Roger's lemma (I.8.6(2)), the six Rogers simplices V_S have density less than δ_{oct} and $\text{vor}(S, V_S) < 0$ if the circumradius of the three faces is at least $1.207(\delta(1, 1.207, 1.3045) < \delta_{oct})$. This condition on the circum-radius of the faces holds whenever there are two edges longer than 2.51 at the origin $(\eta(2.51, 2, 2) > 1.207)$ or whenever there are two oppositely arranged edges longer than 2.51.

Thus, to show that $vor(S) < 0$ for small simplices of Voronoi type, we must consider the following cases: (1) one edge longer than 2.51, (2) two adjacent edges longer than 2.51, and (3) three edges longer than 2.51 meeting at a vertex. These cases are covered by Calculation 4.5.2. In (1), we may assume that at least one of the conditions for compression scoring in Section 3.1 fails to hold. In Calculation 4.5.2(2), we may make the stronger assumption $rad(S) < 1.39$ for otherwise, the Rogers simplices at the origin have density at most $\delta(1, \eta(2, 2, 2.06), 1.39) < \delta_{oct}$ so that $vor(S) < 0$.

We rely on Calculation 4.5.1 for small simplices of compression type. The appendix proves the result for simplices in an explicit neighborhood of $S(2\sqrt{2}, 2, 2, 2, 2, 2)$. These calculations are established by methods of interval arithmetic described in Part I. Source code appears in [6].

Calculation 4.5.1. If S is a small simplex that is not a quasi-regular tetrahedron, then $\Gamma(S) \leq 0$. If equality is attained, then the simplex S is congruent to $S(2\sqrt{2}, 2, 2, 2, 2, 2)$ or to the simplex of zero volume ($S(2\sqrt{2}, 2, 2, 2\sqrt{2}, 2, 2)$.

Calculation 4.5.2. Assume S is small. $vor(S(y_1, \ldots, y_6)) < 0$ if $y_1, \ldots, y_6)$ belongs to any of the cells (1)–(11). Let I denote the interval $[2, 2.51]$ and $L = [2.51, 2\sqrt{2}]$.

(1) $L[2.06, 2.51]I^4$.
(2) $LI^2[2.06, 2.51]I^2$.
(3) $LI^3[2.08, 2.51]I$.
(4) $[2.06, 2.51]I^2LI^2$.
(5) $I[2.08, 2.51]ILI^2$.
(6) $I^3L[2.2, 2.51]I$.
(7) $I^3[2.58, 2\sqrt{2}][2.12, 2.51]I$.
(8) LI^3LI.
(9) LI^3L^2.
(10) I^3L^2I.
(11) I^3L^3.

Type 2. The set $\{tx : 0 \leq t \leq 1, x \in P_2\}$. In this case the density is increased by intersecting the set with a ball of radius 1.18 centered at the origin. The resulting intersection has density $1/1.18^2 < \delta_{oct}$, as required.

Type 3. The set $\{tx : 0 \leq t \leq 1, x \in P_3\}$. The bounding circular arc of P_3 has distance $\sqrt{2}$ from the origin. The set has the same density as a right circular cone, with base a disk of radius $\sqrt{2 - h^2}$ and height h. This cone has volume $\pi(2 - h^2)h/3$. The solid angle at the apex of the cone is $2\pi(1 - \cos\theta)$, where $\cos\theta = h/\sqrt{2}$. This gives a density of $\sqrt{2}/(h^2 + h\sqrt{2})$. This function is maximized over the interval $[1, 1.18]$ at $h = 1$. The density is then at most $2 - \sqrt{2} < \delta_{oct}$.

Type 4. A Rogers simplex $R_1 = R(a, b, \sqrt{2})$, where $1 \le a \le 1.18$ and $\frac{4}{3} \le b^2 \le 2$.

By Lemma I.8.6(2), the density of this simplex is at most that of the Rogers simplex $R_2 = R(1, 2\sqrt{3}/3, \sqrt{2})$. This simplex has the density δ_{oct} of a regular octahedron. (In fact, the regular octahedron may be partitioned into simplices congruent to R_2 and its mirror.) We see that the original simplex R_1 has density δ_{oct} exactly when, in the notation of Lemma I.8.6(2), $|s_1| = |s_2|$, for all λ_1, λ_2, and λ_3 as above. This implies that $a = 1$ and $b = 2\sqrt{3}/3$. This completes the proof of Theorem 4.1.   □

Proposition 4.6. *A cluster other than a quasi-regular tetrahedron attains a score of* 0 *pt if and only if it is made up of simplices congruent to* $S(2, 2, 2, 2, 2, 2\sqrt{2})$, *and possibly some additional simplices of zero volume.*

Proof. Types 2 and 3 always give strictly negative scores for regions of positive volume. According to Calculation 4.5, a region of the first type with positive volume gives a strictly negative score unless it is congruent to $S(2, 2, 2, 2, 2, 2\sqrt{2})$.

Consider a region of the fourth type with score 0 *pt*. We must have $a = 1$ and $b = 2\sqrt{3}/3$. The circumradius of the faces F_1 and F_2 of Diagram 4.3 is then $2\sqrt{3}/3$. This forces the faces F_1 and F_2 to be equilateral triangles of edge length 2. The arc α in Diagram 4.3 must reduce to a point. The edges e_1 and e_2 in Diagram 4.3—if they have positive length—must then meet at a point at distance $\sqrt{2}$ from the origin. This point is a vertex of a Voronoi cell and the circumcenter of a Delaunay simplex S (of circumradius $\sqrt{2}$). The only simplex with two equilateral faces of side 2 and $\mathrm{rad}(S) = \sqrt{2}$ is the wedge of an octahedron $S = S(2, 2, 2, 2, 2, 2\sqrt{2})$. This is a small simplex.

The other possibility is that both the arc α and an edge (say e_2) degenerate to length 0. In this case, Lemma 4.4 shows that the restricted cell belongs to a small Delaunay simplex or a quasi-regular tetrahedron. These cases have already been treated.   □

Appendix

We give a direct argument that $\Gamma(S) \le 0$ *pt*, when the lengths of a small simplex S are within 0.001 of $S_0 = S(2\sqrt{2}, 2, 2, 2, 2, 2)$. Set $S = S(y_1, y_2, y_3, y_4, y_5, y_6)$. Write $y_1 = 2\sqrt{2} - f_1$, and $y_i = 2 + f_i$ for $i > 1$, where $0 \le f_i \le 0.001$. Set $x_1 = y_1^2 = 8 - e_1$ and $x_i = y_i^2 = 4 + e_i$, for $i > 1$. Then $0 \le e_i \le 0.006$. Recall from Section I.8.4 that

$$a(y_1, y_2, \ldots, y_6) = y_1 y_2 y_3 + \tfrac{1}{2} y_1 (y_2^2 + y_3^2 - y_4^2) + \tfrac{1}{2} y_2 (y_1^2 + y_3^2 - y_5^2) + \tfrac{1}{2} y_3 (y_1^2 + y_2^2 - y_6^2).$$

Set

$$
\begin{aligned}
a_0 &= a(y_1, y_2, y_3, y_4, y_5, y_6), & a_{00} &= a(2\sqrt{2}, 2, 2, 2, 2, 2) = 16 + 12\sqrt{2}, \\
a_1 &= a(y_1, y_5, y_6, y_4, y_2, y_3), & a_{10} &= a(2\sqrt{2}, 2, 2, 2, 2, 2) = 16 + 12\sqrt{2}, \\
a_2 &= a(y_4, y_2, y_6, y_1, y_5, y_3), & a_{20} &= a(2, 2, 2, 2\sqrt{2}, 2, 2) = 16, \\
a_3 &= a(y_4, y_5, y_3, y_1, y_2, y_6), & a_{30} &= a(2, 2, 2, 2\sqrt{2}, 2, 2) = 16.
\end{aligned}
$$

Section I.8.4 and the bounds on f_i give $a_i \ge a_i^-$, where $a_0^- = a_1^- = 32.27$ and $a_2^- = a_3^- = 15.3$.

Let Δ be the function of Section I.8.1, and set $\Delta_0 = \Delta(8, 4, 4, 4, 4, 4)$. Set $t = \sqrt{\Delta(x_1, \ldots, x_6)}/2$, and $t_0 = \sqrt{\Delta_0}/2$. A simple calculus exercise shows that

$$\Delta(x_1, \ldots, x_6) \geq \Delta(8, 4, 4, x_4, 4, 4) = 128 - 8e_4^2.$$

This gives $t \geq 5.628$. Let $b_i = 2/(3(1 + t_0^2/a_{i0}^2))$, so that $b_0 = b_1 = (3 + 2\sqrt{2})/9$ and $b_2 = b_3 = 16/27$. Set $c_0 = -\delta_{oct}/6 + \sum_0^3 b_i/a_{i0} \approx -0.00679271$. Then

$$(t - t_0)c_0 = \frac{(\Delta - \Delta_0)c_0}{4(t + t_0)} \leq \frac{-2e_4^2 c_0}{(t + t_0)} \leq 0.002e_4^2 < 0.0002 f_4.$$

We are ready to estimate $\Gamma(S)$. An argument parallel to that of Lemma I.9.1(1) gives

$$\Gamma(S) \leq \Gamma(S_0) + (t - t_0)c_0 + t \sum_{i=0}^{3} \frac{b_i(a_{i0} - a_i)}{a_{i0}^2} + t \sum_{i=0}^{3} \frac{b_i(a_{i0} - a_i)^2}{a_{i0}^2 a_i^-}. \tag{1}$$

The two sums on the right-hand side are polynomials in f_i with no constant terms. To give an upper bound on these polynomials, write them as a sum of monomials, and discard the negative monomials of order greater than 2. The positive monomials of order greater than 2 are dominated by

$$f_1^{d_1} f_2^{d_2} \cdots f_6^{d_6} \leq (0.001)^{d_1 + \cdots + d_6 - 1}(f_1 + f_2 + \cdots + f_6).$$

This approximation shows that the first sum in (1) is at most $-0.005 f_1 - 0.04 f_4 - 0.03(f_2 + f_3 + f_5 + f_6)$ and the second sum in (1) is at most $0.00056(f_1 + f_2 + \cdots + f_6)$. The result easily follows.

This argument is easily adapted to a neighborhood of $S_1 = (2\sqrt{2}, 2, 2, 2\sqrt{2}, 2, 2)$. In this case, for $i = 1, \ldots, 4$, we have $a_{i0} = 16 + 8\sqrt{2}$, $b_i = \frac{2}{3}$, $a_i^- = 27$, $t_0 = 0$, $c_0 \approx -0.0225$, and $t \geq 0$. A similar arguments leads to the conclusion that $\Gamma(S) < \Gamma(S_1) = 0\,pt$, if S is a small simplex such that $S \neq S_1$, and the lengths of the edges of S are within 0.01 of those of S_1.

References

1. T. C. Hales, The sphere packing problem, *J. Comput. Appl. Math.* **44** (1992), 41–76.
2. T. C. Hales, Remarks on the density of sphere packings in three dimensions, *Combinatorica* **13**(2) (1993), 181–197.
3. T. C. Hales, The status of the Kepler conjecture, *Math. Intelligencer* (1994).
4. T. C. Hales, Sphere packings, I, *Discrete Comput. Geom.* **17** (1997), 1–51.
5. T. C. Hales, Sphere packings, III, Preprint.
6. T. C. Hales, http://www.math.lsa.umich.edu/~hales/packings.html.
7. W.-Y. Hsiang, On the sphere packing problem and the proof of Kepler's conjecture, *Internat. J. Math.* **4**(5) (1993), 739–831.
8. C. A. Rogers, The packing of equal spheres, *Proc. London Math. Soc.* **8**(3) (1958), 609–620.

Received April 24, 1995, and in revised form April 11, 1996.

The Kepler Conjecture

The Hales-Ferguson Proof

Jeffrey C. Lagarias
Editor

Department of Mathematics, University of Michigan, Ann Arbor, MI 48109-1043, USA

Jeffrey C. Lagarias (ed.) *The Kepler Conjecture: The Hales-Ferguson Proof*, DOI: 10.1007/978-1-4614-1129-1, © T.C. Hales 2012

DOI 10.1007/978-1-4614-1129-1_12

The publisher regrets that the chapters on the following pages incorrectly states Springer Science+Business Media, LLC, as the Copyright holder. Hales and Ferguson should have been the Copyright holders for these chapters.

Part II Proof of the Kepler Conjecture

3

Historical Overview of the Kepler Conjecture,
DOI: 10.1007/978-1-4614-1129-1_3,
pp. 65–82, © by T.C. Hales 2012
301 Thackeray Hall, University of Pittsburgh, Pittsburgh, PA 15260, USA

The online version of the original chapter can be found at
http://dx.doi.org/10.1007/978-1-4614-1129-1_3

4

A Formulation of the Kepler Conjecture, DOI: 10.1007/978-1-4614-1129-1_4,
pp. 83–133, © by T.C. Hales and S.P. Ferguson, 2012
301 Thackeray Hall, University of Pittsburgh, Pittsburgh, PA 15260, USA
5960 Millrace Court B-303, Columbia, MD 21045,USA

The online version of the original chapter can be found at
http://dx.doi.org/10.1007/978-1-4614-1129-1_4

5

Sphere Packings III. Extremal Cases, DOI: 10.1007/978-1-4614-1129-1_5,
pp. 135–176, © by T.C. Hales, 2012
301 Thackeray Hall, University of Pittsburgh, Pittsburgh, PA 15260, USA

The online version of the original chapter can be found at
http://dx.doi.org/10.1007/978-1-4614-1129-1_5

6

Sphere Packings IV. Detailed Bounds, DOI: 10.1007/978-1-4614-1129-1_6,
pp. 177–234, © by T.C. Hales 2012
301 Thackeray Hall, University of Pittsburgh, Pittsburgh, PA 15260, USA

The online version of the original chapter can be found at
http://dx.doi.org/10.1007/978-1-4614-1129-1_6

7
Sphere Packings V. Pentahedral Prisms, DOI: 10.1007/978-1-4614-1129-1_7,
pp. 235–274, © by S.P. Ferguson 2012
5960 Millrace Court B-303, Columbia, MD 21045, USA

The online version of the original chapter can be found at
http://dx.doi.org/10.1007/978-1-4614-1129-1_7

8
Sphere Packings VI. Tame Graphs and Linear Programs,
DOI: 10.1007/978-1-4614-1129-1_8, pp. 275–337, © by T.C. Hales, 2012
301 Thackeray Hall, University of Pittsburgh, Pittsburgh, PA 15260, USA

The online version of the original chapter can be found at
http://dx.doi.org/10.1007/978-1-4614-1129-1_8

Part III A Revision to the Proof of the Kepler Conjecture
9
A Revision of the Proof of the Kepler Conjecture,
DOI: 10.1007/978-1-4614-1129-1_9,
pp. 341–376, © by T.C. Hales, 2012
301 Thackeray Hall, University of Pittsburgh, Pittsburgh, PA 15260, USA

The online version of the original chapter can be found at
http://dx.doi.org/10.1007/978-1-4614-1129-1_9

Part IV Initial Papers of the Hales Program
10
Sphere Packings I, DOI: 10.1007/978-1-4614-1129-1_10,
pp. 379-431, © by T.C. Hales 2012
301 Thackeray Hall, University of Pittsburgh, Pittsburgh, PA 15260, USA

The online version of the original chapter can be found at
http://dx.doi.org/10.1007/978-1-4614-1129-1_10

11

Sphere Packings II, DOI: 10.1007/978-1-4614-1129-1_11,
pp. 433–449, © by T.C. Hales 2012
301 Thackeray Hall, University of Pittsburgh, Pittsburgh, PA 15260, USA

The online version of the original chapter can be found at
http://dx.doi.org/10.1007/978-1-4614-1129-1_11

Index of Symbols

Index of Subjects

Permissions

p. i Image courtesy History of Science Collections, University of Oklahoma Libraries.

p. 4 The Six-Cornered Snowflake translated by Hardie (1966) Title page. By permission of Oxford University Press.

p. 6 Image courtesy of the Division of Rare and Manuscript Collections, Cornell University Libraries.

p. 7 Image courtesy of the Division of Rare and Manuscript Collections, Cornell University Libraries.

Printed in the United States
By Bookmasters